Springer Water

Series Editor

The book series Springer Water comprises a broad portfolio of multi- and interdisciplinary scientific books, aiming at researchers, students, and everyone interested in water-related science. The series includes peer-reviewed monographs, edited volumes, textbooks, and conference proceedings. Its volumes combine all kinds of water-related research areas, such as: the movement, distribution and quality of freshwater; water resources; the quality and pollution of water and its influence on health; the water industry including drinking water, wastewater, and desalination services and technologies; water history; as well as water management and the governmental, political, developmental, and ethical aspects of water.

More information about this series at http://www.springer.com/series/13419

Mohamed Abu-hashim · Faiza Khebour Allouche ·
Abdelazim Negm
Editors

Agro-Environmental Sustainability in MENA Regions

Editors
Mohamed Abu-hashim
Faculty of Agriculture
Zagazig University
Zagazig, Egypt

Faiza Khebour Allouche
Higher Institute of Agronomic
Sciences-Chott Meriem
Sousse, Tunisia

Abdelazim Negm
Faculty of Engineering
Zagazig University
Zagazig, Egypt

ISSN 2364-6934 ISSN 2364-8198 (electronic)
Springer Water
ISBN 978-3-030-78573-4 ISBN 978-3-030-78574-1 (eBook)
https://doi.org/10.1007/978-3-030-78574-1

This Springer imprint is published by the registered company Springer Nature Switzerland AG
The registered company address is: Gewerbestrasse 11, 6330 Cham, Switzerland

Preface

The future of sustainable agriculture in MENA regions faces many challenges. This book captures agro-environmental sustainability in MENA regions and provides ideas extracted from the volume cases. In addition, some (update) findings from a few recently published research work related to the agro-environmental sustainability covered themes.

Environmental sustainability is concerned with the possibility of protecting and maintaining environmental resources for future generations. Developing countries are looking for growth while developed nations are looking for instruments for post-growth and intellectual development to think about sustainable development (economically effective, socially equitable and environmentally sustainable). For any economy that needs progress and development, growth has always been a significant goal. It is based mainly on the growth of factors of production owing to the enhanced use of available resources. The MENA region, however, faces a triple challenge: emerging climate patterns herald a future where water resources lie below sustainable levels; rapid population growth is threatening to imperil food security; and over-reliance on oil is curtailing governments' ability to act. Nevertheless, water crises are more acute than in the MENA region. In addition to food security, water scarcity is the biggest threat to human and environmental security in the region. Droughts, soil salinity and pollution, land subsidence and rural exodus have been triggered by the lack and inefficient use of resources. It has also helped trigger conflicts. A major culprit behind water scarcity in the MENA region is agriculture and food production. The situation needs new thinking on sustainability, efficient implementation of technology and radical transformation of agriculture.

This book holds much promise and potential for agro-environmental sustainability in MENA regions. It addresses the question of how science and technology can be mobilized to make that promise come true. Therefore, the intention of the book is to improve and address the following main theme: water management practices, diagnosis and new farming technologies, practices for sustainable plant and soil production, sustainable industry approach, and tourism activities in agricultural area.

The following are the key features for the book:

Part One is an introduction to agro-environmental sustainability in MENA regions where the editors present a general overview and highlight the technical elements of each chapter.

In **Part Two "Climate Change and Water Management Practices",** three chapters are identified in the book related to water management practices. The first review chapter treats climate change impacts on water balance in Egypt and opportunities for adaptations. The second chapter gives an overview of these technologies and its environmental economics applications. Furthermore, to demonstrate the application of several wastewater treatment technologies with the high-efficiency treatment of municipal wastewater in small- and large-scale using biofilm systems with respect to the application of low-cost wastewater treatment also to reuse the treated wastewater for irrigation purpose as successful case studies from the MENA region. On the other hand, chapter three covers (i) the use of climatic data for estimating water requirements in olive trees (cv Meski) in the semi-arid Tunisian (Enfidha); (ii) the use of the sap flow method to quantify transpiration and water consumption; (iii) the evaluation of the physiological method according to ET0.

Part Three "Diagnosis and New Farming Technologies" contains five chapters, and different approaches are used to delineate and discuss the new farming technologies. The first one presents the agrarian system diagnosis in Kerkennah Archipelago in Tunisia. The second approach represents precision farming technologies to increasing soil and crop productivity. The third treats the importance of implementing an environmental information system in data-scarce countries. The fourth one is related to the green spaces for residential projects as a commitment to environmental concerns and a sustainable development initiative: design of a peri-urban park in Casablanca, Morocco. The last approach emphasizes the environment and sustainable development in the face of coastal artificialization cases of Tunisia, Morocco and Algeria.

In **Part Four "Practices for Sustainable Plant and Soil Production",** several potential practices for sustainable soil production are implemented in this part. The first chapter presents the role of Tunisian medicinal plants as antifungal and antibacterial potentials for plants diseases control and also antioxidants activity of plants. The second presents an overview of sustainable agriculture in some Arab Maghreb countries (Morocco, Algeria and Tunisia) that has become the main economic sector responsible for multiple environmental impacts. The second developed potential practice is the possibilities of mineral fertilizer substitution via bio and organic fertilizers for decreasing the environmental pollution. The third one explains potential practice which is the possibility of mineral fertilizer substitution via bio and organic fertilizers for decreasing environmental pollution and improving of sesame (*Sesamum indicum* L.) vegetative growth. In the fourth chapter, the author gives an overview of the animal and rangeland resources in Shalatin—Abou Ramad—Halaib Triangle Region, Red Sea Governorate, Egypt, to identify the potentialities of animal and rangeland resources; main constraints and problems that would help in planning specific strategies for developing animal production in the region and enhancing local Bedouin's welfare. While in the fifth chapter, the authors we carried out an analysis of the collected bibliographic corpus relating to the evolution of the urban and peri-urban territories in some of the cities of the MENA, including Morocco, Algeria and Tunisia and, essentially, in the littoral regions of Morocco and Algeria.

In **Part Five "Industrial, Landscape, Touristic and Political Approaches for Agro-Environment Sustainability"**, different approaches are presented.

The fifth part of this book introduced six sustainable industry approaches. The first one is the sustainable mining site remediation under (semi) arid climates in the Middle East and in Northern Africa. Then, two study cases are presented in this part emphasizing on the value of using trees as a key element in the sustainability of cities, and the second explains how urban extension can modify the agricultural landscape of a coastal region. In both research studies, the landscape analysis approach is used.

The fourth approach is related to the interchange between agriculture and tourism in Tunisia in the context of sustainability. The fifth approach is the climate factors affecting sustainable human and tourism comfort in Egypt. Sustainable tourism is the application of sustainable development ideas to the tourism sector, that is, tourism that meets the needs of the existing generations without compromising the ability of future generations to meet their own life needs. In the last chapter, authors review the 2030 Agenda as a blueprint for water and ecosystem services in the context of agriculture with an interesting comparison between policies in both EU and Egypt.

The last chapter in this book (**Part Six**) is the conclusions and recommendations of the book. The chapter presents an update of the most recent findings, the most significant conclusions and recommendations of the chapters contained in the volume.

The editors (Mohamed Abu-hashim and Abdelazim Negm) acknowledge the support of the Science, Technology, and Innovation Authority (STIFA) of Egypt in the framework of the grant no. 30771 for the project titled "a novel standalone solar-driven agriculture greenhouse—desalination system: that grows its energy and irrigation water" via the Newton-Mosharafa funding scheme.

Additionally, the editors are happy to acknowledge the contributions of all authors to make this book a great source of knowledge for the MENA regions countries on the level of researchers, graduate students, stakeholders and decision planners. We hope it helps the MENA regions countries to move forward towards sustainable development to achieve the related SDGs goals.

Moreover, the editors want to extend their thanks to the Springer team who worked hard for a long period to produce this unique book and made the authors' and the editors' dream a reality. Special thanks are due to Andrey Kostianoy and Alexis Vizcaino.

Last but not least, The editors love to close this preface by requesting feedback from the researchers' and professionals' communities and from all audiences as well to improve the next editions. Also, new chapters for next editions are welcomed.

Please send your feedback and your constructive comments and/or your new chapter to the editors via email. The emails are posted in the chapters.

Zagazig, Egypt Mohamed Abu-hashim
Sousse, Tunisia Faiza Khebour Allouche
Zagazig, Egypt Abdelazim Negm
October 2020

Contents

Introduction

Introduction to "Agro-Environmental Sustainability in MENA Regions"

Mohamed Abu-Hashim, El-Sayed E. Omran, Faiza Khebour Allouche, and Abdelazim Negm

Abstract In this chapter, the book titled "Agro-Environmental Sustainability in MENA Regions" is introduced. In addition to two chapters in the section Conclusions and this chapter, the book consisted of 19 chapters, which entirely included 17 chapters. The 17 chapters fall under four themes. The themes include (a) Climate Change and Water Management Practices and covered in 3 chapters, (b) Diagnosis and New Farming Technologies and covered in 5 chapters too, (c) Practices For Sustainable Plant and Soil Production which is covered in 3 chapters and (d) Industrial, Landscape, Touristic and political Approaches for Agro-environment sustainability which is covered in 6 chapters.

Keywords Water resources · Agriculture · Environment · Sustainable development · Farming · Best practices · Technologies · Production · Crop

1 Background

Agro-Environmental Sustainability in MENA Regions is one of the main critical issues for enhancing developing countries. Developing countries are looking for

M. Abu-Hashim
Soil Science Department, Faculty of Agriculture, Zagazig University, Zagazig, Egypt
e-mail: dr.mabuhashim@zu.edu.eg

E.-S. E. Omran
Soil and Water Department, Faculty of Agriculture, Suez Canal University, Ismailia 41522, Egypt

Institute of African Research and Studies and Nile Basin Countries, Aswan University, Aswan, Egypt

F. K. Allouche
High School of Agronomic Sciences, Chott Meriem (ISA CM), Sousse University, Sousse, Tunisia

A. Negm (✉)
Water and Water Structures Department, Faculty of Engineering, Zagazig University, Zagazig 44519, Egypt
e-mail: amnegm@zu.edu.eg

M. Abu-hashim et al. (eds.), *Agro-Environmental Sustainability in MENA Regions*, Springer Water, https://doi.org/10.1007/978-3-030-78574-1_1

growth while developed nations are looking for instruments for post-growth and intellectual development to think about sustainable development (economically effective, socially equitable, and environmentally sustainable). Natural agricultural resources are at the heart of sustainable development in MENA regions and are critical to socio-economic growth [1, 2]. This book demonstrates the efforts made in agro-environmental sustainability in the MENA regions (in terms of observations and suggestions) and presents ideas that have been derived from the volume cases. Furthermore, the conclusions from the recently published study and governmental works on agro-environmental sustainability related to the topics discussed in the book (see e.g. [1, 3, 4]). This book presents a collection of guidelines for resource safeguarding to the current problems facing agro-environmental sustainability in MENA regions. Environmental sustainability is concerned with the possibility of protecting and maintaining environmental resources for future generations [5–7]. For the country that needs progress and development, growth has always been a significant goal, which is based mainly on the growth of factors of production owing to the enhanced use of available resources [6, 8, 9]. MENA region faces a triple challenge: Emerging climate patterns herald a future where water resources lie below sustainable levels [10–12]; rapid population growth which threatening to imperil food security [13] and over-reliance on oil that curtailing governments [9] ability to act. In addition to food security, water scarcity is the biggest threat to human and environmental security in the region [10, 14]. Droughts, soil salinity and pollution, land subsidence, and rural exodus have been triggered by the lack and inefficient use of resources [15]. It has also helped trigger conflicts [16]. A major culprit behind water scarcity in the MENA region is agriculture and food production [1, 16]. The issue needs creative thinking in terms of growth, effective technology deployment and revolutionary agricultural transformation. This book offers much promise and potential in MENA Regions for agro-environmental sustainability. This book deals with the issue of how to integrate science and technology to make the promise come true. Environmental sustainability is concerned with the possibility of protecting and maintaining environmental resources for future generations [17]. USDA [2] mentioned that developing countries are looking for growth while developed nations are looking for instruments for post-growth and intellectual development to think about sustainable development (economically effective, socially equitable, and environmentally sustainable). For any economy that needs progress and development, growth has always been a significant goal. It is based mainly on the growth of factors of production owing to the enhanced use of available resources [1]. The MENA region, however, faces a triple challenge: Emerging climate patterns herald a future where water resources lie below sustainable levels, rapid population growth is threatening to imperil food security; and over-reliance on oil is curtailing governments' ability to act [17]. Nowhere else is the water crisis more acute than in the MENA region. In addition to food security, water scarcity represents the biggest threat to human and environmental security in the region [18–21]. Droughts, soil salinity and pollution, land subsidence, and rural exodus have been triggered by the lack and inefficient use of resources [22, 23]. It has also helped trigger conflicts.

The book's intention is, therefore to improve and tackle the following main theme.

- Climate Change and Water Management Practices.
- Diagnosis and New Farming Technologies.
- Practices for Sustainable Plant and Soil Production.
- Industrial, Landscape, Touristic and political Approaches for Agro-environment sustainability.

The summary of the objectives of the chapters under each theme will be presented in the following section.

2 Summaries of the Book Chapters

2.1 Climate Change and Water Management Practices

In this part, three chapters were displayed related to water management practices, policies and climate change. The first chapter assessed the impacts of climate change on water balance in Egypt and opportunities for Adaptations. It presents a review of the findings of recent studies related to the impact of climate change on water balance in Egypt and discusses the proposed adaptation measures for such impacts. The second chapter displayed an overview of the decentralized wastewater treatment technologies using biofilm technologies. The study showed that the implementation of several biofilm technologies with the high-efficiency treatment of municipal wastewater on a small and large scale to reuse the treated wastewater for irrigation purposes was successful. The third chapter in this section is devoted to the estimation of the olive orchards' water requirements using climatic and physiological methods in Tunisia. That is when there is a lack of irrigation water, supplemental or supplemental irrigation is necessary to increase crop production significantly.

2.2 Diagnosis and New Farming Technologies

The second part of this book consists of five methods that are used to delineate and analyze the evolving agricultural developments. The first approach is the agrarian system diagnosis in Kerkennah Archipelago, Tunisia. The agrarian system is the theoretical expression of a form of agriculture that is traditionally constituted and geographically located, consisting of a featured cultivated ecosystem and a given productive social structure. This allows the sustainable exploitation of the corresponding cultivated ecosystem. The second approach is the precision farming technologies that applied to increase soil and crop productivity. In precision farming (PF) or Site-specific land management (SSLM), the farm field is classified into "site-specific management zones" depending on soil pH, yield rates, pest invasion, and other factors that affect soil and crop production. The third potential practical approach is implementing an environmental information system in data-scarce countries: issues and guidelines. Integrating the environmental dimension into socio-economic development strategies has become a requirement for sustainable development. The aim of implementing such an environmental information system (EIS) is to develop an integrated framework for the storage, production, management, and exchange of environmental information in a decision-making perspective.

The fourth approach is related to the green spaces for residential projects as a commitment to environmental concerns and a sustainable development initiative in Morocco: Design of a periurban park in Casablanca, Morocco. Morocco has embarked on a long process of modernization of the territorial administrative organization and its adaptation to the current political, economic and social context while gradually introducing a decentralization that will allow a rapprochement between the administration and the local authorities. The fifth approach is the environment and sustainable development in the face of coastal artificialisation: Case of **Tunisia**, Morocco and Algeria.

2.3 Practices for Sustainable Plant and Soil Production

The third part of this book introduces **three p**otential practices for sustainable soil production. The first potential practice is sustainable agriculture in some Arab Maghreb countries (Morocco, Algeria, and Tunisia). Agriculture has changed fundamentally and overtime has become the main economic sector responsible for multiple environmental impacts. The second potential practice is the possibilities of mineral fertilizer substitution via bio and organic fertilizers for decreasing environmental pollution and improving of sesame (*Sesamum indicum* L.) vegetative growth. Mineral fertilizers are expensive, not easily available, and increasing its doses about plant needs leading to environmental pollution. Possibilities of partial or complete substitution of mineral fertilizer via bio and organic fertilizers were assessed for improving the characteristics of sesame vegetative growth and contributing to decreasing environmental pollution.

The third potential practice gives an overview of the animal and Rangeland resources in Shalatin—Abou Ramad—Halaib Triangle Region (SAHTR), Red Sea governorate, Egypt. Shalatin—Abou Ramad—Halaib Triangle Region (SAHTR) has vital and strategic importance to Egypt. Due to a combination of severing long-lasting drought and loss of access to traditional grazing area has dramatically undetermined the continued viability of their nomadic pastoralist way of life. The aim of this chapter is to present the author's observations and findings on animal and natural resources in the Shalatin—Abou Ramad—Halaib Triangle Region (SAHTR) to identify the potentialities of animal and rangeland resources; key constraints and problems that would help to prepare specific strategies for improving the region's animal production and enhancing the welfare of local Bedouins.

2.4 *Industrial, Landscape, Touristic and Political Approaches for Agro-Environment Sustainability*

The fourth part of this book introduced six sustainable industry approaches. The first approach is the sustainable mining site remediation under (Semi) Arid Climates in the Middle East and in Northern Africa. The second approach is the agricultural landscape of Sahel Bizertin: A Heritage in Peril. The landscape with harmonious and consistent relations between human activity and preserved nature is an essential competitive advantage. The Sahel of Bizerte is rich in heritage and landscape identity reflecting the evolution of the human occupation of the territory. The heritage value of the landscape is mainly due to time. The third approach is the urban tree: a key element for the sustainable development of Tunisian cities. Green spaces are an essential element for the aesthetics, setting and quality of life of a city. They help to ventilate the cities and must be considered as the lungs of the city. They are places of relaxation, walk, rest, games for the youngest, sports fields for all ages. It was about to see the short history of the introduction of the tree in urban areas in Tunisia and especially in the city of Tunis. The tree that was having a role only for now is having an ecological ornamental role and even symbolic.

The fourth approach is related to the question: What prospects for sustainable tourism development in Hergla (Tunisia)? Hergla is an old farming village located on the eastern coast. Tourism is affected by a tight transformation aimed at a development that is inevitably more sustainable. Tourism development is part of global land management, which integrates local actors. Indeed, "the concept of sustainable land management (SLM) offers solutions that go beyond technologic recommendations by including aspects of social participation and policy dialogue". The fifth approach is the climate factors affecting sustainable human and tourism comfort in Aswan Governorate. Sustainable tourism is the application of sustainable development ideas to the tourism sector. The tourism industry, which is mainly an outdoor economy, can be particularly vulnerable to climate and weather effects. Climate and its various components influence directly on the human body and its social and psychological life. Because the place chosen by the tourist must have adequate climatic and environmental conditions for his movements and needs, which is different from the place where he lives, therefore, the elements of climate are one of the most important natural geographical components on which tourism is based.

The six approaches is the agri-environmental policies: Comparison and critical evaluation between EU and Egyptian structure. The transformation of the agri-food system has become a priority for achieving the Sustainable Development Goals. Agricultural transformation is believed to be vital for achieving a sustainable future and being able to contribute to the achievement of several SDGs. In the European Union (EU), 39% of the total land area were being used for agriculture in 2016, yet only 4.2% of the population worked in the agricultural sector at the same time. In 2017, agriculture contributed to just 1.7% of the gross domestic product (GDP). In Egypt, however, agriculture is the largest employer providing income-generating activities to more than 30% of the workforce. The Egyptian agricultural sector plays a central role in Egypt's economy as it accounts for 14.5% of the GDP., although the estimated land under agricultural use consists of only about 2.5% of the country´s total area (24,960 km^2 of 1 million km^2).

The book ends with the conclusion section which summarizes the most important conclusions and recommendations of the book and a concise update of the book themes.

Acknowledgements Mohamed Abu-Hashim and Negm acknowledge the support of the Science, Technology, and Innovation Authority (STIFA) of Egypt in the framework of the grant no. 30771 for the project titled "a novel standalone solar-driven agriculture greenhouse—desalination system: that grows its energy and irrigation water" via the Newton-Mosharafa funding scheme.

References

1. FAO (2014) Adapting to climate change through land and water management in Eastern Africa results of pilot projects in Ethiopia, Kenya and Tanzania. ISBN 978-92-5-108354-3 (print), E-ISBN 978-92-5-108355-0 (PDF)
2. USDA (2013) Climate change and conflict in Africa and Latin America: findings and preliminary lessons from Uganda, Ethiopia, and Peru, African and Latin American resilience to climate change (ARCC), July 2013
3. Wrachien DD, Mambretti S (2015) Irrigation and drainage systems in flood-prone areas: the role of mathematical models. Austin J Irrig 1(1):1002
4. Yannopoulos SI, Lyberatos G, Theodossiou N, Li W, Valipour M, Tamburrino A, Angelakis AN (2015) Evolution of water lifting devices (pumps) over the centuries worldwide. Water 7(9):5031–5060
5. Cochet H, Devienne S, Dufumier M (2007) Comparative agriculture, a synthetic discipline? Rural Econ 297–298. http://economierurale.revues.org/index2043.html
6. Djemalia M, Bedhiaf-Romdhani S, Iniguezc L, Inounouc I (2009) Saving threatened native breeds by autonomous production, involvement of farmers organization, research and policy makers: the case of the Sicilo-Sarde breed in Tunisia. North Afr Spec Issue Anim Genet Resour 120(3):213–217
7. Sacklokham S, Baudran E (2005) Using agrarian system analysis to understand agriculture. Improving livelihoods in the uplands of the Lao PDR was produced in 2005 by NAFRI, NAFES and NUOL, pp 222–229
8. Kayouli C, Jouany JP, Dardillat C, Tisserand JL (1995) Physiological peculiarities of the dromedary: consequences for its feeding. Mediterranean Options, pp 143–155 (French)
9. Ministry of Equipment and Environment (2011) Study of the development of the management scheme. The economic region of the center east. General direction of spatial planning, 311 p (French)
10. Abu-Hashim M, Negm A (2018) Deficit irrigation management as strategy under water scarcity; potential application at North Sinai, Egypt. In: The handbook of environmental chemistry. Springer, Berlin, Heidelberg. https://doi.org/10.1007/698_2018_292
11. Göll E (2017). Future challenges of climate change in the Mena region. Future Notes, No. 7, July 2017. Project of Middle East and North Africa Regional Architecture: Mapping geopolitical shifts, regional order and domestic transformations (MENARA)
12. Hussain I (2005) Pro-poor intervention strategies in irrigated agriculture in Asia: poverty in irrigated agriculture: issues, lessons, options and guidelines. Final synthesis report submitted to the Asian Development Bank, International Water Management Institute (IWMI), Colombo

13. Emerton L, Bos E (2004) Value. Counting ecosystems as an economic part of water infrastructure. IUCN, Gland, Cambridge, 88pp
14. El Sayed LM (2012) Determining an optimum cropping pattern for Egypt. The American University in Cairo
15. FAO, Water Report 36 (2011) Reference: climate change water and food security. ISBN 978-92-5-106795-6
16. FAOSTAT (2015) Food and Agriculture Organization of the United Nations statistics. Retrieved 25 Aug 2015 from http://faostat.fao.org/site/291/default.aspx
17. Elbehri A, Sadiddin A (2016) Climate change adaptation solutions for the green sectors of selected zones in the MENA region. Future Food J Food Agric Soc 4(3):39–54
18. Abu-Hashim M, Shaban K, Sallam A, Negm A (2019) Effect of water deficit on food productivity under saline conditions: case study—North Sinai, Egypt. In: Negm AM, Abu-Hashim M (eds) Sustainability of agricultural environment. The handbook of environmental chemistry. Springer, Berlin, Heidelberg. https://doi.org/10.1007/698_2018_316
19. Egyptian Ministry of Agriculture and Land Reclamation (EMALR) (2014) Bulletin of agricultural statistics. Egyptian Ministry of Agriculture and Land Reclamation
20. El-Ramady HR, El-Marsafawy SM, Lewis LN (2013) Sustainable agriculture and climate changes in Egypt. In: Lichtfouse E (ed) Sustainable agriculture reviews: volume 12. Springer Netherlands, Dordrecht
21. Omran ESE (2018) Hydrological simulation of a rainfed agricultural watershed using the soil and water assessment tool (SWAT). In: Negm AM, Abu-Hashim M (eds) Sustainability of agricultural environment in Egypt: part I. The handbook of environmental chemistry, vol 76. Springer, Cham
22. Abu-Hashim M, Mohamed E, Belal A (2017) Land-use changes and site variables on the soil organic carbon pool: the potential application for the MENA region. Adv Environ Res 55. ISSN: 2158-4717
23. Mohamed ES, Abu-Hashim M, Belal AAA (2018) Sustainable indicators in arid region: case study—Egypt. In: The handbook of environmental chemistry. Springer, Berlin, Heidelberg. https://doi.org/10.1007/698_2018_243

Climate Change and Water Management Practices

Climate Change Impacts on Water Balance in Egypt and Opportunities for Adaptations

Tamer A. Gado and Doaa E. El-Agha

Abstract Climate change has altered the hydrological cycle by increasing temperature, rising sea level, and changing precipitation patterns in many regions of the world. Egypt faces several serious risks from climate change that could reduce water supplies and adversely impact agriculture, economy, human health, and ecosystems. Thus, a dynamic adaptation strategy should be considered and updated periodically according to acquired knowledge to reveal current uncertainties. In this work, recent studies of climate change impacts on Nile water resources, sea level rise, groundwater, precipitation, evaporation, evapotranspiration, and temperature are revisited for providing helpful literature for future studies in Egypt. Furthermore, the proposed adaptation measures for such climate change impacts are discussed to come up with a robust adaptation strategy for Egypt. The results of the various studies concerning the climate change impact on Nile flows indicate considerable contradictory projections about the future availability of Nile water. Thus, studies for proposing and evaluating several adaptation measures for both the optimistic and pessimistic scenarios of the Nile flow still needed as most of the adaptation measures mentioned in the literature are general, descriptive, and not quantified. Studies about the climate change impacts on rainfall, evaporation, evapotranspiration, and groundwater quality and quantity are still limited in Egypt. In addition, research related to improved early warning and prediction, rainfall harvesting techniques, seawater agriculture, salt and tolerant crop varieties, and modern irrigation techniques are essential to cope with future challenges caused by climate change. The review suggests that further investigation is needed to evaluate the possible climate change scenarios of different hydrometeorological processes affecting Egypt's water resources.

Keywords Climate change · Water resources management · Rainfall · Nile River · Adaptation measures · Egypt

---

T. A. Gado (✉)
Department of Irrigation and Hydraulics Engineering, Faculty of Engineering, Tanta University, Sibirbay Campus, Tanta 31733, Egypt
e-mail: tamer.gado@f-eng.tanta.edu.eg

D. E. El-Agha
Department of Civil Engineering, Higher Institute of Engineering and Technology, Kafr El-Sheikh 33651, Egypt

M. Abu-hashim et al. (eds.), *Agro-Environmental Sustainability in MENA Regions*, Springer Water, https://doi.org/10.1007/978-3-030-78574-1_2

1 Introduction

In the past few decades, the evidence on the rapid warming clearly noted through monitoring the historical global temperature data by several international science institutions (Fig. 1). Most climate scientists indicated that the current global warming trend has become significant, as the planet's average surface temperature has increased about 1.62 °F since the late nineteenth century due to the human expansion of the "greenhouse effect". Although industrialized nations are primarily responsible for this phenomenon, the costs of climate change (CC) will be borne most directly by the developing countries.

Temperature is the most impacted climate variable by CC. Increasing temperature affects water vapor concentrations, humidity, wind speed, cloud characteristics, evapotranspiration rates, precipitation, soil moisture, and snowfall and snowmelt regimes. Meanwhile, changes in precipitation affect the rates of groundwater recharge, the size and timing of floods and droughts, and surface runoff regimes. All these factors are crucial for water resources management; accordingly, CC has severe effects on water resources in many regions in the world.

As shown in Fig. 2, the Middle East and North Africa (MENA) region is among the most water-scarce regions worldwide [1, 2]. The freshwater availability declined from 990 m^3 in 2005 to 800 m^3 in 2015 [3] and expected to reach 600 m^3 per capita by 2050 which is much below the poverty index (1000 m^3/capita) [4]. The weather in MENA varies predominantly from arid to semi-arid, as vast zone characterized by very low and highly variable annual rainfall, and a high degree of aridity [5].

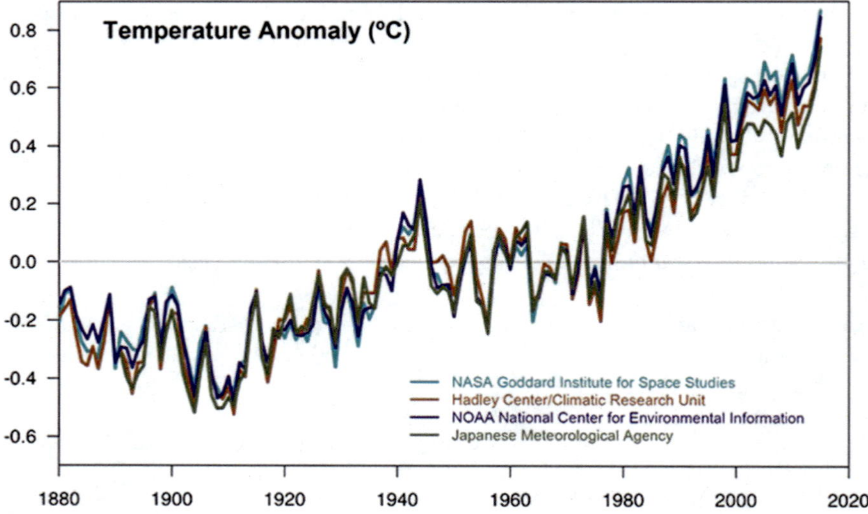

Fig. 1 Global temperature data from four international science institutions (http://climate.nasa.gov/evidence/)

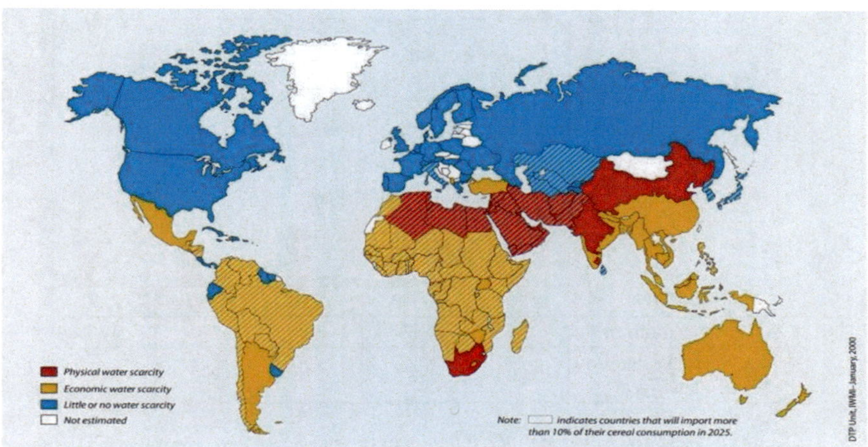

Fig. 2 Projected water scarcity in 2025 [7]

Moreover, the water scarcity in the region will be exacerbated by vastly increasing population as it is expected to double by 2050, to nearly 450 million [6].

The MENA region is highly vulnerable to CC due to social, economic, and environmental conditions [8], which puts more stress on water resources that affect both the agriculture sector and the environment. According to World Bank [9], most agricultural areas in MENA depend mainly on rain-fed agriculture and 60% of the population lives in surface water-scarce areas, which means that the region is extremely vulnerable to changes in temperature and precipitation. Thus, understanding the threats from CC is crucial to formulate policies and rectify strategies to lessen the risks [10, 11]. Modeling studies recently reported that the MENA region will face an increase in the surface temperature of 2–5.5 °C along with a decrease in precipitation by less than 20% by the end of the twenty-first century, that will cause high rates of heatwaves and extreme weather events [9].

Egypt is facing major challenges because of its limited water resources and rapidly growing population. Water use per capita has decreased from 2200 m^3/capita/year in the 1960s to 570 m^3/capita/year in 2018, and it is projected to be only 324 m^3/capita/year in 2050 (Fig. 3). Thereby, CC is projected to cause even greater concerns throughout the country, with severe environmental, social, and economic implications. The current water availability in Egypt is insufficient to address the required water demand for agriculture, industry, domestic use, and others, where transboundary water agreements commitments do exist. Moreover, the threats of reducing the share of the Nile in Egypt arise due to the development plans in the upper Nile Basin countries. It is worth mention that CC can provide challenges and opportunities for the Nile River Basin (NRB) countries to collaborate for reducing the adverse impacts of CC.

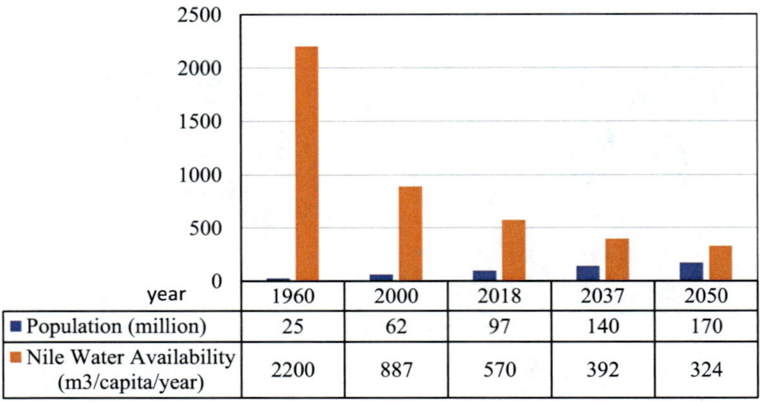

Fig. 3 Decline of freshwater availability per capita in Egypt (*Data source* MWRI [12])

In view of the previously mentioned issues, the purposes of this chapter are to (1) review the findings of recent CC impact studies on water balance in Egypt; and (2) discuss the proposed adaptation measures for such CC impacts.

2 Study Area (Egypt)

Egypt is in the North East of Africa, between 22° and 32° latitude and 24° and 37° longitude (Fig. 4), with an overall area of approximately 1×10^6 km^2, and a population of 99 million until 5 September 2019, as stated by the central agency for public mobilization and statistics (http://www.capmas.gov.eg). The altitude varies from −133 m (the Qattara Depression) to 2629 m above the sea level (Catherine Mountain in Southern Sinai), with an average of 306 m above the sea level [13]. Egypt's coastline extends for around 1200 km along the Mediterranean Sea in the North and about 1000 km along the Red Sea in the East [14]. Geomorphologically, Egypt can be classified into four regions [13]: the western desert, which represents about 68% of the total area of Egypt, the Nile valley and the delta (4%) which considers the country's most populated lands; the eastern desert (22%); and Sinai Peninsula (6%). Major urban cities, commercial center, and industrial activities are confined to the narrow corridor along the Nile and its delta and a narrow strip along the Suez Canal and Red sea Coast. Therefore, Egypt is extremely vulnerable to any potential impacts on its water resources and coastal zones due to very special distribution of its population, land-use and agriculture, and economic activity.

Egypt has four seasons: summer (21 June–22 September), autumn (23 September–20 December), winter (21 December–20 March), and spring (21 March–20 June). Generally, the climate of Egypt, as a semi-arid country, is characterized by a hot dry

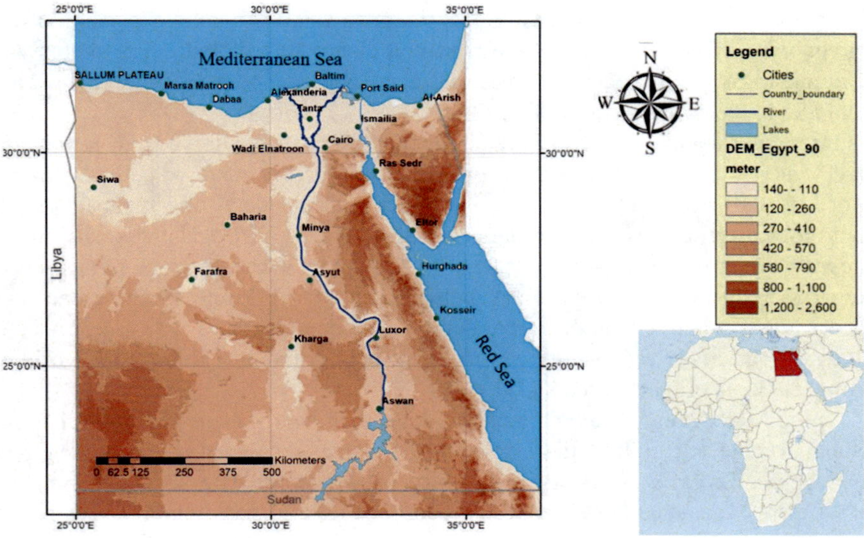

Fig. 4 Map of Egypt with DEM (modified after Gado and El-Agha [15])

summer and a mild winter with little rainfall. According to the Egyptian Meteorological Authority (https://www.emaeg.org), the temperature in the coastal regions, located in the north, ranges between around a minimum of 14 °C in winter and a maximum of 30 °C in summer. Going to the south, temperature varies widely, especially in the desert in summer, when it ranges from 7 °C at night to 43 °C during the day. In winter, temperatures vary less significantly, but they can reach 0 °C at night and as high as 18 °C during the day. Egypt has areas with strong winds, particularly along the coasts, with an average annual wind speed of 8.0–10.0 m/s along the Red Sea coast, and about 6.0–6.5 m/s along the Mediterranean coast [16]. The annual rainfall ranges between 200 mm in the north coast to almost zero in Upper Egypt [16, 17].

3 Climate Change

The greenhouse gases (GHGs) act as a thermal blanket for the earth absorbing heat and warming the surface to an average of 15 °C [18]. It consists of water vapor, carbon dioxide, nitrous oxide, and methane. Too much carbon dioxide (CO_2) and other GHGs in the air are making the greenhouse effect stronger; and thus, the earth gets warmer. Indeed, human activities (e.g., the burning of fossil fuels like coal and oil) have raised the level of CO_2 in the atmosphere from 280 to 400 parts per million in the last 150 years; and the average surface temperature of the planet increased by 0.60 °C during the twentieth century [8]. Hence, most climate scientists conclude that the main cause of the current global warming trend is human expansion of the "greenhouse effect", and they have high confidence that global temperatures will continue to rise for decades to come, i.e., the average temperature increase by 2100 could be in the range of 1.4–5.8 °C [8].

Intergovernmental Panel on Climate Change (IPCC) has observed that any change in climate variables related to global warming is affecting the regional hydrological processes [19]. Precipitation and temperature are considered the main factors of hydrological processes that are anticipated to change [20], and the variations of these two factors are the key measures of CC. The temperature increase has severe impacts on different aspects such as melting ice, sea level rise, extreme events, and diseases associated with high temperatures. Consequently, there is an urgent need to study the various effects of temperature changes in different regions in the world by using available climate models [21].

In the MENA region, the temperature has significantly increased more than the global average, e.g., in Iraq, the temperature is increasing much faster than the global average rise [22]. Egypt, as an arid country, is extremely vulnerable to the impacts of CC, thus, many studies have been recently concerned with spatial and temporal temperature changes in Egypt, as shown in Table 1. Most of the studies indicated a substantial increase in temperature over the past few decades. For example, Hasanean and Abdel Basset [23] investigated trends of the average temperature at 19 stations for the period of (1905–2000), and the results revealed that summer temperature had striking increasing trends during the last 20 years. Nashwan et al. [24] indicated rising trends in some temperature extremes from 1948 to 2010. By investigating temperature trends at 40 stations, El Kenawy et al. [25] showed that the frequency, intensity, and persistence of warm extremes have increased significantly during the period (1983–2015). However, further investigation is essential to evaluate the IPCC scenarios in order to have a comprehensive assessment of the impacts of the CC on different hydrometeorological processes in Egypt.

3.1 Climate Models

An important step in CC impact studies is to predict climate data in the future via Global Climate Models (GCMs) or Regional Climate Models (RCMs). Then, the outputs of these climate models are used in hydrological models, which are calibrated by using historical hydro-meteorological data, to predict future water availability. Thus, GCMs and RCMs have become very important tools to describe and predict different hydrological processes (e.g., precipitation and evapotranspiration). GCMs are numerical models that reasonably reproduce the global and continental-scale climate [34] by representing the physical processes in the atmosphere, ocean, cryosphere, and land surface [35]. Table 2 summarizes some of the most common GCMs worldwide. Although these models are the most sophisticated and advanced tool for simulating the global climate system, they have problems resulting from their coarse spatial resolution [36, 37]. The spatial resolution of GCMs ranges between

Table 1 Summary of recent studies on temperature in Egypt

Study	Region	No. of stations	Observation period	Temperature index	Time scale	Trend test/model	Main findings
Nasrallah and Balling [26]	Middle East		1950–1990 (41 years)	Mean	Month, year	Simple and stepwise multiple regression	A linear significant temperature increase of 0.07 °C per decade was expected
Aesawy and Hasanean [27]	Southern Mediterranean	6	1823–1991 (169 years)	Mean	Month	Sequential version of the Mann–Kendall, first order linear Markov	Most of the stations showed evidence of abrupt climatic changes for all seasons
Hasanean [28]	Egypt	18	1905–2000 (96 years)	Mean	Month	Mann–Kendall, correlation analysis	Wintertime temperature increased at most stations, while decreasing trends were observed over Upper Egypt
Domroes, and El-Tantawi [29]	Egypt	9	1941–2000 (60 years)	Mean, min, max	Year, season	Least-squares, Mann–Kendall, principal component analysis	Decreasing trends of the mean annual temperature in northern Egypt and increasing trends in southern Egypt

(continued)

Table 1 (continued)

Study	Region	No. of stations	Observation period	Temperature index	Time scale	Trend test/model	Main findings
Hasanean and Abdel Basset [23]	Egypt	19	1905–2000 (96 years)	Mean	Month	Mann–Kendall	Summer temperature had striking positive trends during the last 20 years
Shaltout et al. [30]	Egyptian Mediterranean coast	7	1997–2009 (13 years)	Mean	Hour, day		The region would experience significant warming and substantial droughts at the end of the current century
Donat et al. [31]	Arab region	61	1940–2011 (72 years)	Min, max	Day	Least squares regression	Consistent warming trends were found since the middle of the twentieth century across the region
Sayad et al. [32]	Alexandria, Egypt	1	1980–2009 (30 years)	Min, max	Day	Statistical downscaling	RCP 4.5 was the best with increasing in both T_{max} and T_{min}

(continued)

Table 1 (continued)

Study	Region	No. of stations	Observation period	Temperature index	Time scale	Trend test/model	Main findings
Nashwan et al. [24]	Egypt		1948–2010 (63 years)	Min, max, DTR	Day	Mann–Kendall and modified Mann–Kendall	Increasing trends in temperature and temperature extremes
El Kenawy et al. [25]	Egypt	40	1983–2015 (32 years)	Min, max	Day	Least squares regression, modified Mann–Kendall	The frequency, intensity, and persistence of warm extremes have significantly increased
Gado et al. [33]	Egypt	24	2001–2005 (5 years)	Mean	Day	Linear and variance scaling, quantile mapping	RCM simulated temperature data are significantly biased from observed data at most stations

Table 2 Summary of the most common GCMs in the world

Modeling centre	Country	Acronym	Resolution (lon. × lat.)
National Center for Atmospheric Research	USA	NCAR-CCSM3	1.41° × 1.41°
Commonwealth Scientific and Industrial Research Organization	Australia	CSIRO-Mk3.0 CSIRO-Mk-3.5	1.88° × 1.88° 1.88° × 1.88°
Geophysical Fluid Dynamics Laboratory	USA	GFDL-CM2.0 GFDL-CM2.1	2.50° × 2.00° 2.50° × 2.00°
NASA/Goddard Institute for Space Studies	USA	GISS-EH GISS-ER	5.00° × 3.91° 5.00° × 3.91°
Hadley Centre Climate Model	England	UKMO-HadCM3	3.80° × 2.50°
Institute Pierre Simon Laplace	France	IPSL-CM5A-LR	1.89° × 3.75°
Center for Climate System Research (The University of Tokyo), National Institute for Environmental Studies, and Frontier Research Center for Global Change (JAMSTEC)	Japan	MIROC 3.2 (medres)	2.81° × 2.81°
Max Planck Institute for Meteorology	Germany	MPI-ECHAM5	1.88° × 1.88°

250 and 600 km, which is still not suitable to conduct simulation of the regional climate features required by hydrological and national scale impact studies [8, 38]. This necessitates downscaling GCM outputs, to transform their coarse scales to finer scales [39], before they can be used in hydrological impact studies. In general, two downscaling techniques can be applied: statistical and dynamical approaches. Statistical approaches depend on building statistical relationships between large-scale climate variables and regional climate variables [40] to efficiently downscale the outputs GCMs to a finer time and spatial scale [39]. Dynamical downscaling approaches, on the other hand, lead to regional climate models (RCMs) which have finer resolution than GCMs, by considering physical and climate conditions of an area such as clouds, orography, and volcanic eruption to reach a resolution ranges between 1 and 50 km [41]. Table 3 shows some of the most popular RCMs that covered the MENA region.

3.2 Emissions Scenarios

To study the possible impact of future CC, the IPCC proposed long-term emissions scenarios for the twenty-first century (six scenarios, IS92A to F), which were published in the supplementary report to IPCC [42]. These scenarios reflect a wide range of assumptions that influence the future evolution of GHG emissions in the

Table 3 Summary of the most common RCMs that covered the MENA region

Modeling centre	GCM	RCM	Resolution (lon. × lat.)
Swedish Meteorological and Hydrological Institute	ICHEC-EC-EARTH	SMHI-RCA4	0.44° × 0.44°
Max Planck Institute for Meteorology	ICHEC-EC-EARTH	MPI-CSC-REMO2009	0.44° × 0.44°
Max Planck Institute for Meteorology	MPI-M-MPI-ESM-LR	MPI-CSC-REMO2009	0.44° × 0.44°
Swedish Meteorological and Hydrological Institute	MOHC-HadGEM2-ES	SMHI-RCA4	0.44° × 0.44°
Swedish Meteorological and Hydrological Institute	MPI-M-MPI-ESM-LR	SMHI-RCA4	0.44° × 0.44°
Canadian Centre for Climate Modelling and Analysis	CCCma-CanESM2	CCCma-CanRCCM4	0.44° × 0.44°

absence of climate policies that have already been adopted subsequently. Then, the IPCC developed different scenarios that were first published on Special Report on Emissions Scenarios (SRES) in 2000, and then in the Fourth Assessment Report (AR4) of IPCC [43]. The SRES consists of four storylines (A1, A2, B1, and B2), which represent different demographic, social, economic, technological, and environmental developments. The purpose of the SRES is to explore future development in the global environment based on GHGs emissions.

A new set of four scenarios called the "Representative Concentration Pathways" (RCPs) has been launched in the Fifth Assessment Report (AR5) of IPCC [44]. RCPs are defined by estimated total radiative forcing at the end of the century (2100) compared to the year 1750, which are 2.6 (RCP 2.6), 4.5 (RCP 4.5), 6.0 (RCP 6), and 8.5 (RCP 8.5) W/m^2 [45], as shown in Table 4. The radiative forcing, defined as the difference between sunlight absorbed by the earth and the radiant energy back into space, causes the temperature to rise or fall and is measured in CO_2 equivalents [46]. The four RCPs consist of a mitigation scenario (RCP 2.6) that leads to a low forcing level, two equilibrium scenarios (RCP 4.5 and RCP 6), and the worst-case scenario (RCP 8.5) with very high GHG emissions [44]. Therefore, the RCPs can represent a range of climate policies during the twenty-first century, as compared with the no-climate policy of the SRES [44]. These emission scenarios help current understanding of possible future climate uncertainties, which could be useful for mitigation and/or adaptation strategies to cope with CC effects.

The Shared Socioeconomic Pathways (SSPs) have recently been established to provide a socio-economic dimension to the integrative work initiated by RCPs [47]. They include five scenarios (Fig. 5): sustainability, middle-of-the-road, regional rivalry, fossil-fueled development, and inequality [47, 48]. It is worth mention that

Table 4 Impact of different RCPs on both global temperature and sea level rise during twenty-first century (*Data source* [44])

Scenario	Description	Global warming (°C) mean and likely range		Global SLR (cm) mean and likely range	
		2046–2065	2081–2100	2046–2065	2081–2100
RCP 8.5	Rising radiative forcing pathway leading to 8.5 W/m^2 (~1370 ppm CO_2 equivalent) by 2100	2.0 (1.4–2.6)	3.7 (2.6–4.8)	30 (22–38)	63 (45–82)
RCP 6.0	Stabilization without overshoot pathway to 6.0 W/m^2 (~850 ppm CO_2 equivalent) at stabilization after 2100	1.3 (0.8–1.8)	2.2 (1.4–3.1)	25 (18–32)	48 (33–63)
RCP 4.5	Stabilization without overshoot pathway to 4.5 W/m^2 (~650 ppm CO_2 equivalent) at stabilization after 2100	1.4 (0.9–2.0)	1.8 (1.1–2.6)	26 (19–33)	47 (32–63)
RCP 2.6	Peak in radiative forcing at ~3 W/m^2 (~490 ppm CO_2 equivalent) before 2100 and then decline (the selected pathway declines to 2.6 W/m^2 by 2100)	1.0 (0.4–1.6)	1.0 (0.3–1.7)	24 (17–32)	40 (26–55)

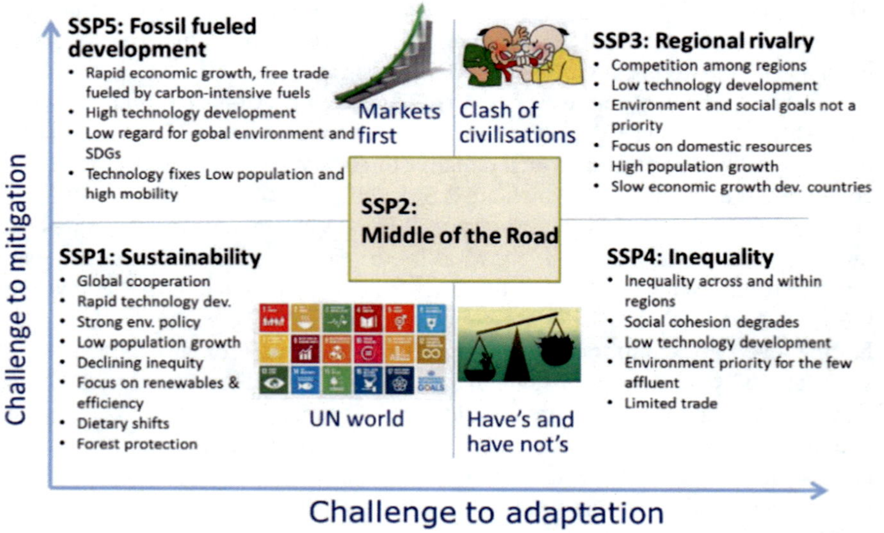

Fig. 5 Summary of drivers under the different SSPs [49]

the sixth assessment report of global warming by IPCC (due in 2021) will use these scenarios.

4 Water Balance in Egypt in the Context of Climate Change

Egypt's total water resources are 60 BCM/year according to the Egyptian Ministry of Water Resources and Irrigation [50]. The Nile River provides 55.5 BCM/year (93.7% of the total water resources) based on the Nile water-sharing agreement between Sudan and Egypt in 1959. The rainfall contributes about 1.3 BCM/year and the uses of deep groundwater and seawater desalination are around 2.1 and 0.35 BCM/year, respectively (Fig. 6). Meanwhile, the total water consumptions are 80.25 BCM/year, divided as 61.1 BCM/year (76.1%) for the agriculture sector, 10.75 BCM/year (13.4%) for the domestic use, 5.4 BCM/year (6.7%) for the industry, and 2.5 BCM/year (3.1%) losses due evaporation from water surface areas (Fig. 6).

Reuse of drainage water and taping from shallow groundwater of the Nile Delta aquifer are closing the gap between demand and supply. This gap may intensify under CC; and thus, understanding its impact on water balance parameters (input and output) becomes essential for having a complete picture of the upcoming situation and help in planning and management of water resources for balancing water demands. CC not only affect water resources but also water consumption. It may impact Nile River flow, rainfall, groundwater recharge, agriculture water consumption, evaporation from water surface areas (e.g., Lake Nasser), environmental needs (e.g., maintaining the water quality of canals and lakes), municipal water needs, and industry (e.g., cooling purposes).

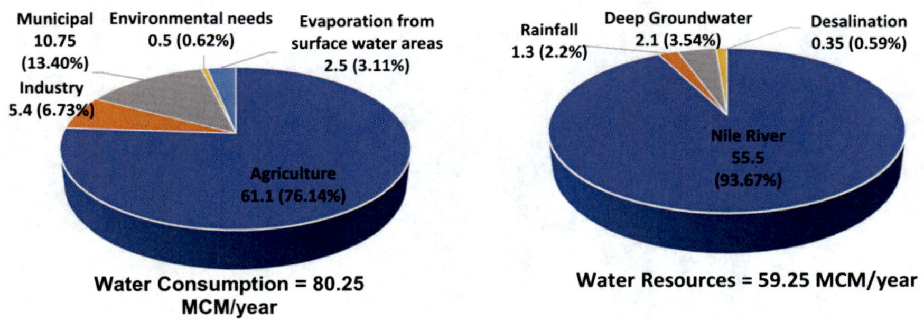

Fig. 6 Water resources and water consumption in Egypt (*Data source* [50])

4.1 Rainfall

Egypt has recently experienced an increase in rainfall extremes, which in many cases caused severe flooding, property damage, and many deaths in various regions of Egypt [51, 52]. Thus, different studies have investigated the rainfall behavior in the country (e.g., [52, 53]). The northern coast region has a relatively high annual total rainfall, in contrast to the central and southern regions of the country, where precipitation is sparce [15]. The rainy season runs from October to March, with rainfall peaks from December to February [17]. The spatial variation of the average annual total precipitation over Egypt is shown in Fig. 7.

Globally, the occurrence of severe rainfall events has increased, in line with rising temperatures and atmospheric water vapor [8]. Accordingly, studying the changes in rainfall patterns in Egypt is vital to developing adaptation and mitigation measures for the potential CC impacts. However, research on this topic is still limited, and nearly all researchers either used a few gauged sites or gridded data. For instance, spatial trend patterns were investigated for gridded rainfall data for the period 1948–2010 by Nashwan et al. [24], when no change was indicated in rainfall extremes. Based on historical data at 31 gauged stations, Gado et al. [52] studied the variability of rainfall indices to reveal the expected CC impacts on rainfall characteristics in Egypt. Significant trends were detected in annual maximum precipitation (AMP), annual total precipitation (ATP), annual number of rainy days (ANRD), and simple

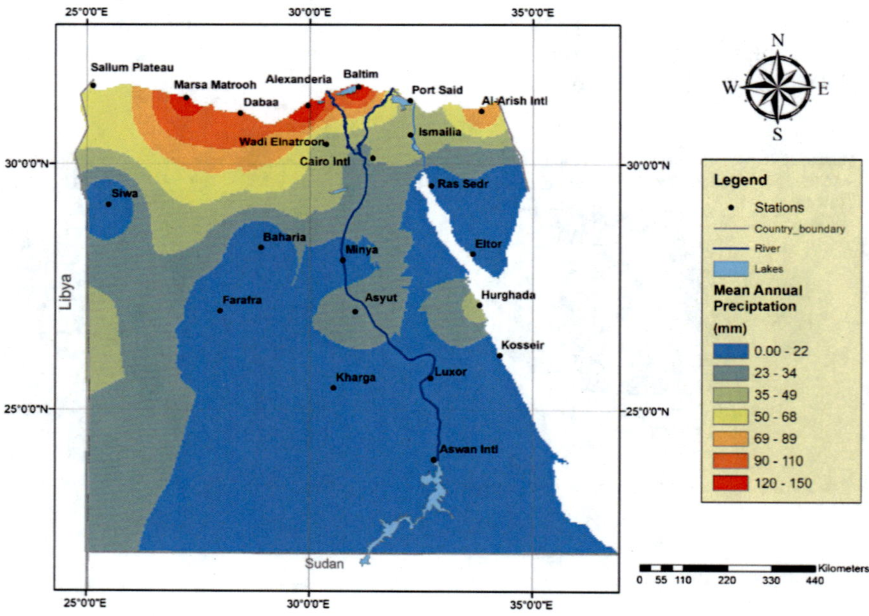

Fig. 7 Spatial distribution of mean annual precipitation in Egypt [15]

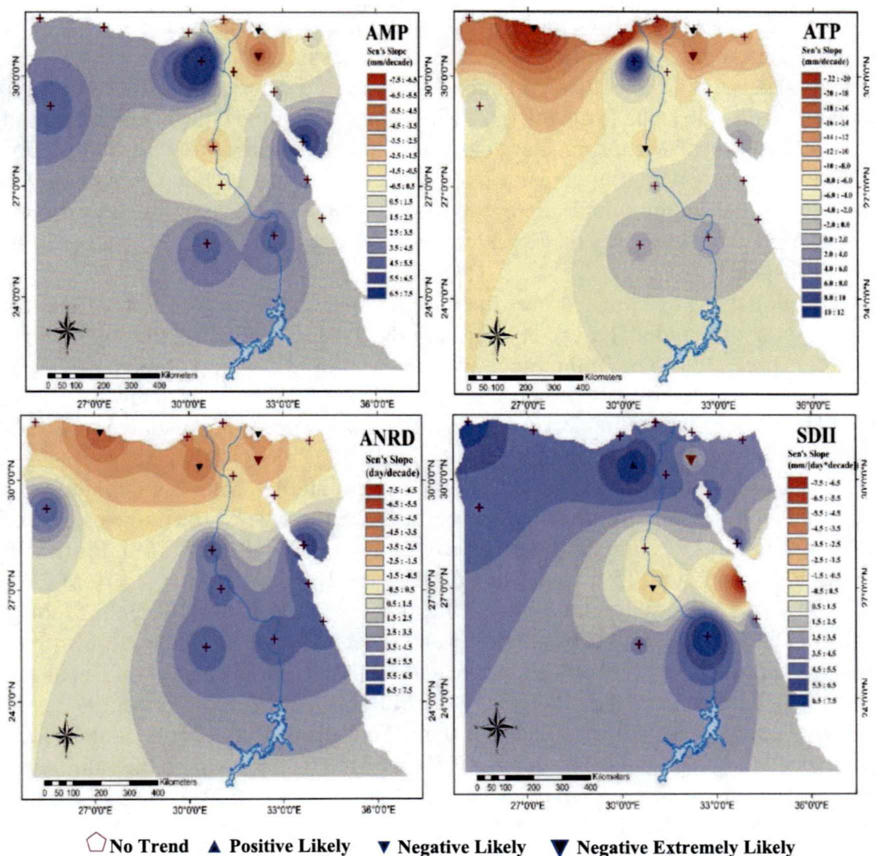

Fig. 8 Spatial distribution of trends by Sen's slope method in annual rainfall indices: AMP (mm/decade), ATP (mm/decade), ANRD (day/decade), and SDII [mm/(day × decade)], for the period (1990–2016) [52]

daily intensity index (SDII) at 29, 19, 13, and 19% of sites, respectively (Fig. 8). Most of the detected trends were negative, which could lead to a decrease in the amount of precipitation in Egypt. They recommended locating more stations, with a recent common period of record, to have more reliable results of the possible effects of CC on rainfall characteristics in Egypt.

4.2 *Groundwater*

Groundwater is an important source that attracts more attention nowadays due to the limited surface water resources and increasing demands. There are six groundwater aquifers in Egypt (Fig. 9) including the Nile aquifer which accounts for 87% of the country's groundwater abstraction [12]. It recharges mainly from the infiltration of irrigation networks and cultivated lands. Two third of the agricultural lands of Egypt located in the Nile Delta [54] which is the terminal part of the Nile River and extends around 240 km of Mediterranean coastline, from Alexandria to Port Said. About 50% of Egypt's population density lives in the Nile Delta and depends mainly on the agricultural sector, which represents about 15% of the Gross Domestic Product (GDP) [55]. This increases the expected vulnerability to CC impacts, especially the rise in the sea level.

The global sea level raised over the past 100 years by about 0.10–0.25 m due to the increase of the global mean temperature [8]. By the year 2100, the sea level rise (SLR) is predicted to be in the range of 0.18–0.59 m [8]. This will increase coastal erosion, seawater intrusion, and flooding of wetlands and lowlands in the delta. Also, it will have a significant impact on the availability of groundwater resources in coastal plains and recharge and discharge patterns [56].

Given the importance of the Nile Delta, several studies have been conducted to predict the SLR using different climate models. They predicted the range of the SLR in the Delta for the coming 100 years between 0.30 and 1.50 m [57, 58]. From 1990 to 2016, Dawod et al. [59] found that SLR in the Delta region ranged from 2.6 to 4.3 mm/year. Other studies investigated the impact of SLR on the Nile Delta (Table 5) and concluded that it will cause flooding and severe damage of lands, salinization in the river, dramatic impacts on agriculture and economy. For instance, Nofal et al. [60] investigated the impact of SLR on the Nile Delta aquifer at the coastal zone by testing three scenarios of SLR (0.25, 0.5 and 1.0 m). They predicted a change in the aquifer static head ranging from 0.1 to 0.5 m and defined the location where the salinity will be changed (through seven km from the sea). Mabrouk et al. [61] assessed the impact of both SLR and groundwater extraction on Nile Delta aquifer. A reduction of fresh groundwater by 18.7% in 90 years is predicted for the extreme scenario with SLR of 1.5 m and an increase in groundwater extraction to 12 BCM/year. Consequently,

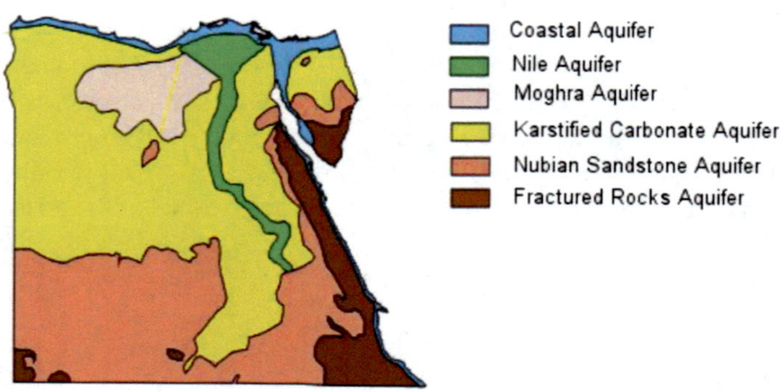

Fig. 9 Groundwater aquifers in Egypt [12]

Table 5 Summary of recent studies on the impact of SLR on the Nile Delta aquifer in Egypt

Study	Region	Model	Time scale	Future scenarios	Main findings
Nofal et al. [60]	15 km along the sea from Ras El-Bar to Gamasa	Finite difference flow and transport simulation model (SEAWAT)	2011–2041	SLR = 0.25, 0.5, and 1.0 m	Increase in groundwater head from 0.1 to 0.5 m and change in the GW salinity at 7 km from the sea (southwards)
Attwa et al. [62]	Northeast Nile Delta (El Sharkia area)	1D model generation using a hybrid genetic algorithm	1983–2007		A dramatic increase of saltwater intrusion was observed close to the low-lying coastal area of El Manzala Lake
Abd-Elhamid et al. [63]	Nile Delta aquifer	Coupled transient finite element model (2D-FEST)	100 years	SLR = 1 m, decline of piezometric head, and combination SLR and decline of piezometric head	The combination of reduction in piezometric head due to over pumping and SLR will cause an intrusion of 15 km
Nofal et al. [64]	Nile Delta aquifer	3D regional model by SEAWAT (MODFLOW and MT3DMS)	2013–2063	SLR = 0.5 and 1 m	The salinity of groundwater in the northern portion of coastal aquifer will be increased from 10,000 to 18,000 ppm
Nofal et al. [65]	Western Delta	Numerical simulations using the software ArcMap and FeFlow	1990–2100	RCP 2.6, RCP 8.5, and no SLR	A shift of the fresh/saline water interface of the Nile Delta aquifer is expected by a maximum from 43 to 57 km covering an area from 1980 to 2870 km^2

(continued)

Table 5 (continued)

Study	Region	Model	Time scale	Future scenarios	Main findings
Mabrouk et al. [61]	Nile Delta aquifer	3D variable-density groundwater flow model and salt transport by SEAWAT	2010–2100	6 scenarios of a combination of SLR and groundwater extraction rates	The fresh water will significantly decrease by 18.7% in 90 years, according to the extreme scenario

adaptation measures should take place to reduce the possible threats of CC to the groundwater in the Delta.

4.3 *Evaporation and Evapotranspiration*

The evaporation losses from open waterways and Lake Nasser were estimated to be 2.5 and 10 BCM/year, respectively [50]. Several studies investigated the impact of CC on the evaporation from Lake Nasser (e.g., [35]). El-Sawwaf et al. [66] and El-Sawwaf and Willems [67] estimated the mean evaporation rate for Lake Nasser to be 5.88 mm/day by analyzing 14 days periods from 1995 to 2004. Evaporation from the lake is expected to increase by nearly 0.47, 0.88, and 1.66 km^3 for the years 2030, 2050, and 2100, respectively [68]. Badway [69] expected that the increase by 2050 will be 0.29% and can be neglected due to the increase in humidity and the reduction in wind speed.

The consumptive use of agricultural lands through evapotranspiration is the largest terms of outflow from the system [12]. However, the spatial variation of evapotranspiration (ET) under the effect of CC was investigated in Egypt in a few studies, as shown in Table 6. For instance, Khalil [70] applied the Penman–Monteith equation to estimate the reference evapotranspiration (ET_o) by using data from 20 stations in the Nile valley and Delta. He employed the GCM (HadCM3) under the emission scenarios IPCC SRES (A1, A2, B1, and B2) to predict ET_o. The results showed significant increases in evapotranspiration under the considered CC scenarios. Abdrabbo and El-Afandi [71] compared between four ET methods (Blaney–Criddle, Hargreaves, Thornthwaite, and FAO-56 Penman–Monteith) in three different regions (Nile Delta in the north, middle, and Upper Egypt) under the different RCP scenarios. They concluded that the Blaney–Criddle and Thornthwaite equations produced the lowest ET_o values compared to the Hargreaves and FAO-56 Penman–Monteith equations under all RCPs. For the same three regions, Abdrabbo et al. [72] used the FAO-56 Penman–Monteith equation to project changes in evapotranspiration under the impact

Table 6 Summary of recent studies on climate change impact on evapotranspiration in Egypt

Study	Region	Stations	Period	Method	GCM	Scenario	Main findings
Khalil [70]	Nile valley and Delta	20	2000–2009	Penman–Monteith	HadCM3	IPCC SRES: A1, A2, B1, B2	A2 scenario provided the highest ET_0, while B1 scenario gave the lowest ET_0
Kamel [73]	Egypt	21	2000–2010	Penman–Monteith			A general increase in ET_0 till years 2005, then a slight decrease till 2010
Abdrabbo and El-Afandi [71]	Egypt	3	1998–2007	Blaney–Criddle, Hargreaves, Thornthwaite, FAO-56 Penman–Monteith	HadGEM2	RCP 2.6, RCP 4.5, RCP 6.0, RCP 8.5	The Blaney–Criddle and Thornthwaite equations had the lowest ET_0 values under different RCPs scenarios
Abdrabbo et al. [72]	Egypt	3	1971–2000	FAO-56 Penman–Monteith		RCP 2.6, RCP 4.5, RCP 6.0, RCP 8.5	The annual ET_0 would significantly increase for all different zones by uneven values

of the RCP scenarios. The results indicated that the annual ET would significantly increase for all different zones by uneven values.

5 Climate Change Impacts on the Nile River Flow

The Nile River Basin (NRB) is one of the most complex basins worldwide, shared by 11 countries (Fig. 10). It has an area of 2.9 M km^2 (10% of African continent landmass) and a living population of 190 million (25% of Africa population) [74]. Nevertheless, due to the varying climate and topography, the river's discharge (0.98 l/s/km^2) is the lowest compared with other rivers having similar basin area [75]. The available water resources governance is already under pressure and affected by the social, political, and legislative conditions of the NRB. Despite the impacts of CC, previous studies have concluded that population growth and land-use changes may have the uppermost impact (e.g., [76, 77]).

As mentioned earlier, water has become a limiting factor for promising development plans in Egypt, especially agriculture expansion. Most of the riparian countries encounter development problems due to frequent events of floods and droughts. Different water management policies are needed in NRB countries, depending on the requirements for drought mitigation, crop irrigation, and hydropower. Thus, any future changes in the Nile water (quantitative or qualitative) may have significant impacts on water management strategies in NRB countries.

Some of the recent studies that investigated both existing and future impacts of CC on NRB were reviewed in Table 7. These studies utilized different approaches ranging from GCM outputs directly (e.g., [79]) to RCMs with statistical bias correction downscaling (e.g., [80, 81]). The studies can be classified into three types: (1) evaluating the sensitivity of the Nile discharge to changes in climate variables, (2) constructing the climate scenarios based on the outputs of GCMs, and (3) obtaining more accurate results using downscaling techniques (statistical or dynamic).

5.1 Sensitivity Studies

The sensitivity of the Nile discharge to changes in temperature and precipitation has been evaluated in various studies for the whole basin (e.g., [95]) and for different sub-basins (e.g., [83, 86, 96]). First, Kite and Waititu [96] investigated the flow of a tributary of Lake Victoria (Nzoia River) and indicated that a 10% increase in precipitation resulted in a 40% increase in runoff. Sene [95] studied the whole basin and indicated an increase of 4–7% inflows if the rainfall increased by 1%. Conway and Hulme [83] assessed the sensitivity of runoff regarding changes in ET_o and precipitation and indicated that changes in temperature in the range of ± 1 °C cause changes in ET_o by $\pm 4\%$. They also concluded that changes in precipitation yielded larger changes in the runoff than that of ET_o. Sayed [86] indicated increases of the

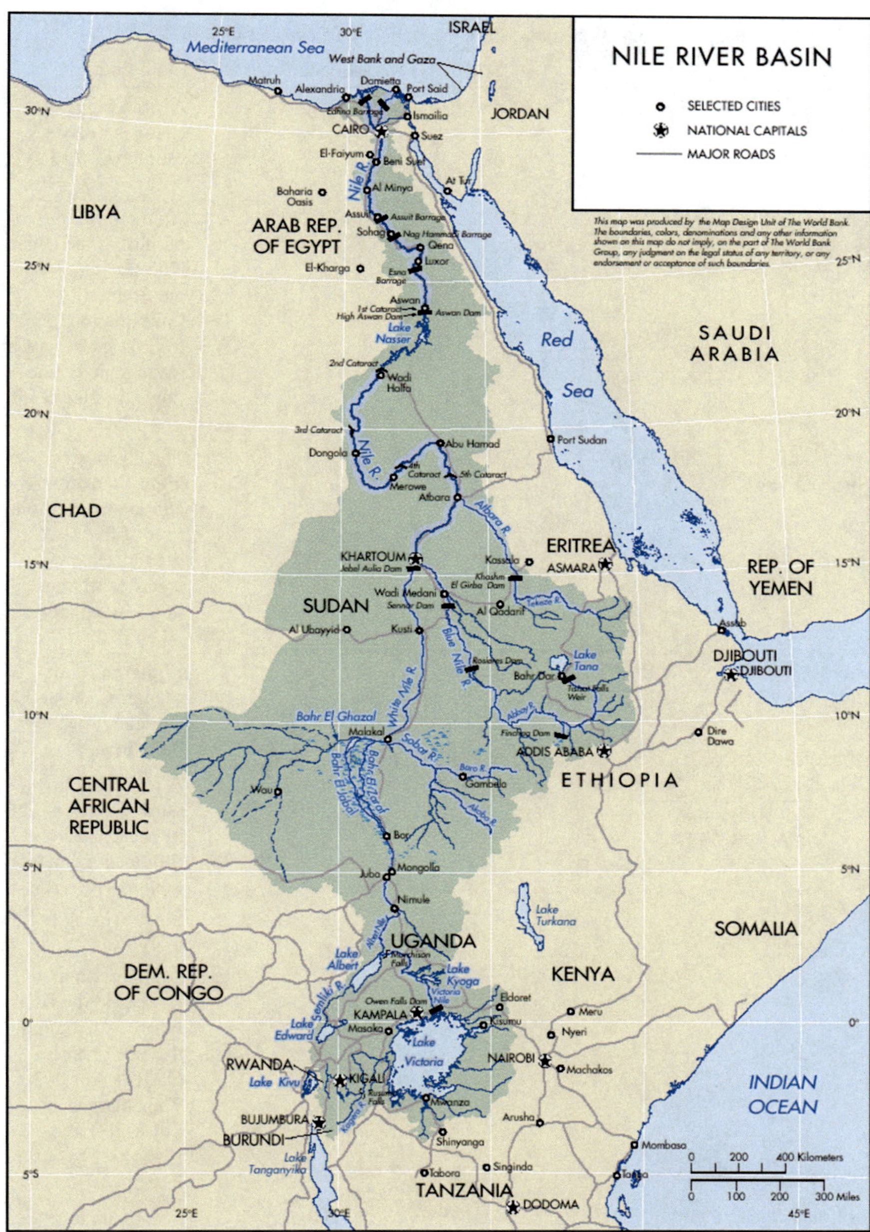

Fig. 10 The Nile River Basin. *Source* Map Design Unit of the World Bank, Gotur [78]

Table 7 Summary of recent studies on future projections of flows for the NRB

Study	Region	GCM/RCM	Scenario	Main findings
Strzepek et al. [82]	Nile Basin	3 GCMs		A wide range of changes in Nile supply (−77 to + 30%)
Conway and Hulme [83]	Nile Basin	3 GCMs		An increase in rainfall in the Blue Nile Basin (2%) and the Equatorial Lakes region (5%) for 2025; changes in mean annual Nile flows for 2025 (−9 to +12%)
Yates and Strzepek [84]	Nile Basin	5 GCMs		The basin was highly sensitive to climate fluctuations, four of five GCMs predicted significantly larger flows in equatorial lakes
Strzepek et al. [79]	Nile Basin	9 GCMs		Eight out of nine scenarios showed a tendency for lower Nile flows
Tate et al. [85]	Lake Victoria	HadCM3 GCM	A2, B2	A reduction in outflow of 2.6–4.2% by the 2050s followed by an increase between 6.3 and 9.7% in the 2080s
Sayed [86]	Nile Basin	4 GCMs		An increase in rainfall in the Blue Nile (2–11%) and the White Nile (1–10%) for 2030. The changes in inflow to Lake Nasser was (−14 to 32%)
Soliman et al. [81]	Blue Nile	ECHAM5 RCM	A1B	Future annual increases in the Blue Nile flow at Diem of about 1.5%

(continued)

Table 7 (continued)

Study	Region	GCM/RCM	Scenario	Main findings
Elshamy et al. [87]	Nile Basin	3 GCMs	A2, B2	The uncertainty across GCMs was larger than that across emission scenarios especially till 2050
Elshamy et al. [88]	Blue Nile	17 GCMs	A1B	−15 to +14% change in precipitation −15% mean change in runoff
Beyene et al. [35]	Ethiopian Highlands	11 GCMs	A2, B1	The Nile River flow increased for the period (2010–2039) and decreased during mid- (2040–2069) and late (2070–2099) century
Nawaz et al. [89]	Upper Blue Nile	3 GCMs	A2, B2	An increased in future mean flows, but this would be rather offset by rising evapotranspiration
Taye et al. [90]	Nyando River and Lake Tana	17 GCMs	A1B, B1	An increase in mean runoff for Nyando catchment for the 2050s
Chen et al. [91]	Sudan and South Sudan	7 GCMs	SRA2	A decreasing (an increasing) trend in rainy (dry) season in the future in Sudan. An increasing trend in most seasons in South Sudan
Dile et al. [92]	Upper Blue Nile	HadCM3 GCM	A2, B2	An annual increase in flow for the river at the end of the century
Gebre and Ludwig [93]	Upper Blue Nile	5 GCMs	RCP 4.5, RCP 8.5	The average annual runoff of the river will significantly increase in the future

(continued)

Table 7 (continued)

Study	Region	GCM/RCM	Scenario	Main findings
Mekonnen and Disse [94]	Blue Nile	6 GCMs	A1B, B1, A2	An increasing trend for precipitation in the range of 1.0–14.4%

outflow by about 5.7%, 34% and 32% of the whole basin, upper Blue Nile (BN), and Atbara sub-basins, respectively, due to a 10% increase in rainfall. Contrary, a reduction in rainfall by 10% would decrease the outflows by 24%, 24%, and 4% for the Atabra, BN, and Lake Victoria sub-basins, respectively. He indicated that these sub-basins are much more sensitive to climatic changes than Lake Victoria sub-basin. The balance of these changes in Dongola results in an increase (a decrease) in the mean annual flow of 30% (−25%) for a 10% increase (reduction) of rainfall over the whole NRB due to the dominance of the flows of the Ethiopian plateau, through Atbara and the BN. Therefore, the NRB is vulnerable to climate variations and that would result in considerable impacts of its water resources [97]. Hence, it is necessary to study the likely changes in water resources of the different sub-basins and the basin under the changing climatic conditions.

5.2 *Direct Use of GCMs Outputs*

Using the results of three GCMs, Strzepek et al. [82] showed a very wide range of changes in the Nile supply (−77 to +30%). Conway and Hulme [83] predicted an increase in rainfall of 2% in the BN basin and 5% in the Equatorial Lakes region for the 2025. They also concluded that changes in mean annual Nile flow for 2025 ranged between −9 and +12%. According to the 1959 Nile Waters Agreement, they estimated a mean annual flow for the Egyptian Nile varying between 50 and 65 BCM, compared to the current share of 55.5 BCM.

Yates and Strzepek [84] divided the NRB into 12 sub-catchments and assessed potential CC impacts on Nile runoff using a monthly water balance model of the NRB. They used mean monthly climate variables of precipitation, temperature, relative humidity, and sunshine hours from five GCMs and showed that the basin was highly sensitive to climate fluctuations. Considerably, larger flows in the equatorial lakes indicated by four GCMs, whereas the results varied for the different GCMs in the Ethiopian highlands of the BN and Atbara basins. Strzepek et al. [79] showed a propensity for lower Nile flows, in contrast to the results of their earlier study. Using a macro-scale hydrological model fed by GCMs data (HadCM2 and HadCM3), Arnell [98] showed that precipitation will generally increase over the NRB but the resulting increase in runoff will be offset by increases in evaporation losses due to temperature rise. Thus, he expected a little change in the mean Nile flows by 2050. Sayed [86] used four GCMs and concluded that the increase in rainfall in the BN and the White Nile would be 2–11% and 1–10%, respectively, in 2030, and the associated flow changes in Lake Nasser ranged from −14 to 32%.

In the Ethiopian Highlands, the potential impact of CC on the hydrological extremes was investigated in the catchments of the Nyando River and Lake Tana [90], which are located in two NRB source regions. They used 17 GCMs to produce CC scenarios for rainfall and potential evapotranspiration and applied two conceptual hydrological models to predict runoff changes for two scenarios (A1B and B1) for the 2050s. The results revealed increasing mean runoff and peak flows of the Nyando catchment, in contrast to the catchment of Lake Tana where no trend of such flows observed. For Lake Victoria, Tate et al. [85] used two GCMs and showed a reduction in the outflow of 2.6–4.2% by the 2050s and then an increase of 6.3–9.7% in the 2080s compared by 1961–90 as a baseline.

5.3 *Downscaling Techniques of GCMs*

The statistical downscaling approach was applied to the main Nile flow in Dongola [87] and the BN flow in Diem [89]. Both studies downscaled the coarse spatial resolution outputs from three GCMs, representing two scenarios, into fine resolution. Although their results differed for both the main Nile and the BN, they concluded that the uncertainty through general circulation models was greater than that of future scenarios (Fig. 11). In another study, a bias-correction approach was applied to the simulations of 17 GCMs, using the A1B emission scenario, in the upper BN at Diem station [88]. The results revealed that the changes in the total annual precipitation ranged from −15 to +14%, while the predicted annual flow at the station declined by 15%.

In Ethiopia, changes in precipitation were assessed by Beyene et al. [35], using bias-corrected data from 11 GCMs. Although the Nile flow is expected to increase during the period (2010–2039), the results also revealed that the flow would decline afterward due to increased evaporation and lower precipitation. To evaluate CC impacts of the Upper Blue Nile Basin (UBNB), Gebre and Ludwig [93] applied five adjusted GCMs for two scenarios (RCP 4.5 and 8.5) and indicated that the average annual runoff of the river would increase significantly in the future. Recently, Mekonnen and Disse [94] downscaled the outputs from six GCMs of the UBNB, considering three scenarios (A1B, B1, and A2), where four GCMs reported an increasing trend for precipitation in the range of 1.0–14.4% (Fig. 12).

The first study to apply the RCM in NRB was probably the study of Mohamed et al. [99], who assessed the impact of the Sudd wetlands on the hydroclimatology of the Nile. Following, some studies utilized different RCMs to investigate the hydrological response to climate change in BN (e.g., [81, 100]). Soliman et al. [81] used ECHAM5 RCM for the A1B scenario and revealed prospective annual increases in BN flow in Diem of nearly 1.5%. Although the increase in flow was expected to be greater at the start of the flood season, a decrease in flow was also predicted at the end of the

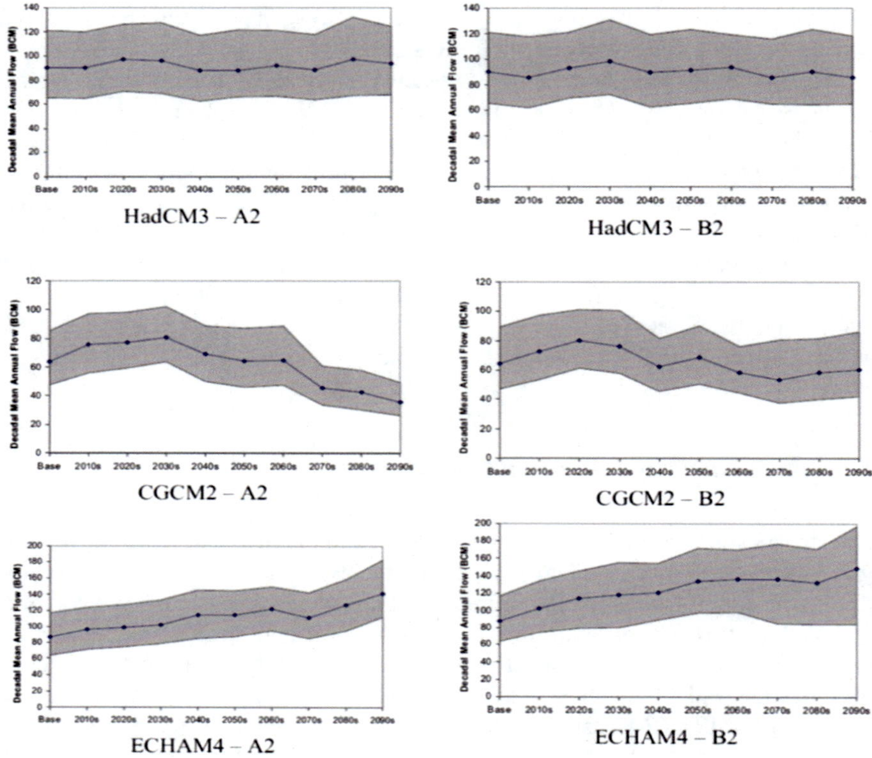

Fig. 11 Decadal mean annual flow at Dongola from three GCMs with two scenarios [87]

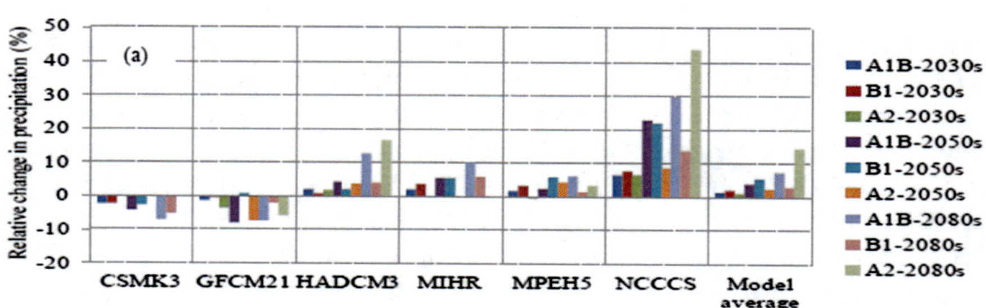

Fig. 12 Relative change of mean annual precipitation from six GCMs of Upper Blue Nile Basin compared from the reference period of 1948–2011 [94]

rainy season and in the dry season. The results of these studies lead to a wide range of uncertainties about flow projections for the NRB.

6 Climate Change Adaptation Measures in Egypt

The challenge of climate change can be addressed in two ways: mitigation and adaptation. The idea of mitigation is to tackle the problem's roots by reducing, or even stabilizing, GHGs concentrations in the atmosphere. Unfortunately, the international efforts to mitigation by controlling greenhouse emissions have had limited success. On the other hand, adaptation can be defined as initiatives and measures to decrease ecosystems' vulnerability to the current and prospective impacts of CC [101]. Adaptation has two main goals: reducing the adverse impacts of CC and taking advantage of any potential beneficial effects [8]. A wide variety of adaptation options for both supply and demand sides have been listed in the literature (e.g., [102]). Although the high cost of mitigation and adaptation impedes the implementation of the measures required, especially in developing countries, Egypt must prepare for the unavoidable risks expected from CC through a robust adaptation policy, as proposed here.

As mentioned earlier, the uncertainty in determining the direction and value of CC impact on the Nile flow is rather high, which requires higher flexibility operation and management of the water resources system. For Egypt, the increases in the flow are of course better than the reductions, as the High Aswan Dam (HAD) allows Egypt to exploit any benefits from the increases in the Nile River flows for both irrigation and hydropower, but the situation will be critical if the flow decreases. Egypt could adapt to a less 15% reduction in Nile flows; otherwise, it would have severe social and economic impacts [103]. To accommodate current and future climate variability, the current operating rules need to be adaptive to such changes to reduce the risks of both flood and drought, which requires a robust management policy for the HAD and Lake Nasser.

The Ministry of Water Resources and Irrigation in Egypt (MWRI) has launched an integrated national water resources plane for the year 2037 (NWRP 2037), which includes coordination between nine ministries with the aim of improving planning and implementation of integrated water resources management, increasing the availability of fresh water, improving water quality, and enhancing water use management [50]. The NWRP 2037 included three main pillars for the CC adaptation measures [104]. First, the MWRI planned to achieve optimal use of available resources by improving the irrigation system, applying modern irrigation systems (drip, sprinkler, etc.), changing cropping pattern by growing tolerant and short duration crops, promoting water-saving technologies in the domestic and industrial sectors, expand the reuse of treated wastewater, and studying alternative operations of HAD. Second, developing water resources would be through adding more conventional and non-conventional resources through continuing the collaboration with riparian states of the NRB, developing deep groundwater, rainfall harvesting, desalination of seawater and brackish groundwater, and increasing pumping from shallow groundwater. Third, the water quality improvement will be achieved through monitoring and preventing pollution from industries activities and promote treatment of effluents, controlling the use of agriculture pesticides and fertilizers and promoting environmentally friendly and organic agriculture practices, expand wastewater treatment and sanitation services, initiate cost recovery for drinking water and sanitary services, and protect wells from pollution.

Two main approaches to flood risk adaptation are available: (1) operating HAD at lower levels to allow more room to receive higher floods and (2) constructing additional storage structures upstream HAD to reduce the risk of dam overtopping. In addition, MWRI is working to implement a plan to rehabilitate and replace major regulators on the Nile, such as New Esna, Naga Hammadi, Assiut Barrages, and the current Dairout group of regulators. Also, several flood protection projects are being implemented in the Nile Delta region (e.g., Rosetta Branch, Baltim, Ras El Bar, and Alexandria, and Northern Lakes) [105].

In Egypt, several studies have recently been conducted on CC adaptation. Blanken [106] provided a systematic overview of the CC impact and adaptation measures based on the initial national communication reports [107–109] for water resources and agricultural sectors. SNC [109] categorized key adaptation measures for the water resources sector to adapt to uncertainty, adapt to a possible increase in flow, adequately confront reductions in flow, develop new water resources, and finally adapt to soft intervention measures. Nour El-Din [110] commented that most of these adaptation measures are general, not quantified, and unrealistic, such as the proposal to increase the Nile flow, which ignores the current promising development plans for upstream countries. Based on the fuzzy logic theory, Batisha [111] proposed a decision support model to assess measures to adapt to sea-level rise in the Nile Delta. Nofal et al. [60] evaluated the mitigation measures for three scenarios of SLR for the coastal region of the Nile Delta. They suggested artificial recharge to the Nile Aquifer through injection wells, implementing impervious barriers, and limiting the abstraction of groundwater in the northern part of the Nile Delta.

7 Conclusions and Recommendations

Recent studies of the water balance in Egypt emphasized that any reduction in water resources, due to CC, can have serious concerns in the agriculture sector, the main consumer (85%) of water resources, which will affect livelihoods and economy. This chapter focused on the Nile water resources, SLR, groundwater, precipitation, evaporation, evapotranspiration, and temperature to examine their vulnerability to CC.

Rainfall patterns and evaporation rates over the NRB are among the main factors that effectively affect Egypt's water resources. Therefore, CC impact studies for the NRB and Egypt were reviewed and discussed in this chapter to provide useful literature for future studies in Egypt. There is a consensus among the regional climate projections performed in the literature indicating a general increase in precipitation during the main rainy season in the Ethiopian Plateau, where the source of both the BN and the Atbara River (the main source of freshwater in Egypt). For other seasons, no significant model agreement was found. For the Nile Equatorial Lakes region, however, most of the models indicate a significant drying during the summer and a general increase in precipitation during the winter. Furthermore, most studies concluded that the uncertainties in predicting CC impacts on the Nile flows are very high, which complicates the proposed plans of water resources in the basin countries.

For precipitation in Egypt, most studies have identified significant decreasing trends in some locations, in contrast to the temperatures that will witness a certain increase due to CC. This will significantly increase evapotranspiration under the different CC scenarios, as investigated by limited studies. For groundwater, recent studies concluded that SLR will cause severe problems in coastal areas (e.g., inundation of lowlands, seawater intrusion, and coastal erosion), which may negatively affect the quantity and quality of groundwater resources.

To address water related problems in Egypt due to CC, the following recommendations are suggested based on the present study:

1. Improvement of the existing planning and hydrologic methodologies to deal with the significant uncertainties involved in predicting future CC impacts.
2. Encouraging data exchange between NRB countries.
3. Improving networks of precipitation gauged stations in NRB countries (e.g., early warning systems).
4. Considering climate change effects in national projects.
5. Rehabilitation of canals and drains and maintenance of hydraulic structures.
6. Using advanced irrigation networks instead of traditional ones.
7. Improvement of the distribution systems for municipal and industrial water use.
8. Implementation of rainwater harvesting that includes recharge wells in urban areas and low-cost techniques on the north coast to assist rain-fed agriculture.
9. Researching for new technologies of water desalination to reduce costs.
10. Strengthen the capacity building of the institutions working on studying CC and its impacts.
11. Involving Stakeholders in adaptation planning.
12. Exploring the vulnerability of both structural and non-structural water systems to possible future CC.
13. Co-operation between the water authorities and scientific communities to help in exchanging information on the state-of-the-art on CC and its impact on water resources.
14. Public awareness campaigns.
15. Enforcing environmental laws.

8 Recommendations for Future Studies

The following are suggestions for possible future research related to climate change impacts on water resources in Egypt:

1. Evaluating the updated IPCC scenarios to have a comprehensive assessment of CC impacts on different hydrometeorological processes (e.g., evapotranspiration and precipitation) in Egypt.
2. Updating Regional Climate Models capable of predicting the impact of CC on the local (Egypt) and regional (NRB) water resources.
3. Studying the basic characteristics of stream networks that contribute to flash floods by using advanced tools (e.g., GIS and remote sensing).
4. Investigating the impact of change in both land use and climate on hydrological processes and water resources.
5. Conducting comprehensive research on the joint impact of CC and new water projects on the water resources of the BN.
6. Assessing CC impacts on food security, energy, and water resources.

References

1. Berkoff J (1994) A strategy for managing water in the Middle East and North Africa. The International Bank for Reconstruction and Development, The World Bank, Washington, DC
2. Bou-Zeid E, El-Fadel M (2002) Climate change and water resources in Lebanon and the Middle East. J Water Resour Plan Manag 128(5):343
3. AFED (2017) Arab environment in 10 years. Annual report of Arab forum for environment and development. In: Saab N (ed) Technical Publications, Beirut
4. Abouelnaga M (2019) Why the MENA region needs to better prepare for climate change. https://www.atlanticcouncil.org/blogs/menasource/why-the-mena-region-needs-to-better-prepare-for-climate-change. Accessed 7 May 2019
5. FAO (2002) The state of food and agriculture 2002. Food and Agriculture Organization at the United Nation, Rome
6. World Bank (2019) Database. https://data.worldbank.org/region/middle-east-and-north-africa?view=chart
7. IWMI (2000) World water supply and demand. International Water Management Institute, Colombo. ISBN: 92-9090-400-3
8. IPCC (2014) Geneva, Switzerland. In: Solomon S, Qin D, Manning M, Chen Z, Marquis M, Averyt KB, Tignor M, Miller HL (eds) Contribution of working group I to the fourth assessment report of the Intergovernmental Panel on Climate Change. Cambridge University Press, Cambridge, New York, NY
9. World Bank (2018) Beyond scarcity: water security in the Middle East and North Africa. https://openknowledge.worldbank.org/handle/10986/2168
10. Abou-Hadid A (2006) Assessment of impacts, adaptation, and vulnerability to climate change in North Africa: food production and water resources, project no. AF 90. Assessments of Impacts and Adaptations to Climate Change (AIACC), Washington, District of Columbia
11. Elsharkawy H, Rashed H, Rached I (2009) Climate change: the impact of sea level rise on Egypt. In: 45th ISOCARP congress
12. MWRI (2005) Water for the future. National water resources plan for Egypt. Planning sector of Ministry of Water Resources and Irrigation
13. Hereher ME (2016) Time series trends of land surface temperatures in Egypt: a signal for global warming. Environ Earth Sci 75
14. El-Raey M (1997) Vulnerability assessment of the coastal zone of the Nile Delta of Egypt to the impacts of sea level rise. Ocean Coast Manag 37(1):29–40

15. Gado TA, El-Agha DE (2020) Feasibility of rainwater harvesting for sustainable water management in urban areas of Egypt. Environ Sci Pollut Res. https://doi.org/10.1007/s11356-019-06529-5
16. El-Menoufy HM, Morsy M, Eid MM, El Ganzoury A, El-Hussainy FM, Abdel Wahab MM (2017) Towards enhancing rainfall projection using bias correction method: case study Egypt. IJSRSET 3(6):187–194
17. Gado TA (2017) Statistical characteristics of extreme rainfall events in Egypt. In: Twentieth international water technology conference, IWTC20 2017, May, pp 18–20. Available at http://iwtc.info/wp-content/uploads/2017/05/44.pdf
18. EPA (2017) Multi-model framework for quantitative sectoral impacts analysis: a technical report for the fourth national climate assessment. U.S. Environmental Protection Agency, Washington, DC
19. Mpelasoka FS, Chiew FH (2009) Influence of rainfall scenario construction methods on runoff projections. J Hydrometeorol 10:1168–1183
20. Teutschbein C, Seibert J (2010) Regional climate models for hydrological impact studies at the catchment scale: a review of recent modeling strategies. Geogr Compass 4:834–860
21. Teutschbein C, Seibert J (2012) Bias correction of regional climate model simulations for hydrological climate-change impact studies: review and evaluation of different methods. J Hydrol 456:12–29
22. Salman SA, Shahid S, Ismail T, Chung E-S, Al-Abadi AM (2017) Long-term trends in daily temperature extremes in Iraq. Atmos Res 2017(198):97–107
23. Hasanean HM, Abdel BH (2006) Variability of summer temperature over Egypt. Int J Climatol 26:1619–1634. https://doi.org/10.1002/joc.1321
24. Nashwan MS, Shahid S, Abd Rahim N (2018) Unidirectional trends in annual and seasonal climate and extremes in Egypt. Theoret Appl Climatol 18 pp
25. El Kenawy AM, Lopez-Moreno JI, McCabe MF, Robaa SM, Domínguez-Castro F, Peña GM, Trigo RM, Hereher ME, Al-Awadhi T, Vicente-Serrano SM (2019) Daily temperature extremes over Egypt: spatial patterns, temporal trends, and driving forces. Atmos Res 226:219–239
26. Nasrallah HA, Balling RC (1993) Spatial and temporal analysis of Middle Eastern temperature changes. Clim Change 25:153–161
27. Aesawy AM, Hasanean HM (1998) Annual and seasonal climatic analysis of surface air temperature variations at six southern Mediterranean stations. Theor Appl Climatol 61:55–68
28. Hasanean HM (2004) Wintertime surface temperature in Egypt in relation to the associated atmospheric circulation. Int J Climatol 24:985–999. https://doi.org/10.1002/joc.1043
29. Domroes M, El-Tantawi A (2005) Recent temporal and spatial temperature changes in Egypt. Int J Climatol 25(1):51–63
30. Shaltout M, El-Gindy A, Omstedt A (2013) Recent climate trends and future scenarios along the Egyptian Mediterranean coast. Geofizika 30(1):19–41
31. Donat MG, Peterson TC, Brunet M, King AD, Almazroui M, Kolli RK, Boucherf D, Al-Mulla AY, Nour AY, Aly AA, Nada TAA, Semawi MM, Al Dashti HA, Salhab TG, El Fadli KI, Muftah MK, Eida SD, Badi W, Driouech F, El Rhaz K, Abubaker MJY, Ghulam AS, Sanhouri Erayah AS, Mansour MB, Alabdouli WO, Al Dhanhani JS, Al-Shekaili MN (2014) Changes in extreme temperature and precipitation in the Arab region: long-term trends and variability related to ENSO and NAO. Int J Climatol 34:581–592. https://doi.org/10.1002/joc.3707
32. Sayad TA, Ali AM, Kamel AM (2016) Study the impact of climate change on maximum and minimum temperature over Alexandria, Egypt using statistical downscaling model (SDSM). Glob J Adv Res 3(8):694–712. ISSN: 2394-5788
33. Gado TA, Mohameden MB, Rashwan IMH (2019b) Bias correction of regional climate model simulations of temperature in Egypt. In: SWINDON/EXCEED workshop on water resources and climate change: impacts, mitigation and adaptation, Amman, 3–7 Nov 2019
34. Hewitson BC, Crane RG (1996) Climate downscaling: techniques and application. Clim Res 7(2):85–95

35. Beyene T, Lettenmair DP, Kabat P (2010) Hydrologic impacts of climate change on the Nile River Basin: implication of 2007 IPCC scenarios. Clim Change 100:433–461
36. Christensen JH, Boberg F, Christensen OB, Lucas-Picher P (2008) On the need for bias correction of regional climate change projections of temperature and precipitation. Geophys Res Lett 35(20):L20709
37. Maraun D, Wetterhall F, Ireson AM, Chandler RE, Kendon EJ, Widmann M, Brienen S, Rust HW, Sauter T, Themessl M, Venema VKC, Chun KP, Goodess CM, Jones RG, Onof C, Vrac M, Thiele-Eich I (2010) Precipitation downscaling under climate change: recent developments to bridge the gap between dynamical models and the end user. Rev Geophys 48:RG3003
38. Wilby RL, Wigley TM (1997) Downscaling general circulation model output: a review of methods and limitations. Prog Phys Geogr 21:530–548
39. Ahmed KF, Wang G, Silander J, Wilson AM, Allen JM, Horton R, Anyah R (2013) Statistical downscaling and bias correction of climate model outputs for climate change impact assessment in the U.S. northeast. Glob Planet Change 100:320–332. https://doi.org/10.1016/j.gloplacha.2012.11.003
40. Chandniha SK, Kansal ML (2016) Rainfall estimation using multiple regression based statistical downscaling for Piperiya watershed in Chhattisgarh. J Agrometeorol 18(1):106–112
41. Xu Z, Yang ZL (2012) An improved dynamical downscaling method with GCM bias corrections and its validation with 30 years of climate simulations. J Clim 25:6271–6286
42. IPCC (1994) Climate change: radiative forcing of climate change and an evaluation of the IPCCIS92 emission scenarios. In: Houghton JT, Meira Filho LG, Bruce J, Lee H, Callander BA, Haites E, Harris N, Maskell K (eds) Reports of working groups I and III of the Intergovernmental Panel on Climate Change, forming part of the IPCC special report to the first session of the Conference of the Parties to the UN Framework Convention on Climate Change. Cambridge University Press, Cambridge
43. Trenberth KE, Jones PD, Ambenje P, Bojariu R, Easterling D, Klein Tank D, Parker D, Rahimzadeh F, Renwick JA, Rusticucci M, Soden B, Zhai P (2007) Observations: surface and atmospheric climate change. In: Solomon S, Qin D, Manning M, Chen Z, Marquis M, Averyt KB, Tignor M, Miller HL (eds) Climate change 2007: the physical science basis. Contribution of working group I to the fourth assessment report of the Intergovernmental Panel on Climate Change. Cambridge University Press, Cambridge, New York, NY
44. IPCC (2013) Climate change: the physical science basis working group I contribution to the fifth assessment report of the Intergovernmental Panel on Climate Change. In: Stocker TF, Qin D, Plattner G-K, Tignor M, Allen SK, Boschung J, Nauels A, Xia Y, Bex V, Midgley PM (eds) Cambridge University Press, Cambridge, New York, NY, 1535 pp
45. Van Vuuren DP, Edmonds JA, Kainuma M, Riahi K, Weyant J (2011) A special issue on the RCPs. Clim Change 109:1–4. https://doi.org/10.1007/s10584-011-0157-y
46. Shindell DT, Lamarque JF, Schulz M, Flanner M, Jiao C, Chin M, Young PJ, Lee YH, Rotstayn L, Mahowald N, Milly G, Faluvegi G, Balkanski Y, Collins WJ, Conley AJ, Dalsoren S, Easter R, Ghan S, Horowitz L, Liu X, Myhre G, Nagashima T, Naik V, Rumbold ST, Skeie R, Sudo K, Szopa S, Takemura T, Voulgarakis A, Yoon JH, Lo F (2013) Radiative forcing in the ACCMIP historical and future climate simulations. Atmos Chem Phys 13:2939–2974. https://doi.org/10.5194/acp-13-2939-2013
47. Van Vuuren DP, Stehfest E, Gernaat DEHJ, Doelman JC, van den Berg M, Harmsen M, Tabeau A (2017) Energy, land-use and greenhouse gas emissions trajectories under a green growth paradigm. Glob Environ Change 42:237–250. https://doi.org/10.1016/j.gloenvcha.2016.05.008
48. Fujimori S, Hasegawa T, Masui T, Takahashi K, Herran DS, Dai H, Kainuma M (2017) SSP3: AIM implementation of shared socioeconomic pathways. Glob Environ Change 42:268–283. https://doi.org/10.1016/j.gloenvcha.2016.06.009
49. UNECE (2019) https://www.unece.org/fileadmin/DAM/energy/se/pdfs/CSE/PATHWAYS/2019/ws_Consult_14_15.May.2019/supp_doc/SSP2_Overview.pdf
50. MWRI (2017) Water security for all, the national water resources plan 2017–2030–2037. Planning sector of Ministry of Water Resources and Irrigation

51. Ahram Online (2018) Egypt's top prosecutor orders report into 'failure to manage' heavy rain in New Cairo. Saturday, 28 Apr 2018
52. Gado TA, El-Hagrsy RM, Rashwan IMH (2019a) Spatial and temporal rainfall changes in Egypt. Environ Sci Pollut Res. https://doi.org/10.1007/s11356-019-06039-4
53. Salama AM, Gado TA, Zeidan BA (2018) On selection of probability distributions for annual extreme rainfall series in Egypt. In: Twenty-first international water technology conference, IWTC21, Ismailia, 28–30 June 2018
54. MALR (Ministry of Agriculture and Land Reclamation) (2011) Egypt—country profile. FAO, Rome. http://www.fao.org/ag/AGP/AGPC/doc/Counprof/Egypt/Egypt.html#intro
55. FAO (2000) Water and agriculture in the Nile basin. Nile basin initiative report to ICCON. Background paper prepared by FAO, Rome
56. FAO (2013) Monitoring of climate change risk impacts of sea level rise on groundwater and agriculture in the Nile Delta. TCP/EGY/3301. FAO, Rome
57. Eid H, El-Marsafawy S, Ouda S (2007) Assessing the economic impacts of climate change on agriculture in Egypt. Policy research working paper 4293. The World Bank Development Research Group
58. Stanley DJ (1990) Recent subsidence and northeast tilting of the Nile Delta, Egypt. J Mar Geol 94:147–154
59. Dawod GM, Mohamed HF, Haggag GG (2019) Relative and absolute sea level rise based on recent heterogeneous geospatial data: a case study in the Nile Delta, Egypt. J Sci Eng Res 6(6):55–64
60. Nofal ER, Fekry AF, El-Didy SM (2014) Adaptation to the impact of sea level rise in the Nile Delta coastal zone, Egypt. Am Sci J 10:17–29
61. Mabrouk M, Jonoski A, Oude Essink GHP, Uhlenbrook S (2018) Impacts of sea level rise and increasing freshwater demand on sustainable groundwater management. Water 10(1690):1–14. https://doi.org/10.3390/w10111690
62. Attwa M, Gemail KS, Eleraki M (2016) Use of salinity and resistivity measurements to study the coastal aquifer salinization in a semi-arid region: a case study in northeast Nile Delta, Egypt. Environ Earth Sci 75:784. https://doi.org/10.1007/s12665-016-5585-6
63. Abd-Elhamid H, Javadi A, Abdelaty I, Sherif M (2016) Simulation of seawater intrusion in the Nile Delta aquifer under the conditions of climate change. Hydrol Res 47(6):1198–1210. https://doi.org/10.2166/nh.2016.157
64. Nofal E, Amer A, El-Didy S, Fekry A (2016) Modeling the impact of anticipated sea level rise on the Nile Delta aquifer system in Egypt. In: 2nd world irrigation forum (WIF2), Chiang Mai, 6–8 Nov 2016
65. Wassef R, Schüttrumpf H (2016) Impact of sea-level rise on groundwater salinity at the development area western delta, Egypt. Groundw Sustain Dev 2–3:85–103. ISSN 2352-801X. https://doi.org/10.1016/j.gsd.2016.06.001
66. El-Sawwaf M, Williams P, Pagano A, Berlamont J (2010) Evaporation estimates from Nasser Lake, Egypt, based on three floating station data and Bowen ratio energy budget. Theor Appl Climatol 100:439–465
67. El-Sawwaf M, Willems P (2012) Analysis of the climate variability on Lake Nasser evaporation based on the Bowen ratio energy budget method. J Environ Biol 33(2 Suppl):475–485
68. LNFDC (2008) Climate change and its effects on water resources management in Egypt. Lake Nasser flood and drought control (LNFDC) project reports. Ministry of Water Resources and Irrigation—Planning Sector, Giza
69. Badway HA (2009) Effect of expected climate changes on evaporation losses from Aswan High Dam Reservoir (AHDR). In: Thirteenth international water technology conference, IWTC 13, Hurghada
70. Khalil AA (2013) Effect of climate change on evapotranspiration in Egypt. Researcher 5(1):7–12
71. Abdrabbo MA, El-Afandi G (2015) Comparison of reference evapotranspiration equations under climate change conditions. Glob J Adv Res 2:556–568

72. Abdrabbo MA, Farag AA, El-Desokey WMS (2015) Implementing of RCPs scenarios for the prediction of evapotranspiration in Egypt. Int J Plant Soil Sci 6(1):50–63
73. Kamel ST (2015) Variation of evapotranspiration during 2000 to 2010 in Egypt due to consequences of climate change. In: Eighteenth international water technology conference, IWTC, Sharm ElSheikh
74. NBI (2006) Baseline and needs assessment of national water policies of the Nile Basin countries: a regional synthesis. Water resources planning & management project, water policy component, Addis Ababa, Ethiopia
75. El-Ganzori (2012) Towards a climate change adaptation strategy for the water sector of Egypt. In: Climate change risk management in Egypt integrated water resources management and forecasting component
76. Koutsoyiannis D, Huaming Y, Georgakakos A (2008) Medium-range flow prediction for the Nile: a comparison of stochastic and deterministic methods. Hydrol Sci J 53(1):132–164. https://doi.org/10.1623/hysj.53.1.142
77. Wilby RL, Beven KJ, Reynard NS (2008) Climate change and fluvial flood risk in the UK: more of the same? Hydrol Process 22:2511–2523
78. Gotur M (ed) (2001) The World Bank annual report 2000: annual review and summary financial information (English). World Bank, Washington, DC. http://documents.worldbank.org/curated/en/931281468741326669/Annual-review-and-summary-financial-information
79. Strzepek K, Yates D, Yohe G, Tol R, Mader N (2001) Constructing 'not implausible' climate and economic scenarios for Egypt. Integr Assess 2:139–157
80. Di Baldassarre G, Elshamy M, Griensven A, Soliman E, Kigobe M, Ndomba P, Mutemi J, Mutua F, Moges S, Xuan Y, Solomatine D, Uhlenbrook S (2011) Future hydrology and climate in the River Nile basin: a review. Hydrol Sci J 56(2):199–211
81. Soliman ESA, Sayed MAA, Jeuland M (2009) Impact assessment of future climate change for the Blue Nile basin using an RCM nested in a GCM. Nile Water Sci Eng Spec Issue Water Clim 2:31–38
82. Strzepek KM, Yates DN, El-Quosy DE (1996) Vulnerability assessment of water resources in Egypt to climatic change in the Nile Basin. Clim Res 6:89–95
83. Conway D, Hulme M (1996) The impacts of climate variability and future climate change in the Nile Basin on water resources in Egypt. Int J Water Resour Dev 12(3):277–296
84. Yates DN, Strzepek KM (1998) Modeling the Nile Basin under climatic change. J Hydrol Eng. https://doi.org/10.1061/(ASCE)1084-0699(1998)3:2(98)
85. Tate E, Sutcliffe J, Conway D, Farquharson F (2004) Water balance of Lake Victoria: update to 2000 and climate change modelling to 2100. Hydrol Sci J 49(4):563–574
86. Sayed MAA (2004) Impacts of climate change on the Nile flows. PhD thesis, Ain Shams University, Cairo
87. Elshamy ME, Sayed MAA, Badawy B (2009a) Impacts of climate change on the Nile flows at Dongola using statistical downscaled GC. Nile Basin Water Eng Sci Mag 2
88. Elshamy ME, Seierstad IA, Sorteberg A (2009b) Impacts of climate change on Blue Nile flows using bias-corrected GCM scenarios. Hydrol Earth Syst Sci 13:551–565. https://doi.org/10.5194/hess-13-551-2009
89. Nawaz R, Bellerby TJ, Sayed MAA, Elshamy ME (2010) Blue Nile runoff sensitivity to climate change. Open Hydrol J 4
90. Taye MT, Ntegeka V, Ogiramoi NP, Willems P (2011) Assessment of climate change impact on hydrological extremes in two source regions of the Nile River Basin. Hydrol Earth Syst Sci 15:209–222
91. Chen H, Guo J, Zhang Z, Xu CY (2013) Prediction of temperature and precipitation in Sudan and South Sudan by using LARSWG in future. Theor Appl Climatol 113:363–375
92. Dile YT, Berndtsson R, Setegn SG (2013) Hydrological response to climate change for Gilgel Abay River, in the Lake Tana Basin—upper Blue Nile Basin of Ethiopia. PLoS ONE 8(10):e79296. https://doi.org/10.1371/journal.pone.0079296
93. Gebre SL, Ludwig F (2015) Hydrological response to climate change of the upper Blue Nile River Basin: based on IPCC fifth assessment report (AR5). J Climatol Weather Forecast 3:121. https://doi.org/10.4172/2332-2594.1000121

94. Mekonnen DF, Disse M (2018) Analyzing the future climate change of upper Blue Nile River Basin using statistical downscaling techniques. Hydrol Earth Syst Sci 22:2391–2408. https://doi.org/10.5194/hess-22-2391-2018
95. Sene KJ (2000) Theoretical estimates for the influence of Lake Victoria on flows in the upper White Nile. Hydrol Sci J 45(1):125–145
96. Kite GW, Waititu JK (1981) Contribution to world climate programme, Nairobi (unpublished)
97. Kim U, Kaluarachchi JJ (2009) Climate change impacts on water resources in the upper Blue Nile River Basin, Ethiopia. J Am Water Res Assoc 46(6):1361–1378
98. Arnell NW (1999) Climate change and global water resources. Glob Environ Change 9(Supplement1):S31–S49
99. Mohamed YA, van den Hurk B, Savenije HHG, Bastiaanssen WGM (2005) Impact of the Sudd wetland on the Nile hydroclimatology. Water Resour Res 41(8):W08420
100. Soliman ESA, Sayed MAA, Nour E-D, Samy G (2008) Integration of NFS with regional climate model to simulate the Nile basin hydro-climatology. Nile Basin Water Eng Sci Mag 1:75–85
101. Sowers J, Weinthal E, Vengosh A (2011) Climate change, water resources, and the politics of adaptation in the Middle East and North Africa. Clim Change 104:599–627
102. Malashkhia N (2003) Social and environmental constraints to the irrigation water conservation measures in Egypt. M.Sc. thesis, Lund University, Sweden
103. Conway D, Hulme M (1993) Recent fluctuations in precipitation and runoff over the Nile sub-basins and their impact on main Nile discharge. Clim Change 25:127–151
104. MWRI (2013) NWRP measures. National water resources plan, coordination project (NWRP-CP). Technical report 44. Ministry of Water Resources and Irrigation
105. NWRI (2018) Water management strategies for the Nile Basin, a better world: volume 3. http://digital.tudor-rose.co.uk/a-better-world-vol-3/index.html#65
106. Blanken J (2012) Mainstreaming of MDGF projects (FAO component) and provision of tools for strategic decisions for the adaptation to climate change in the Egyptian agricultural sector. Final mission report—UNJP/EGY/022/SPA, climate change risk management program
107. INC, Egyptian Environmental Affairs Agency (EEAA) (1999) Initial national communication on climate change. Prepared for the United Nations Framework Convention on Climate Change (UNFCCC), The Arab Republic of Egypt
108. NSACC (2011) Egypt's national strategy for adaptation to climate change and disaster risk reduction. Egyptian Cabinet/Information and Decision Support Center/UNDP
109. SNC (2010) Egypt's second national communication, Egyptian Environmental Affairs Agency (EEAA-May 2010), under the United Nations framework convention on climate change on climate change
110. Nour El-Din MM (2013) Climate change risk management in Egypt proposed climate change adaptation strategy for the Ministry of Water Resources & Irrigation in Egypt. Prepared for UNESCO, Cairo
111. Batisha F (2012) Adaptation of sea level rise in Nile Delta due to climate change. Earth Sci Clim Change 3:2. https://doi.org/10.4172/2157-7617.1000114

Decentralized Wastewater Treatment Using Biofilm Technologies as Cost Effective Applications

Case Studies—MENA Region

Noama Shareef

Abstract This study was conducted to demonstrate and evaluate the economic effects of biofilm application in municipal wastewater treatment in the MENA region. These wastewater treatment pilots were designed and operated of different Biofilm application in tow MENA countries (Tunis and Jordan) as decentralized wastewater treatment applications in small communities. A submerged fixed biofilm was utilized in Tunis, as the first pilot in this study, in the aerobic biological reactor (SFBB) and the moving bed process in an anaerobic denitrification reactor has been designed to treat municipal wastewater treatment for a small community 200 PE. This application aims to achieve full treatment and high-quality outlet before discharge into water bodies. This was requested to protect the Mediterranean Sea including reuse options in agriculture. While the second pilot plant was implemented in Jordan and designed as two Modified Septic Tanks to treat municipal wastewater just to secondary treatment level for reuse in agriculture and to test the biofilm efficiency treatment. Septic Tanks were operated in two systems (suspended and attached growth bacteria). The two pilot plants in (Tunis and Jordan) were implemented, tested, and evaluated as part of different research projects focused on low-cost treatment due to reducing energy requirements by using biofilm technologies. High removal of the organic load was achieved in both systems pilot plants in Tunis and Jordan. Effluent BOD5 concentration was found to be less than 10 mg/l for the pilot in Tunis and 15 mg/l for the pilot plant in Jordan, high nitrogen removal for more than 95% in Pilot-Tunis but it was at a lower rate in Pilot-Jordan as well as *Escherishia coli* removal, nitrifications, denitrification were achieved in a high rate both Pilots. The application of different biofilm system technologies at the two pilots were proved to be a cost-effective technology with low energy consumption, low-cost implementation, and O&M requirements. Therefore, the Effluent quality met secondary wastewater quality criteria as well as WHO Guide and local standards for wastewater reuse in agriculture.

Keywords Biofilm technologies · Onsite decentralized wastewater treatment · Septic tank · Aerobic · Anaerobic · Bacteria suspended/attached · Submerged fixed film · Moving bed · Hybrid-reactor

N. Shareef (✉)
Water Policy Officer, European Water Association-Bonn, Reuter street 153, 53113 Bonn, Germany

M. Abu-hashim et al. (eds.), *Agro-Environmental Sustainability in MENA Regions*, Springer Water, https://doi.org/10.1007/978-3-030-78574-1_3

1 Introduction

Centralized wastewater management faced many challenges as high-cost collection and treatment system due to many factors like the high cost of intensive operation, using high-tech and costly investment in maintenance [1, 2].

In small communities, a decentralized wastewater treatment management DWWM as a solution for rural and small communities is needed to avoid the challenges of a centralized system [3, 4]. Therefore, supporting wastewater activates and services management in rural areas addresses serious concerns accordingly to water scarcity, protecting the environment from pollution, and public health. Moreover, the application of decentralized wastewater treatment technologies has been recognized as a cost-effective solution [5, 6]. However, a long time ago, Biofilms technologies were used in wastewater treatment plants, to degrade organic particles also to improve the nitrification and denitrification processes in wastewater treatment units. Different fixed growth biofilm reactors are commercially used for the treatment of municipal as well as industrial wastewater. Therefore, the awareness of biofilms has increased enormously in recent years due to the impact of biofilms on natural and industrial systems, as well as human health. Biofilm technologies are used widely to develop decentralized wastewater technologies which conceders in many applications as cost-effective solutions for rural-small communities [6–8].

Biofilm processes require less space as they keep consistently the active sludge concentration in the biological reactor at a high level [9]. Biofilm processes do not need sludge recirculation thus making the biological treatment process independent from sedimentation tank performance and sludge characteristics. This is especially of great importance for applications where the type of waste implies the risk of high sludge volume indices. The increase of the demands on efficiency and cost of wastewater treatment leads to new interest in biofilm technology which leads to an important positive impact that can affect the treatment cost, operation, and treatment efficiency in several countries in the region. The analysis of the incremental scenarios has shown that the change in precipitation and temperature will highly affect the amounts of monthly surface runoff and the mean annual temperature increased by 1, 2, 3, and 4 °C [10].

This chapter gives some experience about the application of several Biofilm technologies in the MENA region like submerged fixed-film, Moving bed, and MBBR technologies as a good example of a biofilm system applications in arid to semi-arid areas with several specifications as hot weather, high organic load, and other related factors effects. As shown in some studies on Innovations in wastewater applications in Jordan in adaption to climate change [11].

The aim of this chapter: Decentralized wastewater treatment by using Biofilm technologies is to give an overview of these technologies and its environmental economics applications. Furthermore, to demonstrate the application of several wastewater treatment technologies with the high-efficiency treatment of municipal wastewater in small and large scale using biofilm systems with respect to the application of low-cost wastewater treatment also to reuse the treated wastewater for irrigation purpose as successful case studies from the MENA region.

2 Overviews of Wastewater Treatment

Wastewater treatment used to remove pollutions from wastewater like organic and non-organic matters in wastewater treatment plants to the recycle level and reused it again for various purposes like irrigation, recharge the groundwater ... etc. Many factors related to Wastewater and climate change have to be taken into consideration during the above treatment stages and processes [12]. The temperature, season, and type of wastewater can drastically affect wastewater treatment factors. For example, the discharge of treated wastewater affects the treatment stages and level, in addition, it affects the treatment cost. If the treated wastewater is to discharge into water bodies, the biological and chemical load has to be less than its disagree for reuse purposes in agriculture. Wastewater treatment stages are divided normally between three main stages, which are needed according to the treatment levels and discharge. These main stages are preliminary and primary treatment, secondary treatment, and advanced treatment. Every stage has different treatment process, by chemical, Physical and mechanical treatment process in the primary treatment stage the majority of the solids, organic and inorganic material are removed [8, 12]. The sedimentation process is used to remove solids, while grease, fat, and oil are removed by the skimming process. The reduction of (BOD) biochemical oxygen demand is between 25 and 50%, and Approximately 50–70% of (SS) total suspended solids reduction, while the reduction of oil and greases up to 65% during primary treatment stage in wastewater treatment plant" [13].

Secondary treatment of wastewater uses microorganisms that break down the organic in wastewater and remove any organic compounds. The biological treatment process in this stage, simulates exactly the natural process as it happens in the nature. Biofilm treatment system is one of many approaches that can be used as biological treatment technology in this stage: Submerged fixed film system, as a biofilm system where attached microorganisms grow directly on the surfaces of several types of filter media (Rock, Sand, Plastic, textile ... etc.) to create a thin biofilm. In this stage, the treatment process accrues by feeding off the wastewater including its concentration of organic matter and nutrients that flow over the media and the biofilm [14]. While advanced wastewater treatment used to remove pollutions like color, odor, and taste. It refers to all process treatment are used to increase the wastewater quality obtained from other stages primary and secondary treatments before discharge it to water bodies, because it needs an advanced treatment level. A combination of treatment techniques is usually used as an advanced stage for treatment such as UV disinfection, ozonation, and chlorination [15].

2.1 Biofilm Treatment Process

Biofilms technologies are used in wastewater treatment plants, for a very long time, to remove organic compounds and to improve the treatment process producing high nitrification and denitrification levels. Attached bacteria grows on the filter media as a thin biofilm to break down the organic waste to carbon dioxide and water and new biomass. Oxygen diffusion is limited to the upper biofilm layer; lower layers are anoxic or even anaerobic. Many factors are effects the treatment process like biofilm thickness, wastewater load, oxygen diffusion … etc.

Thin biofilm is more active than thick biofilm, because, anaerobic treatment process in absence of oxygen will happened near the surface of the surface media; this builds a problem in the capacity treatment and decrease the treatment efficiency.

Deferent microorganisms created on the media surface to break down the organic matters by the contact between the microorganisms and wastewater. Throw the MSU [16], Fig. 1 shows the combined aerobic and anaerobic biofilm processes.

Where a combination of moisture and nutrients exists on a media surface, it is likely to find a biofilm there. Biofilms can grow on any natural materials above and below ground, on any materials like: Rock, Sand, metals, plastics, medical implant materials, and plant and body tissue. Over time, a biofilm in the appropriate environment will grow and become strongly attached to the media surface to live on it [17].

Microbial biofilms behavior is naturally tolerant of antibiotic doses up to 1000 times greater than doses that kill free-swimming, planktonic bacteria [16].

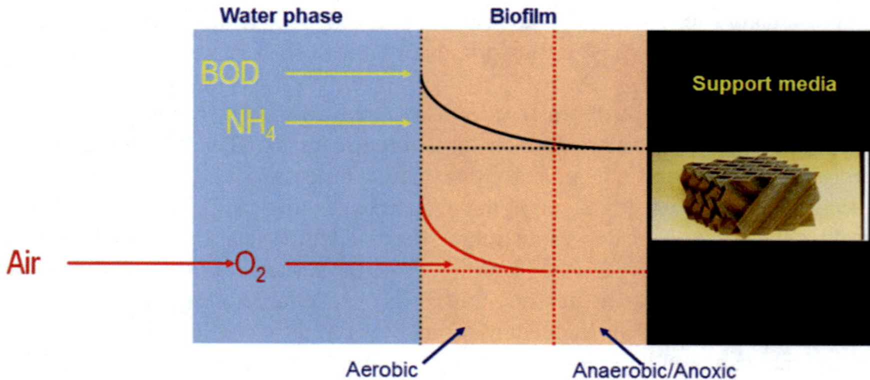

Fig. 1 Combined aerobic and anaerobic biofilm processes

2.1.1 Aerobic Decomposition of Typical Organic Compounds

Decomposition is an aerobic biological process, of degradation of organic matter which is easily degradable by different microorganisms in the presence of oxygen like Organic matter degradation and nitrification.

Organic Matter Degradation

Organic matter degradation is the first stage that simulates the natural process as it happens normally in nature when microorganisms break down organic matters [12].

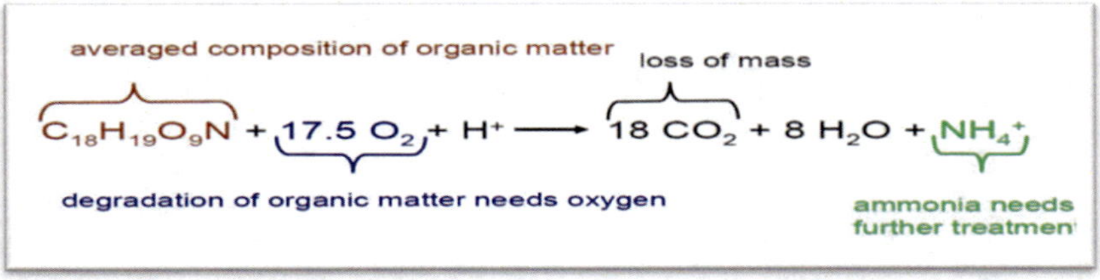

averaged composition of organic matter

loss of mass

$$C_{18}H_{19}O_9N + 17.5\ O_2 + H^+ \longrightarrow 18\ CO_2 + 8\ H_2O + NH_4^+$$

degradation of organic matter needs oxygen

ammonia needs further treatmen

Nitrification (Aerobic Process)

Nitrification is a biological treatment process carried out by microorganism or ammonia-oxidizing bacteria (AOB). This anaerobic biological process converts the organic nitrogen into nitrate-nitrogen in 2 steps. Nitrification is a common method used for the removal of ammonia nitrogen in the wastewater treatment plant. Nitrification occurs in tow steps oxidation reactions, the first step the ammonia oxidizes into nitrites in presence of oxygen, whereby nitrites get oxidized into nitrates as a second step shown by the half equations below: Nitrification is the oxidation of TKN to NO_3 in 2 steps: All nitrification process

$$NH_4^+ + 1.5O_2 \rightarrow NO_2^- + H_2O + H^+$$

$$NO_2^- + 0.5\ O_2 \rightarrow NO_3^-$$

Nitrification is affected by wastewater temperature, for example: with a drop in temperature from 20 to 15 °C resulting in a drop in nitrification effectiveness of 35% [18].

2.1.2 Anaerobic Decomposition of Typical Organic Compounds

An anaerobic biological treatment is a partial conversion of organic matters into simpler organic compounds.

Denitrification (Anaerobic Process)

Denitrification is an anaerobic biological process that uses different microorganisms to convert nitrite and nitrate to elemental nitrogen. It is an anoxic process, it takes place in the absence of oxygen; and heterotrophic processes, requiring organic carbon for cell synthesis and energy generation [19]. Therefor the Denitrification process is converted of $2NO_3$ to N_2 plus $3O_2$ in absence of Oxygen:

$$2NO_3^- + 2H^+ \rightarrow N_2 + H_2O + 2.5O_2$$

2.2 *Design Features*

Biofilm technologies are designed usually based on the specific surface area of the media filter, organic loading rate (gBOD or COD/m^2/day), therefore the size of the carrier media is not be significant as long as the effective surface area is the same. Biofilm application technologies in municipal and industrial wastewater indicate that high-efficiency wastewater treatment can be used to remove soluble BOD in high organic loading rates also improve biological treatment process as Nitrification and denitrification. Organic loading rate in wastewater is defined as the concentration of soluble and particulate organic matter. It is typically expressed as organic load rate BOD per unit area which has to be removed by the treatment process, as organic matter BOD5/m^2/day [20]. Therefore, the biofilm reactor volume calculation depends on the organic load, specific surface area, and wastewater inflow as in the quotation below.

$$OSLR = \frac{Q(\mathrm{m^3/day}) * BOD(\mathrm{g/m^3})}{V_{filter}(\mathrm{m^3}) * A_{specific}(\mathrm{m^2/m^3})}$$

Control of organic loading can be accomplished by pre-treatment to reduce the BOD and TSS concentrations or by increasing the size of the infiltration area to reduce the mass loading per unit area. There are two methods of degradation of organic compounds. The design of most selected case studies in this chapter was to make sure that the reliable operation of the plant does not need highly experienced personnel.

2.3 *Biofilm Application Area*

2.3.1 Municipal Wastewater Treatment

The media material of biofilm technologies can be designed to meet a large specific contact area, which increases the treatment efficiency. Therefore, this kind of treatment process is an ideal process to achieve full secondary treatment. These technologies also are applicable as a cost-effective solution for wastewater treatment.

Bio-film technologies required a very good pre-treatment to decrease the load before it comes into biofilm where the capacity of existing biofilm has to be increased. However, the special treatment process leads the water quality to meet new standards dictated by legislation [21]. The Plastic biofilm-media range enables wastewater treatment plants by this system to increase treatment efficiency be to meet future needs with low-costs and disruptions to the treatment capacity.

2.3.2 Industrial Wastewater Treatment

Industrial wastewater mainly has a variable quantity and quality, it is characterized by special load degree depending on the type of industry producing it, may it includes organic matters easy or complex with different loads, or heavy metals and toxic matters, as the case in many developing countries. The main issue for industrial wastewater is increasing wastewater volume production which may including a high load of synthetic compounds to be directly discharged in nature. Special features like (temperature, different organic load, different PH levels, and Salinity) can be affected the treatment process. Biofilm system is utilizing to treat different types of industrial wastewater with diffrent loads and PH levels to produce high quality effluent. A temperature reduction of the industrial wastewater through the biofilter treatment process is often an additional effect [22].

3 Implementation of Biofilm Technologies (Successful Case Studies)

The implementing of biofilm technologies as (submerged fixed film, moving bed, MBBR … etc.) has been considered to use for BOD reduction, good nitrification in aerobic stages and improve the denitrification efficiency in consideration with the climate change effects at the implementation area.

3.1 *Case Study: Small Municipal Wastewater Treatment Unit for 200 PE (Application in Tunis)*

Aerated Submerged Fixed Biofilm Bed (SFBB) reactors work on the basis of biofilms attached to monolithic plastic supports. Mixing and transport processes within the fixed media are achieved by the aeration in aerobic systems (oxidation of organic compounds and nitrification). Although simultaneous denitrification has been reported by some authors, anoxic conditions cannot be realized in aerated systems.

This study presented a small size, easy to handle wastewater treatment in a compact unit which has been designed and developed primarily for MENA countries. This pilot was implemented in Tunis for municipal wastewater treatment of small community 200 PE and to demonstrate a full treatment efficiency for reuse including discharge in water bodies.

It is an important solution to protect the Mediterranean Sea in such tourist areas including the possibility of reuse. Therefore, this study demonstrates a new concept of decentralized wastewater treatment pilot plant by using biofilm technologies: Submerged fixed biofilm in aerobic nitrification reactor and moving bed in anaerobic denitrification reactor. This kind of small treatment unit by using a simple design makes sure that the reliable operation of the plant does not need experienced personnel so it can be operated simply by the owner or any person with normal knowledge. Figure 2 presents the schematic design concept of reactors.

The pilot plant was designed and tested for wastewater with nitrogen input concentrations of 100 mg/l and higher as it is frequent in arid countries.

Basic Design Data

Unit capacity: 200 PE (Population Equivalent)

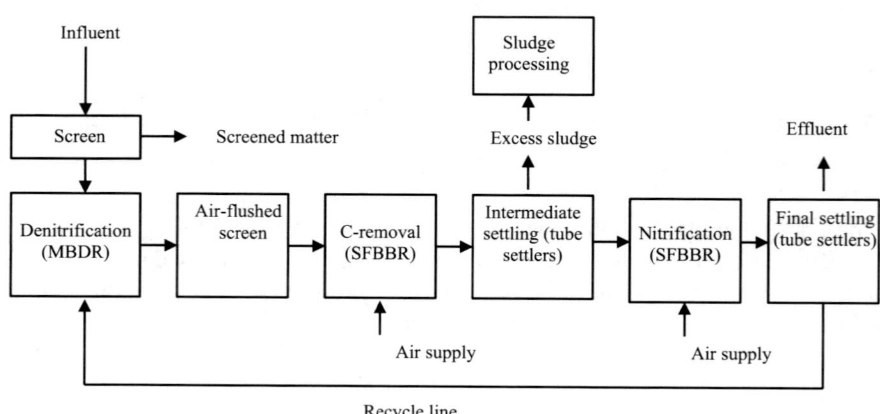

Fig. 2 Displays the flow scheme of the treatment plant 2H-water technologies solution

Specific wastewater load: 20–40 m^3/d
BOD5 load, daily average: 20 kg/d.

Target Quality of Treated Wastewater

BOD5 concentration: $\leq$ 10 mg/l
COD concentration: $\leq$100 mg/l
TSS concentration: $\leq$25 mg/l
NH_4-concentration: $\leq$2 mg N/l
NO_3-concentration: $\leq$10 mg N/l.

Table 1 shows that each compact system (Submerged Fixed Film for nitrification and moving bed for denitrification), was generally used by 2 h GmbH for such applicable technologies, this step is designed as a tailor-made design to favour the growth of specialized bacteria. These specialized bacteria are kept on its place and supplied with its individual needs of oxygen and nutrients.

The plastic foils for the bio-filter media are normally produced using the very specific system for extruding the Polypropylene polymer and directly forming the molten mass into the corrugated foils. The polypropylene polymer material shall be specially formulated for optimum UV stability like the patented 2H Kunststoff GmbH system. A plastic component applied in the experimental above WWTP is presented in Fig. 3.

Table 1 Design data for the projected wastewater plant 2 h GmbH

Unit	gBOD/m^2/d	gNO_3-N/m^2/d	gNH_4-N/m^2/d	Applied surface (m^2)
Denitrification	3.5	1.0	–	834
C-removal	9.0	–	–	770
Nitrification	–	–	1.2	2082
	Max. flow m^3/h	Tube volume m^3	Proj. surface m^2	Surface load m^3/m^2/h
Intermediate settling	7.6	2.53	13.7	0.55
Final settling	2.1	0.76	6.6	0.32

a) Bio carriers

b) Structured media

c) Tube settler module

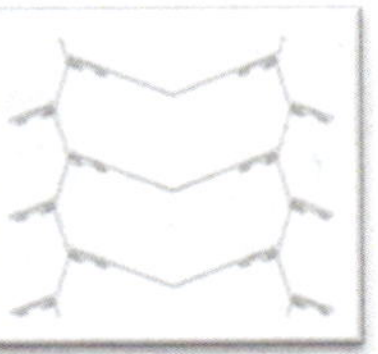
d) Tube settler geometry

Fig. 3 Plastic components applied in the experimental WWTP (2H Kunststoff GmbH) Germany

3.1.1 Results and Analysis

The treated municipal wastewater in the compact unit in Tunis by using both biofilm technologies (Submerged fixed-film for an aerobic reactor. and Moving bed for anaerobic denitrification stage), shows high treatment efficiency in final plant outlet (BOD5 $\leq$10 mg/l, COD $\leq$100 mg/l, TSS: $\leq$25 mg/l, NH_4^-: $\leq$2 mgN/l, NO_3 $\leq$10 mgN/l).

First results of the demonstration plant shows that: halving the loads does not have any effect on mining performance in the load case of this study also during the starting an operation time the COD degrad > 80%, NH_4-N degradation almost 100%, and low COD degradation after changing the operating system, Fig. 4.

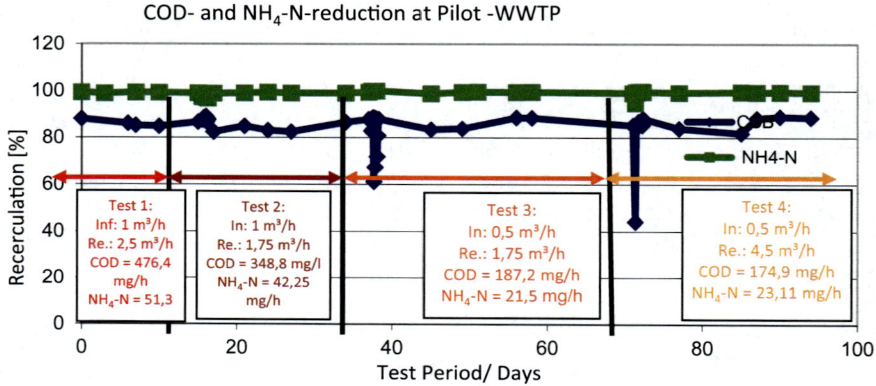

Fig. 4 COD- and NH_4-N reduction at the pilot plant during the operation phase

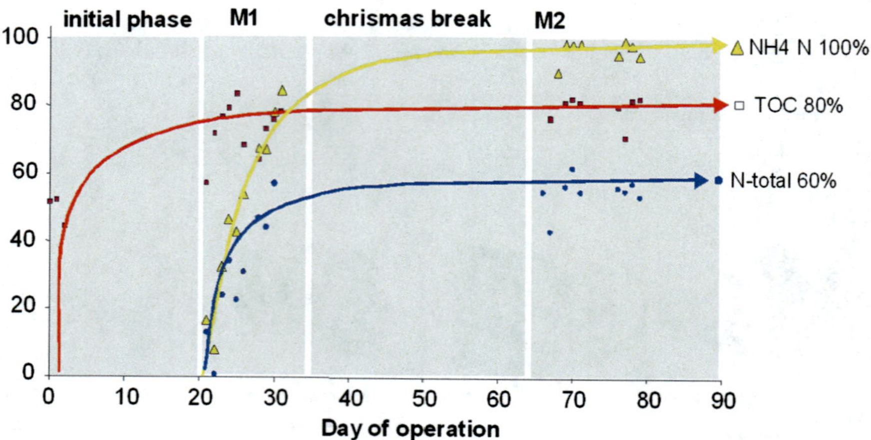

Fig. 5 Reduction of TOC and N-total during the operation phase

Figure 5 shows a high reduction of TOC and N-Total during the first operation phase of the pilot plant which is improved during the operation phases.

Figures 6,7, and 8 also show the high process eliminations of biofilm technology applications (Submerged fixed-film for the aerobic reactor. and Moving bed for anaerobic denitrification stage) and therefore wastewater treatment efficiency.

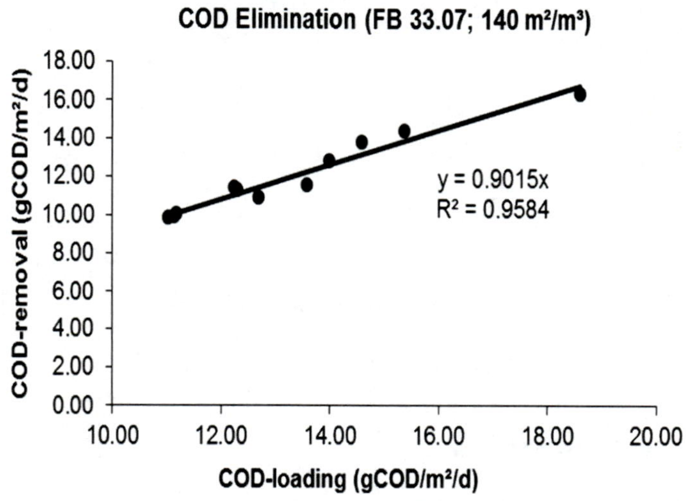

Fig. 6 COD-loading rates versus COD elimination per surface unit

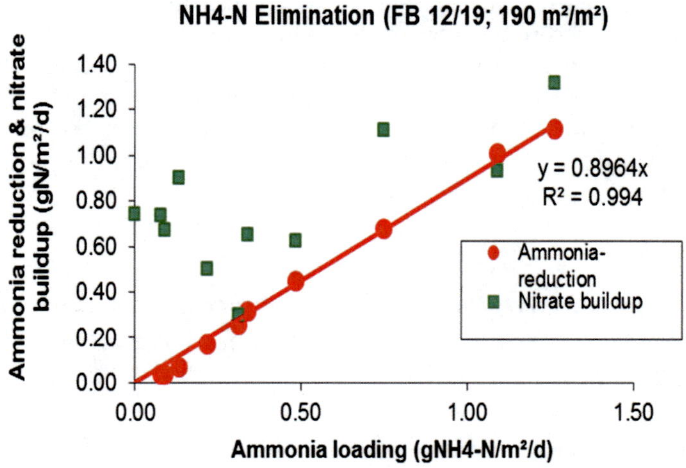

Fig. 7 Ammonia-reduction and nitrate build-up in the nitrification reactor

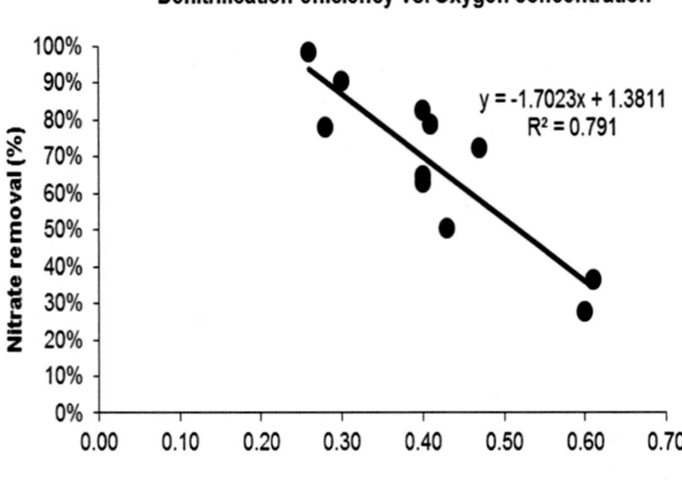

Fig. 8 Denitrification versus oxygen concentration

3.1.2 Economic Aspect

The comparison between the application of the above treatment concept and the activated sludge treatment system for the same 200 PE small unit shows reduction of about 50% of tanks volumes in the system by using the Biofilm concept taking into consideration using lamellar settler in the sedimentation stages.

Therefore, and according to the treatment evaluation, the application of Biofilm technologies by using plastic media showed a clear improvement in the system performance. This study revealed the using (Submerged fixed-film for the aerobic reactor. and Moving bed for anaerobic denitrification stage) can be used as a cost-effective system due to its low energy consumption which leads to low-cost treatment applications.

3.2 Case Study 2: Optimization of Modified Septic Tank Wastewater Treatment Performance for Small Scale (Application in Jordan)

An Application of Biofilm system at Modification septic tank for municipal wastewater treatment in Jordan shows a good treatment efficiency of BOD reduction, good nitrification and denitrification performance.

Two Modified Septic Tanks with dual operational conditions for the anaerobic stage and aerobic stage have been implemented and designed, and operated in parallel

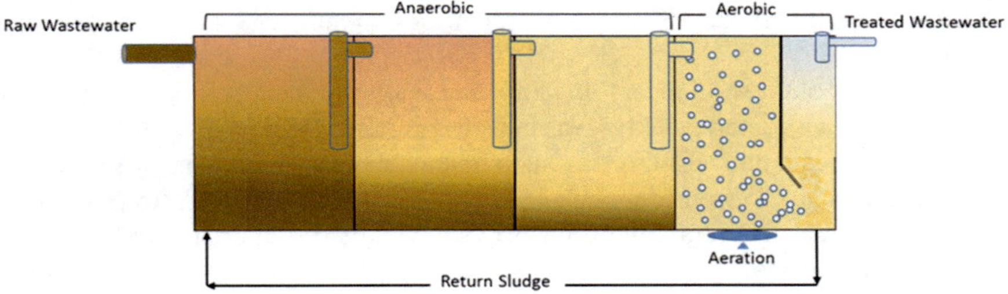

Fig. 9 Schematic diagram modified septic tank—SMART project—Jordan

at the Al Fuhies demonstration site of the SMART (Sustainable Management of Available Water Resources with Innovative Technologies) as one of a potential decentralized system that can be used for the onsite treatment of local domestic wastewater in Jordan [5, 23, 24].

The aim of this research study is to check the treatment effect of biofilm plastic media application and its treatment efficiency. To compare the attached growth system (effect of fixed biofilm) to the conventional suspended growth system in treatment efficiency, some tested under several hydraulic loads were tested online. Septic tank is the primary settlement tank and anaerobic reactor as in Fig. 9. Influential sewage depends through concrete splash baffle and is scattered on the surface of the tank's content disturbing the scum layer and settled sludge.

Each modified septic tank technology suspended growth system (BS), and attached growth system (BA) was utilized and implemented as one anaerobic reactor as the first chamber, and followed by aerobic reactor process, SMART [11, 23]. The anaerobic chamber was designed to be very similar to the traditional septic tank in general, where treated wastewater flows from the first anaerobic part to the aerobic part without pumping (by gravity), for the further treatment process. The Septic tank working by Attached bacteria process (BA) is implemented to be filled with submerged fixed-film plastic media in anaerobic and aerobic chambers to check the treatment efficiency in both and to test the effect of fixed biofilm to compare its treatment capacity with the suspended bacteria system (BS) in the same conditions with the same hydraulic loads. The analyses of outlet data showed that the attached Bacteria reactor system BA achieved high efficiency of biomass reduction and better wastewater treatment efficiency than the Suspended bacteria system. This was because the attached bacteria on the plastic media (support materials); can be more efficient to reduce the organic load and has achieved good nitrification and denitrification levels.

3.2.1 Methods and Measurements

The tow modified septic tanks (BA and BS) have been operated under different hydraulic loadings (phase1 is 1200 l/day, phase 2 is 1600 l/day and phase3 is:

2000 l/day) each hydraulic load was operated for 6 months and with inflow organic load BOD 600 mg/l.

Inflow and outflow of every loading phase during its operational time was controlled and evaluated weekly. Wastewater sampling from each system (BA and BS) were collected and tested onsite such: chemical oxygen demand (COD), Total suspended solids (TSS), Biochemical oxygen demand (BOD5), nitrate (NO_3^-), dissolved oxygen (DO), orthophosphate (PO_4^{3-}) pH value, ammonia (NH_4^+), electrical conductivity (EC), temperature. All the above parameters were tested according to the Standard Methods for Examination of Water and Wastewater [23, 25]. *E. coli* was measured using the IDEXXTM Colilert-18 Quanti-tray method according to the manufacturer's specifications.

3.2.2 The Operation Phases

The systems BA and BS were operated with deferent hydraulic loading of 1.2 m^3/d (Phase 1) and under higher loading rates of 1.6 and 2.0 m^3/d (Phases 2 and 3). Three different hydraulic loadings above were carried out for 26 weeks for every loading level during the operating phases. The modified septic tank systems were designed to accommodate wastewater generated from a single household. Preliminary investigation shows that the average daily wastewater discharge from a single household is about (1.2 m^3). The schematic diagram of the pilot is shown in Fig. 10 [26].

The first chamber (anaerobic chamber) is considered as sludge holding tank, where sludge stabilisation is anticipated to take place. The plant has been equipped with several by-passes, fittings, and valves so that the mode of operation could be easily changed and adjusted.

3.2.3 Results and Analyses

Two modified septic tank systems were implemented for municipal wastewater treatment in a rural place about 17 km south of Amman in Jordan. The tow Septic tank systems (BA and BS) were tested and evaluated. The treatment efficiency evaluation shows a good elimination of several parameters during the biological treatment phases and deferent hydraulic load (Fig. 11). As it shows a reduction of BOD to 95–98% and between 92 and 98% TSS removal. Also, analyses show nitrogen removal of 59% in the BA system, while the nitrogen removal varied between 26 and 33% at the higher hydraulic loading and 0–29% in the BS system [23].

Regarding the WHO guideline and local standard the Effluent quality meets secondary wastewater quality criteria as well as the Jordanian standard for wastewater reuse for all parameters (Table 2), except for the parameter *E. coli*, where further disinfection is required.

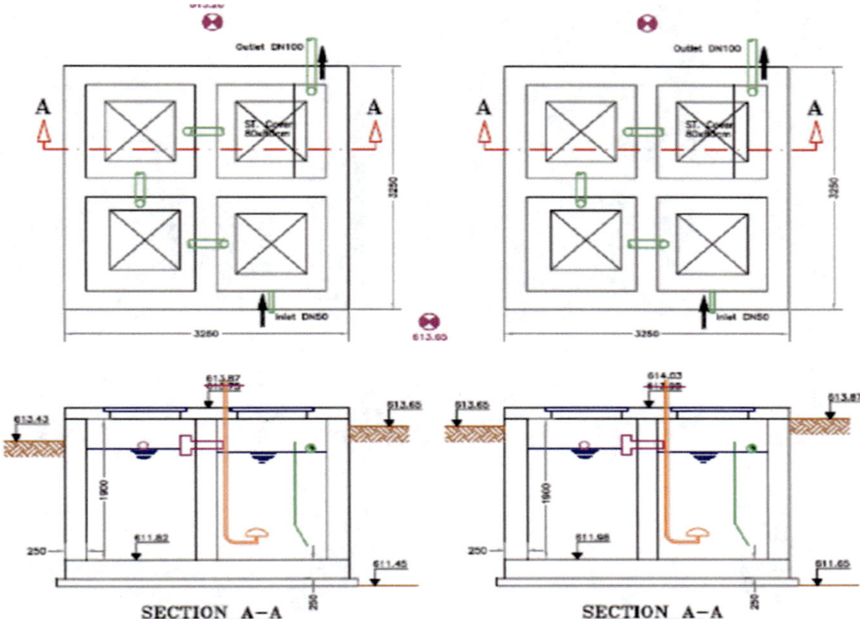

Fig. 10 Two modified septic tanks—process diagram of pilot project SMART

4 The Demand for Capacity Building in the MENA Region

Capacity the building is the main objective of the implementation of a biofilm system as a low-cost, decentralized wastewater treatment technologies in the Middle East and North Africa.

Applying a very specific capacity building for operation and maintenance of biofilm-system is very important because of the demand of this know-how in field experience also the reuse of its end product as an alternative water resource for Irrigation. This could be achieved by providing technical groups in local institutes with required capacity in the biofilm applications as very specific fields. Universities and research centers can play a leading role in raising awareness and technical experience with respect to attached biofilm system as low-cost wastewater treatment technologies because of low energy consumption and reuse treated water for agricultural purposes. SMART project in Fig. 12 was a successful implementation of economic wastewater treatment technologies. This demonstration and research center demonstrates successful applications of several treatment technologies as low cost treatment solution for rural area in the MENA region.

Implementations of several projects in the MENA region demonstrated a successful application of capacity-building activities especially there is a high demand to strengthen experience in water management in the region [28]. During the implementation of the SMART project, a capacity-building program was established

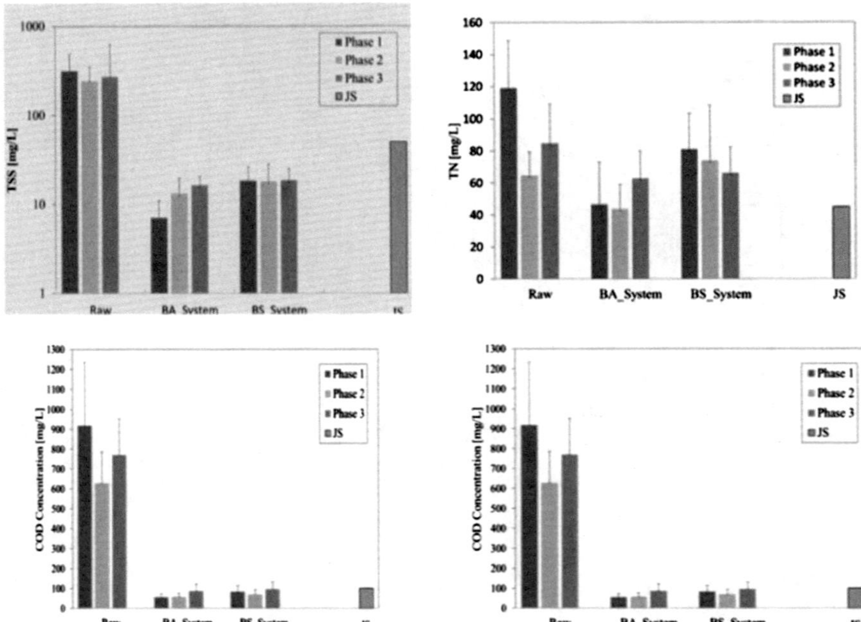

Fig. 11 Process elimination in 3 different phases of BA&BS modified septic tanks

Table 2 The current Jordanian standards for treated wastewater reuse in irrigation [27]

Parameter	Cooked vegetables, parks, playgrounds and sides of roads, within city limits	Fruit trees, sides of roads outside city, limits and landscape	Field crop, industrial, crops and forest, trees	Cut flowers
	A	B	C	D
BOD_5 [mg/L]	30	200	300	15
COD [mg/L]	100	500	500	50
DO [mg/L]	>2	–	–	>2
TSS [mg/L]	50	150	150	15
pH	6–9	6–9	6–9	6–9
Turbidity [NTU]	10	–	–	5
NO_3-N [mg/L]	30	45	45	45
TN [mg/L]	45	70	70	70
E. coli [MPN/100 mL]	100	1000	–	<1.1
Intestinal helminth eggs [egg/L]	< or = 1	< or = 1	< or = 1	< 1
Grease, oils [mg/L]	8	8	8	8

E. coli: *Escherishia coli*

Fig. 12 Research demonstration site for decentralized wastewater treatment—SMART project

by Al-Balqa applied university to support master students, to improving teaching skills and lecturing output, and developing lines of research in their corresponding institutes by designing some related regional workshops. In addition, the exchange of basic research experience at this level was enhanced by coordination between several stakeholders and institutions that carried out research in the counterpart institutes. Research findings were disseminated via scientific publications and a number of technical papers that were produced. These publications established a position for the partner institutes in the field of anaerobic treatment and agricultural reuse of treated wastewater, and most were directly related to the findings of the doctoral students. The doctoral students also participated in international conferences to present their work on the research demonstration center in Jordan to share their experience with other researchers worldwide.The recognition of BAU-University encouraged by the participation of external high-level experts as lecturers in an intensive specialized course in several levels on wastewater treatment as field experience in anaerobic/aerobic process and in agricultural reuse held in Amman, Berlin, and the Dead Sea via the project period. These indirect targets consisted of professionals from the academic sector, water authorities, ministries, and private sectors. Professionals were trained during specialized courses held in Germany and Jordan. The course material covered the operation and maintenance of decentralized/centralized wastewater treatment plants, environmental technologies, emphasizing low-cost anaerobic, aerobic treatment methods, and aspects of wastewater re-use in agriculture. The trainees are expected to positively influence the national agenda related to MENA sanitation policy.

5 Conclusions

The applications of wastewater treatment by biofilm technologies show the successful implementation of wastewater treatment technologies to improve wastewater treatment efficiency and low energy requirement.

Case studies in this chapter revealed that the application of biofilm technologies in wastewater treatment in the region can be used as a low-cost system due to its low energy requirement.

Effluent quality met secondary wastewater quality criteria as well as the WHO and local standard for wastewater reuse in agriculture.

The generated excess sludge in the tow above pilot projects was exposed for drying in sludge drying bed for composting and sludge reduces from the wastewater treatment stages, and also the sludge as an end product. This is after ensuring that it meets the local reuse standard which can be reused safely in agriculture or for other disposals purposes.

6 Recommendation

Experience in the operation of biofilm technologies shows that: Biofilm system applications faced several operations problems in the MENA region. This is related to the lack of experience in biofilm system operation and in know-how of its applications. Some recommendations are need to be taken into consideration by application of biofilm technologies under several climates conditions but not limited to:

- Fats, oils, Solids and Suspended solids, and greases must be removed or minimized in the influent by very good pre-treatment efficiency to protect the biofilm and its specific surface area this leads to an increase in the Biofilm treatment efficiency.
- A good pre-treatment is required to reduce high concentrations of BOD5, TSS in inflow, this protects the biofilm treatment capacity in the biological stage.
- Researches demonstrate that biofilm technologies should be designed based on the surface area loading rate (gCOD/m^2/d), Media material, wastewater type.
- In Tricking filter, Pressure distribution is required for the design option in order to strive for equal distribution of effluent at the design organic loading rate and also to keep the attached biofilm wet.
- Intermediate clarifier tank is useful to build between treatment stages to draw out the sludge produced and to keep the attached bacteria working in its full capacity.

Challenges faced the implementation of DWWM

Decentralized wastewater management is an important solution in wastewater treatment for rural areas in the MENA region, but there are many challenges faced the implementation of Decentralized Wastewater Management System (DWWM) but not limited to:

- Financial and Institutional Problems.
- Non-linked areas on drainage systems.
- Technical Problems.
- The Inefficient Water and Wastewater Networks.
- Small Capacities of the centralized wastewater treatment plants.
- Lack of know-how of new wastewater technologies.
- Unskilled Operation and Maintenance team especially in biofilm technologies (O&M).
- Lack of Master Plan in Governorates.
- Local ministries support in (investments, regulations, control, laws, etc.) still in discussion in most MENA countries.
- Increase the water demand in some hosting countries because of internal and external migration related to the political issue in the MENA region.
- High Water consumption in agriculture (for Jordan: 75% of water use in agriculture).
- The absence in Plans to Preserve Sector Investments.

Acknowledgements The research case studies were funded by a grant from the German Ministry of Research and Education (BMBF) in different period of time within the framework of project "export-oriented research and development for waste water treatment" Application in Tunis, and "SMART project" Application in Jordan. Author contribution was for the first case study in design, calculation, implementation, operation, data evaluation during her working for GEA Water technologies in Germany and she supported in in data analyses, unit operation for the second study, as a senior wastewater engineering for CIM-Germany in cooperation with Al Balqa applied university, as it was a part of Abuharb master thesis.

References

1. Mazumder D, Roy B (2000) Low cost options for treatment and reuse of municipal wastewater. Indian J Environ Prot 20:529–532
2. Nasr A, Mikhaeil B (2015) Treatment of domestic wastewater using modified septic tank. Desalin Water Treat 56:2073–2081
3. Capodaglio AG (2017) Integrated decentralized wastewater management for resource recovery in rural and peri-urban areas. Resources 6:22
4. Withers PJ, Jordan P, May L, Jarvie HP, Do Deal NE (2014) Septic tank systems pose a hidden threat to water quality? Front Ecol Environ 12:123–130
5. Flasche K (2002) Possible applications and performance of small-scale wastewater treatment plants. Ph.D. thesis, University of Hanover, Hanover
6. Wilderer PA (2005) UN water action decade: a unique challenge and chance for water engineers. Water Sci Technol 51:99–107
7. Anil R, Neera AL (2016) Modified septic tank treatment system. Proc Technol 24:240–247
8. Sharma MK, Kazmi AA (2015) Anaerobic onsite treatment of black water using filter-based packaged system as an alternative of conventional septic tank. Ecol Eng 75:457–461
9. Boltz JP, La Motta EJ (2007) The kinetics of particulate organic matter removal as a response to bio flocculation in aerobic biofilm reactors. Water Environ Res 79:725

10. IPCC (2006) IPCC guidelines for national greenhouse gas inventories. In: Intergovernmental Panel on Climate Change in Jordan, pp 102–105
11. Shareef N (2019) Innovation technologies in wastewater treatment: a cost-effective solution. In: Waste management. Springer, pp 30–37. ISBN 978-3-030-18350-9
12. Metcalf E (2003) Wastewater engineering: treatment and reuse, 4th edn. McGraw-Hill, Boston, MA, pp 602–607
13. Pescod M, Rome B (1992) Wastewater treatment. In: Wastewater treatment and use in agriculture, FAO
14. Manc LK (2009) Wastewater treatment principles and regulations. AEX-768-96. Ohioline, Ohio State University. http://ohioline.osu.edu/aexfact/0768.html. Accessed 24 Jan 2012
15. Hogan M (2012) Sewage treatment. Encyclopedia of earth. The Encyclopedia of Earth, FAO, Rome. http://www.eoearth.org/article/Sewage_treatment. Accessed 27 May 2012
16. MSU Center for Biofilm Engineering (2008) Center for biofilm engineering: biofilm research & education relevant to industry, health, and the environment | center for biofilm engineering. Montana State University. Accessed 18 Jan 2012
17. Deibel N (2003) Develop a defense against biofilms. Food Safety Magazine. http://www.biofilm.montana.edu/biofilm-basics.html. Accessed 18 Jan 2012
18. Cogan G (2004) Wastewater treatment. In: Wastewater treatment and use in agriculture. FAO
19. Proal A (2008) Understanding biofilms. Bacteriality—exploring chronic disease, 26 May 2008. http://bacteriality.com/2008/05/26/biofilm/. Accessed 23 Jan 2012
20. Otis J (2001) Boundary design: a strategy for subsurface wastewater infiltration system design and rehabilitation. On-site wastewater treatment procedures. In: Proceedings of the ninth national symposium on individual and small community sewage systems. ASAE, St. Joseph, MI, pp 245–260
21. Manual on sewage and sewerage treatment (2013) Central Public Health and Environmental Engineering Organization, Ministry of Urban Development, New Delhi. Retrieved http://www.cmamp.com/CP/FDocument/ManualonSewerageTreatment.pdf. Accessed 30 May 2020
22. DWA-Arbeitsbericht IG-5.6 (2007) Aerobe Biofilmverfahren in der Industrieabwasserreinigung Beispielanlagen
23. Abu Harb R (2017) Optimization of modified septic tank performance in treating domestic wastewater in Jordan. Ph.D. thesis, University of BAU, Salt, Jordan, pp 25–37
24. Klinger J, Goldscheider N, Hötzl H (2015) SMART-IWRM—integrated water resources management in the lower Jordan Rift Valley. Project report 2nd phase; KIT scientific reports. KIT Scientific Publishing, Karlsruhe. ISBN 978-3-7315-0393-4
25. APHA, AWWA, WEF (2005) Standard methods for examination of water and wastewater, 21st edn. American Public Health Association; American Water Works Association, Water Environment Federation, New York, NY
26. Van Afferden M, Mueller RA (2007) "Design of an training and exhibition trail" for decentralized wastewater treatment and reuse at the SMART-research and demonstration site in Fuheis, Jordan. University of Karlsruhe, Karlsruhe, p 28. Available at http://www.iwrm-smart.org
27. JISM (2006) Department of standards and measures. Jordanian Standards No. JS893/2006, Regulation and Standards for Reclaimed Domestic Wastewater. Amman, Department of Standards and Measures
28. Shareef N, Abdelazim N (2020) Waste management in MENA region: a review. In: Shareef N (ed) Handbook efficient management of solid waste in MENA countries, 1st edn. Springer, Germany, pp 50–66. ISBN 978-3-030-18350-9

Estimation of the Olive Orchards Water Requirements Using Climatic and Physiological Methods: Case Study (Tunisian Semi-arid)

A. Bchir, S. Ben Mansour-Gueddes, R. Lemeur, J. M. Escalona, H. Medrano, F. Ben Mariem, W. Gariani, N. Boukherissa, and M. Braham

Abstract Studding olive trees water requirements allows a better water efficiency management. That is why more and more efforts are being made to develop new techniques for more efficient irrigation. In this respect, the estimation of the water needs was carried out using precise methods integrating the maximum parameters of the soil-plant-atmosphere continuum. Some researchers have been based on the use of the climatic method, which is summed up by estimating reference evapotranspiration. Other studies have been based on the direct measurement of tree transpiration (physiological method) by xylem sap flow measurements that are more representative of the tree's water status. Each of these methods is based on a set of climatic, edaphic and physiological parameters of the olive tree. The present work is based on an experimental study carried out on the table olive cultivar "Meski" conducted in intensive. This study aims to estimate the water requirements of the olive tree for a possible optimization of irrigation. To do this, a calculation of the water requirements was carried out by two methods: (1) physiological methods based on Sap flow measured at stem level (T1) and (2) climatic method based on climatic parameters and the water balance (T2). These results allowed us to state that the physiological method allows a better estimation of water requirements. This method also improves the profitability of the olive tree with better optimization of the use of water that arrives up to save 15% water.

Keywords Olivier · Water needs · Optimize irrigation · Climatic method · Reference evapotranspiration · Physiological method · Sap flow

---

A. Bchir (✉) · S. Ben Mansour-Gueddes · F. Ben Mariem · W. Gariani · N. Boukherissa · M. Braham
Institut de l'Olivier, Université de Sfax, Sfax, Tunisia

R. Lemeur
Laboratoire d'Ecologie des Plantes, Université de Ghent, Ghent, Belgium

J. M. Escalona · H. Medrano
Grup de Recerca en Biologia de les Plantes en Condicions Mediterrànies, Departament de Biologia (UIB-INAGEA), Universitat de les Illes Balears, Palma, Spain

M. Abu-hashim et al. (eds.), *Agro-Environmental Sustainability in MENA Regions*, Springer Water, https://doi.org/10.1007/978-3-030-78574-1_4

1 Introduction

Water is the essential element for maintaining plant turgescence, stomatal control, gas exchange also as cell elongation and plant growth. The olive trees need water especially at the instant of the pre-blooming-maturing phases, which unfortunately coincides with the driest and hottest period of the year. During these phases, the olive tree requires high quantities of water supply in order to be able to complete the stages of differentiation of floral and reproductive growth [1]. To accelerate its growth and its setting in fruit, olive tree needs water. It also plays a regulatory role within the various physiological processes of the plant by conditioning its productivity and quantity [2–4]. Irrigation water management should be done with methods to save lots of water and maximize productivity. Water productivity increases with deficit irrigation. This technique aims to save water and control vegetative growth in orchards without any adverse effects on production [5]. During the first years after planting, the water requirements needs should be compensated by irrigation (Full irrigation) to establish tree growth as soon as possible [6]. Laster, a deficit irrigation strategy is the best option for most olive orchards [6–13]. When there's a shortage of water for irrigation, supplementary or supplementary irrigation is sufficient to significantly increase crop performance [14, 15].

Several methods were used to determine water requirements. Some researchers have used the climatic method, which consists on estimating ET_0 and ETc, by several equations such as that of Penman–Monteith, to estimate the daily water requirements of the olive orchards [16–22]. Others have used xylem sap flow measurements, which also allow estimation of water consumption and transpiration [17, 22–26].

The objectives of this chapter are: (i) the use of climatic data for estimating water requirements in olive trees (cv. Meski) in the semi-arid Tunisian (Enfidha); (ii) the use of the sap flow method to quantify transpiration and water consumption; (iii) the evaluation of the physiological method according to ET_0.

2 Soil-Plant-Atmosphere Relations

2.1 *Water in the Soil*

In the soil, the water moves gradually, from wet areas to the driest areas. Its displacement is multidirectional. It is influenced by soil characteristics (depth, texture, porosity), those of the subsoil and those of the plant (root and aerial systems). In packed, compact and high clay soils, water flow is slow, unlike light-textured soils with higher filtration capacity [27].

The water balance is based on the soil physical characteristics. This is a method that can predict the changing water requirements of olive trees. The water available for the olive tree corresponds to the reserves easily usable in the ground. The easily usable reserve in water of a soil, expressed in millimeters of water, corresponds to the upper fraction of the useful reserve (UR) for which the plant is not brought to regulate its evapotranspiration by the stomata. The reserves easily usable are difficult to evaluate because it varies between 30 and 60% of the UR depending on the type of soil encountered [27–30]. To characterize the water status of soil, it is necessary to know the soil moisture at different depths and to be able to follow their spatiotemporal evolution [27].

2.2 Soil Water Balance

The different components of the soil water balance allow the determination of the actual evapotranspiration of the crop, and to evaluate the water supply and the irrigation water requirements. The soil moisture balance Eq. (1) is expressed as follows:

$$\Delta S = Re + I + C - ETc - D - R \tag{1}$$

With: ΔS: variation of soil water stock; Re: effective rain; I: irrigation; C: capillary rise; ETc: evapotranspiration culture; D: drainage and R: runoff.

In the case of localized irrigation, the evapotranspiration culture (ETc) of the olive tree is estimated by the following Formula (2) [31, 32]:

$$ETc = Kc \times Kr \times ET_0 \tag{2}$$

ET_0: reference evapotranspiration measured at meteorological stations. Kc: a cultural coefficient applied to the ET_0 to determine the actual evapotranspiration of the olive tree [1, 32] (Table 1).

Kr is the recovery coefficient that represents the land cover ratio (3). It takes in concideration the area of the ground covered by tree canopy and planting density [32].

$$Kr = 2 \times Sc/100 \tag{3}$$

With: Sc as the surface of the ground covered by the foliage (Sc = n × Π × d 2/4 × 100) and n is the planting density and the diameter of the foliage [32].

Table 1 Monthly Kc values for adult olive trees planted at a density of 286 plants/ha, in southern Spain [32]

	Jan	Feb	March	April	May	June	July	August	Sep	Oct	Nov	Dec
Kc	0.50	0.50	0.65	0.55	0.55	0.50	0.45	0.45	0.55	0.60	0.65	0.50

3 Estimation of Water Requirements

3.1 Climatic Method: Determination of Reference Evapotranspiration (ET_0)

According to Elsayed-Farag [7], the pan evaporation or automatic meteorological stations are the most used methods to determine ET0. The variability of evapotranspiration values, in most case, is controlled by sun radiation, air temperature, air humidity and wind speed [33–35]. To calculate ET_0, authors used diverse empirical combinations of climatic parameters in their formulas [16, 20, 21, 31] (Table 2). Some of them use Air Temperature like Blaney and Criddle [36]. Other formulas are based on Air Temperature and Solar Radiation, for example, Hargreaves Samani (1985). Hargreaves Radiation and Stephens and Stewart (1965) methods. Ivanov (1954), Eagleman (1965), Turc (1961–1965), Chirstiansen–Hargreaves (1969) and Priestley–Taylor formulas also use Relative Humidity with Air Temperature and Solar Radiation parameters. The most complicated formulas are the Penman Original (1963) and the Penman–Monteith (1998), which include the use of the Wind Speed with all the previously cited climatic parameters (Table 2). The use of the different empirical formulas depend on available weather data. Many authors studied the use of limited weather data to determine ET_0 [16, 20, 21, 37–39].

3.2 Physiological Method: Direct Measurement of Transpiration in Olive Trees (Sap Flow)

The different irrigation management techniques associated directly to the soil and partially to the plant only allow the determination of the moment to start irrigation and not the quantity of water consumed by the plant. The sap flow method is an alternative for direct measurement of the quantity of water conveyed through the trunk, branches and tree roots under open field conditions. It allowed to study the physiological behavior of several species of plants and the movement of water in the soil-plant-atmosphere continuum, to estimate the water needs and to evaluate the water balance.

According to Köstner et al. [41] and Saitta et al. [22], sap flow measurements are the most common method to determine transpiration at the tree and the forest canopy level. This kind of measurement represents an integrated measure of tree transpiration. It allows separating between the physiologically controlled plant process and physical evaporation from soil and interception.

The technology was used across a wide range of applications on diverse plant types:

- Natural and urban forest trees [42, 43],
- Woody horticultural trees and vines [7, 22, 44–51],

Table 2 Methods for ET_0 calculation, corresponding formulas and climatic parameters used

Methods	Equations	Climatic parameters used for the calculation
Blaney and Criddle [36] modified by Doorenbos and Pruitt (1977)	$ET_0 = (8 + 0.46T_{average}) \times n/N/i$	$T_{average}$, n, N
Hagreaves-Samani (1985)	Original: $ET_0 = 0.0023\ R_a \Delta T^{0.5} \times (T_{average} + 17.8)$ Modified: $ET_0 = 0.0035\ R_a \Delta T^{0.5} \times (T_{average} + 12.54)$	R_a, $T_{average}$, T_{max}, T_{min}
Christiansen-Hargreaves (1969)	Original: $ET_0 = 0.492\ R_s \times C_{TT} \times C_{WT} \times C_{HT}$ Modified: $ET_0 = 0.492\ R_a \times C_{TT} \times C_{WT} \times C_{HT} \times C_{ST} \times C_R$	R_s or R_a, $T_{average}$, U_2, $RH_{average}$, E
Ivanov (1954)	$ET_0 = 0.0018 \times (T + 25)^2 \times (100 - e/e^0\ 100)$	$T_{average}$, e, e^0
Eagleman (1967)	$ET_0 = 0.035 \times e^0 \times (100 - RH_{average})^{0.5}$	e^0, $RH_{average}$
Stephens and Stewart (1965)	$ET_0 = (0.014\ T_{average} - 0.37) \times R_s/1500/0.039$ With $Rs = (0.25 + 0.5\ n/N) \times R_a$	R_a, $T_{average}$, n, N
Turc (1961–1965)	$RH > 50\%$ $ET_0 = 0.40\ (R_s + 50)\ T_{average}/(T_{average} + 15)$ $RH < 50\%$ $ET_0 = 0.40\ (R_s + 50)\ T_{average}/(T_{average} + 15)\ (1 + 50 - RH_{average})/70$ With $R_s = (0.25 + 0.5\ n/N) \times R_a$	$T_{average}$, $RH_{average}$, R_a, n, N
Priestley and Taylor [40]	$ET_0 = \alpha \frac{\Delta}{\Delta+\gamma}(Rn - G)$	Rn, G, $T_{average}$, $RH_{average}\gamma$, α
Penman Original (1963)	$ET_0 = \left[\frac{\Delta}{\Delta+\gamma}(Rn - G)\right] + \left[\frac{\Delta}{\Delta+\gamma} 15.36\left(1 + 0.0062\, U_2(e^0 - e)\right)\right]$	Rn, G, $T_{average}$, $RH_{average}$, U_2, e, e^0, γ
Penman Monteith (1998)	$ET_0 = \frac{[0.408\,\Delta\,(Rn-G)]+\left[900\frac{\gamma}{T+273}(e^0-e)U_2\right]}{\Delta+[\gamma(1+0.34)U_2]}$	Rn, G, $T_{average}$, $RH_{average}$, U_2, e, e^0, γ

Masmoudi-Charfi and Habaieb [31]

Where: T_{max}: Maximum air temperature (°C), T_{min}: Minimum air temperature (°C), $T_{average}$: Mean daily air temperature (°C), ΔT: $T_{max} - T_{min}$ (°C), Δ: Slope of the saturated vapor pressure curve (KPa°C^{-1}), γ: Psychrometric constant (KPa°C^{-1}), U2: Wind speed measured at 2 m height (m/s), RH_{max} and RH_{min}: Maximum and minimum relative humidity of the air (%), es: Saturated vapor pressure (KPa), $e_s = 0.5[e_0(T_{max}) + e_0(T_{min})]$, $e_0(T_{max})$: Saturation pressure at T_{max}, $e_0(T_{min})$: Saturation pressure at T_{min}, e_a: Average value of vapor pressure or actual vapor pressure (KPa), $e_a = \left[e_{0T_{max}} RH_{min} + e_{0T_{min}} RH_{max}\right]/200$, $e_0 - e_a$: Saturated vapor pressure deficit of the air (KPa), Ra and Rs: atmospheric and solar radiation (MJ/m^2/month), Rn: net radiation at the crop surface (MJ/m^2 day), N: Maximum sunshine duration (Hours/day), n: Actual sunshine duration (Hours/day), n/N = p: daylight hours monthly/annual daylight hours, G: Soil heat flux density (Mj/m^2/day), C_{TT}: temperature function, C_{WT}: wind function, C_{HT}: air relative humidity function, C_{ST}: sunshine function, C_R: elevation function

– Agricultural crop species [52–54].

Several sap flow methods are available to calculate the amount of water flowing through the conductive organ of the plant (www.wgsapflow.com; [48]). According to Elsayed-Farag [7], some methods provide information for calculating the sap flow at different depths below the cambium, while others integrate sap flows throughout the sapwood. Some methods are invasive since the sensors are located inside the sapwood; others are non-invasive, the sensors being located outside, well in close contact with the conductive member. Some methods are suitable used for small diameter stems, while others may be used for bigger trees [7]. In this sense, several techniques have been developed, in particular, heat dissipation methods, heat balance and heat pulse methods.

3.2.1 Heat Pulse Method

This method is based on the measurement of xylemmic sap temperatures at distances upstream and downstream of a heat pulse source (Fig. 1). This method was first used by Moreno et al. [55] at the root level and by Fernandez et al. [56] at the trunk level. According to Fernández and Moreno [57], this method has the advantage of determining the kinetics of the sap flow in four locations at the xylem level. This method is widely used because of its low cost, low sensitivity to thermal gradients and low demand for operating energy [22, 53, 58–60].

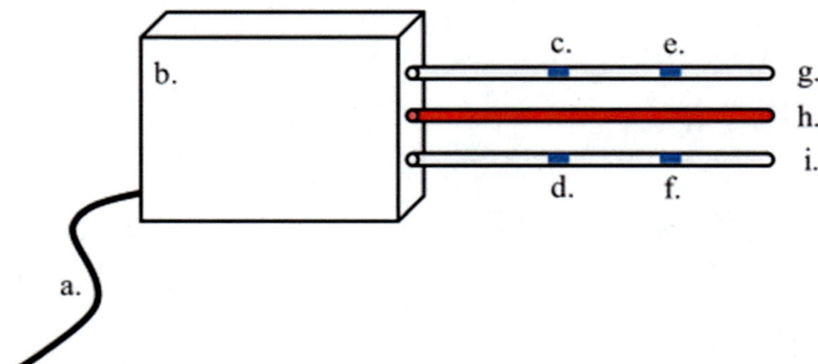

Fig. 1 A schematic for a heat pulse probe to measure heat velocity in xylem. **a** Cable to data acquisition system; **b** epoxy body, PCB or analog–digital interface; **c** downsteam, outer thermistor (temperature) sensor; **d** upsteam outer thermistor; **e** downsteam inner thermistor; **f** upstream inner thermistor; **g** downstream temperature probe; **h** heater probe with an internal heating element; **i** upstream temperature probe. The nominal distance between (**g**) and (**h**) and (**i**) and (**h**) is 0.006 m. The nominal length of probes is 0.003 m [58]

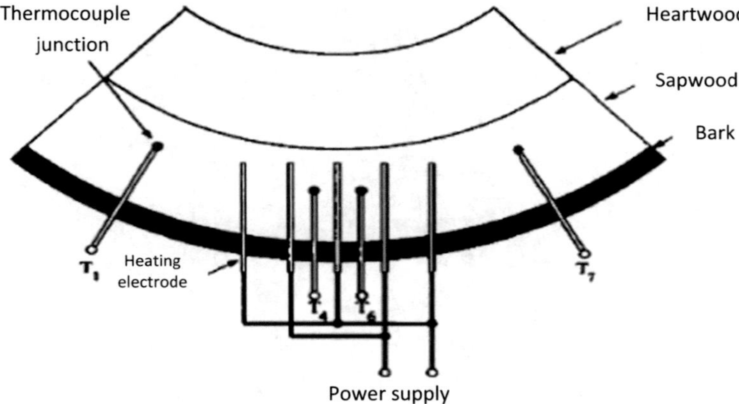

Fig. 2 Schematic representation of the thermal balance method of a trunk area [61]

3.2.2 Heat Balance Method of a Trunk Area

This method is proposed to be used on tree trunks with diameters larger than 120 mm. The sap flow can be calculated from the thermal equilibrium of the heated stem tissues (Fig. 2). However, in this method, heat is applied within a single segment of the trunk, rather than superficially throughout the circumference [61].

3.2.3 Heat Balance Method

This method is employed for various kind of stems (woody and herbaceous) and for different trunk diameters, small, branches and tree trunks [62–66]. According to Smith and Allen [61], the heat balance method employs gauges suitable for different rod diameters ranging from 2 to 125 mm (Fig. 3). The entire stem circumference is heated [62, 67].

3.2.4 Deformation Heat Field (HFD) Method

The current sensor includes radial heater, installed in the sapwood, and two thermocouples made from copper [42, 68–70]. The reference thermocouples measuring the temperature are inserted in a common needle placed below the heater. In the axial direction, the thermocouples of the needles symmetrical pair, are equidistant from the heater. They measure the difference in temperature, allowing bidirectional measurement of very low flow rates [68]. The magnitude of the sap flow's average and high values is due to the asymmetric temperature difference measured with the pair of asymmetric differential thermocouples [71] (Fig. 4).

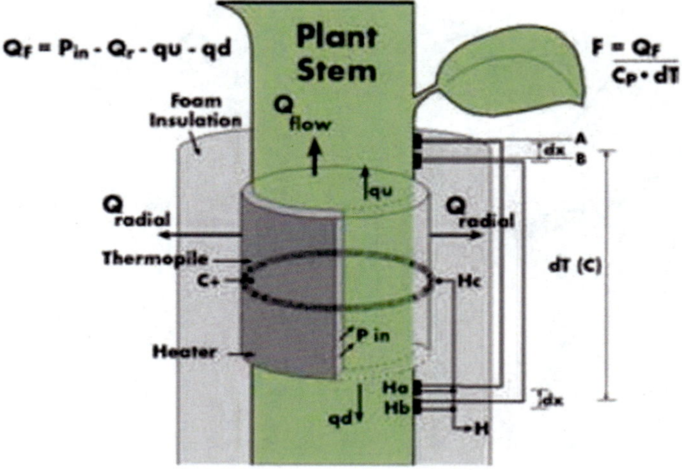

Fig. 3 Schematic representation of the sensor measuring raw sap flow from heat balance [61]

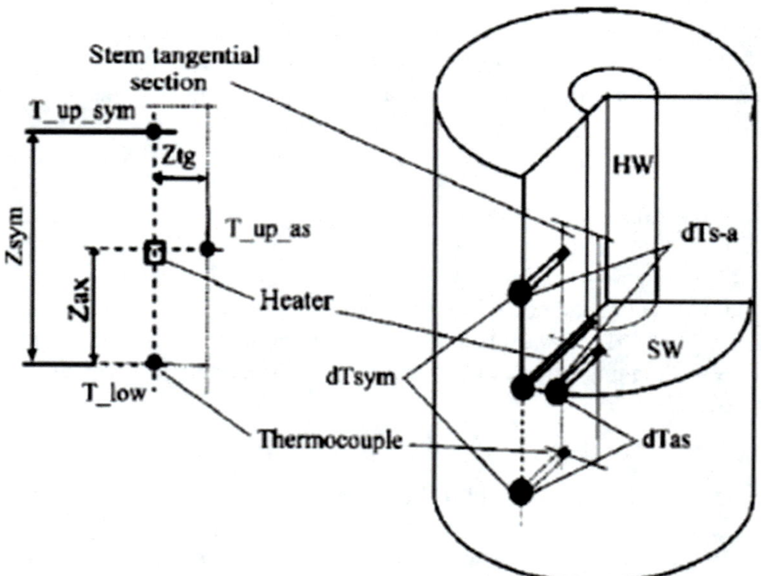

Fig. 4 Schematic representation of the tangential section of stem xylem with arrangements of the thermocouples around the heater of the HFD-sensor installed in the sapwood of a stem [71]

3.2.5 Heat Dissipation Method

The heat dissipation method introduced by Granier [72, 73] is based on the convective heat dissipation following the sap flow (Fig. 5). It consists of measuring the temperature at a heating probe and another non-heating (at wood temperature) inserted radially into the conductive tissue of the raw sap. The difference in temperature between the two probes is strongly influenced by the movement of xylemic sap in the trunk. At zero transpiration, this difference is maximal; the passage of a sap flow transports a quantity of heat by the convective transfer and the temperature in the heating probe decreases. The difference intemperature is inversely proportional to the flux density. It is estimated by an empirical formula expressed per unit area of sap-conducting wood [51, 72, 73]. According to Abid Karray [74], this method seems to be the best adapted to the olive tree since it allows to integrate the sap flow over a length of 2 cm (the length of the sensor). It allows for long periods of measurement since it has a low drift over time [73]. Moreover, this technique is simple to install, its calculation principle is simple, and its cost is low which makes this technique the most used [61].

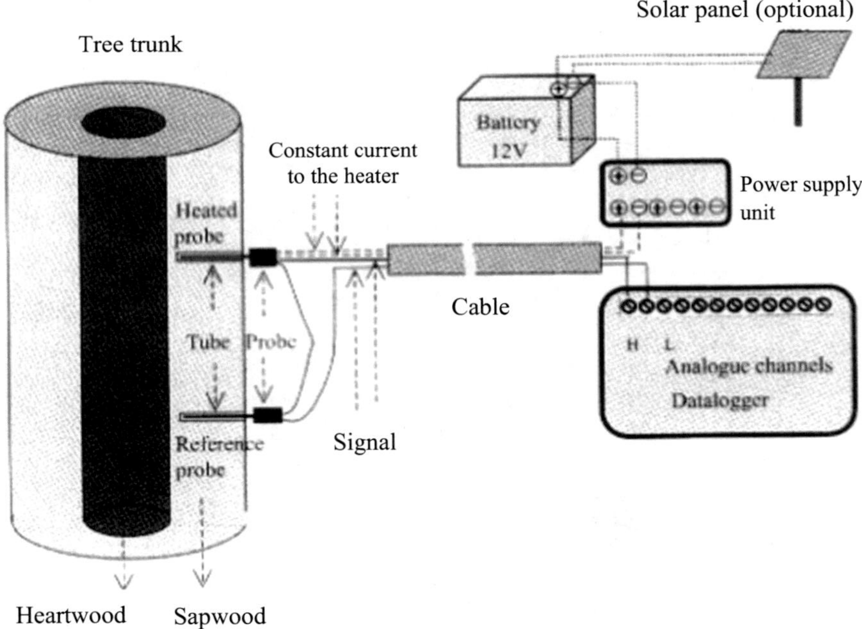

Fig. 5 Schematic representation of the Granier system for sap flow measurement. Each probe contains a thermocouple [50]

4 Case Study: Estimation of the Olive Orchards Water Requirements Using Climatic and Physiological Methods in the Tunisian Semi-arid

In olive growing, it is essential to optimize irrigation by a real estimate of the water needs. To achieve this objective, it is necessary to control the terms of the soil water balance, the energetic balance and eco-physiological parameters related to the olive tree. These terms give us the possibility to better manage water supplies and increase the water use efficiency. Indeed, there is a growing interest in new techniques designed for more accurate estimation of irrigation doses.

The objectives of this study case are: (i) the use of climatic data for estimating water requirements in olive trees (cv. Meski) in the semi-arid Tunisian (Enfidha); (ii) the use of the sap flow method to measure transpiration and water consumption; (iii) the expression of the physiological method according to ET_0.

4.1 Materials and Methods

See Fig. 6.

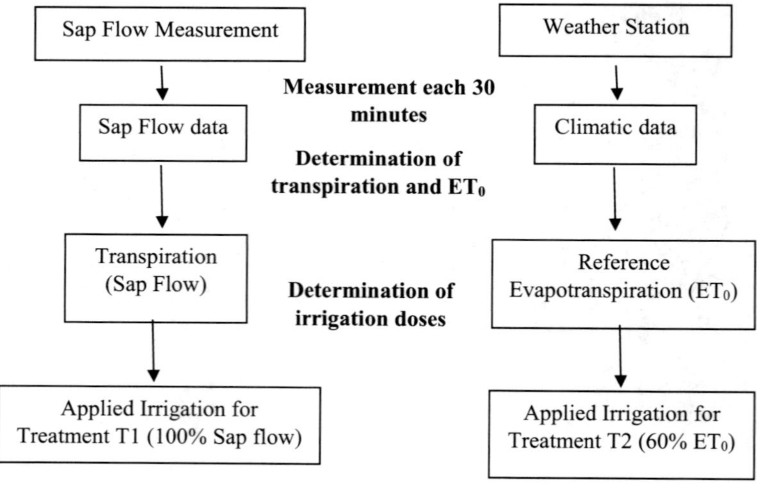

Fig. 6 Methodology flowchart

4.1.1 Description of Experimental Field

The experimentwas conducted in an irrigated olive orchards of 22 ha in Enfidha, central Tunisia (36° 08′ N, 10° 22′ E, 23 m), which belongs to "Office of Dominion Land of Enfidha" (OTD), on 2009. The orchard was planted in 1985 with *Olea europaea* L. cv. Meski trees at 7 × 7 m spacing. In this orchard, one plot of 10 trees, with a trunk diameter of 18 ± 3.4 cm at breast height and a similar height (3.5 ± 0.4 m), was selected to perform the experiments. These trees were irrigated by a drip irrigation system with, four drip nozzles for each tree, "located at a distance of 1.0 m to the north, east, south and west of the trunk, respectively".

The determination of olive tree water requirements was done according to the physiological method (T1: irrigation with a 100% of sap flow) based on the determination of xylem sap flow and the climatic method (T2: irrigation with a 60% of ET_0), based on the determination of reference evapotranspiration according to several approaches.

T1: irrigation according to the physiological method (100% of sap flow);
T2: irrigation according to the climatic method (60% of ET_0).

4.1.2 Sap Flow Measurements

Heat dissipation probes (TDP 30, Dynamax Inc., Houston, TX, USA) were used to measure sap flow (see discription in Sect. 3.2.5). Two probes were installed on each tree in two directions (South and North). Details are given by Bchir et al. [46].

4.1.3 Reference and Crop Evapotranspiration

The In-situ measured climatic paramiters were used to determine the reference evapotranspiration (ET0, mm/day) using the Penman Monteith formula [28], cited in Table 2 (Sect. 3.1).

4.2 *Results and Discussion*

4.2.1 Comparison of Applied Doses During the Experimental Period

The annual irrigation doses applied during the experimental period of the T1 and T2 treatments are respectively 1589 and 3080 m^3/ha (see Table 3). The calculated doses accordingto both treatments were variable and depended on physiological and climatic conditions. For the physiological method, the applied dose ranges from 84 to 285 m^3/ha/month. While the climatic method dose varies from 141 to 608 m^3/ha/month (see Table 3).

Table 3 Irrigation rates applied during the experimental period for the three treatments during the 2009 season (T1: 100% sap flow and T2: 60% ET_0)

	T1	T2
	m^3/ha	
February	84	141
March	143	245
April	180	313
May	187	421
June	262	557
July	283	608
August	247	409
September	203	386
Total	1589	3080

These results confirmed those found by Chehab et al. [75]. These authors have shown that according to the physiological method (100% FS) and the climatic method (100% ETc), the irrigation needs were respectively 1600 and 2600 m^3/ha applied from April to August 2008. According to Elsayed-Farag [7], in the Mediterranean regions with an ET_0 of around 1200 mm/year and an average annual rainfall of 500 mm, the orchards of Mature olive trees irrigated drip with planting densities of 100–300 trees/ha, require about 7000 m^3/ha/year to account for ETc. Among them, between 3000 and 4000 m^3/ha/year are supplemented by irrigations, according to the precipitation collected [57, 76]. According to Fernández et al. [77], for high-density olive orchards planting (about 2000 trees/ha), irrigation needs can increase to 5000 m^3/ha/year. In similar climatic condition in Spain, López Bernal et al. [11] had applied, from late spring to early autumn different irrigation doses ranged between 35 and 70% of the ETmax. Fernández et al. [9], Hernandez-Santana et al. [10] and Fernández [78] had applied three different irrigation treatments 100%, 60% and 30% ETc.

4.2.2 Evolution and Comparison of Transpiration According to the Applied Doses

The transpiration measured by the sap flow method during the experimental period showed that the highest value of transpiration was recorded for climatic method T2 (Table 4). These values vary from 73 m^3/ha (February 2009) to more than 400 m^3/ha (June and July 2009), this can be explained by the fact that this treatment receives the highest irrigation dose. The transpiration recorded for physiologic method (T1) is in a range of 72 m^3/ha (February 2009) and 245 m^3/ha (July 2009). Chehab et al. [75], found that during the months of high demand (June and July), olive trees irrigated by the climatic method transpired 23% more than those irrigated by the physiological method.

Table 4 Transpiration, sol evaporation and water requirements determined by the physiological method (T1) and the climatic method (T2)

	ET_0	Climatic method (T2)			Physiological method (T1)			T1 = %ET_0
		Dose	Transpiration	Sol evaporation	Dose	Transpiration	Sol evaporation	
	(m^3/ha)							
February	235	141	73	21	84	73	11	36
March	409	245	135	37	143	124	19	35
April	521	313	181	47	180	156	23	34
May	701	421	293	63	187	163	24	27
June	929	557	424	84	262	228	34	28
July	1014	608	417	91	283	246	37	28
August	682	409	318	61	247	215	32	36
September	644	386	286	58	203	176	26	32
Total	5134	3080	2127	462	1589	1381	207	
Average								32

The use of micro-lysimeters allowed us to estimate soil evaporation, which averages 15% of the doses applied for T1 and T2 treatment (Table 4).

For both treatments T1 and T2, the minimum evaporation of soil (Es) is recorded in February. It is equal to 11 m^3/ha and 21 m^3/ha respectively. The highest values of Es are observed in July, for both irrigation methods.

During the experimental period, we observed that for T2 treatment (60% of ET_0), transpiration was equivalent to an average of 69% of the applied dose. The sum of transpiration and soil evaporation (on average 15%) is less than the dose. This indicates that this method overestimates the irrigation needs and that leads to significant losses through evaporation and drainage.

In order to optimize the irrigation dose, the physiological method was converted according to ET_0 (Table 4). The irrigation dose applied by the physiological method represents on average about 32% of the ET_0. During May, June and July, irrigation needs to be applied for T1 is variable between 27 and 28% of the ET_0. During the rest of the year, the percentages were about are approximately 35%. Many authors have found similar results. Laouar [79] has shown that under the climatic conditions of Tunisia (arid and semi-arid), the water requirements of the olive tree do not exceed 75% of the ET_0. The same author indicated in 1978 that the olive tree could reduce its needs up to 35% of ET_0 without affecting its productivity [80]. According to Bchir [23] and Bchir et al. [47], the physiological method based on xylem sap flow allows a better estimation of the water needs of the olive tree as well as more efficient management of limited water resources by referring to the set of eco-physiological measures analyzed in the olive tree table variety 'Meski'. Ben Ahmed et al. [81] showed that the contribution of 600 mm of water per year (33% of ETc) could cover the needs of the Chemlali variety grown in semi-arid regions without decreasing photosynthetic and productive activity.

According to Chehab [82], in order to maximize the eco-physiological and reproductive parameters in the 'Meski' variety, a continuous 100% ETc irrigation is necessary, whereas the 'Picholine' variety can better value the respective water intake from 75 to 50% ETc during the fruiting stage. Haouari [83] found that 60% ET_0 irrigation leads to improved biochemical activity and ecophysiological parameters in table olive. According to García et al. [84], the application of regulated deficit irrigation strategy supplying 30% of the total irrigation needs for 'Arbequina' olive trees induces an increase in natural antioxidants in virgin olive oil. Neither yield, nor the rest of the quality parameters were affected by the reduced irrigation.

Hernandez-Santana et al. [85], showed that regulated deficit irrigation (RDI) has many advantages for olive tree orchards, such as increasing their long-term productive life, by reducing leaf area and without affecting single fruit weight nor total fruit yield. Also, the same authors found that the lowest RDI levels (30 and 45%) lead to greater water savings than 60%, with the best balance between crop water consumption and fruit yield.

5 Conclusions

Water has particular importance in the agricultural sector, particularly in arid and semi-arid regions, but water resources are increasingly scarce. Having reached a high level of mobilization of these hydraulic resources, Tunisia is now confronted with the imperative to better manageand valorizethem.

The study of the olive water needs allowed more efficient management of the water supplies and saving in irrigation water and this by an estimation of the needs in water by using more precise methods integrating the maximum of parameters from the Soil-Plant-Atmosphere continuum. Each of these methods is based on a set of climatic, edaphic and/or physiological parameters of the olive tree. The study of the water requirements of the olive tree in a semi-arid region has allowed us to manage water supplies more efficiently and save on irrigation water using the climatic method and/or the physiological method.

The evaluation of the different irrigation doses, applied according to the climatic (T2 = 60% ET_0) and physiological (T1 = 100% sap flow) methods and in relation to the transpiration of the trees measured by the sap flow method, allows to note that the higher the irrigation dose is, the highest is loses by transpiration. The highest transpiration values are recorded in the climatic method.

6 Recommendations

Deficit irrigation strategies, applying climatic or physiological methods, allow better water use efficient for olive orchards without affecting yield or fruit and oil quality. The physiological method (T1), equivalent to 32% of the ET_0, allowed us a saving in water of the order of 50% in comparison with the climatic method. So the physiological method allows formore efficient management of water resources.

To apply the physiological method in different bioclimatic stages, the use of transpiration modelingin relation to climatic parameters is necessary to meet the needs of the economy and the management of water supplies.

Acknowledgements The authors wish to thank the Flemish Interuniversity Council (VLIR) for supporting this study as part of a larger Own Initiatives project (ZEIN2006PR326).

References

1. Masmoudi-Charfi C (2006) Gestion des eaux d'irrigation dans les vergers d'olivier, document technique de l'Institut de l'olivier, 13 pp
2. Ben Rouina B, Trigui A, Boukhris M (1998) Effect of the climate and the soil condition on crops performance of the Chemlali de Sfax olive trees. ISHS Acta Horticult 586(1):285–289

3. Braham M (1997) Activité éco-physiologique, état nutritif et croissance de l'Olivier (*Olea europaea* L.) soumis à une contrainte hydrique. Thèse d'Etat de la Faculté des Sciences Agronomiques de Gand, Belgique, 160 pp
4. Trigui A (1993) Le secteur oléicole: Potentialité, production et évolution à paraître. Sérieétudes 1/93
5. Gòmez-del-Campo M (2011) Summer deficit–irrigation strategies in a hedgerow olive orchard cv. 'Arbequina': effect on fruit characteristics and yield. Irrig Sci. https://doi.org/10.1007/s00271-0299-8
6. Fernández JE, Diaz-Espejo A, Romero R, Hernandez-Santana V, García JM, Padilla-Díaz CM, Cuevas MV (2018) Precision irrigation in olive (*Olea europaea* L.) tree orchards. In: Water scarcity and sustainable agriculture in semi-arid environment. Tools, strategies, and challenges for woody crops, Chap 9, pp 179–217
7. Elsayed-Farag S (2014) Irrigation scheduling from plant-based measurements in hedgerow olive orchards. University of Seville, 218 pp
8. Fernández JE (2014) Understanding olive adaptation to abiotic stresses as a tool to increase crop performance. Environ Exp Bot 103:158–179
9. Fernández JE, Moreno F, Martín-Palomo MJ, Cuevas MV, Torres-Ruiz JM, Moriana A (2011) Combining sap flow and trunk diameter measurements to assess water needs in mature olive orchards. Environ Exp Bot 72:330–338
10. Hernandez-Santana V, Fernández JE, Rodriguez-Dominguez CM, Romeroa R, Diaz-Espejo A (2016) The dynamics of radial sap flux density reflects changes in stomatal conductance in response to soil and air-water deficit. Agric For Meteorol 218–219:92–101
11. López Bernal Á, García Tejera O, Vega VA, Hidalgo JC, Testi L, Orgaz F, Villalobos FJ (2015) Using sap flow measurements to estimate net assimilation in olive trees under different irrigation regimes. Irrig Sci 33:357–366
12. Padilla-Díaza CM, Rodriguez-Dominguez CM, Hernandez-Santana V, Perez-Martin A, Fernandes RDM, Montero A, García JM, Fernández JE (2018) Water status, gas exchange and crop performance in a super high density olive orchard under deficit irrigation scheduled from leaf turgor measurements. Agric Water Manag 202:241–252
13. Rallo L, Caruso T, Díez CM, Campisi G (2016) Olive growing in a time of change: from empiricism to genomics. In: Rugini E et al (eds) The olive tree genome. Compendium of plant genomes, Chap 4. Springer, pp 55–64
14. Lavee S, Nashef M, Wodner M, Harshemesh H (1990) The effect of complementary irrigation added to old olive trees (*Olea europaea* L.)
15. Proietti P, Nasini L, Ilarioni L (2012) Photosynthetic behavior of Spanish Arbequina and Italian Maurino olive (*Olea europaea* L.) cultivars under super-intensive grove conditions. Photosynthetica 50(2):239–246
16. Bchir A, Lemeur R, Ben Mariem F, Boukherissa N, Gariani W, Sbaii H, Ben Dhiab A, Ben Mansour Gueddes S, Braham M (2019) Estimation and comparison of reference evapotranspiration using different methods to determine olive trees irrigation schedule in different bioclimatic stages of Tunisia. Braz J Biol Sci 6(14):615–628
17. Fernandez JE, Palomo MJ, Diaz-Espejo A, Clothier BE, Green SR, Giron IF, Moreno F (2001) Heat-pulse measurements of sap flow in olives for automating irrigation: tests, root flow and diagnostics of water stress. Agric Water Manag 51:99–123
18. Giorio P, Giorio G (2003) Sap flow of several olive trees estimated with the heat-pulse technique by continuous monitoring of a single gauge. Environ Exp Bot 49:9–20
19. Nicolas E, Torrecillas A, Ortuno MF, Domongo R, Alarcon JJ (2005) Evaluation of transpiration in adult apricot trees from sap flow measurements. Agric Water Manag 72:131–145
20. Paredes P, Pereira LS, Almorox J, Darouich H (2020) Reference grass evapotranspiration with reduced data sets: parameterization of the FAO Penman-Monteith temperature approach and the Hargeaves-Samani equation using local climatic variables. Agric Water Manag 240:106210
21. Pereira LS, Paredes P, Jovanovic N (2020) Soil water balance models for determining crop water and irrigation requirements and irrigation scheduling focusing on the FAO56 method and the dual Kc approach. Agric Water Manag 241(1):106357

22. Saitta D, Vanella D, Ramírez-Cuesta JM, Longo-Minnolo G, Ferlito F, Consoli S (2020) Comparison of orange orchard evapotranspiration by eddy covariance, sap flow, and FAO-56 methods under different irrigation strategies. J Irrig Drain Eng 146(7):05020002
23. Bchir A (2015) Etude de l'évapotranspiration et de la transpiration pour l'estimation des besoins en eau de l'olivier (*Olea europaea* L.) conduit en intensif dans différents étages bioclimatiques. Thèse de doctorat en Sciences Agronomiques, Institut Supérieur Agronomique de Chott-Mariem, 127 pp
24. Fernández JE, Durán PJ, Palomo MJ, Diaz-Espejo A, Chamorro V, Girón IF (2006) Calibration of sap flow estimated by the compensation heat pulse method in olive, plum and orange trees: relationships with xylem anatomy. Tree Physiol 26:719–728
25. Intrigliolo DS, Castel JR (2006) Performance of various water stress indicators for prediction of fruit size response to deficit irrigation in plum. Agric Water Manag 83:173–180
26. Ortuño MF, Garcia-Orellana Y, Conejero W, Ruiz-Sanchez MC, Alarcon JJ, Torrecillas A (2006) Stem and leaf water potentials, gas exchange, sap flow, and trunk diameter fluctuations for detecting water stress in lemon trees. Trees 20:1–8
27. Masmoudi-Charfi C (2012) Manuel d'irrigation de l'olivier Techniques et Applications. Institut de l'Olivier, 206 pp
28. Allen RG, Pereira LS, Raes D, Smith M (1998) Crop evapotranspiration: guidelines for computing crop water requirements. Food and Agriculture Organization (FAO), Rome, 300 pp
29. Food and Agriculture Organization (1976) Les besoins en eau des cultures. Bulletin d'Irrigation et de Drainage 24
30. Gendrier JP, Lichou J, Baudry O, Orts R, Rondeau S, Soing P, Mandrin JF (1999) Outils de pilotage. Bonnes pratiques en arboriculture fruitière. Edition CTIFEL-ACTA, 202 pp. www.ctifel.fr
31. Masmoudi-Charfi C, Habaieb H (2014) Rainfall distribution functions for irrigation scheduling: calculation procedures following site of olive (*Olea europaea* L.) cultivation and growing periods. Am J Plant Sci 5:2094–2133
32. Pastor M, Hidalog J, Vega V, Castro J (1998) Irrigation des cultures oléicoles dans la région de Loma (province de Jaèn). Olivae 17:39–49
33. Khoshravesh M, Gholami Sefidkouhi MA, Valipour M (2017) Estimation of reference evapotranspiration using multivariate fractional polynomial, Bayesian regression, and robust regression models in three arid environments. Appl Water Sci 7:1911–1922
34. Wrachien DD, Mambretti S (2015) Irrigation and drainage systems in flood-prone areas: the role of mathematical models. Austin J Irrig 1(1):1002
35. Yannopoulos SI, Lyberatos G, Theodossiou N, Li W, Valipour M, Tamburrino A, Angelakis AN (2015) Evolution of water lifting devices (pumps) over the centuries worldwide. Water 7(9):5031–5060
36. Blaney HF, Criddle WD (1950) Determining water requirements in irrigated areas from climatologically and irrigation data. USDA (SCS) TP 96, p 48
37. Feng Y, Jia Y, Cui N, Zhao L, Li C, Gong D (2017) Calibration of Hargreaves model for reference evapotranspiration estimation in Sichuan basin of southwest China. Agric Water Manag 181:1–9
38. Landeras G, Bekoe E, Ampofo J, Logah F, Diop M, Cisse M, Shiri J (2018) New alternatives for reference evapotranspiration estimation in West Africa using limited weather data and ancillary data supply strategies. Theor Appl Climatol 132:701–716
39. Valipour M (2017) Analysis of potential evapotranspiration using limited weather data. Appl Water Sci 7:187–197
40. Priestley CHB, Taylor RJ (1972) On the assessment of surface heat flux and evaporation using large-scale parameters. Mon Weather Rev 100:81–92
41. Köstner B, Falge E, Alsheimer M (2017) Sap flow measurements. In: Foken T (ed) Energy and matter fluxes of a spruce forest ecosystem. Ecological studies (analysis and synthesis), vol 229. Springer, Cham

42. Cermák J, Kucera J, Nadezhdina N (2004) Sap flow measurements with some thermodynamic methods, flow integration within trees and scaling up from sample trees to entire forest stands. Trees Struct Funct 18:529–546
43. Pataki DE, McCarthy HR, Litvak E, Pincetl S (2011) Transpiration of urban forests in the Los Angeles metropolitan area. Ecol Appl 21(3):661–677
44. Alarcón J, Domingo R, Green S, Nicolás E, Torrecillas A (2003) Estimation of hydraulic conductance within field-grown apricot using sap flow measurements. Plant Soil 251(1):125–135
45. Bauerle WL, Whitlow TH, Pollock CR, Frongillo EA (2002) A laser-diode-based system for measuring sap flow by the heat-pulse method. Agric For Meteorol 110(4):275–284
46. Bchir A, Boussadia O, Lemeur R, Braham M (2013a) Comparison between sap flow measurements and two prediction climate formulas to estimate transpiration in olive orchards (*Olea europaea* L. cv. Chemlali). Eur Sci J 9(21):161–167
47. Bchir A, Boussadia O, Lemeur R, Braham M (2013b) Water use in olive orchards estimated by physiologic and climatic methods in Tunisia. Eur Sci J 9(24):374–385
48. Escalona JM, Ribas-Carbó M (2010) Methodologies for the measurement of water flow in gravevines. In: Methodologies and results in grapevine research, Chap 5, pp 57–58
49. Fernández JE, Green S, Caspari H, Diaz-Espejo A, Cuevas M (2008) The use of sap flow measurements for scheduling irrigation in olive, apple and Asian pear trees and in grapevines. Plant Soil 305(1–2):91–104
50. Lu P, Urban L, Zhao P (2004) Granier's thermal dissipation probe (TDP) method for measuring sap flow in trees: theory and practice. Acta Bot Sin 46(6):631–646
51. Siqueira JM, Paço TA, da Silva JM, Silvestre JC (2020) Biot-Granier sensor: a novel strategy to measuring sap flow in trees. Sensors 20:3538
52. Cohen Y, Li Y (1996) Validating sap flow measurement in field-grown sunflower and corn. J Exp Bot 47(304):1699–1707
53. Miner GL, Ham JM, Kluitenberg GJ (2017) A heat-pulse method for measuring sap flow in corn and sunflower using 3D-printed sensor bodies and low-cost electronics. Agric For Meteorol 246:86–97
54. Myburgh P (2016) Estimating transpiration of whole grapevines under field conditions. S Afr J Enol Vitic 37(1):47–60
55. Moreno F, Fernandez JE, Clothier BE, Green SR (1996) Transpiration and root water uptake by olive trees. Plant Soil 184:85–96
56. Fernandez JE, Moreno F, Clothier BE, Green SR (1996) Aplicasion de la técnica de componsacion de pulso de calor a la medida del flujo de savia en olivo. In: Proceedings of the XIV Congreso Nacional de Riegos, Aguadulcev (Almeria), pp 1–7
57. Fernández JE, Moreno F (1999) Water use by the olive tree. J Crop Prod 22:101–162
58. Forster MA (2020) The importance of conduction versus convection in heat pulse sap flow methods. Tree Physiol 40:683–694. https://doi.org/10.1093/treephys/tpaa009
59. Kostner B, Granier A, Cermak J (1998) Sap flow measurement in forest stands: methods and uncertainties. Ann Sci For 55:13–27
60. Larsen EK, Palau JL, Valiente JA, Chirino E, Bellot J (2020) Technical note: long-term probe misalignment and proposed quality control using the heat pulse method for transpiration estimations. Hydrol Earth Syst Sci 24:2755–2767. https://doi.org/10.5194/hess-24-2755-2020
61. Smith D, Allen S (1996) Measurement of sap flow in plant stems. J Exp Bot 47:1833–1844
62. Baker JM, Van Bavel CHM (1987) Measurement of mass flow of water in stem of herbaceous plant. Plant Cell Environ 10:777–782
63. Groot A, King KM (1992) Measurement of sap flow by the heat balance method: numerical analysis and application to coniferous seedlings. Agric For Meteorol 59:289–308
64. Hoelscher MT, Martin Andreas Kern MA, Wessolek G, Nehls T (2018) A new consistent sap flow baseline-correction approach for the stem heat balance method using nocturnal water vapour pressure deficits and its application in the measurements of urban climbing plant transpiration. Agric For Meteorol 248:169–176

65. Steinberg SL, van Bavel CHM, McFarland MJ (1989) A gauge to measure mass flow rate of sap in stems and trunks of woody plants. J Am Soc Hortic Sci 114:466–472
66. Steinberg SL, McFarland MJ, Worthington JW (1990) Comparison of trunk and brunch sap flow with canopy transpiration in Pecan. J Exp Bot 41(227):653–659
67. Sakuratani T (1981) A heat balance method for measuring water flux in the stem of intact plants. J Agric Meteorol 37:9–17
68. Nadezhdina N, Cermak J, Nadezhdin V (1998) Heat field deformation method for sap flow measurements. In: 4th international workshop on measuring sap flow in intact plants, Zilochovice. IUFRO Publications, Publishing House of Mendel University, Brno
69. Nadezhdina N, Vandegehuchte MW, Steppe K (2012) Sap flux density measurements based on the heat field deformation method. Trees 26:1439–1448
70. Nadezhdina N (2018) Revisiting the Heat Field Deformation (HFD) method for measuring sap flow. Forest 11:118–130. https://doi.org/10.3832/ifor2381-011
71. Nadezhdina N, Ferreira MI, Silva R, Pacheco CA (2008) Seasonal variation of water uptake of a Quercus suber tree in Central Portugal. Plant Soil 305:105–119
72. Granier A (1985) Une nouvelle méthode pour la mesure du flux de sève brute dans les troncs des arbres. Ann Sci For 42:193–200
73. Granier A (1987) Evaluation of transpiration in a Douglas-fir stand by means of sap flow measurements. Tree Physiol 3:309–320
74. Abid Karray J (2006) Bilan hydrique d'un système de cultures intercalaires (Olivier-Culture maraîchère) en Tunisie Centrale: Approche expérimentale et essai de modélisation. Thèse de Doctorat, Université de Montpellier II, 180 pp
75. Chehab H, Mechri B, Haouari A, Mahjoub Z, Brahem M, Boujnah D (2014) Effects of two drip-irrigation regimes on sap flow, water potential and leaf photosynthetic activity of mature olive trees, pp 2446–2447
76. Gucci R, Goldhamer DA, Fereres E (2012) Olive. In: Steduto P, Hsiao TC, Fereres E, Raes D (eds) Crop yield response to water. Irrigation & drainage paper no. 66, FAO, Rome, pp 298–313
77. Fernández JE, Perez-Martin A, Torres-Ruiz JM, Cuevas MV, Rodriguez-Dominguez CM, Elsayed-Farag S, Morales-Sillero A, García JM, Hernandez-Santana V, Diaz-Espejo A (2013) A regulated deficit irrigation strategy for hedgerow olive orchards with high plant density. Plant Soil 372:279–295
78. Fernández JE (2017) Plant-based methods for irrigation scheduling of woody crops. Horticulturae 3:35. https://doi.org/10.3390/horticulturae3020035
79. Laouar S (1977) Caractéristiques écophysiologiques et aspects de l'économie de l'olivier (*Olea europaea* L.) et de l'orange (*Citrus sinensis* L.). Thèse de doctorat d'état Es Science, Université Paris VII, 247 pp
80. Laouar S (1978) Quelques aspects de l'économie de l'eau de l'Olivier (*Olea europaea* L.) et de l'oranger (*Citrus sinensis* L. Osbeck). Thèse de Doctorat d'Etat Es Sciences. Université Paris VII, 247 pp
81. Ben Ahmed C, Ben Rouina B, Boukhris M (2007) Effects of water deficit on olive trees cv. Chemlali under field conditions in arid region in Tunisia. Sci Horticult 113:267–277
82. Chehab H (2007) Etude éco-physiologique, agronomique de production et relation source-puits chez l'Olivier de table en rapport avec les besoins en eau. Thèse de Doctorat en Sciences Agronomiques, INAT, 175 pp
83. Haouari A (2009) Etude de certains paramètres éco-physiologiques, biochimiques et nutritionnels chez l'Olivier (*Olea europaea* L. cv. Meski) soumis, à trois régimes hydriques. Mastère en biologie et écophysiologie des organismes végétaux, 166 pp
84. García JM, Morales-Sillero A, Pérez-Rubio AG, Diaz-Espejo A, Montero A, Fernández JE (2017) Virgin olive oil quality of hedgerow 'Arbequina' olive trees under deficit irrigation. J Sci Food Agric 97:1018–1026
85. Hernandez-Santana V, Fernández JE, Cuevas MV, Perez-Martin A, Diaz-Espejo A (2017) Photosynthetic limitations by water deficit: effect on fruit and olive oil yield, leaf area and trunk diameter and its potential use to control vegetative growth of super-high density olive orchards. Agric Water Manag 184:9–18

Diagnosis and New Farming Technologies

Agrarian System Diagnosis in Kerkennah Archipelago, Tunisia

Faiza Khebour Allouche, Arwa Hamaideh, Khouloud Khachlouf, Habib Ben Ckhikha, Ali Hanafi, and Youssef M'sadak

Abstract The application of a detailed diagnosis on the agrarian systems in MENA region, more precisely in a Tunisian archipelago through a social survey allowed us to describe the characteristics of cropping systems in general and detailed farming systems. The interaction between different agricultural and social indicators helped to identify gaps in the evolution of such system. Results prove that the breeding sector is averagely developed in the Archipelago. However, the closure of the collected center since 2011 has strongly contributed to the reduction of the application of the bovine breeding. However, attention is paid to rangelands since 65% of breeders apply transhumance. Mapping the transhumance circuits adopted by the breeders of the Archipelago shows the interest of this technic, but it is necessary to study these displacements to ensure the conservation of the natural resources of the Archipelago ensuring the sustainability of natural resources.

Keywords Diagnosis · Agrarian system · Mapping · Archipelago

1 Introduction

Given the small size of its arable land, the impact climate change, and the evolution of demand towards quality products, Tunisia needs to develop agricultural practices of its producers. The purpose of a diagnostic analysis of the agrarian system in a small

F. K. Allouche (✉) · K. Khachlouf · Y. M'sadak
High School of Agronomic Sciences, Chott Meriem (ISA CM), Sousse University, Sousse, Tunisia

F. K. Allouche
GREEN TEAM Laboratory, Carthage University, Tunis, Tunisia

A. Hamaideh
Water, Energy, Environment Research and Study Center (WEEC), University of Jordan, Amman, Jordan

H. B. Ckhikha
Territorial Unit of Agricultural Development of Kerkennah (CTV), Kerkennah-Sfax, Tunisia

A. Hanafi
Lab. GEOMAG, FLAHM, University of Manouba, Manouba, Tunisia

M. Abu-hashim et al. (eds.), *Agro-Environmental Sustainability in MENA Regions*, Springer Water, https://doi.org/10.1007/978-3-030-78574-1_5

region is to identify the factors that influence the evolution of agricultural production systems [1]. From this definition, we have tried to make a diagnosis of the agrarian systems of the Kerkennah Archipelago through a survey on a total sample of 100 farmer-breeders. A questionaire was distributed including 50 questions regarding livestock sector and its characterization. It includes: mode of performance, structure, etc. In this research, a comprehensive approach is considered who the beginning of the process where cropping system, animal husbandry concepts are applied to the plot and animal breeding on the farm rather than to a crop or animals. However, this approach was applied by different researchers, from example, in 2012, Barral et al., have applied it to understand the complex conditions under which farmers work. Then, in 2017, it was applied for determining the conditions necessary for the Niayes to better contribute to the country's agricultural production. More recently, in 2020, Vankovych et al. [2] and Li et al. [3] have used respectively the concept to identify patterns between state regulatory measures of aggregate support for the agricultural sector of Ukraine and its development and to study the assurance of agricultural sustainability and global food security by treating plant diseases diagnosis.

2 Agrarian and Production System Concepts

According to Jouve [4], an agrarian system can be defined as "the historically constituted, sustainable mode of organization used by a rural society to use its area and manage its resources.

It results in interactions between the bio-physical, socio-economic and technical factors" (cited in [5]): "*Agrarian system is the theoretical expression of a historically constituted and geographically located type of agriculture, consisting of a characteristic cultivated ecosystem and a defined productive social system. This allows sustainable exploitation of the corresponding cultivated ecosystem*". Conceiving and analyzing farming as it is practiced at a given moment in a given place as an agro-system "consists in breaking the system down into its main sub-systems and studying the organization and functioning of each, and their inter-relations" [6]. The concept of the agrarian system encompasses both the mode of exploitation and reproduction of one or more ecosystems, the social relations of production and exchange which have contributed to its establishment and development, and overall economic and social conditions [7]. It can be defined as the way farmers exploit the environment by using the relations and interactions that occur between all of its social and physical components. The system also takes into account the limits of the environment and its ability to reproduce [8]. Relationships can be established between [9]:

- The three main basic components (biophysic, human and technics states) of the system make it possible to determine a number of functional characteristics which will then serve as benchmarks and guides for the study of any agrarian system.

- The elements of the physical environment, the elements of the human environment and the technical elements will inform the way in which the environment is exploited by a rural society.
- The physical and human environment are studied through the social organization that determines the allocation of resources: land, water, vegetation.
- The human environment and technical elements cover the organization of work (duration, productivity, distribution of tasks in production units, and in society itself) and the technical means used (equipment, tools, inputs, varieties, etc.).
- Technical components and physical environment are understood by studying the culture and breeding systems (breed or type of crop, practices, and inputs used).

Analytical back and forth between the objects of study that are agrarian systems, production systems, with cropping and breeding systems will be at the heart of our iterative approach to build and validate assumptions about project impacts [10].

The agrarian system cannot then be regarded as a simple technical system of agricultural practices. The idea is to analyze the transformations of agricultural techniques and the changes in social relations, not only at the local level but also at the national and global levels. However, in contrast to many research carried out in agricultural economics or rural sociology, comparative agriculture clearly shows the ambition to take very finely into consideration the technical processes of agricultural production, given the special nature of agro-ecosystems, which are the real objects of work of farmers [11]. Resistance to change, and in particular to agricultural intensification, demonstrated by some rural societies despite the imperatives and incentives of national policy, can be interpreted more adequately if the concept of the agrarian system and the laws of its dynamics are used.

This resistance may indeed have its origin in the absence of a need for local populations to change their mode of exploitation of the environment to the extent that it satisfies their basic needs, at generally lower labor costs than would be required for intensification [12].

An agricultural production system is a combination done by farmers of the production techniques and means of production of the farms. It includes a number of interrelated subsystems: cropping system, forage system, livestock system…" [13]:

Livestock system: Vallerand and Richer [14] cited that "the livestock system is a tool for analyzing preferred livestock activities: on the one hand, the linkages between the major components of this activity and in the other hand the relationships of these components with stakeholders. The Breeding System can be defined in a very general way as the combination of resources, animal species, techniques and practices implemented by a community or by a breeder, to satisfy its needs by valuing natural resources by animals [15]. Following to the same author, the breeding system can itself be broken down into two subsystems. The first one is the management or steering system, where objectives and information on the environment and the structure and functioning of the system are defined. The second one is biotechnology system of production, where production processes and driving patterns are defined to understand the purpose of producers' practices and strategies. The breeding system contains three main components, namely man, animal, and land [16]. Breeding practices are the individual practices of ranchers that can be observed in the field. Thereto, it's possible to determine categories or typologies of practices presented: practices of aggregation, conduct, reproduction, renewal, and territorial practices that will be well developed in the analysis section.

Extensive Systems: Djemalia [17] characterizes this system by the use of natural vegetation from mountainous areas, pastures, and marginal land. This system most often involves small-sized herds of small ruminants, camelids, cattle, equidae, and native poultry. It is characterized by the use of family labor, low productivity, and almost no health coverage. These kinds of systems are economically unprofitable but are very well adapted to the environment and very ecologically efficient [18].

Grassland systems: These systems are based on fencing and cultivation of grass. A grassland system is even more extensive as grass plays a larger role in the animal diet and that the used surfaces receive less input, and undergo less cultural operations.

They are often described as extensive, using little or no chemical fertilizer and complementary foods [19].

Pastoral Systems: Pastoral systems use courses frequently for the guarding of herds. They are also extremely varied, particularly in relation to the diversity and often the heterogeneity of the routes they exploit: forests, alpine pastures, etc. [20].

Integrated Intensive System: According to Kayouli [21], this system mainly concerns dairy farming and fattening activities integrated on farms that grow, exclusively or partially, fodder, etc.

"Landless" or above ground intensive systems: It concerns family-type (all species combined) and industrial (poultry, rabbit breeding and pig farming) farms in irrigated and peri-urban areas. Investments are minimized, with limited external funding.

System of culture: The plant system or the system of culture is a "*whole succession of cultures and techniques implemented on a certain surface of ground handled in a homogeneous way, to obtain a plant production incompatible conditions with the objectives of the farmer*" [22].

Fodder system: At the level of exploitation, the fodder system is placed in the system of production. It assures correspondence between the system of culture and the system of breeding and to link with the system of breeding. There is the process of distribution of various rations of animals [23]. Define the forage system as being the set of the operations which have for objective to organize the food of the herd from the fodder resources produced on the exploitation or bought outside [22]. Clear some global characteristics of systems foragers' such the performance level, the dependence level, and the safety level. The elaboration or the study of a fodder system depends on the consideration of the constraints appropriate set to the exploitation, and the use of all the available economic and technical references. The constraints concern essentially the agro-climatic aspects, the availability in the workforce, the needs for the herd, and the economic environment. The feed balance allows the farmer to compare available feedstocks and feed requirements during the year (winter, summer, or both). It is always accompanied by a feeding plan which studies the use of feed in the form of a ration adapted to each of the categories of animals concerned. This plan provides for a basic and supplementary ration [24]. Keeping a grazing schedule helps the farmer to better manage the pasture on a daily basis. However, it makes it possible to carry out global assessment of this period and to draw lessons for the future.

3 Study Area

The islands of Kerkennah are located almost 20 km from the coast of Sfax occupying an area of 15.7 thousand hectares. The archipelago extends from the southwest to the northeast over a length of nearly 42 km (Fig. 1). The archipelago of Kerkennah constitutes a delegation attached to the governorate of Sfax. It is composed of two main islands: Chergui and Gharbi and twelve islets: Gremdi, Roummadia, ErRakadia, Sefnou, Charmadia, Lazdad, and six islets of HajHmida [25]. Gharbi Island, named «Mellita» is the first island encountered when arriving from Sfax. It is triangular, covered by a large palm grove and has only one village. An ancient 600 m long "El Kantara" dike, built on the site of a Roman bridge, connects it to the other end of the archipelago. Chargui Island or "Grandes Kerkennah" occupies the seat of the Delegation and the main administrative services stelErramla. The total area of the

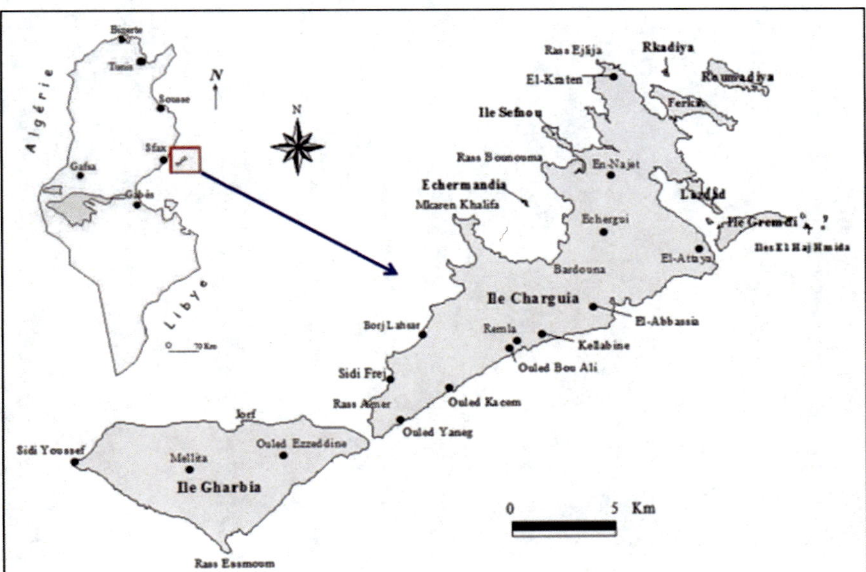

Fig. 1 Location of the Kerkennah Islands (feuilles topographiques au 1/25,000 En-Najet SO, En-Najet SE, Er-Ramla NE, Er-Ramla SO et Sfax SE)

islands of Kerkennah is 14.5 thousand hectares, the most important is that of the imadat of Mellita with 4643 ha. While the imadatd of OuledYaneg occupies the smallest area of 274 ha.

The Kerkennah archipelago belongs to the arid bioclimatic stage, and it enjoys a relatively pleasant arid Mediterranean climate with a hot and dry summer and a mild winter. The topography of the archipelago is characterized by low slopes and an almost flat, monotonous relief. The maximum altitude is 13 m above the hill of Ouled Ezzeddine. Its profile is marked by a succession of depressions from 1 to 5 m of depth "Bh'Iret" which do not exceed 3 m of depth. These depressions are separated by strings of muddy sand called "Dhaar" in which important deep oaks depressions connecting the "Bh'Iret" to the high seas were cut, depending on the direction of the tidal currents [26]. Atlas of the governorate of Sfax. General Direction of Spatial Planning. 105 p (French). The sebkhats correspond to the depressed areas which cover an area of almost 62 km^2 and develop in the coastal areas and extend into the interior of the archipelago. They correspond to soils without vegetation because of the high salinity. These lands are bordered in places by sea marshes [27] (Fig. 2).

The grounds of the archipelago are generally light and poor in organic matter. From the sector of Ramla northward grounds are characterized by calcareous crusting. In this sector, a big part of lands became salaine. In the region of Mallita towards Sidi Youssef, we meet sandy grounds and muddy sablo. The low surface of the Kerkennah

Fig. 2 Depressions separated by strings of muddy sand [27]

Islands and its flat topography does not favor the establishment of important river systems. The streaming is essentially diffuse.

The drainage of grounds is made through small soil erosions passing by towards sebkhas which communicate with the sea during the high swamps. Groundwater aquifers are especially accommodated in marly sands dating the Quaternary. They are often situated unless 3 m of depth. The salinity is generally lower à3 g/1, whereas the thickness of the water lens is very low. They are of the order of 1 m, and it exceeds it only locally. The vegetation of Kerkennah belongs to a mediterranean-steppic domain, with a steppic series of the Juniper of Oleaster. The steppe of white Artemisia (Armoisa herba alba) who distinguishes steppes of crusts limestones and firmly packed, fine grounds in muddyand the steppe of rural artemisia (Artemisia campestris) who characterizes the steppes of the sandy grounds are present with farming [28]. The Kerkennah delegation has a total area of 14,582 ha, including 8472 ha of agricultural land. It is characterised by an extensive system based mainly on sheep breeding. This is in addition to the emergence of irrigated crops on an area of 400 ha. Most of the cultivated land is planted by olive trees, associated in some farms with pomegranates or vines, in addition to palm trees growing spontaneously and which are scattered throughout the islands. Some farmers cultivate fodder crops in the middle with plantations but on relatively limited areas depending on the annual rainfall. The total area of agricultural land is 8300 ha, more than half of which are pasture land.

Sheep breeding is the most widespread in the islands, with a total population of 10,107 head. The average number of head per farmer is 20.336 head for goat breeding and 76 head for cattle. Concerning poultry, there are two chicken breeding farms, with 8000 farm chicken breeding poultry. The feeding of the herd is based on grazing the areas of the range bordering the houses and complementing by the purchase of straw and hay. Farmers preserve a kind of harvested grain barley from their plots to make up for food shortages in their herds. Recently, dairy cattle production has declined due to the constraints faced by farmers, in particular, the increase in feed loads and the difficulties of disposal [29]. Fishing in Kerkennah is a very ancient activity and is based on centuries-indigenous. Currently, it remains the main activity of the archipelago, providing more than 40% of the jobs. The main fishing ports of the island are those of El Ataya and Kraten. Along the coast, there are also several others more or less important landing sites. Fishing techniques in the Kerkennah Islands are distinguished by the practice of cherfias and come in various forms: fixed fisheries (fish fishing); gillnets (fish fishing); nasses (fishing for fish); gargoletlonglines (octopus fishing) and harpoons (sponge fishing). Fixed fisheries mainly develop in the south-west of Gharbi Island (between Sidi Youssef and El Kantara) and between El attaya and El Kantara. Average production in the years 2000–2006 is just half that of the 1980s [30].

4 Methodological Approach

The survey questionnaire was designed to be adaptable to the diversity of experiences but with a fairly strict framework so that the data could be analyzed more easily. We, therefore, chose to use as many closed questions as possible, often with multiple choices, but by giving, when thought it necessary, the opportunity to make a free comment to complete the answer. The preparation phase is used to acquire the knowledge and information necessary for the implementation of the survey method. The survey questionnaire was designed to be adaptable to the diversity of experiences but with a fairly strict framework so that the data could be analyzed more easily. Initially, data was collected based on the number of breeders: number of sheep/head, number of cattle/head, number of goats/head and number of households/imadat for each designation. Then we calculated the percentage of the total number of breeders/imadat. For each designation, we took the 10% of breeders to collect 100 breeders (Table 1).

The design of the survey is based on the principle of asking general and then specific questions. More than 20 questions were asked, grouped into four parts: farmer identification; general characteristics of the production system, production system, and breeding system. The information collected is recorded in the current Excel tables. In view of the large volume of information collected, four tables have been designed in the: identification of farmers; general characteristics of cultural system; livestock system and agropastoral relating to the previous organizational charts. Each table constitutes a pre-analysis matrix. In each matrix, the sum and the percentage per product are calculated.

Table 1 Breeders' choice by imadat

Designation	% of breeders/nbr of animals by imadat	10% of breeders
Mellita	14.7	23
Ouledezzidine	50	6
OuledYenig	16.7	5
OuledKacem	34.9	10
OuledBouali	11.2	6
Ramla	9.9	6
Kallebine	15.4	6
Charki	30.3	6
Ataya	11.5	11
Najet	33.7	13
Kraten	19.4	8
	Total	100

5 Results

5.1 Identification of Farmer

5.1.1 Key Activity and Socio-professional Category

Figure 3 shows that the survey has been done with the participation of 38% women and 62% of men. However, Fig. 4 highlighted that fishing is the main activity of the Archipelago (21%), followed by agricultural activity (10%). The percentage of housewives contains part of these two part-time activities. The rest of the socio-professional categories vary between different full-time or part-time activities. With regard to off-farm activities, fishing occupies a significant or even important place by presenting 21% of extra-agricultural activities. The main source of income for half of farmers is the benefits of agricultural products. This result confirms the results obtained previously, fishing is a socio-economic activity that is considered as an additional source of income and to improve the standard of living.

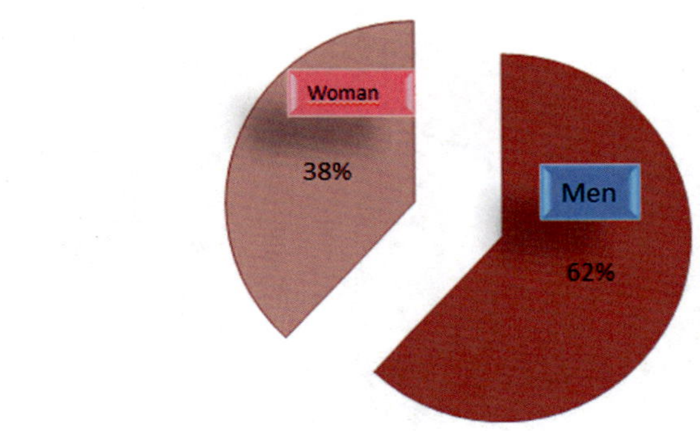

Fig. 3 Gender of investigated populations

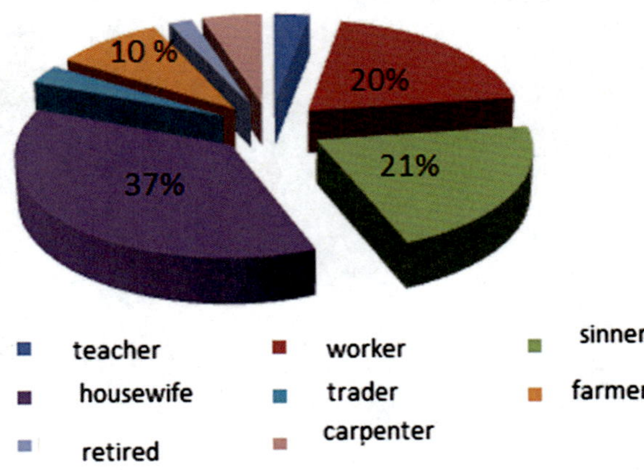

Fig. 4 Socio-professional category

5.1.2 Family Structure and Farmer's Intellectual Level

The majority of farmers are between the ages of 46 and 60, which can be explained by the attachment of farmers to their land and the fact that young people are not interested in farming (Fig. 5). The family structure is characterized by more or less numerous families with 5 to more than 7 individuals. This explains the importance of human resources on the farms.

Nearly half of the inhabitants of Kerkennah are illiterate. The intellectual level of the rest of the sample is divided between primary, secondary, and third cycle (Fig. 6). The majority of farmers joining Agriculture and Fisheries Development Group (GDA) and the Tunisian Union of Agriculture and Fisheries (UTAP) are important among large farmers who would like to increase their productivity while being involved in decision-making within these different structures. This explains the lack of extension and the farmer's bad conscience.

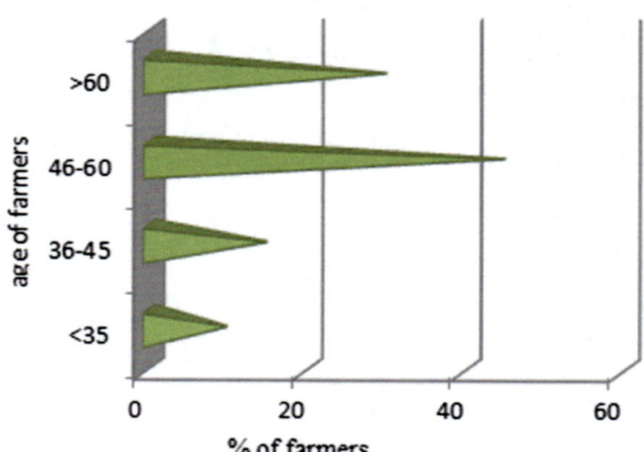

Fig. 5 Farmers' age

Fig. 6 Farmers' educational level

5.1.3 Geographic Distribution of Livestock Producers

The distribution of breeders and their housing is based on the number of households and breeders per imadat. Figure 7 shows Mellita is the busiest space occupied by

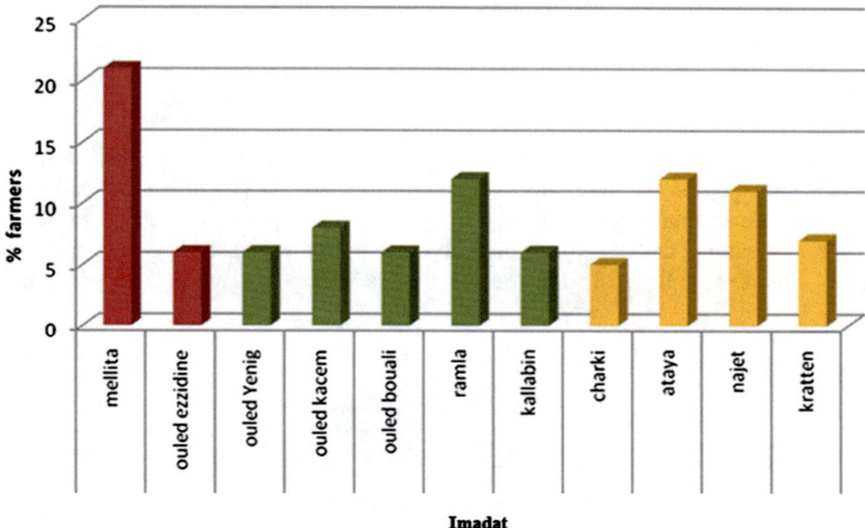

Fig. 7 Geographical distribution of breeders-farmers

herders (20% of farmers). In second place are the regions of Ramla, Ataya, and Najet.

5.2 *General Characteristics of the Production System*

The analysis in Fig. 8 show that the breeding system dominates with a percentage of 43% compared to the other areas associated with breeding such as arboriculture, irrigated, and dry crops. This is due to the high rate of family breeding in the archipelago. The distribution of production system areas varies from a minimum of 1 ha to a maximum of more than 5 ha. As for the spatial distribution, there is a variety of trees, fallow land, and grazing crops. There is a dominance of arboriculture, of which 57% are farmers who own their agricultural land (private or collective) and who practice field crops. The others (43%) only practice livestock. The most widespread form of direct exploitation is the most important in the region. Several factors, including the fragmentation of land ownership: the predominance of large families and the low level of agricultural investment, explain the development of direct marketing (owned by the operator) in the Kerkennah area.

As far as extra-agricultural activities, fisheries play an important role, accounting for 21% of extra-agricultural activities. The main source of income for 57% of farmers is the profits from agricultural products. This result confirms the results obtained previously, fishing is a socio-economic activity which is considered as an additional source of income, and in order to improve the standard of living.

5.3 Livestock System

The livestock of the farmers consists mainly of sheep, cattle, and goats. There is a predominance of sheep breeding with 73% of the types of breeding practiced in the island of Kerkennah, 19% mixed breeding (sheep and goats), 4% cattle breeding, 3% goat breeding, and 1% mixed breeding (cattle, goats, and sheep). Transhumance herds with a percentage of 61% compared to sedentary herds with 39% of the herd (Fig. 9).

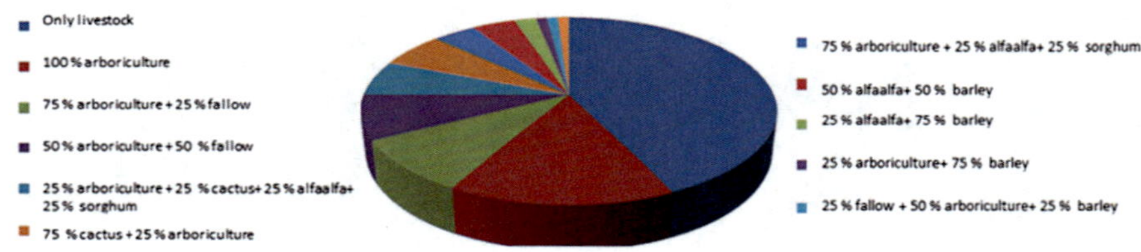

Fig. 8 Characteristic of Kerkennah Archipelago's production system

Fig. 9 Mixed breeding in Kratten

5.3.1 Characteristics of Cattle Farming

Structure of Cattle Farming

The cattle farming is intended only for the dairy production of Holstein/black pie cattle. The livestock may from Sfax origin (40%) or Kerkennah origin (60%). However, 80% of cattle herders have a barn with a family workforce (Fig. 10).

Fig. 10 Cattle breeding in Kallebine

Figure 11 shows that before 2011, more than half of the cattle livestock of the archipelago contains more than 20 heads. While this proportion is totally reversed after the years 2011 indicating a significant reduction in the number of cattle heads, less than 5. Indeed, according to the breeders, the reduction in the number of cattle is accentuated more and more because of the closure of the collection center. Hence, milk production is generally oriented towards self-consumption with a small quantity marketed.

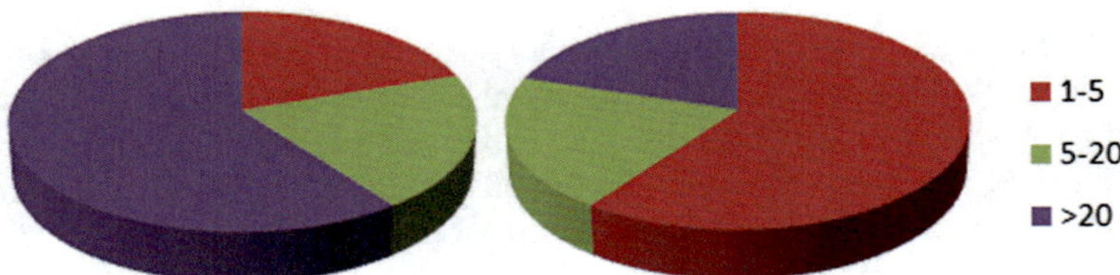

Fig. 11 Cattle numbers before 2011 (on the left) and after 2011 (on the right) in Kerkennah

Conduct of Cattle Farming

Only 20% of farmers rely on a full diet of hay, concentrates, and forage crops. Other farmers (80%) use above-ground feed. This is due to the fact that farmers do not cultivate fodder crops due to the lack of irrigated in the Kerkennah region with the exception of Mellita, Ramla, and ouled-Ezzedine.

The distribution of food differs from one breeder to another. Usually, it is done 3 times/day, with a frequent amount of concentrate distributed. For those with small numbers, the quantity is 1.5–3 kg cc/d. For the rest of the farmers, they distribute a small amount because of the high food prices. Thus, 80% of breeders have an intensive (landless) or above-ground breeding. Due to the increase in concentrate prices, smallholder farmers cannot provide an adequate ration for dairy cattle to meet their daily needs. They use certain complementary foods such as bran and barley, believing that it constitutes a complete ration. However, this will greatly affect cow productivity, production, and health. However, a mortality rate of 1–3% is noted in cattle of 60% of the surveyed breeders. In addition, the majority of newborns who are dead have diarrhoea problems.

Beef Farming Practice

The free and compulsory vaccines applied for cattle breeding are foot-and-mouth disease, tuberculosis, and brucellosis. All breeders benefit from these vaccines without exception. With regard to prepaid vaccines, only 80% of breeders apply them for their flocks. These vaccines are presented as a means of prevention against antiparasites, enterotoxemia, and browning. In addition, 80% of breeders have problems with the level of driving: no milking parlor, no milking machine, no storage room, no water tank, and electricity group for water intake, watering of cows and non-practice of the drying technique.

5.3.2 Characteristics of Sheep Farming

Flock Structure

In the archipelago, most farmers breed sheep from a cross between the Barbarine breed and the Western Tail Fine breed is known locally as Cherki with a percentage of more than 30% of the total. Generally, flocks are small and large flocks fairly common in the past have become very rare. The decrease in the numbers of small ruminants is mainly due to the decrease in agricultural products and by-products of the farm and the increase in prices of staple food and the ignorance of the farmer, breeding conditions (watering, veterinary care, construction of sheepfolds, etc.). Most of the herd has only one sheepfold, except in large farms; there are several sheepfolds that separate the herd according to their different physiological state and their different driving technique.

Conduct of Sheep Breeding

The distributed basic ration differs from one herd to another, but the majority of herders (68%) use hay, concentrate and agropastoral foods as a complete basic ration. Indeed, the feeding of the herd is based on a system of breeding above ground and cultivated, other breeders insist on their conduct only on the food of purchase (except ground). The amount of concentrate feed distributed/head/day varies between 100 and 700 g/head/day with 74% of farmers using 100–500 g/t/d of concentrate. The foods of complementation are very varied in the archipelago.

However, the return to production of planted olive groves produces large quantities of food (olive cake, olive leaves) used by farmers for the nutrition of their herds. This development of arboriculture has allowed, in some agro-pastoral farms, integration between livestock and agriculture which has resulted in an intensification of the breeding system which becomes more productive and more profitable. In these farms, the coverage of supplementation with fodder produced on the farm is relatively low compared to the cropping system (Fig. 12).

Fig. 12 Complementary feeds used for sheep by Kerkennah breeders: a by-product of the palm (on the left), olive leaves (on the right)

The main complementation foods are barley and bran, which represent respectively 28% and 18% of feed units given to animals. The importance of the role played by barley in the complementation explains the continuous extension of cereal plantings in the rangelands, despite their low yield and their very negative impact on the conservation of resources in arid zones. The distribution time of food is generally 2 times/day (54% of farmers) taking into account that the majority of herds are transhumant. Nevertheless, 36% of the farmers make the distribution 3 times/day and the rest for 4 times/day (Fig. 13).

Fig. 13 Manual distribution of concentrate feeds to Gremdi Island

Sheep Farming Practice

93% of the breeders do not apply the flushing technique, 90% of the breeders do not apply the steaming technique, and 82% of the breeders do not apply the technique of creep feeding. This can be explained by the lack of a precise breeding technique. The free and obligatory vaccines are blue tangue, enterotoxemia, foot-and-mouth disease, sheep pox, and brucellosis (Fig. 14). 86% of breeders use these vaccines, the others who are from Ataya imadat refuse to contact the veterinary services. Regarding pre-paid vaccine, 59% of farmers apply this vaccine to their flock. Shearing sheep is done in May, April or June. However, most farmers apply this technique in May with a percentage of 49%. Anti-squamous bath is a mode of fight against external parasites.

As a performance criterion; prolificity that has a zootechnical performance was used. According to the surveyed breeders, the number of lambing/year in 65% of breeders is presented by a single newborn/year, 31% of breeders reaches 2 lambing/head/year. The others may have a newborn/2 year. Analysis of the breeding of sheep shows that 94% of practice and the lack of extension. Except for ranchers who use this technique to improve the size of their herds. In fact, 6% of breeders apply the ram effect taking into account the number of existing sheepfolds. Many breeders use palm pollen as a means of ram sexual activity during the mating period. 62% of breeders use the reform of rams at the age of 1–2 years, 32% at the age of 2–5 years, and the others for more than 5 years. However, 63% of breeders have a mortality rate of 1–3%.

The sorting criteria vary between breeders, 14% of breeders consist of the conformation of the animal, 10% on the conformation and criteria of the sheep mother and

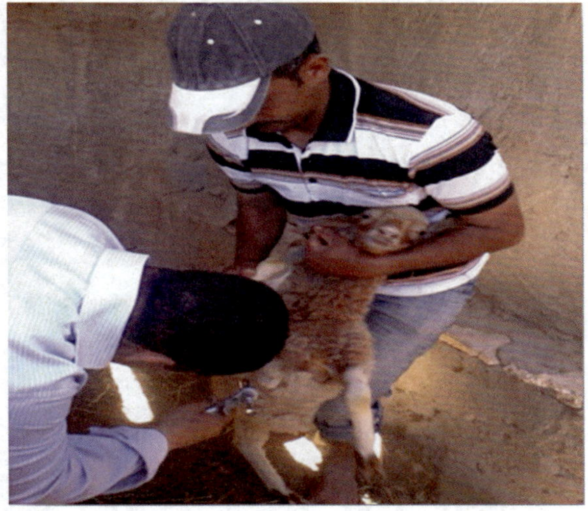

Fig. 14 Treatment of the free vaccine by the veterinarian of the archipelago

77% of breeders do not practice this sorting technique. The products sold are sheep meat and wool; 79% of the products are only meat and the rest are wool and meat.

5.3.3 Characteristics of Goat Farming

Goat Structure

The production destination of goat breeding is estimated as following 100% fattening compared to milk production which is very negligible since in Tunisia the consumer prefers cow's milk, despite the benefits of goat's milk. Speaking about the breeds present in Kerkennah, we find two main breeds which are: (i) local (Tunisian) breed with 49% of the herd, (ii) Damasquine breed with 20% of the herd and (iii) alpine breed (imported) with 31% of the herd.

Nevertheless, the majority of breeders have sheepfolds with a percentage of 89%, which indicates that the livestock sector has a very important role in the agricultural economy of the Archipelago (Fig. 15).

Conduct in Goat Breeding

The number of lambing in the goat exceeds 1 new born/year the mating season is not against the season. There is also a lack of triage within the goat herd.

Fig. 15 Goat farming in Kallebine imadat

Practices in Goat Breeding

The type of mobility encountered is transhumance which is a seasonal, cyclical movement of the breeder and his herd in search of better food conditions. Neither Flushing nor steaming nor creep-feeding is applied by goat breeders. Free and compulsory vaccines are applied to all goats. However, the prepaid vaccine is applied only by 21% of breeders. This is the parasite vaccine. 21% of breeders use the services of the veterinarian alone in case of an emergency. The scab bath is a technique applied in August in the sea by 21% of goat breeders.

5.4 Agro-pastoral Forage System

According to data analysis, 76% of farmers have agro-pastoral systems whose type of crop system has the following percentages: 65% of farmers rely on grazing land, 11% of farmers have forage crops, and the rest of the farmers rely on two types of crops.

5.4.1 Parcels Cultivated in Forage Crops

The spatial distribution of plots grown in forage crops is shown in Fig. 16. The largest percentage of parcels are in Ramla (36.10%). The area of the plots varies: 70% of the plots have an area between 0.5 and 1 ha, 20% of them have an area greater than 0.5 ha, and the rest of plots have an area less than 1 ha. The distance of these plots from the head office of the holding varies from 10 m to 1 km or for 90% of the farmers. This distance is greater than 150 m. On the other hand, these plots are generally private (85%), the others are collective (13%) or public (2%).

Based on the synthesized forage calendar, alfalfa is sown throughout the winter. Whereas feed barley is usually sown only once a year (46% of the plots are grown in alfalfa and barley and 31% in barley and alfalfa). Harvest products are either grown and given green to the animals or dried and kept until winter. This helps to address food deficiencies in winter. Sorghum is often grown only once a year, and some farmers spread their production over two years (Table 2).

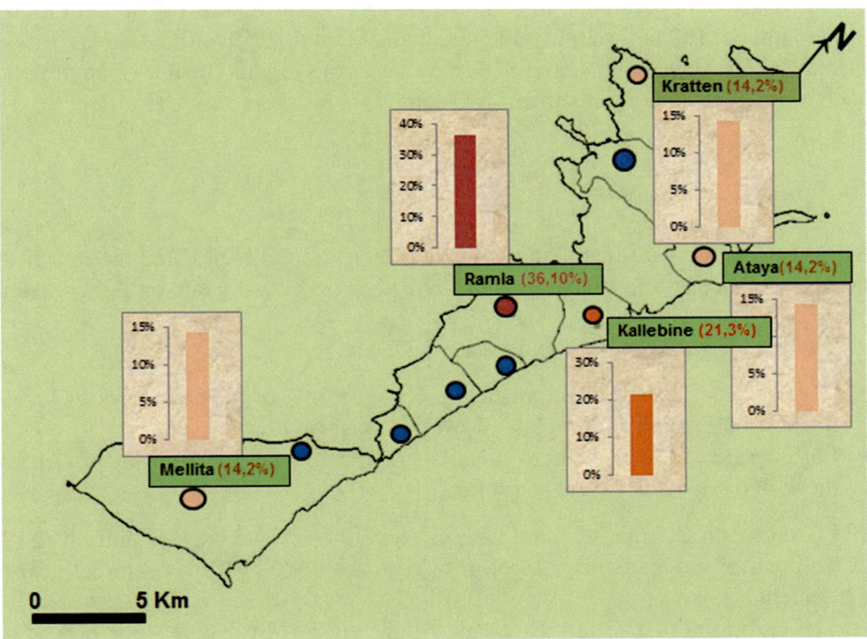

Fig. 16 Share of forage crops by Imadat in Kerkennah

Table 2 Feed schedule applied by farmers in the archipelago

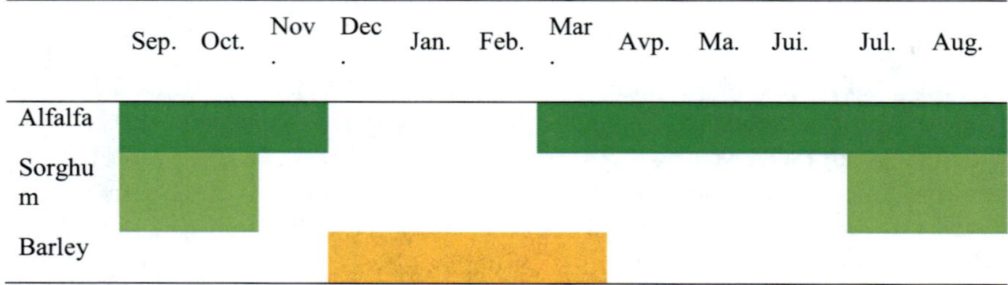

	Sep.	Oct.	Nov.	Dec.	Jan.	Feb.	Mar.	Avp.	Ma.	Jui.	Jul.	Aug.
Alfalfa												
Sorghum												
Barley												

Cultural Techniques

The method of irrigation for dry plots accounts for 72% compared to 28% for irrigated plots. The type of irrigation can be either in the form of a water valve (94%) or a water basin (6%).

The distance between the plot and the water points is about 100–500 m for 50% of the farmers, 100 m for 36%, and more than 500 m for 24% of the farmers 14% of farmers apply amendments, and 81% have access to organic matter (manure) after each crop. Phytosanitary treatment is absent.

Grazing and Cutting Action

84% of the herders estimated that 20–50 sheep graze in the plot and the rest of the herders practice mixed breeding of 20–50 sheep and 20–0 goats on the cultivated plot.

The time left per day for the flock is divided as follows:

- For 2 weeks (87% of the herd stays 1–3 h, and 13% of the herd stays 4–7 h and those in barley and alfalfa plots (flowering stage);
- For 3 weeks (87% of the herd stays 1–3 h, and 13% of the herd stays 4–7 h and those in the plots grown in sorghum.

Farmers estimate that the herd can be kept either by the breeder himself (24%) or the breeder and his family (14%) or by a worker (62%). The workforce is 100% permanent.

The cutting period differs from one forage species to another. For barley, the cutting period is 1–4 months with 1 cut/4 months. For alfalfa, the cutting period varies from 1 to 2 months, with 5 cuts/year for 5 years. Whereas for sorghum, 6 months are recorded as a cut period with 1 cut/year. Thus, the food value of the cut food of the plot presents for 80% of the breeders as a basic ration and for the others is considered as a supplement. Finally, the destination of the sold crop can be either for self-consumption (96%) or for sale (4% of breeders).

5.4.2 The Forage Course

The map of the transhumance circuits adopted by the breeders of the Archipelago is illustrated in Fig. 17. It is noted that the majority of Kerkennah breeders prac-

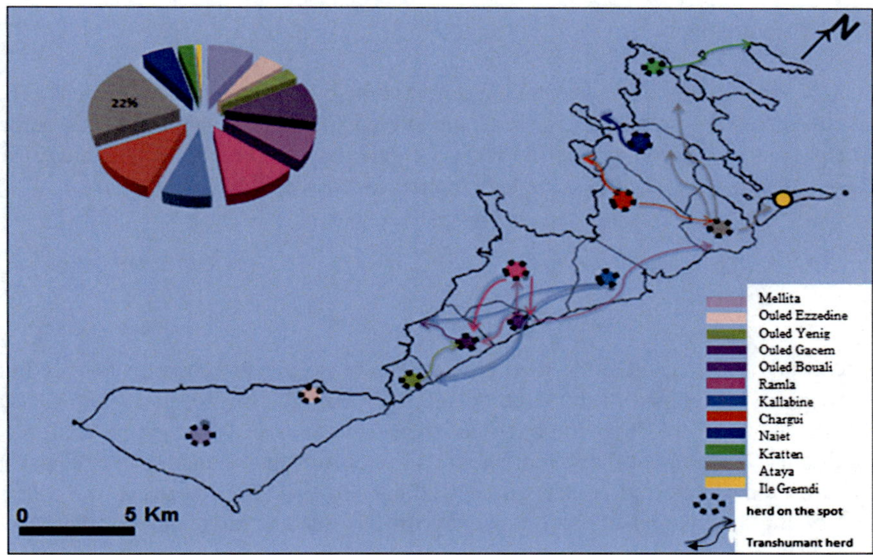

Fig. 17 Transhumance circuits adopted by the Archipelago breeders

tice transhumance while those of imadats of Mellita and OuledEzzdine do not apply this technique. It is mainly attaya breeders (22%) who practice transhumance, which according to them, this technique has important benefits on health and animal productivity.

Pasture on the Forage Course

96% of the breeders exploit routes all year long. The reason of the access to the pasture to the majority of the breeders is to assure the basic ration of the livestock, for the 11% of the breeders it is complementation.

Type of Vegetation

Animals are led on routes to steppic vegetation. This plant place setting is subjected since a few decades to overexploitation which is translated by the rarefaction the best pastoral species. To cover the fodder deficit of routes, complementation, often with barley, is given to animals. The duration of the complementation varies according to the years from 3 to 12 months by the years. Widely state-subsidized feeds at the beginning of the years, allowed the intensification of the breeding and the increase of staff on routes.

Cuts of Route Plants

35% of the breeders make one cut per day, and the others apply 2–4 cuts/week. The taken species are thyme, artemisia champetre, atriplexe, artrocnemum and cactus. The people who cut Plants on the route are generally the breeders. The interest of these plants is to be used as complementation for 87% of the breeders.

6 Discussion

According to the results presented previously, we note that the sector of breeding is averagely developed in the Archipelago. It is only the big breeders who follow the standards of breeding; the oldest of them are convinced by the application of traditional conduct, but the young one tries to improve their conduct. However, the closure of the collected center since 2011 has strongly contributed to the reduction of the application of the bovine breeding in the Archipelago. So, there is a clear need for agricultural extension services to support education and awareness of breeders. In the Archipelago, a dominance of the ovine breeding (73%) based on an aboveground farming system is noted. The majority of the breeders use the hay, the concentrated, and the agropastoral food as basic ration completes what establishes a constraint. Furthermore, the analysis of the ovine reproduction system shows that 94% of the breeders do not practice the mating against the season, which explains the lack of knowledge of the breeding conduct. Also, more than half of the sheep and goat breeders do not apply the techniques of flushing, desteaming and of creep-feeding. This ignorance establishes a constraint for the development of the sheep and goat breeding in the Archipelago, from where the importance of the popularization. According to the future strategy planned by the Archipelago farmers, they provide the increase of the cultivated areas in barley and olive grove and the introduction of the smooth cactus to ensure the success of their breeding. The increase in goat numbers is expected. However, attention is paid to rangelands since 65% of breeders apply transhumance. Mapping the transhumance circuits adopted by the breeders of the Archipelago shows the interest of this technic, but it is necessary to study these displacements to ensure the conservation of the natural resources of the Archipelago ensuring the sustainability of natural resources. This first diagnosis of its kind applied mainly to the breeding system of Kerkennah's archipelago has allowed a detailed characterization. However, this work deserves to be more thorough and completed to identify the agrarian system of the whole archipelago. Few works have been followed in this context but others have treated biodiversity [31], sustainable management [32], ecotourism [33] and [34], plant diseases [3] etc.

7 Conclusions

In Tunisia, the agricultural sector remains of great economic and socio-political importance because of its contribution to the achievement of national objectives in terms of food security, income generation, employment, regional balance and management of natural resource [35]. However, the scarcity of water resources, soil salinity and the frequency of high winds are all factors that did not favor the development of agricultural and livestock activities that remained subsistence activities [36]. Thus a good diagnosis will allow a better analysis of the agrarian system. Moreover, the applications of new technologies such the use of high resolution images, GIS spatial analysis and mapping will better improve this diagnosis.

According to results found in this research, this agrarian system can be classified from the major MENA farming systems to the Dry land Mixed System one identified by Dixon et al. [37]. These authors have mentioned that during the period 2000–2030, the population of MENA region is projected to almost double from its present. This could have a considerable negative impact in areas with fragile or vulnerable soils and sloping land, and will certainly be of importance for water resources everywhere. Then, some priorities proposed by them for this kind of system can be recommended to our case study.

- Priorities should be focused on regulatory measures for access and use of land and water resources, as well as on technology development.
- A new approach to research is called for in order to develop crop varieties with a shorter growing period, drought resistance, and improved grain and straw quality.
- Pastoral areas could be managed sustainably through the revival of, and support for, older institutions for control of communal grazing areas.
- Monitor management systems, including both urban-based and pastoral stakeholder groups.
- Apply a new legislation that will protect the steppe environment and ensure sound long-term management of soil and pasture resources.
- Introduce the concept of agricultural sustainability and global food security.

8 Recommendations

The development of the extension to the farmers by integrating the approach of the participative management will play a crucial role in the development of the agrarian system and will ensure sustainable conservation of the agro-pastoral resources of the Archipelago, and then it is recommended to support education and awareness of breeders.

Likewise, we recommend decision-makers to use the agrarian diagnosis method merged to new agronomic technologies for planning a future strategic plan [38]. Has proposed the introduction of this technology to improve the quality production, with the concern to save water in irrigable areas and to cultivate varieties adapted to the weakness and the irregularity of the pluviometry in rain-fed areas, using energy-saving techniques energy and inputs from oil. The recommendation can be applied to different MENA agrarian systems.

References

1. Fare Y, Dufumier M, Loloum M, Miss F, Pouye A, Khastalani A, Fall A (2017) Analysis and diagnosis of the agrarian system in the Niayes Region, Northwest Senegal (West Africa). Agriculture MDPI 7(759). https://doi.org/10.3390/agriculture7070059
2. Vankovych D, Kulchytskyy V, Zamaslo MI, Boichuk OT, Ruslan M (2020) Diagnosis of the impact of state financial support on the functioning of the agricultural sector in Ukraine. ESPACIOS 41(15):10
3. Li Z, Yu T, Paul R, Fan J, Yang Y, Wei Q (2020) Agricultural nanodiagnostics for plant diseases: recent advances and challenges. Nanoscale Adv J 8. https://doi.org/10.1039/C9NA00724E
4. Jouve PH (1992). The diagnosis of the rural environment. From the region to the parcel. Systemic approach of the farming modes of the environment. CNEARC, Montpellier, 39 p (Studies and works of CNEARC, n.6) (French)
5. Lhopitallier L, Caron P (1999) Diversity and recomposition of rural areas in Amatola District, Eastern Cape Province. Geogr Space 28–2:170–183 (French)
6. Mazoyer M, Roudart L (1997) History of the agriculture of the world. From the Neolithic to the contemporary crisis, 705 p (French)
7. Cochet H (2001) Comparative agriculture. Quae Editions, Paris
8. Sacklokham S, Baudran E (2005) Using agrarian system analysis to understand agriculture improving livelihoods in the Uplands of the Lao PDR was produced in 2005 by NAFRI, NAFES and NUOL, pp 222–229
9. Cochet H, Devienne S, Dufumier M (2007) Comparative agriculture, a synthetic discipline? Rural Econ 297–298. http://economierurale.revues.org/index2043.html
10. Cochet H (2005) Comparative agriculture. Genesis and formalization of a scientific discipline. National Institute of Agronomy of Paris-Grignon (Ina-PG), Paris
11. Griffon M (2006) Feeding the planet: for a doubly green revolution, Odile Jacob edition, 463 p (French)
12. Boserup E (1970) Agrarian evolution and demographic pressure. Flammarion, coll. "New Scientific Library", Paris, 218 p (French)
13. Metge G (1991) Contribution to the ecological study of Anopheles labranchiae in Morocco: Imagos activity and dynamics of pre-imaginal stages in the sidi bettache region. Bull School 22:419–426 (French)
14. Vallerand RJ, Richer F (1988) On the use of the Causal Dimension Scale in a field setting: a test with confirmatory factor analysis in success and failure situations. J Pers Soc Psychol 54:704–712 (French)
15. Lhoste (2001) The study and the DiaNostiCe of Breeding Systems, Training Workshop for agronomists SCV Madagascar, 13–23 Mar 2001, 32 p, Document obtained on the Cirad site of the network, http://agroecologie.cirad.fr (French)
16. Landais E, Balent G (1993) Introduction to the study of extensive farming systems. In: Studies and research on agrarian systems and development, INRA Editions, pp 13–36 (French)
17. Djemalia M, Bedhiaf-Romdhani S, Iniguezc L, Inounouc I (2009) Saving threatened native breeds by autonomous production, involvement of farmers organization, research and policy

makers: The case of the Sicilo-Sarde breed in Tunisia, North Africa. Special Issue: Animal Genetic Resour 120(3):213–217

18. Kayouli C, Jouany JP, Dardillat C, Tisserand JL (1995) Physiological peculiarities of the dromedary: consequences for its feeding. Mediterranean Options 143–155 (French)
19. Landais E, Balent G (1995) Introduction to the study of extensive farming systems. In: Extensive Livestock Practices. Identify, Model Evaluate, Landais E (éd) INRA, Studies and Research on Agrarian Systems and Development, no 27, pp 13–35 (French)
20. Meuret M, Landais E (1997) What's new on livestock systems? In: Blanc-Pamard C (coord.), Boutrais Jean (coord.). Theme and variations: new rural research in the south. ORSTOM, Paris, 323–355. (Colloquia and Seminars). Dynamics of Agrarian Systems: Seminar, Paris (FRA), 1995. ISBN 2-7099-1375-5
21. Kayouli C, Djemali M, Belhadj MT (1989) Situation of intensive production in Tunisia. Mediterranean Options, Sér. Séminaires-n (French)
22. Hnatyszyn M, Guais A (1988) Fodder and breeder. Ed. Lavoisier, Coll. Tec & Doc, Agriculture today—Science Techniqes Applications, Paris (French)
23. Bonneviale JR, Jussiau R, Marshal E (1989) Global approach to farming. National Institute for Research and Educational Applications (Inrap)-Foucher, Dijon
24. Mike S, Ole O (1992) Indigenous integrated farming systems in the Sahel, World Bank technical paper number 179 Africa technical department series (French)
25. Oueslati MA, Ksontini M, Haddad M, Charbonnel Y (1995) Composting Acacia Branches Cyanopyylla and Fresh Sludge from Wastewater Treatment Plants, Rev. For. XLVII-5-1995 (French)
26. Ministry of Equipment and Environment (2013) Atlas of the governorate of Sfax. General Direction of Spatial Planning, 105 p (French)
27. Ministry of Equipment and Environment (2011) Study of the development of the management scheme. The economic region of the center east. General Direction of Spatial Planning, 311 p (French)
28. Ghrabi-Gammar Z, Georgec DR, Daoud-Bouattour A, Ben Haj Jilani I, Ben Saad-Limamb S, Sparaganoc O (2009) Screening of essential oils from wild-growing plants in Tunisia for their yield and toxicity to the poultry red mite, Dermanyssus gallinae. Ind Crops Prod 30(2009):441–443
29. CRDASfax (2008) Agricultural activity report of gouversnorate of sfax, 75 p
30. Rhouma A, Labidi A (2006) Participatory diagnosis of the state of traditional fishing in the Kerkennah Islands United Nations Development Program, 42 p (French)
31. Abdelmajid R, Noureddine N, Ben Salah M, Allala M (2005) Analysis of the genetic diversity of the date palm in the Kerkennah Islands. CRDA Sfax, 55 p
32. Allouche Khebour F, Benbelgacem W (2015) Developpement of a regional park for better protection and management of Kerkennah's archipelago Environment (Tunisia). Int J Manage IT Eng 5(3):94–106
33. Hellal M, Jarraya M (2012) Perspectives of sustainable tourism development in a vulnerable island territory: the case of the Kerkennah Islands. Worlds Dev 157:111–118
34. Abulhawa T, Cummings T (2017) Rapid cultural inventories of wetlands in six Arab States, including Ramsar Sites and World Heritage Properties. Better understand the contribution of cultural values and practices to successful conservation of natural resources, 120 p. https://www.ramsar.org/sites/default/files/2017-11-15_arab_states_report_fr_0.pdf
35. CHEBBI H (2019) Synthesis report on agriculture in Tunisia. Project to support the mediterranean enpard initiative. CIHEAM, 99 p
36. Touzi S, Ben Zakour M (2015) Tunisian experience to deal with the Variability and the Climatic Change in Coastal Areas. National report. As part of the ClimVar (MedPartnership) project "Integrating variability and climate change into national GIZC strategies", 40 p

37. Dixon J, Gulliver A, Gibbon D, Malcolm H (2001) Farming systems and poverty: improving farmers' livelihoods in a changing world (English). World Bank, Washington, DC. http://documents.worldbank.org/curated/en/126251468331211716/Farming-systems-and-poverty-improving-farmers-livelihoods-in-a-changing-world
38. FAO (2013) Tunisia Financing the agricultural sector, 325 p

Precision Farming Technologies to Increase Soil and Crop Productivity

Abdelaziz A. Belal, Hassan EL-Ramady, Mohamed Jalhoum, Abdalla Gad, and Elsayed Said Mohamed

Abstract Excessive information, and practical experience, in crop and soil management, and new agriculture technology, have been accumulated over time. To meet the increasing need of humanity for food, a new farming concept known as Precision Farming (PF) or Precision Agriculture (PA) or Site-Specific Land Management (SSLM), is developed in recent years. This concept is a new approach leads to develop the agricultural processes for increasing soil and crop productivity, while saving efforts and costs. Precision farming includes many techniques such as Global Positioning Systems (GPS), Geographic Information Systems (GIS), Remote Sensing (RS), Yield Monitoring, Variable Rate Application (VRA), Yield Mapping, Site-Specific Management Zones (SSMZ) and Crop Modeling. SSMZ is considered as an important factor of PF application. SSMZ delineation can be improved based on the accuracy of determined of soil and crop characteristics which are used for managing and sustaining soil functions. The improvement of land management, at the field scale, will be based on better characterization of soil variability and crop properties within-field. This can be carried out through mapping soil and crop properties with high resolutions, compared with the traditional way. The goal of delineating site-specific management zones (SSMZ) is to get a better explanation of the actual variation within the field. Soil sampling by traditional methods and laboratory analysis are not cost efficient or timely enough. The Grid soil sampling and management zone are found the most important methods for precision farming to collect soil samples. The grid soil sampling is elaborated by dividing a field into a square of cells. However, recent research estimated single soil chemical and physical attribute by using various sensors order to reduce costs and improve management zone delineation. SSMZ map can be produced from a single layer of data, or combinations of different data layers including, topographic attributes map, yield map, soil maps, and soil nutrients maps. The farmer can use SSMZ map to select which production and management strategies plans are required and where they should be placed.

A. A. Belal (✉) · M. Jalhoum · A. Gad · E. S. Mohamed
National Authority for Remote Sensing and Space Sciences (NARSS), Cairo, Egypt

H. EL-Ramady
Soil and Water Department, Faculty of Agriculture, Kafrelsheikh University, Kafr El-Shaikh, Egypt

M. Abu-hashim et al. (eds.), *Agro-Environmental Sustainability in MENA Regions*, Springer Water, https://doi.org/10.1007/978-3-030-78574-1_6

Therefore, Implementation of this technique under the status of developing countries is necessary to increase soil and crop production, reduce costs, increase farm profitability and reduce environmental risks and desertification processes (Mohamed in Arab J Geosci 6:4647–4659, 2013 [1]).

Keywords Precision farming · Management zones · RS · GIS · GPS · VRT

1 Introduction

Crop cultivation, of course, dates back to the earliest age of humankind. The human has been cultivated and managed soil and crops since the ancient ages. The soil and crop management ideas have been developing with the civilization. Although most of the efforts of ancient people mainly in collecting foods for their life, their efforts increased by the time about soil and crop management as well as the development of new advanced farming techniques [2, 3]. The few examples of such activities are shifting cultivation, crop rotation, irrigation and manure their fields. With the time goes, the people become more and more worried about crop and soil management practices which have the main effect on increase soil and crop production. Identification of site-specific for crop requirements, e.g., soil and climate, fertilizer and manuring, information on the monitoring of pest and disease and their control measures give the farmers' ability to plan and manage their field and other agriculture operations in an efficient method to increase farm income [4–6]. Conventional farming management plans use a whole-field, in which each field is treated as a homogeneous area, and the variability in soil properties, topography attributes, local weather conditions, and land use and land cover are not considered [7–9].

In precision farming (PF) or Site-specific land management (SSLM), the farm field is classified into "site-specific management zones" depending on soil pH, yield rates, pest invasion, and other factors that affect soil and crop production. Management decisions become in need for each zone and SSLM tools [10, 11], for example, remote sensing, GPS, and GIS are used to observe zone variable rate inputs. This method differs from traditional farming due to the traditional farming used a "whole field" approach where the field is considered as a homogeneous area, but SSLM method classifies the field into zones based on the field variability [12, 13]. Management decisions in traditional farming depend on field averages and inputs, which applied uniformly across a field. The advantage of PF or SSLM includes management zones with a higher possibility for an economic return receiving more inputs if needed than less productive areas. Therefore, the maximum economic return can be achieved for each input through the field [14].

The application of new advanced technology in order to increase agricultural productivity becomes an urgent solution to meet the food demand of the growing population [15]. The rapidly changing in the world and decreasing arable land due to urbanization and industrialization, the agricultural productivity needs a 70% increase in production levels. There are many new innovations in information and communications technologies that can be used in the agricultural sector in different areas of precision farming, including the use of farm management software, wireless sensors and the use of the agricultural machinery. Remote sensing technique is an important factor in precision farming [16].

Different components of technologies used in precision farming include remote sensings such as satellites, aerial photography, UAVs, GIS, GPS, the variable rate application, and geostatistics. This geostatistics technique is an associated technology with precision farming, and it is known as a part of applied statistics. It used for calculates the spatial dependence and spatial structure of measured properties and in data transformation, uses that spatial structure to detect values of the properties at unsampled locations on the field. The Interpolation is a method normally used for predicting unknown values of neighboring locations. Also, spatial modeling can be considered associated techniques used for precision farming and also is known as variography, whereas spatial interpolation is called kriging [17, 18].

There are several important applications for precision farming in arable land, including increasing the efficiency of fertilizer use, depending on the three macronutrients nitrogen, phosphorus, and potassium. In the traditional agriculture, the rates of these fertilizers are inputted uniformly distributed everywhere on the field at specific times within the year. This precision farming is allowed to increase different applications in some locations and reduced applications in others ones. Many environmental risks are generally based on the increasing agricultural inputs, which it leads to nitrogen and phosphorus leaching from the field into the surface and ground waters. By using PF technology, fertilizers could be applied in more accurate quantity in different spatial and temporal component to improve the variable rate application [19]. This technique is called Variable Rate Application (VRA). It allows the farmer to be able to control the quantity of fertilizer rate inputs in agriculture lands, which it contains a variable-rate (VR) control system with application equipment to apply variable rate inputs at an accurate time and location to understand the site-specific management variable rate application. These VRs are designed on the basis of previous measurements from remote sensing or machine-mounted sensors. Complete Variable Rate Technology (VRT) systems built from the integration of Differential Global Positioning System (DGPS) receiver, computer, VR software and the controller to make VRA work [20, 21].

2 Definition of Precision Farming

Precision farming or precision agriculture has many definitions, and these definitions are often affected by the commercial equipment or technology currently in mode. Precision farming can be defined as a management system, which it depends upon the information and technology or is site-specific management and it uses one or more of the following data sources: soils properties, crop growth parameters, nutrient maps, pests, soil moisture content and yield map, for maximization of the economics, sustainability and conservation of the environmental farm [22]. On the other hand, precision farming refers to the systems that assess variability in soil and crops through the field. Then, Information, which was collected in these assessments, could be then used to develop site-specific management practices for optimizing soil and crop yield production [23, 24]. The term of precision agriculture can also be known as the precision farming systems, Site-Specific Crop Management (SSCM) or Site-Specific Land Management (SSLM). Moreover, SSLM is defined as the method for managing cropland based on the local environment using the current field variability techniques. It can also be defined as the management of local variability in SSLM plans [5]. Meantime, site-specific crop management is the management of soil and crop production inputs such as water requirements, fertilizer rates, gypsum requirements, seeds, herbicides, insecticides in the soil on field variability basis such that it would support to reduce waste, increase profits and protect the environment from risks [25–27]. SSLM seems to be one of the advantage technologies in soil and crop production in a new century. Panda et al. [28] defined site-specific crop management as one way of precision agriculture, which involves geospatial reference, soil, crop and climate monitoring, variability mapping, decision support systems and differential action to increase soil and crop production with minimal input.

3 Benefits of Precision Farming Technology

According to Folnovic [29], the goal of precision farming is to increase agricultural yield production and decrease the environmental risks. However, the benefits could include the following points:

- detecting soil properties and plant physicochemical parameters including electromagnetic conductivity inductions, nitrates, temperature, evapotranspiration, radiation, leaf area index and soil moisture,
- getting data in real time by installing the remote sensing instruments in the fields, allowing continuous monitoring of the selected attributes and will offer real-time data, ensuring an updated status soil and plant parameters at all-time as well as getting better information for management decisions and farming planning,
- saving time and decreasing the costs through reducing fertilizer costs and other agrochemical applications by reducing the use of chemicals as well as reducing tillage operation,
- supplying the farmers with good farm information and databases, which it is essential for sale and succession and
- integrating farm management software, like Agrivi, to make all farm activities easier and to increase farm productivity.

The increased efficiency of the management plan will come about through a good understanding of the interaction between environment, soil, crop and more detailed information using of new advanced and current information technologies such as short and long-term crop growth modeling, soil conservation, climate predictions and agro-economics modeling. Precision farming management is a cyclical process (Fig. 1). The farmer can start farming depending upon "site-specific land management" information and needs seasonal planning, the program of data collection, and analysis of data in order to complete the cycle of the precision farming management strategies plan. Soil analysis should be don before planting, and data analysis also should be finished for calculation and mapping of the variability in soil properties, which will be used at any time and by any method for variable rate application. Through crop growing season, the monitoring crop condition will do by starting the work by monitoring different values of seeds based on the data and use the variable rate application of fertilizer, which is determined based on soil, plant and water analysis. Crop growth is accomplished for research on many attributes such as variability of soil properties, water requirement, weeds, pests or diseases. At harvest time, the yield crop monitor system which is installed inside the combine provides the variation in soil and crop growth status based on the geographic location and map of crop growth status across the farm according to its geographical location [30, 31].

In order to collect and use information with high effectively and efficiently, it is significant for the farmers who used precision farming should have a good knowledge of the new tools of precision farming technology. These new precision farming tools are content hardware, software and recommended practices [32].

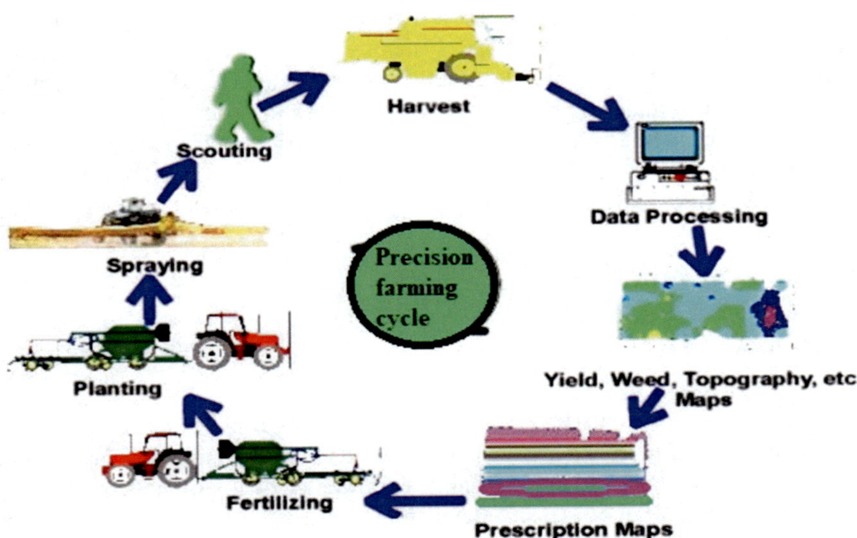

Fig. 1 Precision farming cycle (modified after Goswami et al. [30])

4 Global Position Systems (GPS)

There are many applications for global position systems under precision farming approaches, including mapping of soil and crop variability, farm management planning, soil sampling design, machinery orientation, crop scouting, variable rate technology, application mapping and yield mapping. The farmers can use GPS under different conditions such as low field visibility like rain, dust, fog, and darkness. Several tools could be developed for GPS by equipment manufacturers. They help farmers and agribusinesses to increase soil and crop productivity and effectiveness in their precision farming strategic plans. GPS equipment could be used to collect georeferenced information for delineation field boundaries, collect soil and plant sampling, variable rate application, roads network, irrigation and drainage systems, and crop scouting, e.g. waterlogged, weeds or plant disease. Increasing the accuracy of GPS enables farmers to produce farm application maps with high precise acreage. Also, farmers could use a GPS system to go to precise locations within the field, periodically, to collect plant and soil samples or detect plant growth status. The possibility to collect information about farm activity with Geo-reference gives farmers the ability to analysis, interpretation, map and visually characterization for agricultural processes. These provided vision into both production variability as well as inadequacy in soil and crop production and farm operations [33]. Using a Differential Global Positioning System (DGPS) is the best way to reduce GPS errors. This DGPS is included two things space and ground-based segments that together contain a radio-navigation facility in a DGPS system, where the GPS receiver antenna is placed at a precisely known location. GPS errors will calculate depending upon this receiver station by matching it is with the existing location to the location calculated from the GPS signals. This error information is sent to the machine receiver antenna, which uses it to correct the position information which it is calculated by GPS signals [34]. DGPS corrections can be broadcasted by tower-based or satellite-based systems (Fig. 2).

5 Remote Sensing

The term of remote sensing (RS) is to collect information from a distance for an object without physical contact and analysis, process and interpretation of this information. According to [36], reported that the earth is considered as important to study target by remote sensing. Remote sensing devices are normally collect and measurements of electromagnetic energy. These contain a wide range from the human eye to remote sensing satellite which is used for precision farming.

The most commonly remote sensing devices used for the precision farming application are the color, color infrared aerial photography and videography usually adopt for several applications in precision farming. Also, remote sensing can be used for precision farming through a number of way for providing an application of soil

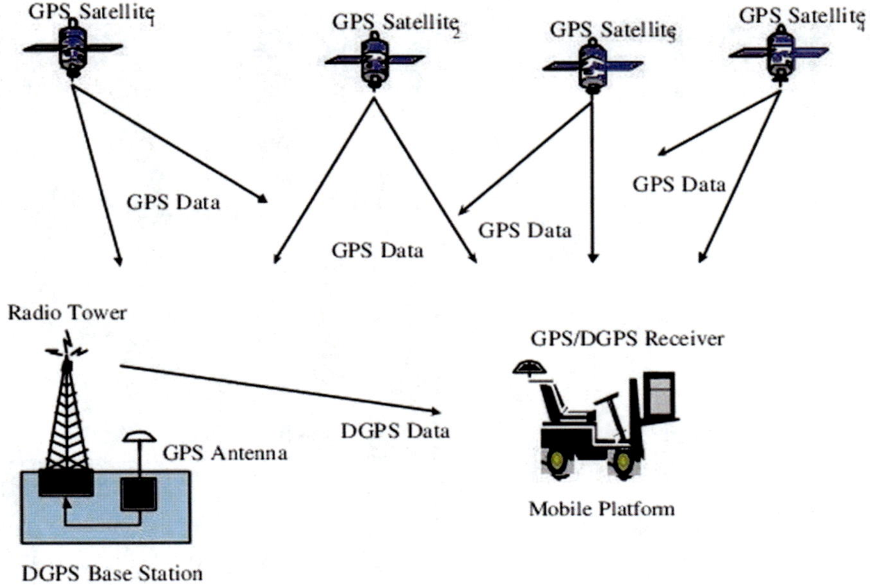

Fig. 2 DGPS tower-based and satellite-based systems (modified after Soares et al. [35])

and plant growth status as well as variability to build decision support system and management plan [37, 38].

5.1 Satellite and Aerial Photography Remote Sensing for Precision Farming

Remote sensing images taken through the growing season are very important to study crop status through the grown season and at its end the growing season (Fig. 3a). Produced yield maps from remote sensing images can be used to give an idea about farm management plan when the farmer does not have yield monitor systems [39]. The same authors mentioned the relationship between yield and the normalized difference vegetation index (NDVI) is highly related. The results showed that yield is highly significantly related to NDVI with a correlation coefficient of 95%. The crop yield can be calculated from NDVI based on the strong relationship between yield and NDVI. Figure 3c shows the crop yield map measure from remote sensing images related to the relationship between yield and NDVI. This map shows the yield variability within the field. Also, this yield map has the same spatial pattern comparing with the colour-infrared image and classification map (Fig. 3b).

Hyperspectral remote sensing device is very important in agriculture application depending upon it collect reflected information in a wide spectral range at small

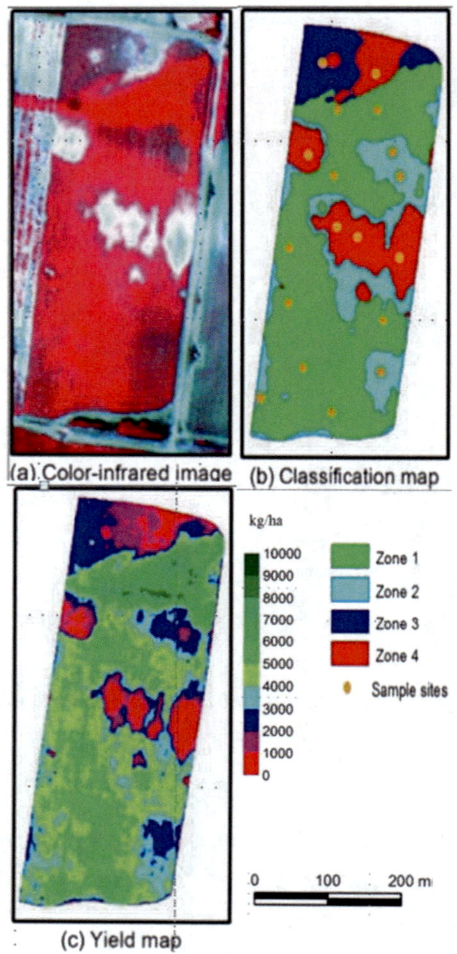

Fig. 3 **a** A colour-infrared digital image, **b** a four-zone classification map, and **c** a yield map (modified after Yang et al. [39])

spectral resolution is around 10 nm. It gives good opportunities to study soil and crop characteristics [33]. Hyper-spectral remote sensing data can be used in precision farming include the following:

(1) Getting images for the bare soil to delineation management zones, (2) detecting weed distribution, (3) monitor N stress in a plant, (4) produce yield mapping and (5) monitoring plant disease and pest. Hyperspectral remote sensing imaging is importance tool for precision farming application through using it in the variable rate application. It can be used to design management plan through the grown season for nitrogen fertilizer depending on spatial variability in chlorophyll content within the field [40].

5.2 Unmanned Aerial Vehicle (UAV) for Precision Farming

The commercial name of Unmanned Aerial Vehicle (UAV) can be called UAV or Unmanned Aircraft System (UAS). UAVs have been recently implemented in the precision farming study in order to solve the limitations of conventional remote sensing systems, e.g., satellites or aerial photo, which demonstrated to be incompetent in addressing and monitoring agriculture problems. The availability and high spatial resolution for satellite images at exactly crop growth status are the most limitations in PF such as the range of time to get the suitable images is usually very small. Based on this situation, UAVs have been normally used for image surveys such as mapping weed in coffee farms, finding the anomaly in the fertilization delivery system and determining maturity analyses [41]. UAVs have also been implemented for analysis of soils variability, pest problems solutions, detecting variation in fruit ripeness, and measurements nighttime temperatures for frost mitigation [42]. The smaller size of electronics such as computers and remote sensing sensors gives a new chance for increasing the efficiency of remote sensing applications. Despite the current restrictive regulations, the use of UAVs equipment with small RGB, hyperspectral, LiDAR, and/or multi-spectral sensors has become as a promising alternative for produce modeling, mapping, and monitoring applications in soil and agriculture management [43]. Increasing continuous technological developments of remote sensing UAVs are increasing the spatial, temporal and radiometric resolution of the image available for study the issue of land and crop management. However, the Low costs and small size of the UAV are benefits in recognizing agricultural monitoring. The three common types of unmanned aerial vehicles include: utilized fixed wings, helicopters, and multi-copter (Fig. 4). The competence and applicability of such systems depend on multiple factors such as the aircraft mass, payload capacity, average dimensions, flying range, average speed, low expenses [44].

Elarab [46] used the UAS acquired imagery in the visual, near infrared and thermal infrared spectra with spatial resolution of less than a meter (15–60 cm) and used applied a machine learning algorithm (relevance vector machine) to the five band imagery red (R), green (G), blue (B), near-infrared (NIR), and thermal infrared (TIR) to estimate plant chlorophyll content, NDVI, LAI and leaf nitrogen content

Fig. 4 Three examples of UAVs: **a** fixed-wing; **b** helicopter; **c** quadrotor (after Valavanis et al. [45], SpringerLink)

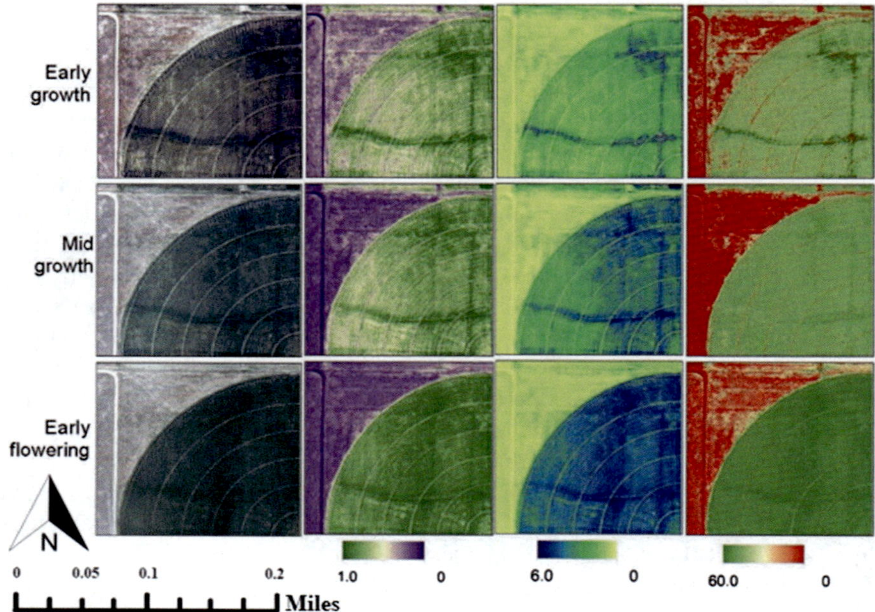

Fig. 5 Show UAS True colour images, NDVI maps, LAI maps, and chlorophyll map for the different growth stages (modified after Elarab [46])

respectively for oat crops in modern center pivot irrigation sprinkler system (Figs. 5 and 6). The high spatial resolution imagery is taken by UAV proximately 15 cm resolutions can be used to measure crop evapotranspiration or ET (see Fig. 7).

Low-cost remote sensing method like UAV is given a good chance for getting images with high spatial resolution over the areas of different size comparing with an aerial photograph and satellite images. The fixed wings or rotor-wing UAV are used in soil and crop survey which it can fly in different locations and conditions. UAV has a wide range application in the geomorphologic and environmental study [47]. d'Oleire-Oltmann et al. [48] the low-cost remote sensing devices are used for detecting and mapping soil erosion by water in Morocco. They studied the gullies erosion due to it has a highly effect on soil degradation in arid and semi-arid regions. Alexakis et al. [49] used UAV to study the soil roughness and soil erosion by water in two different study area in Greece and demonstrated that, gully erosion affect soil loss and soil degradation. Moreover, the assessment of the quantity of soil roughness is essential and influence due to its effect on soil water holding capacity and soil infiltration rate.

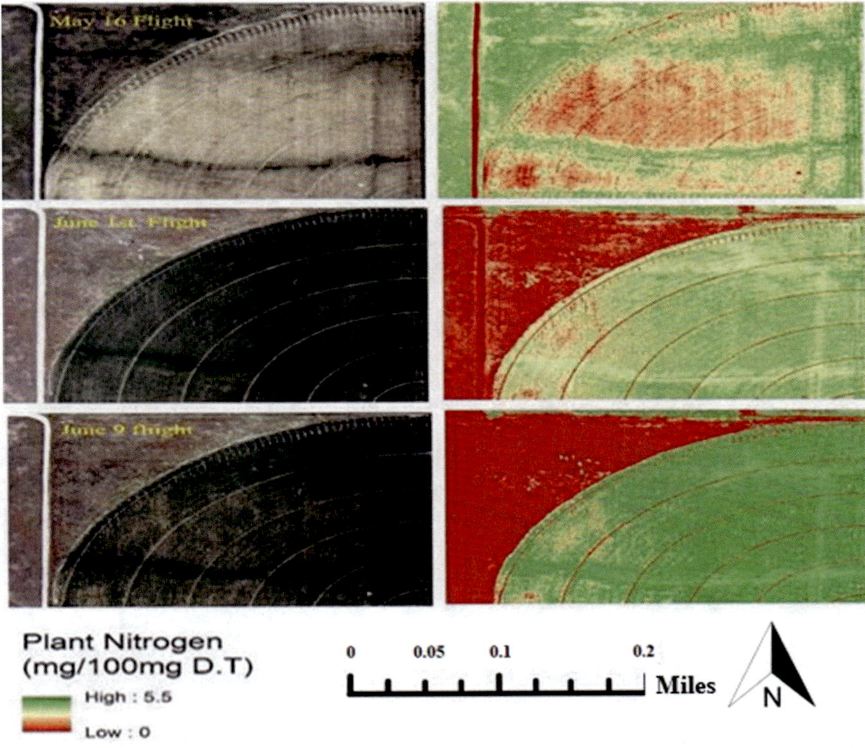

Fig. 6 UAS, True-colour maps (left), the estimated plant leaf nitrogen estimate map (right) (modified after Elarab [46])

5.3 Proximal Soil and Crop Sensors for Precision Farming

Remote sensing techniques include many different sensors system installed on airborne or satellite platforms. Proximal remote sensing sensors record their measurements directly at the field where, the surface and subsurface soil properties can be studied at specific locations [50]. This technology can be used at different levels in the field, and it allows measuring a set of soil characteristics directly. The benefits of single location measurement can use in different applications, produce high-resolution digital soil maps based moving sensors and measurements are taken close to the surface of the landscape [51]. There are two different proximal sensors one of them for soil and the other for crops measurements. Crop proximal sensors have been developed in the last two decades to measure physiological and morphological characteristics such leaf area index, biomass, water stress index, plant height, chlorophyll content, nitrogen content, that show the status of crops [52]. Proximal remote sensing sensors are mounted on ground-based vehicles and its collect the remote sensing data in the same way of the sensors that are used to collect aerial or satellite imagery. The

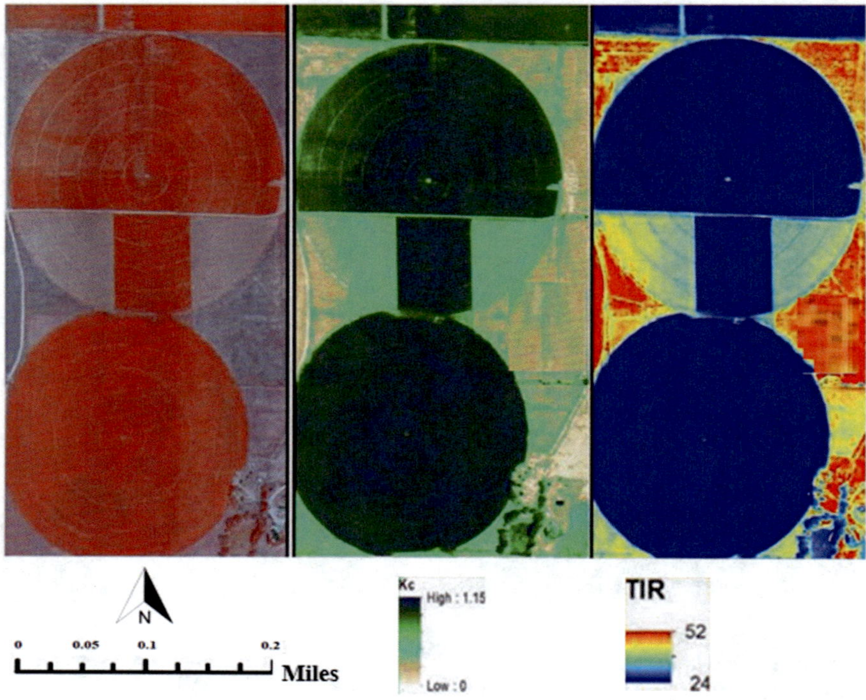

Fig. 7 UAS, reflectance in false colour (left), Mapping Evapotranspiration with Internalized Calibration (METRIC) reference evapotranspiration fraction (ETrf) output (centre), TIR (Celsius) map (right) (modified after Elarab [46])

difference between the ground-based systems and aerial or satellite imagery, it uses their light source to detect the soil and crop situations. Based on speciality, they can be used 24 h a day and any day regardless of its condition. These sensors produce an image in the area of the field, which they pass across. The information collected by the sensor is transformed to a reflectance index directly. This can then be tagged with location information from Global Navigation Satellite System (GNSS) and stored to make a map for later use or used in conjunction with variable-rate technology to control changes in fertilizer or chemical application while driving across the field [53]. Figure 8 shows some of the proximal crop reflectance sensors. Commercially, there are two types of sensors that can be used to measure soil salinity (EC) in the field; contact and non-contact. Both of them give similar results.

The first sensor type is made of electrode coulters that contact with the soil and measure the electric conductivity. This technique has two to three pairs of coulters placed on a toolbar; one provides the electrical current to the soil (transmitting electrodes), and the other (receiving electrodes) measures the voltage drop between them (Fig. 9). The second sensors type is non-contact EC sensors based on electromagnetic induction (EMI). Electromagnetic induction does not contact the soil surface.

Crop circle (A)

Manufacturer	Holland Scientific
Height of operation	0.25m to 2.5m
Field of view	Height x 0.6 (up to 8 sensors on CANbus)
View angle	Nadir
Active light source	Model ACS-220: Yellow (560nm) or Red (650nm) & NIR (770nm) Model 270: 3 user-configurable bands (420 to 800nm)
Data output	Model ACS-220: band information and NDVI or YNDVI Model 270: band information and user-defined index
Calibrations	Crop biomass and nitrogen uptake

Yara N Sensor® ALS (B)

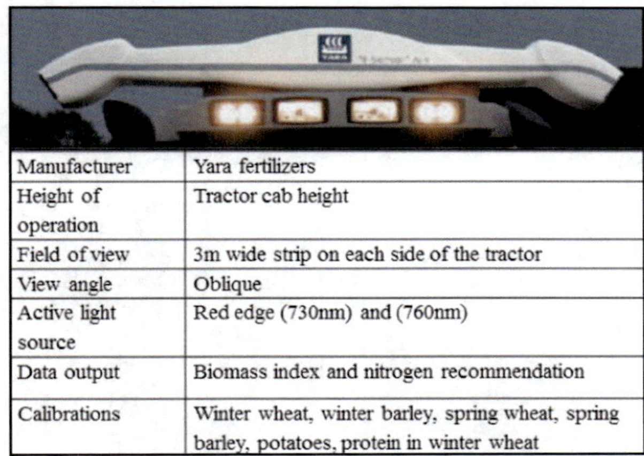

Manufacturer	Yara fertilizers
Height of operation	Tractor cab height
Field of view	3m wide strip on each side of the tractor
View angle	Oblique
Active light source	Red edge (730nm) and (760nm)
Data output	Biomass index and nitrogen recommendation
Calibrations	Winter wheat, winter barley, spring wheat, spring barley, potatoes, protein in winter wheat

Fig. 8 A crop circle sensor and B Yara N sensor (modified after Whelan [53])

It contains two coils; transmitter and another is a receiver (Fig. 10) generally placed at adverse ends of the unit. A sensor measures an electromagnetic field that affected by the concentration soil salinity [54].

The benefit of the on-the-go sensor technology is an effective way to supply rapid measurements, cost-effective, and getting high-resolution soil properties assessments to help site-specific management decisions. The essential soil properties do not have strong absorbance or reflectance lineaments in the visible and near-infrared (VNIR) spectroscopy range (400 and 1400 nm) or do not have correlated with primary soil

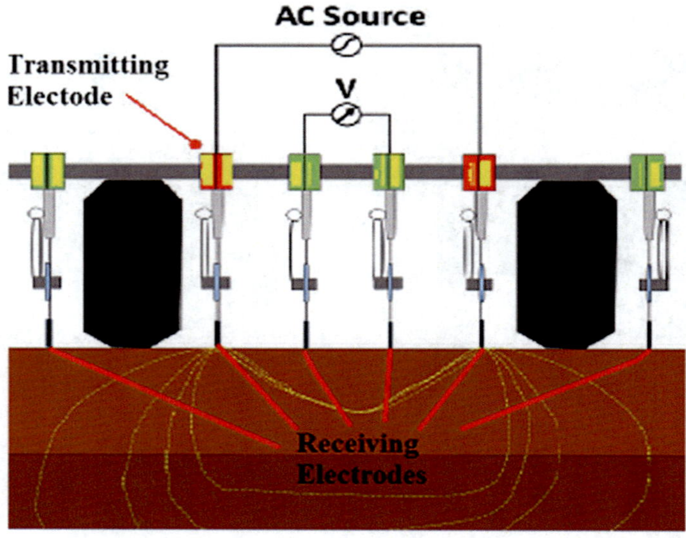

Fig. 9 Contact type EC sensor (modified after Grisso et al. [54])

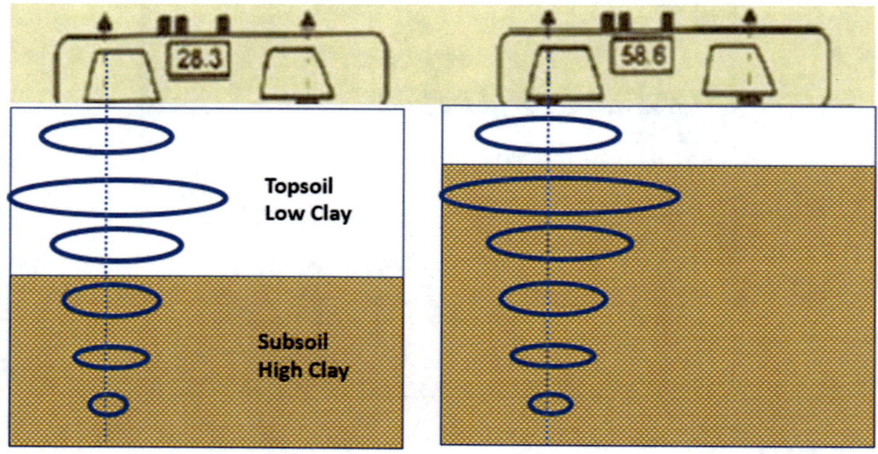

Fig. 10 Non-contact type EC sensor (modified after Grisso et al. [54])

properties that output VNIR lineaments. Therefore, a sensor fusion approach is ideal for in-field assessment of soil health. The use of soil apparent electrical conductivity (ECa), and penetration resistance measured by cone Penetrometer (i.e., cone index, CI) sensors with VNIR to improve assessment of biological and physical aspects of soil health [55]. Overall, proximal sensors or on-the-go sensor technology has the possibility to assessment and inclusive soil health index for improved performance sustainability, profitability, and environmental protection [56, 57].

6 Geographic Information System and Precision Farming

Geographic Information System (GIS) is a tool that analyzes, captures, stores manages and presents the data. A GIS depends on space and time dimensions as the base for all other applications. The basic concept of precision agriculture is collecting, analyzing and interpreting data in addition, making a decision. GIS systems supply interpretation and analysis tools that can be used as requirements for precision farming. It also provides a framework to build the data base and information for researchers, planners, decision-makers and other users. GIS is a term used for characterization of the data and has the ability to storage, manipulate, and present of data geographically [58, 59].

Precision farming is a new technology used to manage and improve the agricultural production, that depends on the field variability, where the application of irrigation, nutrient and seeding, should be considered of a given field. Furthermore, GIS could be used to generate and analyze complex view about the investigated fields and provide agro-technological decisions [60, 61]. Zhang and Taylor [62] Illustrated that field-level geographic information system (FIS) could be used as a research tool for precision farming. FIS is an integrated tool that depends on a raster-based geographic information system that developed for precision farming. Data collected by farmers is then used by GIS software for analyzing, and mapping then can support the farmers for the required information (Fig. 11). GIS is considered the central component of precision farming technology. The data which farmer has the list of coordinates and random data without GIS due to the data cannot be useful information without analyzing and interpretation them using GIS software. Still, when looking to it from a general overview, the needs and essential steps in precision farming may seem difficult. However, its effects and benefits appear to overbalance the necessary dedication on the part of the farmers which is clearly visible both at the level of implementation and the primary data offering [63, 64].

7 Soil Sampling Techniques for Precision Farming

The main purpose of precision agriculture technology is to increase fertilizer application efficiency according to crop requirements. The objectives of precision farming technology improve the efficiency of nutrient application and other inputs for crop production at precise locations where it depends on soil variability.The precision farming technology considered as the main means to success nutrient management plan where it depends on the optimum soil sampling techniques at the field and their variability. Soil sampling technique affects the results of soil analyses and

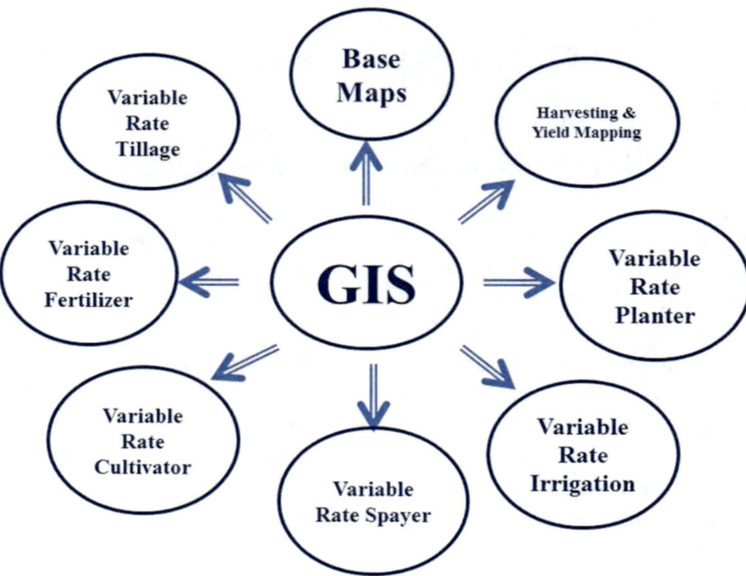

Fig. 11 Some of the components related to GIS in precision farming (modified after Zhang and Taylor [62])

nutrient crops program. The method of soil sampling supposes much greater important when implementation precision farming or site-specific management, because of the precision farming needs to assessment variable rates of nutrient and their applications and maximizing the profits of the precision farming. For developing a soil sampling strategy, the integration of precision farming technology should take into consideration as the most important issue [65]. The Grid soil sampling is the most importance methods for precision farming, and it is subdivided into configuration cells (commonly squares), and the soil sample is taken from each of these cells [66]. Fist technique for soil sampling is the grid soil sampling technique which has several types that include a regular systematic point, staggered start point, systematic unaligned point, and random composite cell [67]. Figure 12 shows an example of a regular, systematic sampling scheme. One soil sample would be taken from each of the center-stratify circles [68].

The random soil sampling is the second technique, which collects soil samples from all across the grid without any direction in the grid center (Fig. 13). The sampling pattern will not be harmonized across the cells, but this approach will ensure good randomization. The disadvantage of this method due to take more time and different sampling points has to be single accessed through the grid area [65]. Both center and random sampling techniques can be produced based on "Feature to Point" and "Create Random Points" tools, respectively [68]. Crozier and Heiniger [69] reported that the third technique for soil sampling is called as direct or management zone sampling. This technique depends on field variability to reduce the number of samples

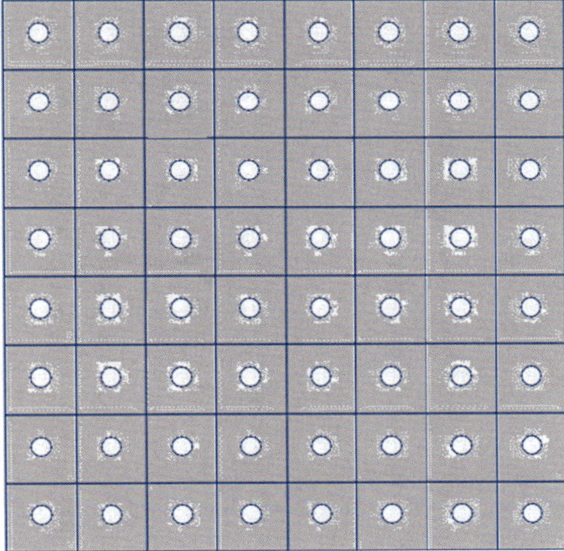

Fig. 12 Grid centered soil sampling (modified after Mylavarapu and Lee [65])

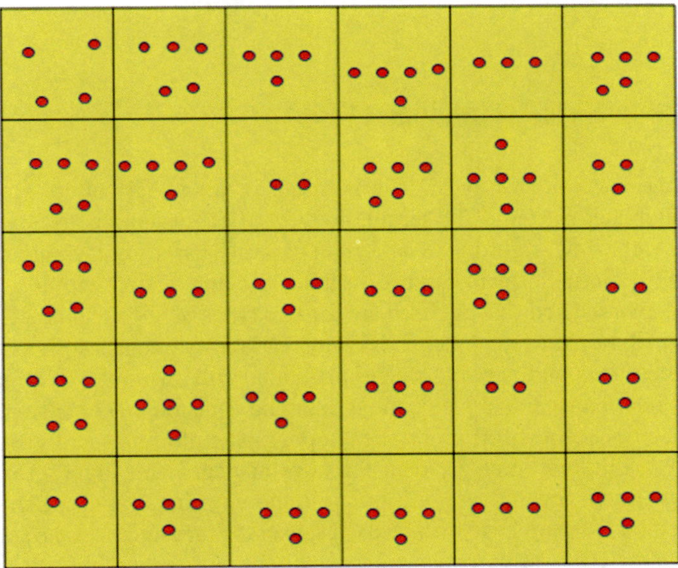

Fig. 13 Random soil sampling schemes [65]

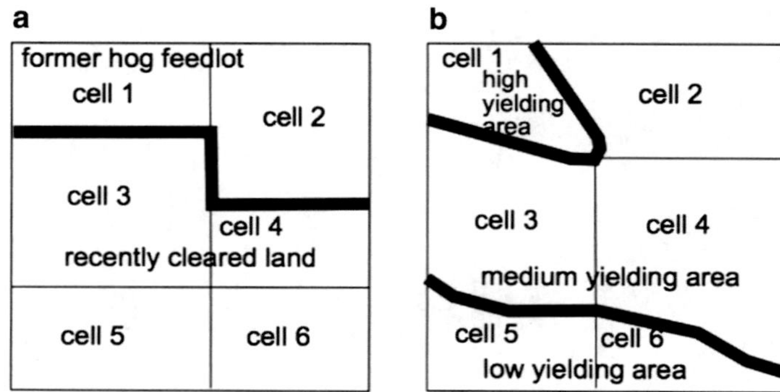

Fig. 14 Directed soil sampling schemes **a** based on previous land use and **b** based on prior crop yields [69]

while still recognizing zones of differing nutrient status as well as crop growth and yield variability based on different factors. Some of these factors related to soil characteristics and other factors are due to the land management program. If a uniform soil factor effect on yield variability in a field, then the distribution of this factor, or the distribution of crop yield, can assist in the design of a soil sampling scheme (Fig. 14).

8 Variable Rate Technology (VRT)

Variable-rate technologies (VRT) can be used for many agriculture practices and it is an automated system. The terminology of VRT systems is to apply variable rate at the right time, right place, and right quantity based on the variation of soil and crop [70]. Some farmers can use variable rate application with a conventional sprayer. Map based and sensor-based are the two types of VRT technology. The first method is map based which it depends on the variability in maps such as crop yield, soil characteristics and variable rate application. This method can be used using many different management plans. the farmer and decision maker have integrations different factors such as soil proprieties, topographic attributes and yield mapping to produce a different strategic plan for farm management. Figure 15 shows the map-based system for the variability of soil and crop input application rates. The second method is sensor based which used sensors to measurement the soil and crop characteristics. The measurements are taken by the sensor then analysis and control variable rate application directly (Fig. 16). This method does not need any DGPS but needs more data analysis to design the variable rate application plan [14, 71].

The sensor based system contains two kinds of sensors (Fig. 17). These sensors are placed on a strip or spray boom then directed in the soil. When the plants have much a chlorophyll content (green), the reflectance signal increase above the threshold,

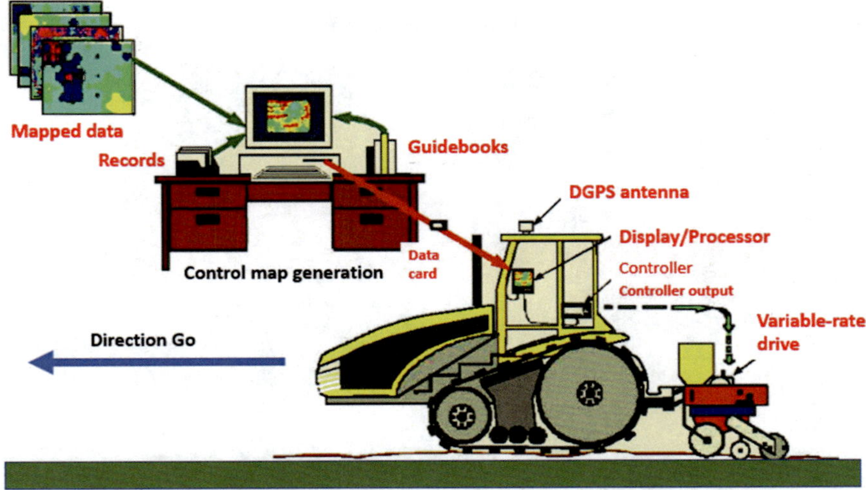

Fig. 15 The map-based system for variable application rates (modified after Ess et al. [71])

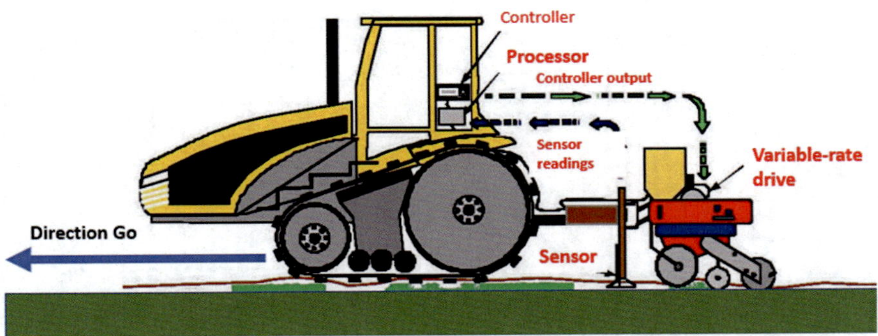

Fig. 16 The sensor-based system for variable application rates (modified after Ess et al. [71])

then the control unit sends a signal to a solenoid to be triggered by the valve to spray the herbicide [14, 72]. Optical sensors record light reflectance to detect crop and soil characterize. These optical sensors, too It can simulate the human eye to recognize different soil properties by using near-infrared, mid-infrared spectroscopy techniques. On the same principle techniques of the aerial photograph or satellites remote sensing [73]. Figure 18 shows the cross-section in optical sensor structure.

Theory of VRT built with the capability of farm field equipment to accurately oversight the variable rate application (VRA) of crop inputs and tillage operations. VRT can be divided into granular, liquid and gaseous form based on the type of fertilizer materials (Fig. 19). VRT system includes a GPS antenna, computer, controller, and geographic information systems map and database. The computer control unit

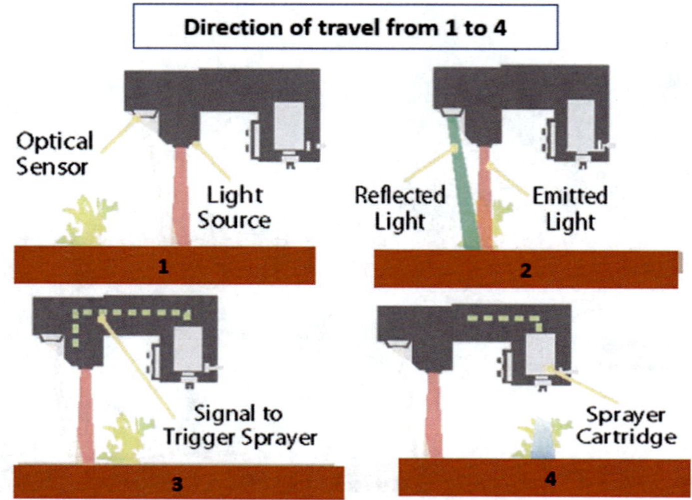

Fig. 17 The optical sensor control of the spray nozzle (modified after Grisso et al. [14])

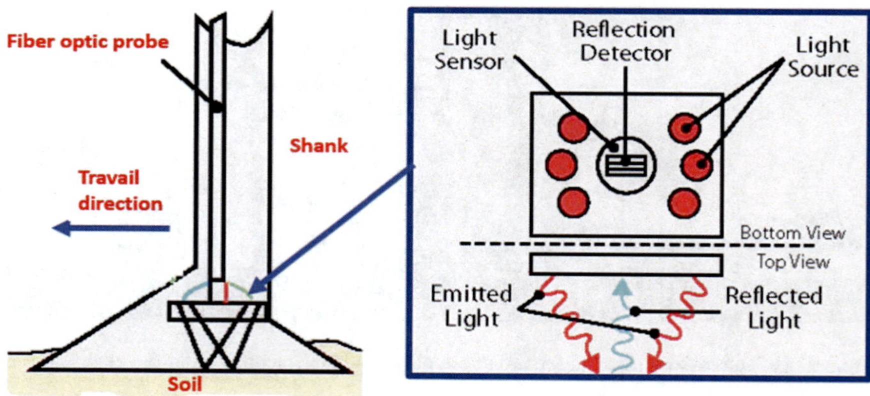

Fig. 18 The cross-section for the optical sensor (modified after Adamchuk and Jasa [73])

determines the interaction of the applied crop variables. The controller includes a spatial database using GIS, along with a precomputed flow rate system. The GPS receiver is integrated with the computer, the equipment location on the map supply by the GIS unit through the computer monitor and used the location coordinates from the GPS unit. The GIS system contains various variables related to crop inputs that are applied to equipment in the field. The actual applied rates at each location will be recorded and stored the information in the GIS computer system by the observer [33].

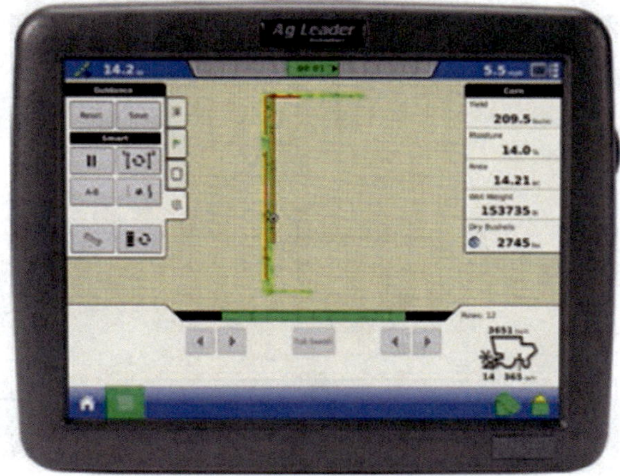

Fig. 19 Variable rate application monitors [74] Sprenger link

9 Yield Monitoring System and Mapping

The most important thing in precision farming is to understand the variability within the field for soil characteristics, crop yield, and crop growth condition. Also, PF has interested in the analysis of the limitation factors affecting plant growth. Finally, it important thing in precision farming to put management factors based on crop needs [75].

The continuing development of yield monitoring systems helps farmers to put the plan for the variable rate of planting and fertilizers application.

Yield mapping also, helps decision maker and farmers to know the different productive areas in the field, and they can evaluate if different planting populations will give a high return on their application. The farmers or producers use the yield map of the previous crop to estimate the fertilizer application rate based on the nutrient depletion from the soil in the previous crop [76, 77]. Figure 20 show combination of yield monitors systems. They contain three parts, user interface (keypad and display), a console located in the combine cab and data storage device. While it has different types of sensors such as a grain flow sensor, grain moisture sensor, separator speed sensor, header height sensor and ground speed sensor. The volume or mass of grain yield is to measure by grain flow sensor [78].

Gaumitz [79] reported that the size of the harvester combine makes a difference in accuracy as a smaller, research combine covers a smaller area and is, therefore, more precise than a regular sized combine. The yield monitor itself requires routine calibration and monitoring by the combine driver to ensure the best results. The yield map is generated from the thousands of points captured by the yield monitor in ArcGIS. The number of points varies greatly depending on the size of the field and harvester combines (Figs. 21 and 22).

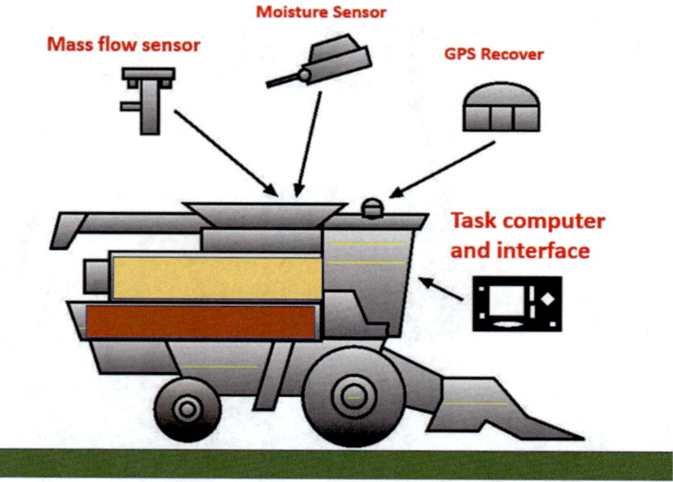

Fig. 20 Yield monitoring systems (modified after Grisso et al. [78])

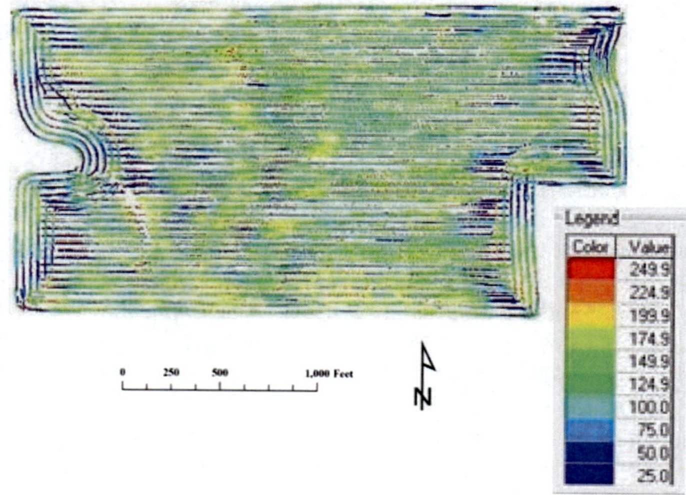

Fig. 21 Example of cleaned yield points generated using Yield Editor Software (modified after Gaumitz [79])

10 Site Specific Land Management (SSLM) and Soil Productivity

Site-specific land management (SSLM) is the technique of managing cropland in its local environment, taking into consideration the current types of the field variable.

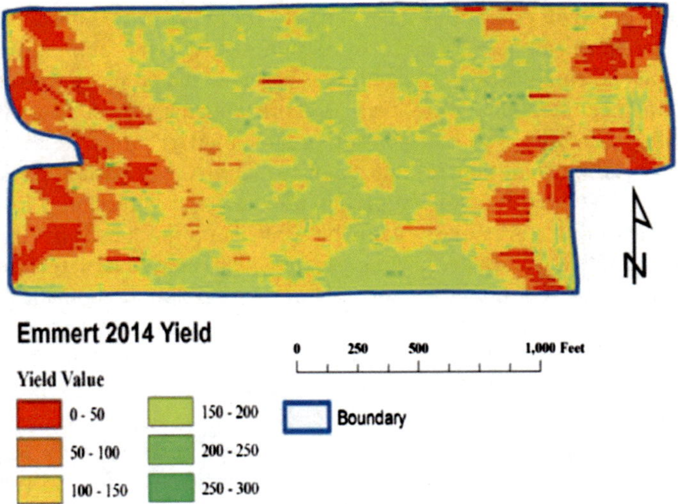

Fig. 22 Yield Interpolations map for Emmert (in bushels per acre) (modified after Gaumitz [79])

It can also show that managing local field variability is the core point in SSLM management plans [5]. Miller et al. [80] reported about three elements that should be achieved in for SSLM to be implemented. These elements include (1), important current factors that have spatial variability within-field that impact on crop yield, (2) source of this variability within—field can be specified and assessment and (3) the collected information from these assessments about field variability can be utilized to adjust the soil and crop management plans to increase farm income or to decrease environmental risks. Many types of research on SSLM have explained that several soil properties have high spatial variability within the field. The evaluation within this field variability and its adequacy play an important role in success SSLM planning, but one should be taken into considerations to its applicability and profitability. Laboratory analysis and spatial interpolation should be involved in this field variability, and this is behind that what can be economically assessed using soil sampling [81]. Spatial variability can occur on a variety of scales, between farmers regions and fields, or within the same fields. Soil chemical characteristics can have spatially varying content of nutrients, pH, salinity and organic matter. The uptake of these three significant macro-nutrients (N, P, K) by crops could be included nitrogen, phosphorus and potassium. Spatial variation in nutrients can happen, based on uptake by plants, filtration or changes such as liming application process.

The most soil physical properties including soil texture and structure also have effects on spatial variability in the field. The term of soil structure is the arranging of soil particles and voids in any specific volume of soil. Soil structure has an impact on the physical characteristics through gas propagation, water movement and plant root mooring, while soil texture impact crop yield by affecting supply nutrient variability and water and gas movement. The spatial variability in soil texture within the field based on the geologic operations cannot easily be influenced by farming activity, while soil structure can be influenced by farming activity through cultivating. The soil structural and compositional variability is based on the property considered [82]. Whelan [83] categorized this overall variability in soil attributes such as soil textural and structural variability; variability in soil organic matter; soil moisture variability; variability in soil nutrient content and their availability and variability in soil pH. Belal et al. [84] studied the soil properties affecting the wheat crop in the center pivot field. Soil salinity was interpolated between sampling locations for the specified sampling depths by ordinary kriging geostatistical interpolation technique. Also, the soil variables, including pH, organic matter (%), calcium carbonate content (%), available nitrogen (mg /kg), available phosphorus (mg /kg) and available potassium (mg /kg) were interpolated and mapped with Inverse Distance Weighted (IDW) interpolation technique for management zone extraction [85]. Figures 23, 24, 25 and 26) show the interpolated soil variables maps.

Soil texture plays an important role in the absorption of oxygen and growth of the potato tuber due to the potato crop is highly sensitive to the soil physical and chemical properties and especially texture. It needs low-soil density and good permeability, well aerated, rich in soil nutrients and good water retention capacity. Also, soil texture is an important soil factor for selecting the suitable agrotechnical and agrochemicals methods to be used to increase and improve soil fertility [86]. He

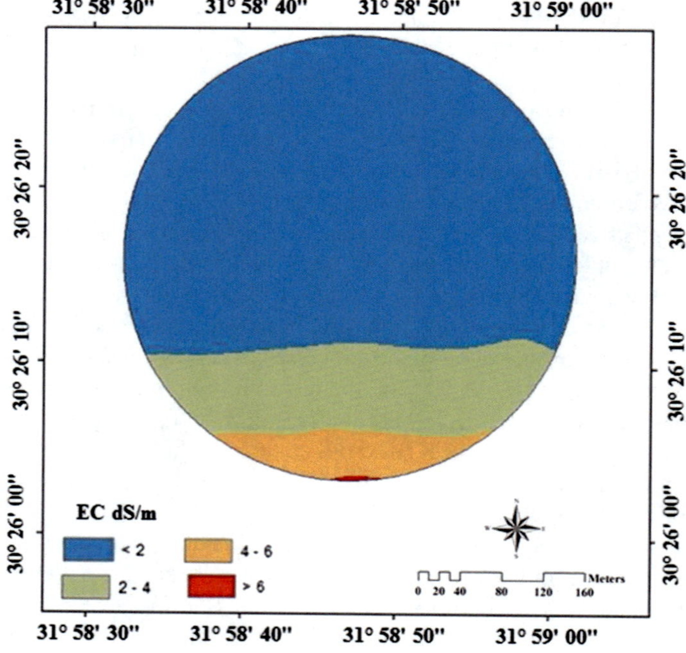

Fig. 23 EM vertical signals map (modified after Belal et al. [84])

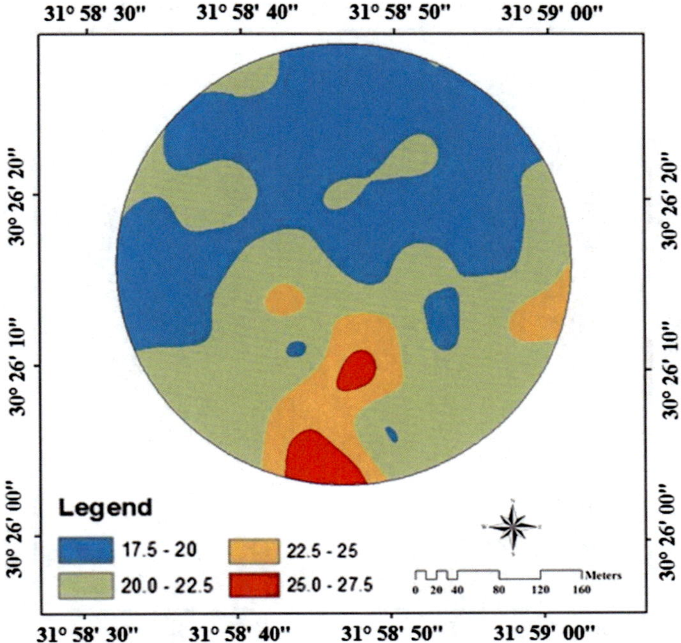

Fig. 24 Soil salinity mapping (modified after Belal et al. [84])

reported that the amount of soil salinity could be used to increase crop production in real time, only if we take into consideration other factors, such as data on yield production, seeds, the biological factors, the fertilizer and agronomic knowledge of the area. In order to achieve this objective of increasing productivity, it is essential to take into consideration both the soil and the plant parameters. By doing the correlating between these two factors will reach to the following objectives: get the good decision for fertilization the plants and the right research regarding the development of the soil's fertility. It could be improvement soil fertility if we used the right technologies as well as supporting cultivated plants and recovering depleted nutrients. These methods will help in the design of the soil fertility plans, in the end, take into consideration the agrochemical indices of the soil. This type of land management has classified the land into different site-specific management zone (SSMZ), which is divided into classes of a field that has a relatively uniform combination of yield-limiting factors. This SSMZ is normally produced from one layer of data or integration of different data layers containing topographic attributes, yield map, soil maps, soil salinity maps and soil fertility maps. The farmer can use the produced SSMZ map as an index to set which type of crop can be planted as well as soil conservation and management strategies are required, and they must be applied. The site-specific management zone is just an accepted method in many developed countries. Therefore, implementation of this new technology under the situation of

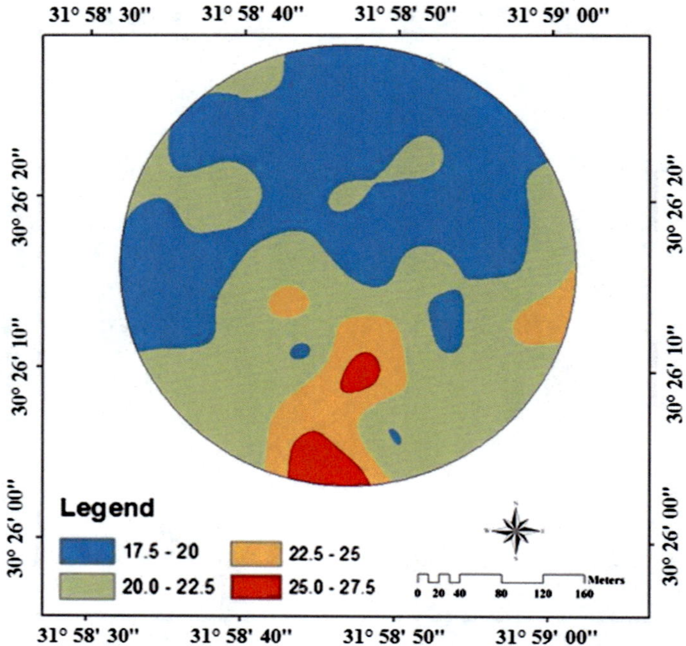

Fig. 25 Soil saturation percent map (modified after Belal et al. [84])

developing countries and trying on experiments in some villages is necessary to increase soil and crop production as well as reducing both cost and environmental risks [87, 88]. A number of methods could be used to delineate effective management zones, including soil-based approach, yield map-based approach and the combination of soil and topographic attributes information to define management zones. Saleh and Belal [89] produced the map for management zones of 154 Fadden of wheat center pivot field at East of Nile Delta in Egypt for use in precision farming technique depending on the spatial variability of soil and topographic attributes. The geostatistics, principal component analysis, and fuzzy logic techniques were used to analyze information and produced soil management zones. These Techniques were demonstrated to be a viable method for the area, lead to the delineation of uniform and distinct regions through them related to soil and topographic attributes. The result represented the five management zones are the suitable number for adequacy used in farm management. The main factors selected for delineation management zones are soil apparent electrical conductivity (EM38h), soil saturation percent (SP), and digital elevation model (DEM). The delineated management zones can be used to explain the spatial variability in soil and topographic properties (Fig. 27).

Belal et al. [58] used integrated approaches of some soil information with crop data to delineate subfield management zones. The soil data included ECa, pH, TDR (Time Domain Reflectrometry) but the plant data were included yield map, NPK,

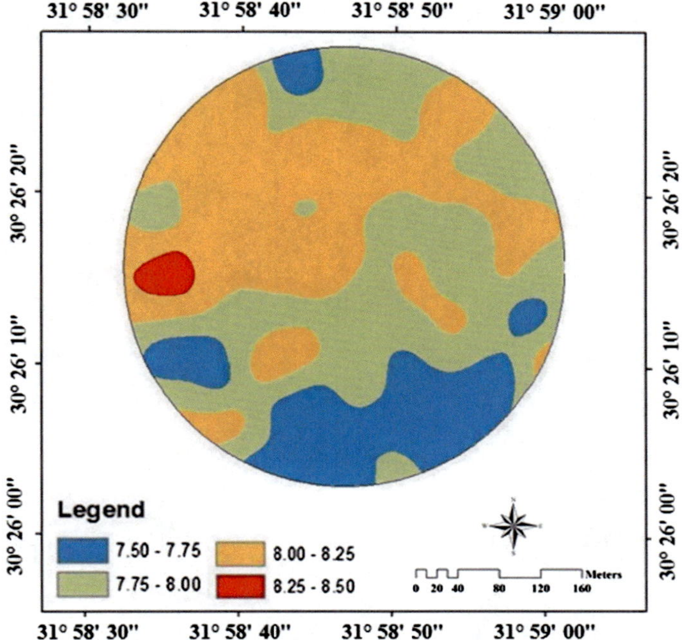

Fig. 26 pH map (modified after Belal et al. [84])

NDVI, chlorophyll A, B. Spatial models were produced based on the Model Builder tool in ArcGIS 10.1. In the Model Builder, a spatial model is displayed as a diagram that looks like a flowchart. The model used to the yield prediction as a base layer and then run to choose only those yield predictions within the base layer. In the final choose prediction map, dark red is related to the areas with the highest yield prediction; light red related to the areas with the lowest yield prediction (Fig. 28). Figure 29 shows the structure of spatial modeling to produce management zones with many layers.

11 Profitability Map of Application Precision Farming Technology

The variability of crop yield is the important issue for farmers to calculate economic analysis within the field to get a precise assessment of risk. The farmers can get a clear idea of choosing new agriculture technology over the conventional method when they compare the net income before and after the application of the SSLM [5]. Profit maps are very important in determining knowledge of the economics of crop production, as they show spatial improvement and compare yield data over time

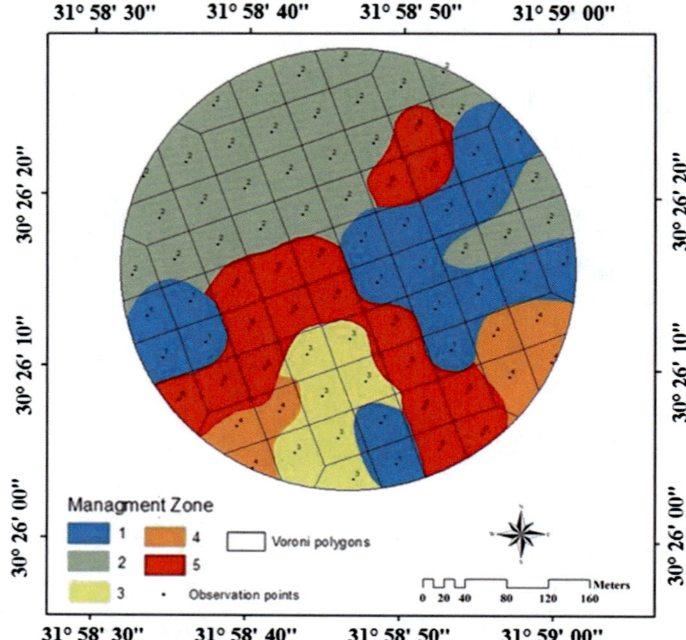

Fig. 27 Classes of management zones for the center pivot (modified after Saleh and Belal [89])

periods (in dollars). Hence, it is considered an effective tool for accurate decisions. Also, profit map for crops and over the years gives a good characterization of farm profitability over time and location. The decision maker can know where areas of the field were above or below economic benchmarks across years depending upon interaction cost and price data. Single crop profit maps let the decision maker assess how specific areas of the field can differ in profitability by crops over the years [90, 91]. Bazzi et al. [92] found that the prices of agricultural crops are constantly variable due to seasonal variation and demand, therefore it is difficult to predict the appropriate times in economic terms that affect farm profitability (equivalent 1) and the difference between total income and total cost.

$$P = Y * PP - Prc \tag{1}$$

where:

P = profit;

Y = yield (ton ha^{-1});

PP = sale price of the product (US$ ton^{-1});

Prc = production cost (US$ ha^{-1}).

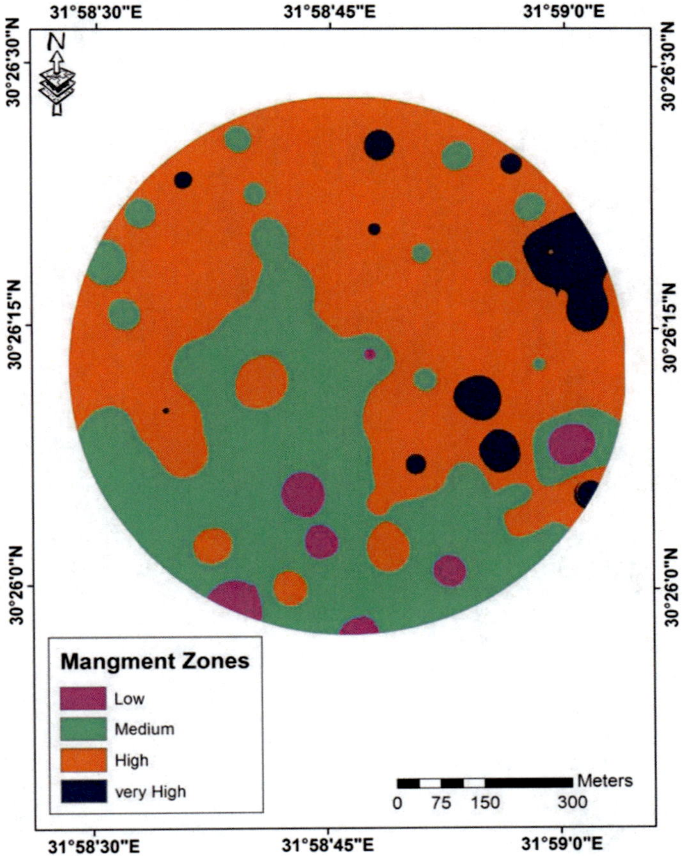

Fig. 28 Management zones map for the study area (modified after Belal et al. [58])

The total cost of the agricultural process includes (depreciation of machinery and tools, depreciation of improvements and formulations, organization and correction of soil, capital, and insurance) as well as operating and improving expenditures for maintenance, labor, seeds, fertilizers, and pesticides. General costs, external transportation, technical assistance, insurance, and benefits). Figures 30 and 31 showed produce yield mapping for soybean and profit (Tables 1, 2 and 3).

12 Conclusions

The precision farming or precision agriculture or site-specific land management is relatively a new farming technology for increase soil and crop production, however,

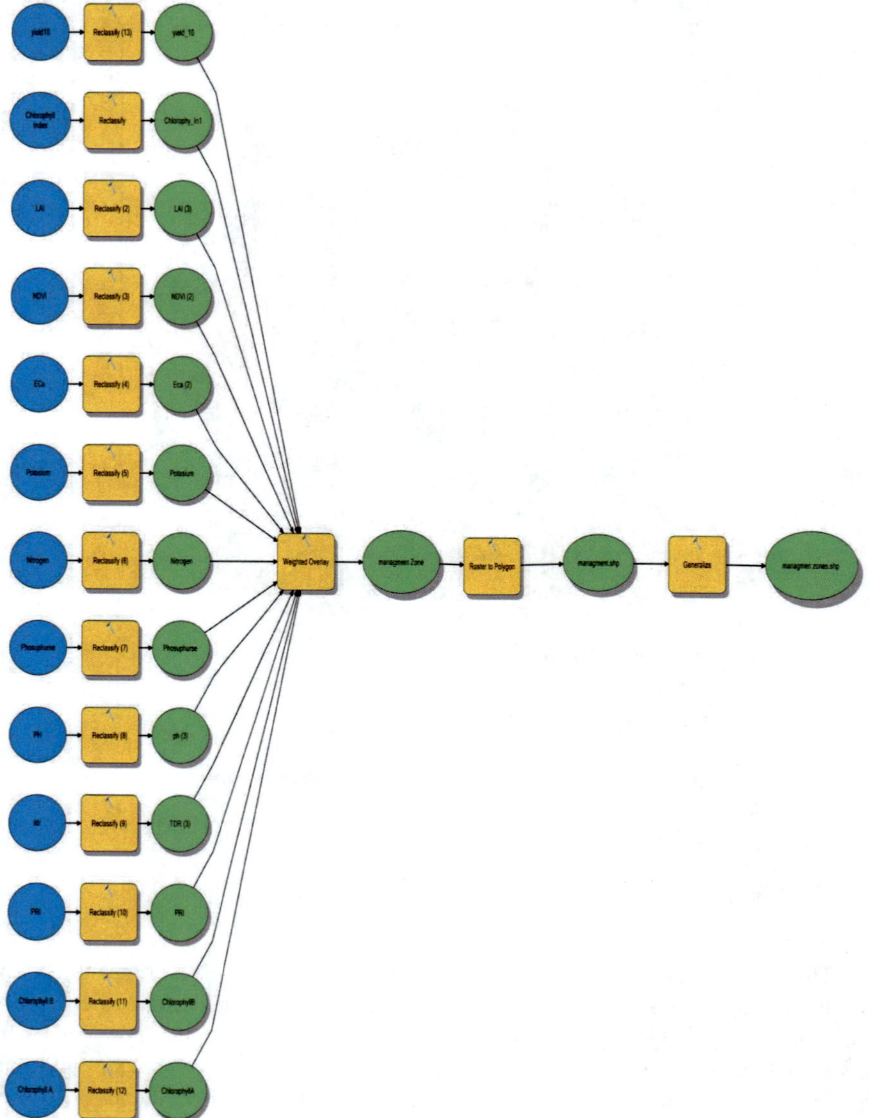

Fig. 29 Diagram of the spatial model developed in the ArcGIS Model Builder for management zones based on soil and plant properties [58]

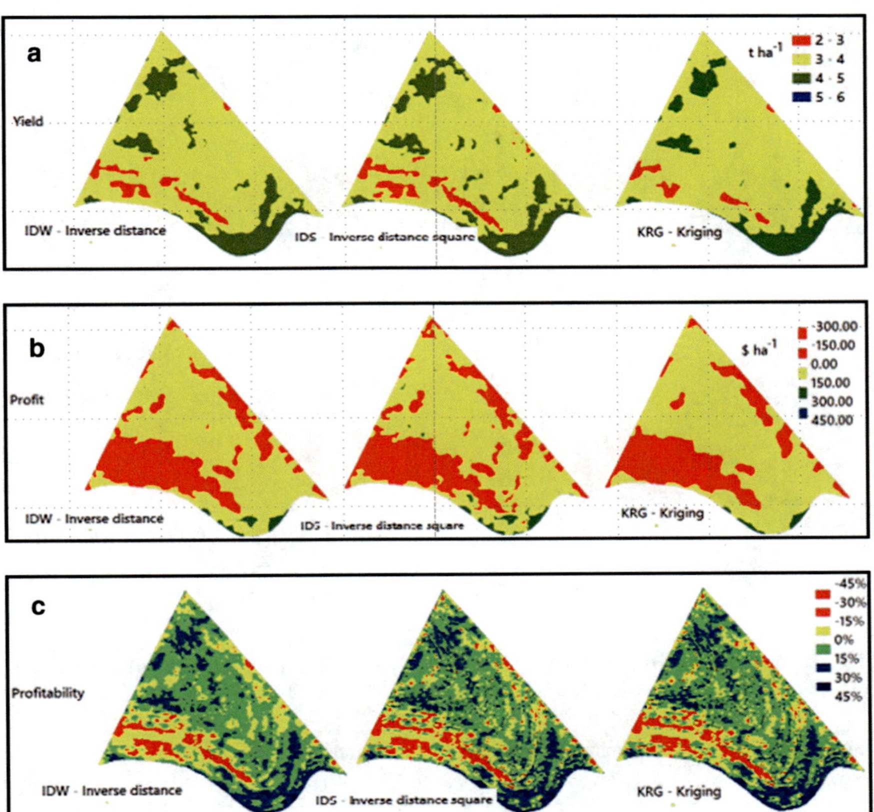

Fig. 30 Spatial distribution maps, **a** Yield, **b** profit and **c** profitability of Soybean in 2006 (modified after Bazzi et al. [92])

precision farming technology includes remote sensing, GIS, DGPS, yield map, variable rate application, soil sampling, and site-specific management zones. The main aim of precision farming include, the good understanding of soil and crop variability within-field and follow up by farm management inputs depending on the main source of the variability Moreover, this precision farming technology is the purpose of assisting management of agriculture resources in farm return and environmentally effective methods in the spatial and temporal field variability. In the last two decades, precision farming technology has been more and more developed, including mechanization and automation. Site-specific management has been an important sector of the agricultural system of the developed countries, but implementation of this high technique of precision farming to farm through the farmers in developing countries is still very slow and needs a lot of extension efforts and farmer motivation.. Due to most farmers do not know new information technologies, e.g. remote sensing, GIS, and GPS. The government should be made to increase

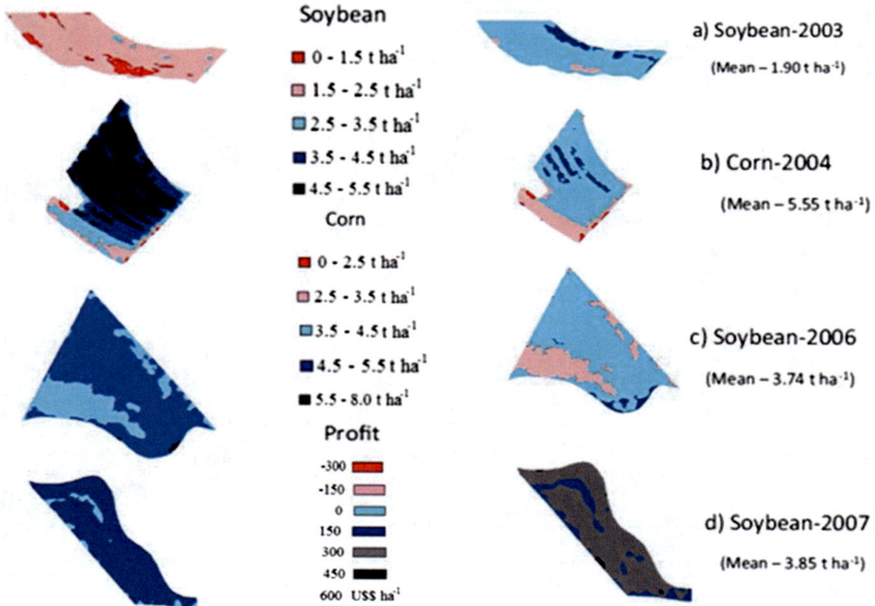

Fig. 31 Yield and profit maps for the harvest of Soybean2003 (**a**), Corn2004 (**b**), Soybean2006 (**c**), and Soybean2007 (**d**) using kriging method (modified after Bazzi et al. [92])

Table 1 Showed production cost of Soybean 03, 06, 07 and Cron crops (US$ ha^{-1})

Culture/harvest	Soybean-03	Cron-2004	Soybean-06	Soybean-07
Production cost	270.90	366.90	641.28	632.35

Source SEAB/PR [93]

Table 2 Showed selling price of corn (US$ ha^{-1})

Year	July	August	September	October	November	December
2004	88.0	85.0	87.67	83.0	82.33	79.83

Source SEAB/PR [93]

Table 3 Showed selling price of soybeans (US$ ha^{-1})

Year	March	April	May	June	July	August
2003	188.50	199.66	189.00	192.83	182.17	184.33
2006	181.83	180.00	173.33	191.00	188.17	186.33
2007	232.50	223.00	233.67	242.33	245.33	156.50

Source SEAB/PR [93]

investments to educate farmers for the application of new methods and technologies in agriculture. The size of the agriculture land is another problem for adopted precision farming technology in the third country like Egypt. Egyptian farmers and experts should use the PF experiences that have been used by developed countries such as USA, Germany, Japan. In conclusion, integrating new technology for farmlands management and provided investment in agriculture are two essential issues that help to improve Egypt's agriculture production.

13 Recommendations

Farmers in developing countries still rely on traditional systems in agricultural operations, as the same procedure is applied to their crops, even in different conditions of climate and soil. Farmers need very high costs to obtain higher production compared to the application of modern technologies aimed at improving and efficient land management. Precision farming systems can be implementation in developing country through collect information about soil and crop using new information technology as the following [94]:

- Spatial management depends on collecting information from the soil that illustrates the spatial variation of crop and soil characteristics using GPS.
- Remote sensing and geographic information systems have increased the efficiency of agricultural operations, such as plowing, planting preparation, seeding rate, fertilizers, pesticides, and the efficiency of irrigation use in modern systems.
- The integration between different data using geographic information systems has led to increased accuracy of mapping and spatial prediction of soil and crop properties across a field, as these maps allow a detailed representation of spatial change within a field by creating a homogeneous map.
- Decision support systems rely on the integration of knowledge about the effects of field variation on crop growth and the degree of responses to appropriate agricultural operations—to achieve a higher rate of accuracy in decision validity.

References

1. Mohamed ES (2013) Spatial assessment of desertification in north Sinai using modified MEDLAUS model. Arab J Geosci 6(12):4647–4659
2. Onyango CM, Nyaga JM, Wetterlind J, Söderström M, Piikki K (2021) Precision agriculture for resource use efficiency in smallholder farming systems in Sub-Saharan Africa: a systematic review. Sustainability 13(3):1158
3. Soropa G, Nyamangara J, Nyakatawa EZ (2019) Nutrient status of sandy soils in smallholder areas of Zimbabwe and the need to develop site-specific fertiliser recommendations for sustainable crop intensification. S Afr J Plant Soil 36:149–151

4. Abu-Hashim M, Mohamed E, Belal AE (2015) Identification of potential soil water retention using hydric numerical model at arid regions by land-use changes. Int Soil Water Conserv Res 3(4):305–315
5. Adhikari K, Carre F, Toth G, Montanarella L (2009) Site specific land management: general concepts and applications. EUR23978 EN. ISBN 978-92-79-13350-3. https://doi.org/10.2788/32619
6. Mohamed ES, Ali A, El-Shirbeny M, Abutaleb K, Shaddad SM (2019) Mapping soil moisture and their correlation with crop pattern using remotely sensed data in arid region. Egypt J Remote Sens Space Sci
7. Hassan AM, Belal AA, Hassan MA, Farag FM, Mohamed ES (2019) Potential of thermal remote sensing techniques in monitoring waterlogged area based on surface soil moisture retrieval. J Afr Earth Sc 155:64–74
8. Mohamed ES, Abu-hashim M, Abdel Rahman MA, Schütt B, Lasaponara R (2019) Evaluating the effects of human activity over the last decades on the soil organic carbon pool using satellite imagery and GIS techniques in the Nile Delta Area, Egypt. Sustainability 11(9):2644
9. Srinivasan A (2006) Handbook of precision agriculture: principles and applications. Food Product Press, New York, p 708P
10. Kayode O, Aizebeokhai A, Odukoya A (2021) Sustainable use of chemical in agricultural soils and implications for precision agriculture. J Environ Treatm Tech 9(2):361–367
11. Lowenberg-DeBoer J, Erickson B (2019) Setting the record straight on precision agriculture adoption. Agron J 111:1552–1569
12. Miao Y, Mulla DJ, Pierre C, Robert PC (2018) An integrated approach to site-specific management zone delineation. Front Agr Sci Eng. https://doi.org/10.15302/J-FASE-2018230
13. Sao Y, Singh G, Jha MK (2018) Site specific nutrient management for crop yield maximization using two soil types of Bilaspur District of C.G. on grain and straw yield. J Pharm Phytochem 7(1):08–10
14. Grisso RB, Alley M, Thomason W, Holshouser D, Roberson GT (2011) Precision farming tools: variable-rate application. Virginia Cooperative Extension, Virginia State University. Publication number, 442–505. https://www.pubs.ext.vt.edu/content/dam/pubs_ext_vt_edu/442/442-505/442-05_PDF.pdf
15. Hendawy E, Belal AA, Mohamed ES, Elfadaly A, Murgante B, Aldosari AA, Lasaponara R (2019) The prediction and assessment of the impacts of soil sealing on agricultural land in the North Nile Delta (Egypt) using satellite data and GIS modeling. Sustainability 11(17):4662
16. Abdullahi HS, Mahieddine F, Sheriff RE (2015) Technology impact on agricultural productivity: a review of precision agriculture using unmanned aerial vehicles. In: Pillai et al (eds) Wireless and satellite systems 7th international conference, WiSATS 2015 Bradford, UK, 6–7 July 2015 Revised Selected Papers. Springer, Cham, pp 388–400. https://doi.org/10.1007/978-3-319-25479-1
17. Davis RJ, Baillie C, Schmidt E (2009) Precision agriculture technologies-relevance and application to sugarcane production, Agric. Technol. a Chang. Clim, pp 114–122
18. Mohamed ES, Baroudy A, El-beshbeshy T, Emam M, Belal AA, Elfadaly A, Aldosari AA, Ali A, Lasaponara R (2020) Vis-NIR spectroscopy and satellite landsat-8 OLI data to map soil nutrients in arid conditions: a case study of the Northwest Coast of Egypt. Remote Sens 12(22):3716
19. Podlasek A, Koda E, Vaverková MD (2021) The variability of nitrogen forms in soils due to traditional and precision agriculture: case studies in Poland. Int J Environ Res Public Health 18(2):465
20. Mani PK, Mandal A, Biswas S, Sarkar B, Mitran T, Meena RS (2021) Remote sensing and geographic information system: a tool for precision farming. In: Geospatial technologies for crops and soils. Springer, Singapore, pp 49–111
21. Zarco-Tejada PJ, Hubbard N, Loudjani P (2014) Precision agriculture: an opportunity for EU Farmers potential support with the cap 201–2020. Policy Department B: Structural and cohesion polices. Agriculture and Rural Development. http://www.europarl.europa.eu/RegData/etudes/note/join/2014/529049/IPOL-AGRI_NT%282014%29529049_EN.pdf

22. USDA (2007) Precision agriculture: NRCS support for emerging technologies. Agronomy Technical Note No 1
23. Baroudy AA, Ali A, Mohamed ES, Moghanm FS, Shokr MS, Savin I, Poddubsky A, Ding Z, Kheir A, Aldosari AA, Elfadaly A (2020) Modeling land suitability for rice crop using remote sensing and soil quality indicators: the case study of the Nile Delta. Sustainability 12(22):9653
24. Paxton KW, Mishra AK, Chintawar S, Roberts RK, Larson JA, English BC, Lambert DM, Marra MC, Larkin SL, Reeves JM, Martin SW (2011) Intensity of precision agriculture technology adoption by cotton producers. Agric Resour Econ Rev 40(1):133–144
25. Abdel-Fattah MK, Abd-Elmabod SK, Aldosari AA, Elrys AS, Mohamed ES (2020) Multivariate analysis for assessing irrigation water quality: a case study of the Bahr Mouise Canal, Eastern Nile Delta. Water 12(9):2537
26. Elarab M (2016) The application of unmanned aerial vehicle to precision agriculture: chlorophyll, nitrogen, and evapotranspiration estimation. Ph.D., Civil and Environmental Engineering, Utah State University, USA
27. Said MES, Ali A, Borin M, Abd-Elmabod SK, Aldosari AA, Khalil M, Abdel-Fattah MK (2020) On the use of multivariate analysis and land evaluation for potential agricultural development of the Northwestern Coast of Egypt. Agronomy 10(9):1318
28. Panda SS, Hoogenboom G, Paz JO (2010) Remote sensing and geospatial technological applications for site-specific management of fruit and nut crops: a review. Remote Sens 2:1973–1997. https://doi.org/10.3390/rs2081973
29. Folnovic T (2015) Benefits of using precision farming. Producing More with Less—See more at: http://blog.agrivi.com/post/benefits-of-using-precision-farming-producing-more-with-less#sthash.N1xXkUyU.dpuf
30. Goswami SB, Matin S, Saxena A, Bairagi GD (2012) A review: the application of remote sensing, GIS and GPS in precision agriculture. Int J Adv Technol Eng Res (IJATER) 2(1). ISSN: 2250-3536
31. Tayari E, Jamshid AR, Goodarzi HR (2015) Role of GPS and GIS in precision agriculture. J Sci Res Dev 2(3):157–162
32. Singh AK (2004) Precision farming. Water Technology Center. I.A.R.I, New Delhi
33. Andreo V (2013) Remote sensing and geographic information systems in precision farming. Maestría en Aplicaciones Espaciales de Alerta y Respuesta Temprana a Emergencias - CONAE – UNC
34. Bakhtiari AA, Hematian A (2013) Precision farming technology, opportunities and difficulty. Int J Sci Emerg Technol Latest Trends 5(1):1–14
35. Soares MG, Malheiro B, Restivo FJDO (2004) Evaluation of a real time DGPS data server. In: First international European conference on the use of modern information and communication technologies (ECUMICT 2004). KU Leuven, pp 105–112
36. Liaghat S, Balasundram SK (2010) A review: the role of remote sensing in precision agriculture. Am J Agric Biol Sci 5(1):50–55
37. Brisco B, Brown RJ, Hirose T, McNairn H, Staenz K (2014) Precision agriculture and the role of remote sensing: a review. Can J Remote Sens 24(3). https://doi.org/10.1080/07038992.1998.10855254
38. Brook A, De Micco V, Battipaglia G, Erbaggio A, Ludeno G, Catapano I, Bonfante A (2020) A smart multiple spatial and temporal resolution system to support precision agriculture from satellite images: proof of concept on Aglianico vineyard. Remote Sens Environ 240:111679
39. Yang C, Everitt JH, Du Q, Luo B, Chanussot J (2013) Using high-resolution airborne and satellite imagery to assess crop growth and yield variably for precision agriculture. Proc IEEE 101(3)
40. Yao HL, Tang L, Tian Brown RL, Bhatnagar D, Cleveland TE (2010) Using hyperspectral data in precision farming applications (Chap 25). In: Thenkabail PS, Lyon JG, Huete A (eds) Hyperspectral remote sensing of vegetation. CRC Press, Boca Raton, p 705
41. Mogili UR, Deepak BV (2018) Review on application of drone systems in precision agriculture. Procedia Comput Sci 133:502–509. In: International Conference on Robotics and Smart Manufacturing (RoSMa2018)

42. Valente J, Sanz D, Barrientos A, del Cerro J, Ribeiro Á, Rossi C (2011) An air-ground wireless sensor network for crop monitoring. Sensors 11:6088–6108. https://doi.org/10.3390/s110606088
43. Salamí E, Barrado C, Pastor E (2014) UAV flight experiments applied to the remote sensing of vegetated areas. Remote Sens 6:11051–11081
44. Marinello F, Pezzuolo A, Chiumenti A, Sartori L (2016) Technical analysis of unmanned aerial vehicles (Drones) for agriculture applications. Engineering for Rural Development, Jelgava, 25.-27.05
45. Valavanis KP, Oh P, Piegl L (2009) Guest editorial for the special volume on unmanned aircraft systems (UAS). J Intell Rob Syst 54(1–3):1–2
46. El-Shirbeny MA, Mohamed ES, Negm A (2018) Estimation of crops water consumptions using remote sensing with case studies from Egypt. In: Negm AM (ed) Conventional water resources and agriculture in Egypt. The handbook of environmental chemistry, vol 74. Springer, Cham. https://doi.org/10.1007/698_2018_305
47. Hackney C, Clayton A (2015) Unmanned Aerial Vehicles (UAVs) and their application in geomorphic mapping. British Society for Geomorphology, Geomorphological Techniques, Chap. 2, Sec. 1.7, ISSN 2047-0371
48. d'Oleire-Oltmanns S, Marzolff I, Peter KD, Johannes B, Ries JB (2012) Unmanned aerial vehicle (UAV) for monitoring soil Erosion in Morocco. Remote Sens 4(11):3390–3416. https://doi.org/10.3390/rs4113390
49. Alexakis D, Seiradakis K, Tsanis I (2016) Using unmanned aerial vehicle (UAV) for spatio-temporal monitoring of soil erosion and roughness in Chania, Crete, Greece. EGU General Assemb 18:EGU2016-11937
50. Viscarra Rossel RA, Adamchuk VI, Sudduth KA, Mckenzie NJ, Lobsey C (2011) Proximal soil sensing: an effective approach for soil measurements in space and time. Adv Agron 113:237–282
51. Adamchuk VI, Viscarra Rossel RA, Sudduth KA, Schulze Lammers P (2011) Sensor fusion for precision agriculture. In: Sensor fusion—foundation and applications. In Tech, pp 27–40
52. Samborski SM, Tremblay N, Fallon E (2009) Strategies to make use of plant sensors based diagnostic information for nitrogen recommendations. Agron J 101:800–816
53. Whelan B (2015) Proximal crop reflectance sensors: a guide to their capabilities and uses. Published by GRDC and the University of Sydney, Australia. ISBN: 978-1-921779-54-1
54. Grisso RB, Alley M, Wysor WG, Holshouser D, Thomason D (2009) Precision farming tools: soil electrical conductivity. Virginia Cooperative Extension, Virginia State University. Publication number, 442–508
55. Mohamed ES, Saleh AM, Belal AB, Gad A (2018) Application of near-infrared reflectance for quantitative assessment of soil properties. Egypt J Remote Sens Space Sci 21(1):1–14
56. Mohamed ES, Ali AM, El Shirbeny MA, Abd El Razek AA, Savin IY (2016) Near infrared spectroscopy techniques for soil contamination assessment in the Nile Delta. Eur Soil Sci 49(6):632–639
57. Veum KS, Sudduth KA, Kitchen NR (2016) Sensor based soil health assessment. In: 13th international conference in precision agriculture, 31 July–4 Aug 2016. St., Louis Missouri, USA
58. Belal AA, Mohamed ES, Abu-hashim MSD (2015) Land evaluation based on GIS-spatial multi-criteria evaluation (SMCE) for agricultural development in dry Wadi, Eastern Desert, Egypt. Int J Soil Sci 10:100–116
59. Earl R, Thomas G, Blackmore BS (2000) The potential role of GIS in autonomous field operations. Comput Electron Agric 25:107–120
60. Abd-Elmabod SK, Mansour H, Hussein AAEF, Mohamed ES, Zhang Z, Anaya-Romero M, Jordán A (2019) Influence of irrigation water quantity on the land capability classification. Plant Arch 2:2253–2561
61. Neményi M, Mesterházi PÁ, Pecze Z, Stépán Z (2003) The role of GIS and GPS in precision farming. Comput Electron Agric 40(1–3):45–55

62. Zhang N, Taylor RK (2001) Applications of a field level geographic information systems (FIS) precision agriculture. Appl Eng Agric 17(6):885–892
63. Abu-hashim M, Elsayed M, Belal AE (2016) Effect of land-use changes and site variables on surface soil organic carbon pool at Mediterranean Region. J Afr Earth Sc 114:78–84
64. Limpisathian P (2011) Geographic information system in agriculture and precision farming. Earth and Mineral Sciences First-year Seminar—EM SC 100S Section 4. The College of Earth and Mineral Sciences the Pennsylvania State University
65. Mylavarapu RS, Lee WD (2014) UF/IFAS Nutrient management series: Soil sampling strategies for precision agriculture. IFAS Extension, University of Florida. http://edis.ifas.ufl.edu/pdffiles/SS/SS40200.pdf. Last accessed 11 Aug 2014
66. Mallarino A, Wittry D (2001) Management zones soil sampling: a Better alternative to grid and soil type sampling? Ames: Iowa State University Extension, 159–164. http://www.agronext.iastate.edu/soilfertility/info/ICM_2001_ZoneSampling_Publ.pdf
67. Franzen DW (2011) Collecting and analyzing soil spatial information using kriging and inverse distance. In: Clay DE, Shanahan JF (eds) GIS applications in agriculture. CRC Press, Boca Raton, pp 61–80
68. Harrell JB (2014) An evaluation of soil sampling methods in support of precision agriculture in Northeastern North Carolina. M.Sc., thesis, Faculty of the USC Graduate School University of Southern California, USA
69. Crozier C, Heiniger R (2015) Soil sampling for precision farming systems. North Carolina State University Extension. https://content.ces.ncsu.edu/soil-sampling-for-precision-farming-systems
70. NESPAL (2005) Benefits of precision agriculture (slides, available at www.nespal.cpes.peachnet.edu/PrecAg/)
71. Ess D, Morgan M, Parson S (2001) Implementing site-specific management: map- versus sensor-based variable-rate application. Technical Report. Available on http://www.ces.purdue.edu/extmedia/AE/SSM-2-W.pdf
72. Ehsani R, Schumann A, Salyani M (2009) Variable rate technology for Florida citrus. UF/IFAS Extension, University of Florida. http://ufdcimages.uflib.ufl.edu/IR/00/00/33/14/00001/AE44400.pdf
73. Adamchuck VI, Jasa P (2002) On-the-go vehicle-based soil sensors. University of NebraskaLincoln Extension EC02-178. www.ianrpubs.unl.edu/epublic/live/ec178/build/ec178.pdf
74. Ahmad L, Mahdi SS (2018) Variable rate technology and variable rate application. In: Satellite farming. Springer, Cham. https://doi.org/10.1007/978-3-030-03448-1_5
75. Zhang M, Li MZ, Liu G, Wang MH (2008) Yield Mapping in precision farming. In: Li D (ed) IFIP international federation for information processing, vol 259; Computer and computing technologies in agriculture, vol 2. Springer, Boston, pp 1407–1410
76. Pathak DUP, Meena MK, Mallikarjun N (2018) Precision farming a promising technology in horticulture: a review. Int J Pure Appl Biosci 6(1):1596–1606. https://doi.org/10.18782/2320-7051.3088
77. Risius NW (2014) Analysis of a combine grain yield monitoring system. MSC, Iowa State University, Ames, Iowa, USA
78. Grisso RB, Alley M, McClellan P (2009) Precision farming tools: yield monitor. Virginia Cooperative Extension, Virginia State University. Publication number, 442–502. https://www.pubs.ext.vt.edu/content/dam/pubs_ext_vt_edu/442/442-502/442-02_pdf.pdf
79. Gaumitz BC (2016) Precision agriculture and GIS: evaluation the use of yield maps combined with LIDAR data. MSC, thesis Faculty of the USC Graduate School University of Southern California, USA
80. Miller RO, Pettygrove S, Denison RF, Jackson L, Cahn M, Plant R, Kearny T (1999) Site-specific relationships among flag leaf nitrogen, SPAD meter values and grain protein in irrigated wheat. In: Robert PC, Rust RH, Larson WE (eds) Proceedings of the fourth international conference on precision agriculture, 19–22 July 1998, St. Paul, USA. Ameri. Soci. of Agron., Madison, WI, pp 113–122

81. McBratney AB, Pringle MJ (1997) Spatial variability in soil: implications for precision agriculture. In: Proceedings of 1st European conference on precision agriculture, vol 1. Warwick, UK, pp 3–31
82. Hellebrand HJ, Umeda M (2004) Soil and plant sensing for precision agriculture. In: 1st Asian conference on precision agriculture, Kuala Lumpur Malaysia, 11–13 May
83. Whelen BM (2003) Precision agriculture, an introduction to concepts, analysis and interpretation. A training course for graduate and industrial professional. Aus, Center for Precision Agriculture, University of Sydney, Australia, pp 11–153
84. Belal AA, Saleh AM, Mohamed E, El baroudy A (2014) Using close-to-ground sensing, advances in spatial Sampling and prediction to characterize soil and wheat crop variability for precision farming. Second progress Project report Funded by STDF, Egypt (Project Number 444)
85. Saleh AM, Belal AA, Mohamed E (2016) Mapping of soil salinity using electromagnetic induction: a case study of East Nile Delta, Egypt. Egypt J Soil Sci (Under Publication)
86. Isabela M (2014) Potato crop monitoring based on spatial variability of recourses. PhD thesis, Faculty of Agriculture Sciences, Ph.D. School University of Agriculture Sciences and Veterinary Medicine CLUJ-NAPOCA
87. Hammam AA, Mohamed ES (2018) Mapping soil salinity in the East Nile Delta using several methodological approaches of salinity assessment. Egypt J Remote Sens Space Sci
88. Tiki L, Kewessa G, Nigatu D (2015) A review on site specific land management as a strategy for sustainable agriculture. Sky J Agric Res 4(8):147–155
89. Saleh AM, Belal AA (2014) Delineation of site-specific management zones by fuzzy clustering of soil and topographic attributes: a case study of East Nile Delta, Egypt. IOP Conf Ser: Earth Environ Sci 18:012046. https://doi.org/10.1088/1755-1315/18/1/012046
90. Gerhards M, Schlerf M, Mallick K, Udelhoven T (2019) Challenges and future perspectives of multi-/hyperspectral thermal infrared remote sensing for crop water-stress detection: a review. Remote Sens 11(10):1240
91. Massey RE, Myers DB, Kitchen NR, Sudduth KA (2008) Profitability maps as an input for site-specific management decision making. Agron J 100(1):50–59
92. Bazzi CL, Souza EG, Khosla R, Uribe-Opazo MA, Schenatto K (2015) Profit maps for precision agriculture. Cien Inv Agr 42(3):385–396
93. SEAB/PR-Secretaria da Agricultura e do Abastecimento do Paraná (2015) Available online at: http://www.agricultura.pr.gov.br/ (Website accessed: 17 Mar 2015)
94. Tran DV, Nguyen NV (2006) The concept and implementation of precision farming and rice integrated crop management systems for sustainable production in the twenty first century. Int Rice Commiss Newslett (FAO) 55:91–102

Implementing an Environmental Information System in Data-Scarce Countries: Issues and Guidelines

Abdelhamid Fadil, Mohamed El Imame Malaainine, and Younes Kharchaf

Abstract Integrating the environmental dimension into socio-economic development strategies has become a requirement for the sustainable development. To reach this goal, it is necessary to monitor the state of the environment in order to identify, manage and supervise the different environmental issues and to determine the appropriate responses for them. This objective cannot be achieved without organizing environmental data in an information system that can be used and updated by all the actors and partners concerned by the environmental questions. The aim of implementing an Environmental Information System (EIS) is to develop an integrated framework for the storage, production, management and exchange of environmental information in a decision-making perspective. However, availability and scarcity of data remain a major obstacle for realizing such systems, especially in developing countries. This study aims to discuss the issues and processes of EIS implementation. It details the environmental monitoring models and the challenges related to building the environmental databases. An overview on the EIS implementation in developed countries was carried out, and an example of EIS in MENA region (Morocco) was presented to extract syntheses and recommendations in order to define a guideline to implement a successful EIS especially in data-scarce countries.

Keywords Environmental information system · Environmental monitoring · Environmental data · Indicators · Data-scarce · Morocco

1 Introduction

The relationship between development and environment was the main focus of the World Charter for Nature of 1982. The World Commission on Environment and Development (WCED), known as the Brundtland Commission, will introduce in its report in 1987 the concept of sustainable development. This report pulls the alarm signal about the need to take into account the preservation of the environment and the

A. Fadil (✉) · M. E. I. Malaainine · Y. Kharchaf
Laboratory of Systems Engineering (LaGes), Hassania School of Public Works, Casablanca, Morocco
e-mail: fadil@ehtp.ac.ma

M. Abu-hashim et al. (eds.), *Agro-Environmental Sustainability in MENA Regions*, Springer Water, https://doi.org/10.1007/978-3-030-78574-1_7

sustainability of natural resources in the halt to growth [1]. It announces also the lack of decision-making structure and institutional tools at the national and international levels to address environmental issues and the requirements of sustainable development. The Earth Summit in Rio de Janeiro in 1992 will highlight this concept and institutionalize it through a set of conventions, protocols and declarations that target the protection of the environment, the safeguarding of ecological balances and the adoption of a model of sustainable economic development [2]. Summits and conferences will follow in order to concretize the principle of sustainable development and insist on the right of future generations to inherit a clean planet and a healthy environment [3].

Guidelines have been issued everywhere to establish a regulatory framework and decision support tools for protecting the environment and conciliating the requirements of socio-economic development with those of the environment preservation [4]. Having information about the environment state and follow her evolution in time was one of the major recommendations of the various earth summits [2]. For this, different countries have embarked on the race to set up their environmental databases and implement tools in order to manage, analyze and synthesize these data. Environmental Information Systems (EIS) have sprung up so all over the world [5]. They were developed to study and manage environmental issues and problems in different thematic dimensions and spatial–temporal scales.

To make a decision, data is required, and processing is needed to transform the data from its raw state into valuable information. Decisions should be based on reliable and accurate information and knowledge [6].

For this purpose, we need a:

- Model: It lets to construct a simple representation of the real world perceived as very complex. When it is simplified, reality can be understood more easily, its components and interactions can be defined and studied in different dimensions.
- Data: It describes the environment (real world). It is collected regarding the model used. The analysis of the environment state and its evolution is done based on data.
- Information system: To store, manage, process and analyze data and generate useful results for decision support systems.

These points are the main components of the process of implementing an information system. Applied to the environment domain, these elements are illustrated in Fig. 1.

EIS implementation is hampered in general by major obstacles, particularly in developing countries that suffer from a cruel lack of data [5].

This study aims to draw up the issues and challenges meeting the EIS implementation, especially in data-scarce countries and provide some answers to overcome these difficulties. The study details the EIS implementation processes illustrated in Fig. 1. It presents an overview of the EIS implementation in northern countries as well as an example of an EIS in MENA region (Morocco). It concludes with a discussion where we present guidelines for deploying operational EIS in the context of data-scarce countries.

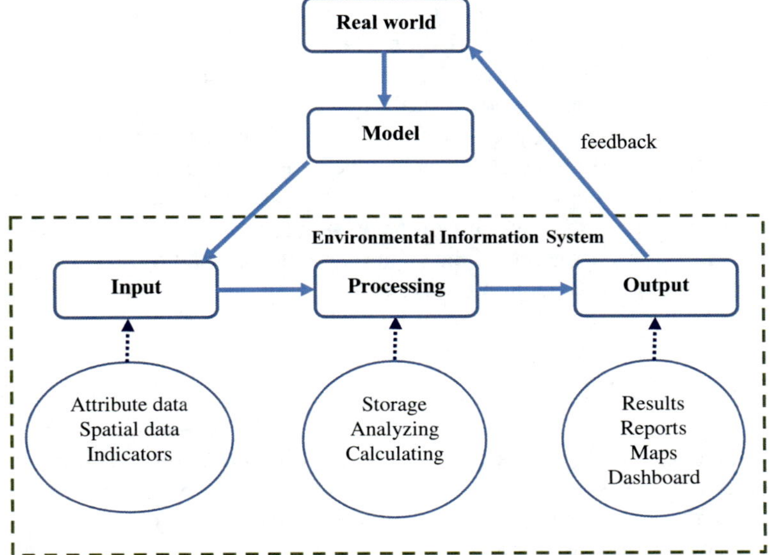

Fig. 1 Environmental information system implementation process

2 Environmental Monitoring Models

The first approaches to measuring the effect of development on the environment are oriented towards description [3]. Their main goal is to create information that describes the changing state of the environment. The following models and those developed primarily from the 1980s refer to the principle of cause and effect [7]. Several approaches, based on this principle, were then conceived and implemented in different zones and contexts. The best known and most used are: PSR (Pressure-State-Response) model, DSR (Driving Force-State-Response) model and DPSIR (Driving Force-Pressure-State-Impact-Response) model [3, 8].

2.1 PSR Model

The PSR (Pressure-State-Response) model was developed by the European Organization for Economic Co-operation and Development (OECD) in the 1980s. It is based on the idea that human activities exert pressures on the environment (pressures) [9]. These activities reduce the quality of the environment and put in danger natural resources (state). Society responds then by trying to provide solutions to these changes by adopting environmental, economic and social policies (Responses) [10].

The PSR model declines these items throw indicators that describe the environmental pressures and state as well as the responses of society [11]:

- Pressure Indicators: They describe the pressures exerted by human activities (energy, transport, industry, agriculture, etc.) on the environment.
- State Indicators: They describe the environmental conditions. They concern the quality of the environment as well as the quality and quantity of natural resources.
- Response Indicators: They reflect the society's involvement in responding to environmental issues. Individual and collective actions and reactions are of several kinds: removing the adverse effects of human activities; stopping the damage already imposed to the environment or seeking to repair it; protecting nature and natural resources. Concretely, the society tries to respond to regulations, taxes, subsidies, pollution reduction rates, etc.

2.2 DSR Model

The DSR model (Driving Force-State-Response) was designed from the PSR model. It replaces the "pressures" component with a broader notion of "driving forces". The aim is to consider the positive as well as the adverse environmental impacts of policies that were missing in the PSR model [12].

The concept of driving forces highlighted by this model describes the activities, processes and human behaviors that can positively or negatively influence the sustainable development. It also targets to identify more effectively the addition of social, economic and institutional indicators [13].

2.3 DPSIR Model

The DPSIR model (Driving Force-Pressure-State-Impact-Response) was set up by the European Environment Agency (EEA) from the OECD PSR and DSR models [14]. This model adds a new component which is "Impact" to describe the complex interactions between Society and Environment [15].

The five elements of the DPSIR model are linked by causal links. The driving forces that drive human activities cause pressure on the environment, which results in a change in the general state of the environment. This will have an impact on nature, humankind, and assets. According to the severity of this impact, it will react and cause a reaction (response) of civil society. This fifth element (reactions) brings together all the measures and political instruments implemented by society to ensure the protection of the environment in an integrated way [14]. They concern the other four elements: preventive measures directed towards the driving forces, curative measures towards Pressures and the State and palliative measures towards the State and the Impact [16].

DPSIR is the model recommended and used by most environmental agencies around the world [17]. It is more oriented towards the evaluation of environmental policies and decision-making since it makes it possible to represent information on

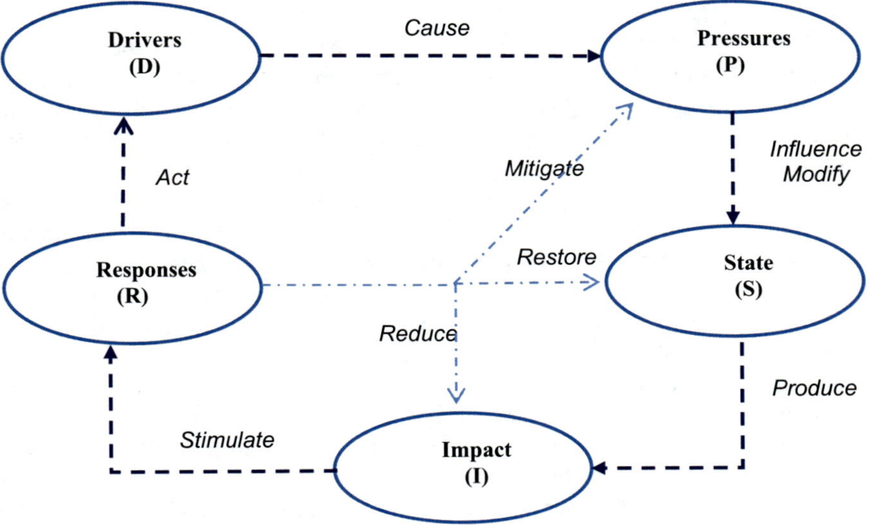

Fig. 2 DPSIR model components [19, 20]

the environment in a simple and efficient way through a multidisciplinary indicators system [18].

The different components and interactions of the DPSIR model are illustrated in Fig. 2 [19, 20].

From the DPSIR, several specific models have been generated and adapted to different requirements and domains. Patricio et al. have identified about thirty adaptations of the DPSIR model to the different environmental issues (Agriculture, marine, water, atmosphere, forest …) [17].

Hence, Environmental models are simplified representations of the real world that make it possible to express the different components and relationships of the terrain considered according to a specific approach [21].

To move from the conceptual level of these models to the concrete and quantified level, the data is required. It is one of the key elements of the information systems.

3 Environmental Data

3.1 Environmental Data Categories

Data is a fundamental prerequisite for monitoring the state of the environment. Without data, no policy or protection strategy can be carried out. The environmental concerns all aspects of human life and its interaction with the natural space. It is in direct or indirect relation with everything that happens on the earth. The assessment of the state of the environment must, therefore, consider all these interactions and fields [22]. Using models makes it possible to formalize and express these different cause-and-effect relationships and present them in the form of indicators calculated from environmental data [16]. Models are often based on a system of indicators that describes the different components of the adopted schema (driving forces, pressures, state, impact, activities, actions, responses…) [12].

The data necessary for monitoring the state of the environment are no longer focused just on the natural space and resources. They cover all the fields and domains describing the relationship between the society and its environment [13].

As an example, the repository of indicators for monitoring the state of the environment in Morocco contains nearly 300 indicators divided into several fields, themes and sub-themes as shown in Table 1 [23, 24].

Access to a wide range of high-quality information is fundamental to understand and protect the environment. However, these data are produced and managed by several institutions and organizations and vary in space and time. Hence, to build useful information and suitable indicators for monitoring the state of the environment, effective data exchange must be ensured between producers and users of the environmental data.

3.2 Environmental Data Exchange

Information and communication technologies were developed extraordinarily so that today we are talking about connected objects, Internet of Things, IP6, etc. The world of computing science has also seen a major change towards adopting standards that promote system interoperability and data exchange [25]. Paradoxically, by moving away from this technological aspect, the reality shows that the level of information exchange remains very limited especially in developing countries. This finding is further intensified in the classic domains where the rate of automation is quite low [26].

The field of the environment is distinguished by its multidisciplinary character. As an example, to study the impact of a landfill on the environment, it is necessary to combine the data of soil, geology, water, air, population, land use, topography, etc. These data come from various institutions and have to be gathered and crossed to carry out the study.

To answer the environmental questions, an effective data exchange has to be established. In this way, many goals can be achieved especially:

- Guarantying access to different environmental data themes by adopting the principle of "offering one type of data and benefiting from all other types of data."
- Using knowledge in a clear and effective way by crossing different data themes.
- Reducing data production and maintenance costs by pooling and sharing investments in data collection and data update.

Table 1 Fields and themes of environmental monitoring in Morocco [23, 24]

Field	Theme	Sub-theme
Natural environment	Climate	
	Water	Water quality, use of resources…
	Air	Air quality…
	Soil	Soil use, quality…
	Biodiversity and ecosystems	Protected sites…
	Climate change	
	Natural risks	Flood, earthquake…
Socio-human environment	Socio-demographics	Population, training, employment…
	Resource consumption	Water, Energy…
	Liquid and solid sanitation	Wastewater, solid waste…
	Health and hygiene	Infrastructure, impact on public health…
	Sustainable development	Actors, policies, expenses, control…
Economic environment	Agriculture	Land use, livestock, actions…
	Industry	Production, rejections…
	Trade	Infrastructure…
	Mines et quarries	Density, rejections…
	Energy	Infrastructure, production, rejections…
	Transport	Density, actions for environment…
	Tourism	
	Craft	
	Technological risks	
Space and territory	Housing et urban systems	Urbanization, regional planning…
	Forest	Density…
	Coastline	

- Ensuring the quality, reliability and integrity of data by assigning responsibility for each type of data to a specific organization.

The benefits and rewards of sharing and exchanging data are so evident that no one can do without today [27]. However, we note that the major obstacle that hinders the development of information systems including the EIS is the lack or scarcity of data [28]. This situation is most evident in developing countries where the problem of data access is pointed out as the major problem obstructing the implementation of information systems. In the context of these countries, several factors may explain this situation:

- Data is sometimes considered as a source of power. Sharing the data is then equivalent to a sharing of power.
- The cost of building the databases is very high. It represents over 80% of the project budget in some cases. Sharing the data with others is therefore seen as a waste of the investment made on creating these data.
- Poor data quality sometimes prevents people from displaying and distributing it. The best way to hide this quality defect is to keep them in-house.
- The framework and policy of data exchange are not often defined and formalized. In the absence of exchange agreements, most stakeholders prefer to keep their data to avoid assuming any responsibility that may arise from the use of these data.

The success of any information system depends mostly on the availability and quality of the input data. The quality and reliability of the results are just a reflection of those of the input data.

4 EIS Purpose and Functions

4.1 EIS Definition and Functions

Bourgeois defines Information systems (IS) as interrelated components that work together to collect, process, store, and disseminate information to support decision making in an organization [29].

EIS is an IS that manipulates environmental data. It follows the same processes like any other IS. The main purpose of EIS is to create a federated and interoperable system where different categories of environmental data can be stored, aggregated, combined, exchanged, updated and assessed. It aims to provide environmental stakeholders with reliable, up-to-date and synthesized information to facilitate the decision-making and the state environmental monitoring [30].

The main functions of an EIS are: acquisition, storage, analysis and visualization of information. These functions are presented in Table 2.

When implementing EIS, two dimensions should be specified:

- Geographical: It defines the spatial extent of the study area covered by the EIS. It can cover the whole earth, continent, country, region or county.
- Thematic: It defines the domains or themes of the environment considered by the EIS. This dimension can be global when all themes are considered or specific when the EIS is dedicated to one or some specific themes (biodiversity, water, air, waste …).

Table 2 Main functions of environmental information system (EIS)

Function	Consistence
Acquisition	Ensure data production and update Ensure data access and exchange
Storage	Organize, structure and archive the environmental database Synchronize, formalize and homogenize data
Analyze and processing	Facilitate data use Meet needs and requests of users Cross and combine different types of data Detect environmental state changes Represent the evolution of environmental parameters on a spatial/temporal scale
Visualization	Visualize and search data, indicators, and different combinations of data and indicators Represent results and summaries in different forms: reports, graphs, statistics, thematic maps, dashboards Distribute environmental state information

The combination of these two dimensions determines the spatial and functional scope of the EIS. They define the range of data that will be required by the EIS as well as their spatial–temporal variation.

4.2 *EIS Architecture*

Like any decision support system, the ultimate goal of an EIS is to support decision-making through the implementation of an environmental information management platform. This makes it possible to collect, organize and analyze in one place, multiple data from different sources.

To set up an EIS, the targeted real-world environmental issues are translated, based on a model (DPSIR for example), into data (input). The data are then processed, analyzed, crossed and aggregated by the information system in order to produce results (output). The outcomes can be represented in different formats: reports, statistical tables, graphs, interactive maps, dashboards.

The core of the EIS is the processing block which is the main function of any information system. This part usually consists of an application server and a mapping engine that allow both the processing and analysis of attribute and spatial data. This core is usually designed as a client–server or web architectures. Data storage is provided by Database Management Systems (DBMS).

5 EIS Benchmarking

5.1 *International EIS Overview*

The aim of analyzing environmental information systems at the international level is to establish a review in terms of uses and approaches and to draw a synthesized overview of the structures and components of these systems. Several systems are studied at the global level (United Nations, Europe) and at the level of certain countries (United States, Canada, Switzerland).

5.1.1 Information System of the United Nations Environment Program (UNEP)

The United Nations Environment Program (UNEP), founded in 1972, is the highest environmental authority of the United Nations. Its mission is to act as a catalyst, advocate, instructor and facilitator to promote the wise use and sustainable development of the global environment [31]. It also aims to be a source of inspiration and information for countries and populations and a facilitation instrument that will enable them to improve the quality of their lives without compromising the future generations' one [32].

In terms of environmental information management, UNEP has established a global network of environmental information centers for managing the global environmental resources database called GRID (Global Resource Information Database). The GRID has a global network of 15 environmental data centers managed by UNEP's Pre-Alert and Evaluation Division (DEWA). Each GRID regional center manages its database and updates it through its partners (FAO, UNESCO, World Bank, regional or national organizations, research laboratories …).

GRID's objective is to provide and facilitate access to environmental data and information for decision-making and environmental policy-making. It also aims to set a solid basis on which UNEP can rely for its study of the world's environmental state [33].

These data are managed through a data portal called "GEO" which is the framework used by UNEP and its partners to manage and assess the environmental issues. This online database contains more than 500 different parameters distributed into several categories:

- Geospatial data (maps) covering different themes such as water, forests, emissions, climate, disasters, health and population.
- Statistics at the national, regional or sub-regional level.

This information can be visualized in the form of maps, graphs and tables. It can also be downloaded in various formats.

5.1.2 Shared Environmental Information System (SEIS) of Europe

The Environmental Information Sharing System (SEIS) was created by the European Environment Agency (EEA) in 2008 in response to the challenges of managing environmental issues that Europe faces. The aim is to improve the collection, exchange and use of environmental data and information throughout Europe [34]. SEIS targets also to set up an online information system to simplify and modernize existing environmental information systems and processes.

SEIS has been designed based on an approach according to which the environmental information must be [35]:

- Managed as close as possible to its source
- Collected once and then shared between partners and users interested in this data
- Accessible to all users to meet data needs and facilitate reporting tasks
- Shared in order to allow comparisons and to draw evolution curves of environmental phenomena
- Available and readily usable for the general public
- Standardized to facilitate its exchange and processing by different information systems.

SEIS is implemented as a web platform consisting of a set of tools and interfaces for displaying and managing data in the tabular, graphical and cartographic form.

It deals with the different themes and issues of the environment at the scale of Europe: Water, Air, Emissions of gases, energy, urban waste, biodiversity …

An SEIS extension to the neighboring countries of East and South has also been put up through support and assistance programs for these countries.

5.1.3 Information System of the European Directive "INSPIRE"

The INSPIRE directive, drawn up by the European Commission through its Directorate General for the Environment, aims to establish a geographic data infrastructure in Europe to ensure interoperability between databases and facilitate the diffusion, availability, use and the reuse of geographical information in Europe [36].

The INSPIRE directive is based on several founding principles of a Spatial Data Infrastructure (SDI), namely the data must be [37]:

- collected once to avoid duplication. It should be then stored, made available and updated by the most competent authority
- combined with other information coming from different sources across Europe
- shared between different users and applications
- shared by all other public units if the original data is collected by a public institution, regardless of their hierarchical or administrative level
- described through metadata to know what information is available and under what conditions they can be acquired and used.

The information system of INSPIRE offers a set of features and services to discover, visualize and query the different types of data managed. It ensures their presentation and manipulation according to the standards that were designed for maximum interoperability between the databases involved in this European directive.

The data managed within the INSPIRE framework concern both general data (topography, administrative boundaries, hydrography, transport networks …) and environmental data (Air, water, ocean, biodiversity, protected areas…).

5.1.4 Information System of the US Environmental Protection Agency (EPA)

The United States Environmental Protection Agency (EPA) was founded in 1970 to address the high concerns of environmental pollution issues. The objective of this federal public institution is to consolidate in a single entity a variety of research, monitoring, management and enforcement activities to ensure environmental protection at the US federal level [38].

Environmental information management at the EPA is achieved through a set of databases and several tools. Access to this information is organized by using a single interface called "Envirofacts" which provides access to all environmental data via a federal web portal for searching information either by study area (county, basin, park …) or by theme (Air, water, soil, waste, toxic materials, radiation …).

The choice of the environmental information leads to an interface that makes it possible to list all the properties associated with this information (metadata), to view the measured values of this data, to locate the measuring network for the parameter, and to download the data in various formats (text files, Excel, KML or Shapefile).

This portal also has a geospatial system entitled "EnviroMapper" that allows to spatially represent all environmental information with a rich cartographic background composed of satellite images, aerial photos, vector layers (roads, hydrographic network, cities …) and address layers.

5.1.5 Environment Canada Information System (EC)

Environment Canada (EC) is the federal organization responsible for the management and protection of the environment in Canada. Its primary role is to ensure a healthy, clean and sustainable environment through credible science, effective regulations and laws, successful partnerships and high-quality service delivery [38]. Its mission is to preserve and enhance the quality of the natural environment, repair damage, enforce legislation, and coordinate environmental policies and programs on behalf of the federal government.

EC environmental information management is conducted through a set of indicators called Canadian Environmental Sustainability Indicators (CESI) that provide data and information to track the performance of the country in environmental sustainability issues, such as climate change, air quality, water quality and nature protection [39]. These environmental indicators are derived from a set of information that identifies environmental trends.

These indicators are prepared by "Environment Canada" with the support of other federal departments as well as provincial and territorial departments. They are designed and developed through methodologies and data regularly obtained from surveys and monitoring networks.

This environmental information is presented through a geoportal that allows display data in different formats: tables, graphs, texts, interactive maps and downloadable files.

5.1.6 Information System of the Swiss Federal Office for the Environment (FOEN)

The Federal Office for the Environment (FOEN) is the Swiss organization responsible for the sustainable exploitation of natural resources, such as soil, water, air and forest. It is responsible for protecting against natural hazards, protecting the environment, preserving health from excessive exposure to pollution and conserving biodiversity. The mission of FOEN is to preserve and manage natural resources according to the principles of sustainable development and eliminate the damage that has already occurred and protected people against pollution and natural hazards [40].

In order to manage environmental information, FOEN has set up a multidisciplinary information system composed of a set of tools based mainly on geographic information systems (GIS) as a method of presenting environmental information. The three main components of this system are:

- set of databases on the different environmental themes: database on the state of surface water (DBGZ), Solid database (thrust volumes in rivers), water resources (GEWISS)…
- EcoGis: Interactive mapping service that allows the display of different information on a cartographic background. This service enables to manage the cartographic rendering, query and search the data as well as download the data in several formats.
- Swiss environmental data catalogue (Envirocat): this project launched by FOEN and the Swiss Federal Statistical Office (FSO) aims to organize, in collaboration with all the others departments, all sets of environmental information in a single system. The purpose of this catalogue is to provide an overview of the environmental observation activities carried out by the Confederation and the cantons. It presents the metadata's properties for all the environmental information sets (source, methods of acquisition, exchange rules, quality …).

5.2 Benchmarking Synthesis

The analysis of the international EIS cited above lead to the following conclusions:

- EIS is often built based on a system of environmental indicators that are calculated from several sets of data
- Environmental data changes over time and varies in space. In most of the systems studied, this information is collected by sensors and telemetry networks
- To ensure the data exchange, the environmental authority prepares partnership agreements with the others institutions managing the environmental data in order to acquire their measurements
- The entity responsible for the environment is usually the one responsible for organizing, updating and distributing the information contained in the EIS
- Most EIS are built based on several levels and tools: a layer for storing data in databases, administration tools, configuring and updating tools and environmental information exchange tools
- For the environmental information diffusion, the geospatial portal (geoportal) is the most widely used technique. Such geoportal is developed either in the form of a very rich web-based GIS or the form of a simple interactive mapping tools
- The data managed by the EIS covers both generic data (topography, administrative boundaries, hydrography, transport networks, etc.) and specific environmental data (Air, water, biodiversity, marine areas, ecological areas, etc.)
- Metadata is a key element of EIS. All the EIS analyzed offer a catalogue system to manage and manipulate the metadata.

6 Use Case: EIS in Morocco

6.1 Environmental Policy in Morocco

Morocco has been experiencing a significant economic and demographic growth for several years, which has led to an improvement in the standard of living but also to strong urbanization and an increase in energy and natural resources needs. This development was not without negative repercussions on the state of the environment. Pollution of the natural environment and overexploitation of natural resources coupled with the hostile impacts of climate change are threatening factors to the environment.

Morocco has put in place a legislative and regulatory framework for the protection of the environment reconciling the imperatives of socio-economic development and those of the preservation of the environment. This vision was concretized by the implementation of the National Charter of Environment and Sustainable Development in 1999 which led to the National Strategy for the Protection of the Environment. Thus, several laws have been promulgated, notably, those relating to the environment, water, waste management and environmental impact studies [24]. Internationally, Morocco has ratified numerous environmental treaties, including the Earth Summit of Rio de Janeiro (1992) and Johannesburg (2002). According to the resolutions of this later summit, all countries had to draw up their National Sustainable Development Strategy [41].

Morocco has developed its 2030 sustainable development strategy [42]. It has defined a list of indicators for assessing the country's sustainable development [23]. Therefore, all specific development projects must consider the environmental dimension in order to guarantee the good management of resources, the protection of ecosystems, the monitoring of the state of the environment and the improvement of citizens' living environment.

6.2 Moroccan EIS

Morocco is engaged to submit periodic reports on the state of the environment at the regional and national levels. In 2010, it decided to set up EIS at the regional level called Regional Environmental Information System (REIS). The objective is to have a platform for monitoring the state of the environment in each region in perspective to consolidate them in a next step to form the national environmental information system.

REIS is designed as a tool for decision support system and information monitoring in the domain of environmental management and sustainable development. It is a platform for generating, sharing and representing region-wide environmental data and indicators using the DPSIR model. The objectives assigned to the REIS are [43]:

- Facilitate and boost the flow of information between environmental data producers and users
- Meet the public needs demanding more transparency in accessing and disseminating environmental information
- Enhance the abilities to produce, manage and distribute the environmental information
- Develop tools to display and summarize the evolution of different environmental indicators and phenomena
- Automate the generation of periodic reports on the state of the environment
- Support the environmental forecasting through simulation tools.

Two significant challenges were encountered during the implementation of these REIS:

- The Data component of the Information System: How to ensure the collect and the regular update of the database?
- The Software component of the information system: What technical architecture should be adopted to meet the needs of different users?

To address the first point, a strategy for the exchange of environmental information between the partners of the region has been set up based on thematic information exchange networks founded by the regional authorities. The second challenge was processed by building the REIS based on a web-oriented architecture that allows each partner (Regional representations of the ministries, urban agencies, water agencies,

Table 3 Roles of regional environmental information system actors and partners [43]

Actor	Role
Regional observatory of the environment	• REIS administrator
Regional partner (data producer)	• Produce and update partner-specific data • Access all other categories of data • Use REIS functions
Regional partner (data user)	• Query data • Use REIS functions
General public	• Query public data

delegation…) to access to the system and databases according to the rights and roles that have been assigned to each entity.

The role of the different entities involved in the REIS is shown in Table 3 [43].

6.3 REIS: Are They Operational?

A few years after the launch of the REIS projects in the different regions of Morocco, the finding is that these systems are struggling to function effectively and to be deployed to the general public. New REIS harmonization and operationalization projects are launched. The question then is why the deployment of EIS encounters problems and fails to move to the operational state. Are these technical difficulties related to information systems, institutional barriers referred to the use of collaborative systems or problems of availability and access to data?

The diagnosis led on some REIS [44] and the conclusions drawn from the implementation of other types of information systems in the Moroccan context, show that the technological aspect of the information systems is no longer a problem. It is rather the weakness of the content, embodied by the lack of data or the poor quality of existing data, that hampers the successful deployment of these systems. This situation is further aggravated when the institutional aspects of data exchange between partners are not defined or formalized.

Solutions and alternatives must be brought in order to overcome the obstacles hindering the implementation of useful, operational and up-to-date EIS, especially in countries that are experiencing problems of lack or scarcity of data.

7 Guidelines for Implementing Operational EIS in Data-Scarce Countries

Following the benchmarking carried out on the EIS of some countries of the Northern Hemisphere and the case study of Morocco, we can draw a set of conclusions and proposals for the deployment of useful, operational and efficient EIS in the context of data-scarce countries.

They are presented in three levels: institutional, technological and related to data problems.

7.1 Institutional Aspects

7.1.1 Public Data is the Business of the State: Need for a National Spatial Data Infrastructure

The core of an EIS is usually a DBMS and a Geographic Information System (GIS) that require spatial data. These data are divided into two parts: global or public data that can be used by all information systems (topography, geomorphology, hydrography, administrative division, roads, demography …) and specific data relating to the field of IS (environment). Establishing a successful EIS requires that it focuses principally on the collection and structuring of specific environmental data. The necessary public data under the responsibility of other departments and entities of the State must be collected and provided by these units. This is the concept known as National Spatial Data Infrastructure (NSDI) that each country needs to implement in order to facilitate the deployment of information system and the reuse of data as is the case in developed countries [45]. Environmental community-based monitoring systems must also be encouraged by public authorities to facilitate citizen participation in the preservation of the environment [46].

7.1.2 Data Exchange is a Necessity and Not a Favor

In multidisciplinary fields such as the environment, data is coming from different sources. It must be cross-referenced in order to have a more real idea of environmental issues. Exchange and sharing of data between institutions become so a necessity and an obligation [26]. A "win–win" policy must be implemented in order to build operational networks for data exchange. Institutionalizing these networks and putting them in a legal and regulatory framework that can give them the strength of application will further reinforce the information flow and address a number of data access and availability issues.

7.1.3 Non-quality Costs More Than Quality

Today, it has become clear that the cost of non-quality or poor quality is higher than that of quality. The use of data including mistakes or biases can potentially produce incorrect conclusions that may lead to decisions with dramatic consequences [47].

Institutions must, therefore, invest from the outset in the creation of reliable and good quality systems and databases. This will ensure the continuity and sustainability of these systems and an effective amortization of the investment made to build the databases.

Non-quality often leads to the abandonment of the system, the creation of the same features in a repetitive manner, the waste of time and money and the waste of opportunities to develop new value-added services that can be achieved if the system was reliable and accurate.

7.2 Technological Aspects

7.2.1 Technology Must Be Adapted to the Needs of Society and Not the Other Way Around

The technology has been developed to facilitate human life and to make it better and safer. It is often the fruit of the need expressed by society. However, its life cycle and change are faster than that of men. While procedures and work processes need to be reviewed and improved in order to benefit from technological development, they must not be jostled and upset to keep up with almost daily changes in technology. The rate and speed of process evolution are more stable and slower than those of technology. It is the latter that must be adapted to business logic and not the other way around.

7.2.2 A Simple and Operational System is Better Than a Pretty Empty Shell

The goal of an information system is to have a platform for managing and transforming input data into information and knowledge to facilitate and guide decision-making. The focus must be on the implementation of operational and useful systems even if we are forced to adopt the simplest and most basic representations. It is apparent that it does not make sense to have a complex and sophisticated but non-functional system.

Therefore, Particular attention must be paid to the step of capturing user needs in order to adapt the functions of the information system to the real needs and expectations of its future users [48].

7.2.3 Interoperable Systems Communicate Better Than Humans

Environmental issues are becoming more complicated and interdependent. The development of information systems must be done by adopting principles of openness and interoperability. Systems must be able to communicate and exchange with each other to facilitate the implementation of data exchange protocols and to ensure the smooth and timely flow of information.

In the last decade, new approaches and techniques, such as service-oriented architectures, have been developed to meet these needs [49, 50]. EIS design should benefit from this evolution to facilitate communication and interoperability of environmental systems.

7.3 Aspects Related to Lack or Scarcity of Data

7.3.1 A Mine of Open Access Satellite Data to Use

To overcome the problems of access and availability of data, other alternatives are to be explored. Earth observation data captured from satellites is one of the promising options in this regard. Indeed, with the free and open access to satellite data of many types and resolutions, these databases become a valuable source that countries suffering from data availability problems can exploit and glamorize [51].

Research conducted on these data sources shows that they can answer the data needs (topography, land use, climate, soil moisture, ocean salinity, etc.) in different areas, especially at small and medium scales. Many environmental studies use already this kind of data such as those relating to water, air, soil, oceans… [52].

However, these data cannot replace in-situ data in most cases and must always be analyzed and validated before use. A combination of these global data with local data collected in-situ would be a suitable and recommended approach to take advantage of the benefits of each of these two categories [53].

7.3.2 Sensors Acquire Data Better Than Humans

The development of Information Technology has allowed the implementation of new modes and techniques of data collection (sensors, drones …). The advantage of these methods is the automation of the information acquisition chain, the increase of the frequency's collection and the access to the data in real time [54]. Delegating data collection tasks to machines limits human intervention in this process and therefore minimizes the sources of error and the dependence on human will and behavior.

The flood warning and water quality monitoring sensors are two concrete examples of the major utility of these techniques and their determining role in the preservation of lives and the environment [55, 56].

With the ongoing reduction in hardware and software costs, it turns out that these techniques are an opportunity that developing countries can seize to build reliable, accurate and up-to-date databases.

7.3.3 Data Availability Issues at the Big Data Era

The amplification and the multitude of techniques of data acquisition (satellites, drones, sensors, cameras, smartphones …) have generated large data, very varied and evolving with tremendous speed. This is the big data, the era where the problem of data is reversed. There is a lot of data, so we do not know which to use and which one has value.

Big data techniques present three significant advantages for the implementation of information systems. Firstly, they provide solutions for the storage of mass data that evolve over time, such as those concerning the environment. They provide algorithms and methods for processing and analyzing extensive data coming from various sources. Finally, they provide tools and interfaces to display information in many ways that make it easy to understand and interpret the phenomena being studied [57].

Developing countries, and especially those suffering from data availability problems, are called upon to formulate their strategy for exploiting and using big data in order to benefit from the advantages and opportunities offered by these techniques. Above all, it is necessary to determine concrete and useful uses of this technology through its contextualization and its adaptation to the concerns of society instead of falling into its charm.

8 Conclusions and Recommendations

Humanity is forced to adopt the concepts and requirements of sustainable development to handle the fragility of ecological equilibrium and the degradation of the environment as a result of the imperatives of socio-economic development. One of the pillars of this requirement is to continuously monitor the state of the environment through an operational information system that can assess and support the actions and policies of society towards the environment.

The implementation of such systems encounters difficulties in developing countries due mainly to problems of availability and scarcity of data. This study provides an overview of the different issues concerning this topic and presents a list of suggestions to overcome these obstacles.

To implement a successful and useful environmental information system, the authors recommend to take into consideration all the elements of the guideline presented in this study. They highlight the need to pay special attention to the problem of lack or scarcity of data. For that, they also recommend studying and analyzing the new opportunities offered by remote sensing techniques and big data technologies.

References

1. Brundtland GH (1987) Our common future: report of the world commission on environment and development. United Nations Commission 4(1):300. https://doi.org/10.1080/07488008808408783
2. Hodge T (1997) Toward a conceptual framework for assessing progress toward sustainability, 5–98. https://doi.org/10.1023/A:1006847209030
3. Vidal C, Marquer P (2002) Towards sustainable European agriculture: tools and methods (Vers Une Agriculture Européenne Durable: Outils et Méthodes). Educagri éditions. https://books.google.co.ma/books?id=qn0SCOIhIWwC&printsec=frontcover&hl=fr#v=onepage&q&f=false
4. Mangalagiu D, Dronin N, Billot M (2019) Implementing the shared environmental information system (SEIS) and environmental policies in Central Asia. Environ Sci Policy 99:29–36. https://doi.org/10.1016/j.envsci.2019.05.017
5. World Health Organization (2005) Implementing environment and health information system in Europe. Regional Office for Europe of the World Health Organization. http://ec.europa.eu/health/ph_projects/2003/action1/docs/2003_1_28_frep_en.pdf
6. Vardon M, Castaneda J, Nagy M, Schenau S (2018) How the system of environmental-economic accounting can improve environmental information systems and data quality for decision making. Environ Sci Policy 89:83–92. https://doi.org/10.1016/j.envsci.2018.07.007
7. Brambila A, Flombaum P (2017) Comparison of environmental indicator sets using a unified indicator classification framework. Ecol Indic 83:96–102. https://doi.org/10.1016/j.ecolind.2017.07.023
8. Levrel H, Kerbiriou C, Couvet D (2009) OECD pressure-state-response indicators for managing biodiversity: a realistic perspective for a French. Biosphere Reserve 1 C 1 C" 18(7):1719–1732. https://doi.org/10.1007/s10531-008-9507-0
9. Hughey KFD, Cullen R, Kerr GN, Cook AJ (2004) Application of the pressure-state-response framework to perceptions reporting of the state of the New Zealand environment. J Environ Manage 70(1):85–93. https://doi.org/10.1016/j.jenvman.2003.09.020
10. Neri AC, Dupin P, Sánchez LE (2016) A pressure-state-response approach to cumulative impact assessment. J Clean Prod 126:288–298. https://doi.org/10.1016/j.jclepro.2016.02.134
11. Organisation for Economic Co-operation and Development (2003) Using the pressure-state-response model to develop indicators of sustainability. Framework for environmental indicators. Agriculture, 1–11. DESA/DSD/2001/3
12. Segnestam L (2002) Indicators of environment and sustainable development: theories and practical experience. World Bank Environment Department. http://orton.catie.ac.cr/cgi-bin/wxis.exe/?IsisScript=IICACR.xis&method=post&formato=2&cantidad=1&expresion=mfn=031489
13. United Nations Department of Economic and Social Affairs (2007) Indicators of sustainable development: guidelines and methodologies. United Nations Publications
14. Carr ER, Wingard PM, Yorty SC, Thompson MC, Jensen NK, Roberson J (2007) Applying DPSIR to sustainable development. Int J Sust Dev World 14(6):543–555. https://doi.org/10.1080/13504500709469753
15. Elliott M (2002) The role of the DPSIR approach and conceptual models in marine environmental management: an example for offshore wind power. Mar Pollut Bull 44(6):iii–vii. https://doi.org/10.1016/S0025-326X(02)00146-7
16. Weigel J, Ird ED, Morand P (2006) Defining criteria and indicators to compare the societal cost of fishing activities in marine protected areas and in unprotected zones. From Criteria to Indicators 1:1–15
17. Patrício J, Elliott M, Mazik K, Papadopoulou K-N, Smith CJ (2016) DPSIR—two decades of trying to develop a unifying framework for marine environmental management? Front Mar Sci 3(September):1–14. https://doi.org/10.3389/fmars.2016.00177

18. Martin DM, Piscopo AN, Chintala MM, Gleason TR, Berry W (2018) Developing qualitative ecosystem service relationships with the driver-pressure-state-impact-response framework: a case study on Cape Cod, Massachusetts. Ecol Indic 84:404–415. https://doi.org/10.1016/j.ecolind.2017.08.047
19. Atkins JP, Burdon D, Elliott M, Gregory AJ (2011) Management of the marine environment: integrating ecosystem services and societal benefits with the DPSIR framework in a systems approach. Mar Pollut Bull 62(2):215–26. https://doi.org/10.1016/j.marpolbul.2010.12.012
20. Kristensen P (2004) The DPSIR framework. In: A comprehensive/detailed assessment of the vulnerability of water resources to environmental change in Africa using river basin approach, pp 1–10. http://enviro.lclark.edu:8002/rid=1145949501662_742777852_522/DPSIROverview.pdf
21. Chandrakumar C, McLaren SJ (2018) Towards a comprehensive absolute sustainability assessment method for effective earth system governance: defining key environmental indicators using an enhanced-DPSIR framework. Ecol Indic 90:577–83. https://doi.org/10.1016/j.ecolind.2018.03.063
22. Harmancioglu NB, Necdet Alpaslan M, Ozkul SD, Singh VP (1997) Integrated approach to environmental data management systems
23. Department of Environment of Morocco (2014) Indicators of sustainable development in Morocco (Indicateurs Du Developpement Durable Au Maroc)
24. Department of Environment of Morocco (2015) State of the Environment of Morocco (L'état de l'environnement Du Maroc)
25. Shu Y, Liu Q, Taylor K (2016) Semantic validation of environmental observations data. Environ Model Softw 79:10–21. https://doi.org/10.1016/j.envsoft.2016.01.004
26. Palle H, Jørgen W (2007) Environmental data exchange network for inland water. Elsevier B.V.
27. Aggestam F (2019) Setting the stage for a shared environmental information system. Environ Sci Policy 92:124–32. https://doi.org/10.1016/j.envsci.2018.11.008
28. Pandeya B, Buytaert W, Zulkafli Z, Karpouzoglou T, Mao F, Hannah DM (2016) A comparative analysis of ecosystem services valuation approaches for application at the local scale and in data scarce regions. Ecosyst Serv 22:250–259. https://doi.org/10.1016/j.ecoser.2016.10.015
29. Bourgeois DT (2014) Information systems for business and beyond. The Saylor Academy
30. Frysinger SP (2001) An integrated environmental information system (IEIS) for corporate environmental management. Adv Environ Res 5(4):361–367. https://doi.org/10.1016/S1093-0191(01)00087-9
31. United Nations Environment Programme (2015) Report of the United Nations Environment Programme (UNEP)
32. Secretariat UNEP (1978) Environment and development. Environ Int 1(3):141–146. https://doi.org/10.1016/S0378-777X(78)80144-9
33. Norgbey S, Perch C (2006) United Nations environment programme division of early warning and assessment evaluation report, no May
34. Bonazountas M, Martirano G, Camilleri T, Trypitsidis A (2007) Design of the shared environmental information system (SEIS) and development of a web-based GIS interface. In: Formosa S (ed) Future preparedness: thematic and spatial issues for the environment and sustainability. University of Malta, pp 85–101. https://www.um.edu.mt/library/oar//handle/123456789/8969
35. Steenmans C (2012) Shared environmental information system (SEIS). https://www.mepa.org.mt/infores-projects
36. European Commission (2006) Consultation paper on a forthcoming EU legal initiative on spatial information for community policy-making and implementation
37. INSPIRE Framework definition support Working Group (2003) Contribution to the Extended Impact Assessment of INSPIRE
38. Information Statistics Environmental and Division (1995) A conceptual framework to support development and use of environmental information in decision-making
39. Environmenet Canada (2006) Canadian environmental sustainability indicators
40. Swiss Federal Department of the Environment (2011) Switzerland's state of the environment in a global context

41. United Nations (2002) Plan of implementation of the world summit on sustainable development contents. Johaneesburg Plan of Implementation
42. Department of Environment of Morocco (2017) National strategy for sustainable development (Stratégie Nationale de Développement Durable): executive summary
43. Department of Environment of Morocco (2012) Establishment of a regional information system on the environment (Etude Relative à La Mise En Place d'un Système d'Information Régional Sur L'Environnement)
44. Candeloro A, Pam P (n.d.) Report of the technical support mission within the framework of the ENPI-SEIS project to the national observatory of the environment of Morocco (ONEM) for the diagnosis and the prospection on the implementation of the regional information systems on the E
45. Rajabifard A, Feeney MEF, Williamson IP (2002) Future directions for SDI development. Int J Appl Earth Observ Geoinf 4(1):11–22. https://doi.org/10.1016/S0303-2434(02)00002-8
46. When U, Almomani A (2019) Incentives and barriers for participation in community-based environmental monitoring and information systems: a critical analysis and integration of the literature. Environ Sci Policy 101:341–357. https://doi.org/10.1016/j.envsci.2019.09.002
47. Gibert K, Horsburgh JS, Athanasiadis IN, Holmes G (2018) Environmental data science. Environ Model Softw 106:4–12. https://doi.org/10.1016/j.envsoft.2018.04.005
48. Walling E (2020) Developing successful environmental decision support systems: challenges and best practices. J Environ Manage 264. https://doi.org/10.1016/j.jenvman.2020.110513
49. Kanevski M, Pozdnoukhov A, Timonin V (2009) Machine learning for spatial environmental data: theory, applications and software. EPFL Press
50. Zafar M (2004) Computer applications in environmental data, 71–81. https://doi.org/10.1002/tqem.20017
51. Guo H, Wang L, Liang D (2016) Big earth data from space: a new engine for earth science. Sci Bull 61(7):505–513. https://doi.org/10.1007/s11434-016-1041-y
52. Fadil A, Rhinane H, Kaoukaya A, Kharchaf Y, Bachir OA (2011) Hydrologic modeling of the bouregreg watershed (Morocco) using GIS and SWAT model. J Geogr Inf Syst 03(04):279–289. https://doi.org/10.4236/jgis.2011.34024
53. Felix K, Powell AM, Fedorov O (2009) Use of satellite and in-situ data to improve sustainability
54. Rennie SC (2016) Providing information on environmental change: data management, Centre. Ecol Indic 1–8. https://doi.org/10.1016/j.ecolind.2016.01.060
55. Horsburgh JS, Reeder SL (2014) Data visualization and analysis within a hydrologic information system: integrating with the R statistical computing environment. Environ Model Softw 52:51–61. https://doi.org/10.1016/j.envsoft.2013.10.016
56. Horsburgh JS, Tarboton DG, Maidment DR, Zaslavsky I (2011) Components of an environmental observatory information system. Comput Geosci 37(2):207–218. https://doi.org/10.1016/j.cageo.2010.07.003
57. Kamilaris A, Kartakoullis A, Prenafeta-Boldú FX (2017) A review on the practice of big data analysis in agriculture. Comput Electron Agric 143:23–37. https://doi.org/10.1016/j.compag.2017.09.037

Green Spaces for Residential Projects as a Commitment to Environmental Concerns and a Sustainable Development Initiative: Design of a Peri-Urban Park in Casablanca, Morocco

Amira Hamdaoui and Mohamed Bsibis

Abstract The need for green space is no longer to be demonstrated in urban areas; it is not only a question of environmental and aesthetic stakes, but also the place of founding experiences and social openness. The presence of this vital space is essential for the well-being of human populations, because its absence presents itself as a phenomenon to be analyzed., Casablanca is a city that is much better known for its buildings than for its green spaces. Places of relaxation in the White City are numerous, but they are poorly maintained. The number of green spaces varies by neighborhood. In fact, the metropolis has less than 2 m^2 per inhabitant on average, far from the 12 m^2 recommended by the World Health Organization and Very far behind the 20 m^2 available to the cities of Paris and New York. An insufficient figure for a metropolis with more than 5 million inhabitants. As a result, real estate developers in Morocco, and especially in Casablanca, used this lack of green space as a marketing charter for housing projects. They present their projects under a natural concept, and the development of green spaces (private garden, park, square) and their interviews are of paramount importance.

Keywords Development · Periurban · High standing · Casablanca · Sustainable park · Environment

1 Introduction

Since independence, Morocco has embarked on a long process of modernization of the territorial administrative organization and its adaptation to the current political, economic and social context while gradually introducing a decentralization that will allow a rapprochement between the administration and the local authorities [1]. Administered, and citizen involvement in the decision-making process and the

A. Hamdaoui (✉)
UR 13 AGR07/Higher Institute of Agronomy—ChottMeriem/University of Sousse, Sousse, Tunisia

M. Bsibis
Higher Institute of Agronomy—ChottMeriem/University of Sousse, Sousse, Tunisia

M. Abu-hashim et al. (eds.), *Agro-Environmental Sustainability in MENA Regions*, Springer Water, https://doi.org/10.1007/978-3-030-78574-1_8

management of local affairs. This will make Morocco one of the most emerging countries in the MENA region [2].

The processes of territorial recomposition in Morocco are characterized by a continuous dynamic, reflecting the major changes that affected the different sectors of the country during the twentieth century. Territorial renewal in Morocco is a process of permanent recomposition, from traditional frames (whether tribal, confessional related to brotherhoods or "seigneurial" caidal) to the most recent territorial meshes in municipalities and regions. The different generations of forms and political and administrative territories born of these processes of recomposition were first the colonial region, then the province, the commune and the postcolonial region, finally many perimeters were distinguished administratively first for the development [3]. And then more recently for development projects and environmental protection. The territorial building reflects the evolution of logic, contexts, strategies, means and approaches to a changing local, national and international reality [4]. Each territorial renewal generates new references and multiplies the forms of the territorial framework. We can even speak of a process of densification and complexity of territorial management in Morocco. The CASADIAA project in Casablanca and the development of its park represents one of the typical examples of what Morocco is experiencing today in terms of sustainable territorial development [5].

2 Landscaping and Land Use Planning in Morocco

A land of art and beauty, a country where, the particular architecture is its history and multiple influences, from war to another and from succession to another, Morocco remains the beauty process fruit and human stupidity. Islam and its influences of the East, mixed with the old Berber strain and spiritual and cultural heritages, although, Morocco has made an original culture.

A country of thirst where rains are infrequent, and streames are limited, the human being has managed to take advantage of these constraints to make visitors explore exceptional gardens contrasting with the drought surrounding climate. Nowadays, Moroccans face several problems, despite the protectorate efforts to deal with the anarchic development on the suburbs, Morocco has to deal with an ever-growing population abandoning the rural world for an urban lifestyle. This current image of both Moroccan land and landscape planning must be the result of a strategy used from the beginning of its history [6].

2.1 Territorial Planning in Morocco

The first form of urbanism to emerge in Morocco is essentially that which exists within traditional cities, which is of Muslim and Arab origin. This organization is dependent on a society deeply oriented towards religious practices. The medina is the historic funds of Morocco city. This term refers to the old city. This is precisely where traditional activities are found. It is defined by a radiating device whose central point is the Great Mosque, the stalemate habitat around the merchant streets and encirclement by the old ramparts. This organization as we know it today would also come from the report that the Arabs have with the economy. Muslims give some importance to the public place, which is reduced to street networks and markets. The urban system consists of buildings that are closely connected to each other that forms a common mass.

From 1912, under the influence of the French protectorate, Morocco undertakes to reorganize its workforce. General Lyautey's plan of action is clear: "To ensure order and security, to promote economic and social development, to ensure justice and equity under the guise of traditional authority." Urbanism is then the main key to the accomplishment of this colonialist policy. It was then that Morocco was imposed Western urbanism as it existed in France at the beginning of the twentieth century. It is precisely this imported urbanism that will be associated with Morocco's entry into the modern foundations. We then move from a model based mainly on human values to a model of structuring based on normative rationality, urbanism of a predictive and secure type [7].

In 1914, Morocco was the first country in the world to have a legislative instrument in urban planning, for example, France does not yet have devices of this kind that:

- defines how to implement the city plan
- defines the terms and conditions for individuals to create groups of dwellings;
- defines the regulation of the act of building

This law allowed the state to create, between 1915 and 1925 a number of cities: a dozen will emerge. Also, the administration has equipped itself with an instrument to put an end to the uncontrolled expansion of cities that had been raging for time in Morocco [7].

The "new cities" or "European cities" are the urbanistic result of the modern thought of the time. In order to preserve the Moroccan culture, these cities were located on the outskirts of the medina and a buffer zone was planned in order to distinguish them. To facilitate the establishment of these new cities, the state has thus acquired a new tool, the Dahir of November 23, 1917 [8].

The state, as mentioned before, which then practices a provisional planning and security type is confronted with the needs of forward planning and with greater control of the urban evolution. Through the adoption of several Dahirs in the 1930s, Morocco is moving to a rather strategic type of urban planning, which seeks greater coherence between the design and execution of these projects. To finally reach the Dahirs of July 30, 1952, which will mark the urbanism until the independence of Morocco. This law is intended to be an extension of the field of application, hitherto reserved to the cities, to other spaces: the suburbs, Peripheral areas, the delimited centers and urban planning groups.

This law states that from now on, only the development plan generates obligations and this plan allows the administration to prepare development plans and zoning plans for each region. It should be mentioned that until now no law has been designed for the rural world. It will be necessary to wait after the independence so that the direction of Morocco fills this gap.

While during the protectorate, Morocco is witnessing an accelerated and constant regulation of its cities, the pace will be somewhat diminished during independence. As well as, until 1992, the 1952 Dahir will remain the main law of reference. However, a legislative law covering the rural world was born on June 25, 1960, which somewhat complemented the Dahir of 1952. This law recognizes and delineates the agglomerations located outside the urban perimeter without, however, reserving any particular regulation. From now on it is compulsory to provide a building permit and the subdivision will be subject to prior authorization. It is only in 1992 that the government will pass new laws to adjust legislation and counter urbanization problems [9] (El Issaoui, 1997).

On June 17, 1992, the government adopts a law which relates to town planning and which applies mainly to Urban Commons, defined centers of rural communes, peripheral areas of urban municipalities, urban planning group, the coastal strip on a depth of 5 km and along the roads at a depth of 1 km.

Until then, it is the state that monopolizes the production of norms and urban planning, and the local authorities take care to implement them. There is a confusion in terms of sharing of responsibilities versus a large number of stakeholders, constituting a hindrance to the implementation of urban plans. Starting in 1998, the government of "alternation", with their vision of integration, will make urban planning the habitat, the environment and regional planning of a single department. The main objective of this government is mainly to establish a real communication with all the actors involved in the areas concerned. In this respect, the meeting of the four poles into a large ministry has the advantage of matching the reality on the ground and of meeting the requirements of coherence and efficiency required by the combined action of the four departments.

So the state makes planning and planning a complementary set because they have the same time scale and the same long-term concern. It is then that the concept of sustainable development is approached. "There is no question of continuing to make the present prevail, to manage the urgency, but to look for the future and for that it requires a medium and long-term vision and values that define and explain the concept of sustainability. The concept of sustainable development is essentially based on four principles:

- Keep in mind the needs of future generations
- Take into account all the environmental efforts of urban activities
- Balance city and rural residents, as well as between current and future generations
- Promote the participation of all citizens in decisions concerning the development of their spaces and territories.

2.2 Landscaping in Morocco

Since the establishment of the protectorate, the first resident general of France in Morocco, Marshal Hubert Lyautey (1854–1934), launched the challenge of creating model cities, symbols of modernity and progress, like the power French colonial; in the middle of the 1914–1918 war, it was a real "war gesture". To achieve its urban program, the resident Lyautey turned to France and more specifically to the Social Museum which was trying to assert the obligation of development plans and extension for major French cities. Member of the Social Museum and Conservator of the Paris walks, landscape gardener Jean Claude Nicolas Forestier proposed to the resident general to implement the concept of "parks system" that he had theorized in 1906 through his book Grandes Villeset Systèmes de parks. This theory, which remained unimplemented in France, consisted of planning new settlements and orienting their development through a hierarchical and continuous network of open spaces, ranging from public gardens to periurban green belts [10].

In 1913, called by Lyautey, Forestier went on a mission a few months in Morocco to draw up a "report of the reserves to be built in and around the capital cities of Morocco …". In this report, Forestier proposed a set of prescriptions to follow in order to implement the principle of the park system in each Moroccan imperial city. To implement the prescriptions of Forestier, Lyautey, on the advice of the landscape architect, brought the architect Henri Prost.

As soon as they arrived in Morocco in 1914, Prost and his team set up development plans for all Moroccan cities. All new settlements would be designed on the basis of the urbanistic principles imposed by the resident Lyautey, namely: the separation of the old and the new city, the creation, outside the medinas, of a non aedificandi area of military protection and hygiene and the division of the new city into differentiated neighborhoods (zoning).

In addition to these principles, Henri Prost will experiment with Forestier's concept of "park system". To understand how the new colonial settlements were built through a system of parks and gardens, we will take the example of two cities: Rabat- the capital—and Marrakech, two ultimate garden cities.

- **Rabat, the prototype of the 1920s landscape city**

Rabat is located on the Atlantic coast, at the mouth of the river Bou-Regreg, facing the town of Salé which occupies the opposite bank. It is naturally bounded on the north by the ocean, on the east and south by the Bou-Regreg valley. The undulating terrain associated with dune cords parallel to the coast offers a succession of high points (belvederes) and low points.

The geographical location and the topography of the site will be decisive in the choice of the distribution of the open spaces of the new agglomeration.

In 1912, at the time of installation of the first in the French capital, the city was concentrated in two urban nuclei: the Kasbah of Oudayas, fortified camp dating from the ninth century, at the mouth of the BouRegreg and medina confined in its Andalusian walls dating from the seventeenth century. Outside the enclosure, some monuments dotted the landscape: the ruins of the Mosque of Hassan Tower—twin sister of the Ghiralda Seville and Koutoubia Marrakech—remained unfinished under the Almohad dynasty (twelfth century) the Es-Sunna mosque, the imperial palace, built in the eighteenth century, and its gardens (l'Aguedal). These elements were delimited by a second enclosure raised in the twelfth century by the Almohads. Beyond these outer ramparts stood, in Belvedere on the valley of Bou-Regreg, the ruins of the Chellah, ancient merinid necropolis of the thirteenth century. The rest of the space between the outdoor enclosure and the inner enclosure—which surrounds the medina—was occupied by gardens, vineyards and orange groves.

– **Forestier's prescriptions for the planning of the new city**

In its 1913 report, Forestier advocated for the future city of Rabat a number of recommendations relating, on the one hand, to the protection of existing gardens and, on the other hand, to the creation of new open spaces. In the first place, he asked that the gardens and orange groves around the medina, inside the Almohad enclosure, be conserved, notably by creating a green zone not aedificandi outside the Andalusian enclosure. Beyond the outer ramparts, he recommended creating, before the city develops, reserves for public gardens and avenues for walks. At the same time, he proposed to link the different neighborhoods by a communication network combining wide roads and narrower streets with a suitable planting system. To anticipate the growth of the city in the medium and long term, Forestier advised the creation of green isolation zones outside the Almohad enclosure.

– **The system of open spaces of Forestier applied to the plan of development of the new city (plan Prost)**

The first action of the architect Prost was to safeguard and classify the monuments of the past, in response to Lyautey's urban doctrine based on respect for the ancient heritage, traditions and morals of the Moroccan people. In parallel with the classification of the urban and architectural patrimony (kasbah of Oudaïas, Hassan tower, ramparts, etc.), Prost imposed nonaedificandi easements around the ramparts of the medina and the tower Hassan and Chellah. New and completely original element: Prost preserved some points of view on the panorama of the ancient city (ramparts, medina, etc.) and its surrounding landscape (ocean, mouth, the silhouette of Salé). The new constructions risk to hide the views, Prost decided to arrange the first plans in gardens. Thus the location of the gardens and open spaces of the future European city was chosen. To establish the development plan for the new agglomeration, Prost drew up:

- A traffic plan in which the main lines of communication appear;
- A distribution plan of the different neighborhoods (administrative district, residential and commercial district, university district, industrial district);
- A plan for open spaces indicating the distribution of the various free spaces (public gardens, residential areas of the General Residence, planted walks).

Prost's plan for open spaces is the perfect illustration of the parks and gardens system outlined by Forestier in his 1913 report. Thus, as he had advocated, before neighborhoods rise, gardens public and open spaces were to be created outside and inside the Almohad enclosure.

The first public garden, the Essais garden, was built in 1914 on the outskirts of the ramparts on about ten hectares. The development plan was designed by Forestier in 1916. The second public garden is the Belvedere garden, built in the continuity of the Essais garden, on one of the highest points of the city. Originally, it was the Franco-Moroccan exhibition fair that took place in September 1917. Prost chose the site because of its privileged location overlooking the city and the ocean. The creation of a fair outside the outer walls was an opportunity to anticipate the development of extra-muros districts (Aguedal district). The site of the fair was indeed preserved and rehabilitated in public garden and surrounded by afforestation of pines and eucalyptus which formed, from 1917, a greenbelt (wood of the Aguedal) around the enclosure of the royal palace. The third public garden is the Garden of the View Triangle, designed in 1924 by the French landscape architect Marcel Zaborski, head of the walks and plantations of Morocco. Located outside the ramparts of the medina, it acts as a buffer between the old and the new city. Why the name "View Triangle"? To respond simply to the desire of General Lyautey to highlight the perspective from the terrace of the Municipality—former General Residence—on the medina and the ocean. Today located in the city center, the garden of the Triangle of View forms a key link in the park system of the city of Rabat. The Garden District of the Residence General is also part of the network of open spaces and privileged points of view that wished to safeguard Lyautey. Built on the highest point of the city, leaning against the outer ramparts, the Residence district was conceived as a garden city in its own right. A veritable "gesture of war", the district gathered all the administrative services of the protectorate and the residential villa with its gardens, all drowned in a green setting (pine and eucalyptus wood) and connected by a judicious system of pedestrian galleries. The gardens of the residential villa, designed in part by Albert Laprade—collaborating architect of Prost—and by Marcel Zaborski around 1920–1925, were realized in the pure French tradition with regular flower beds taking place in front of the main facade of the Residence.

Outside these gardens, others were created like the Oudayas garden laid out in the enclosure of the casbah in the pure Moorish tradition, or the garden of the Mamounia realized on the site of an old orange grove between the ramparts from the medina and the outer enclosure. Let's also mention the gardens of Hassan Tower and Chellah. A set of squares, kindergartens, squares, sports fields and woodlands (green belts) completed the network of parks and gardens of Rabat. To connect the various free spaces of the city, Prost made realize, as Forestier had prescribed, a network of planted walks combining broad avenues and narrower streets.

- **Marrakech, the garden-wintering city**

Marrakech is, like Rabat, an old imperial city. It is located more than four hundred kilometers south of the capital, in the center of a vast plain bounded on the south by the chain of High Atlas, about fifty kilometers, and by the hills of Jbilet, about ten kilometers North. Before the advent of the protectorate, Marrakech had all the elements of a garden city. Although enjoying a semi-arid climate very dry, the city astonishes by its luxuriant vegetation, with its vast palm plantation and its immense gardens. At the beginning of the twentieth century, the city was limited, like Rabat, to its medina. The essential difference is that Marrakech, unlike other Moroccan cities, had many large expanses of gardens within its walls.

The most important are the gardens of Aguedal which extend to the south of the medina, at four hundred and forty hectares—almost the size of the medina which occupies about six hundred hectares. Surrounded by an enclosure, the Aguedal represents the perfect prototype of the Almohad garden of the twelfth century, with its large reservoirs, its squares of olive and fruit plantations and, above all, the magnificent panorama that is offered to the visitor, with on the horizon the silhouette of the High Atlas.

Inside the ramparts, on the edge of the urban fabric, there were also ancient princely gardens named Arsa, dating from the eighteenth century, including Arsat Moulay Abdeslam and Arsat Mamounia, which still exist today.

To these gardens are added the palace gardens, also located inside the enclosure of the medina: the riad of the palace al-Badi realized under the Saadian dynasty in the sixteenth century, and the palace of the Bahia and its Aguedal created at the very beginning of the twentieth century by former vizier Ba Hmed. Outside the ramparts extends, to the west of the medina, another great AlmohadAguedal: the Menara. These gardens, surrounded by high walls, also have a pond with a pavilion, all immersed in an orchard of olive and orange trees. The silhouette of the pavilion, reflected in the water of the basin, with in the background the snow-covered peaks of the High Atlas, is one of the most symbolic images of Marrakech.

Outside this walled garden, the medina was surrounded by a vast palm grove that extended to the north and west of the city. Made up of groves of date palms, the palm grove was punctuated with nourishing gardens (jnan); among those who were rehabilitated was Jnan El Harti. The irrigation of all these gardens was ensured, since the foundation of the city, by a judicious network of khettara (underground draining galleries) and seguias (aqueducts), the khettaras used to capture water from the water table and seguias ensuring the delivery of wadis water.

Thus, when they arrived in 1912, the French discovered a set of centuries-old plantations and gardens and a remarkable irrigation system. Unlike other Moroccan cities, when Forestier went to Marrakech in 1913, he discovered a new city already drawn by the French military. The latter had begun to develop a European district (the district of Guéliz), in the north-west of the medina, between the ramparts and the military camps installed further north, below the hill of Guéliz (name taken again for the new city).

Protective measures had already been taken to preserve the existing palm grove and gardens. In addition to these first measures, Forestier recommended that the whole of the garden and intramural patrimony of the medina (Aguedal, Menara, Arsa, Riad, Jnan, etc.) be conserved and restored. He particularly insisted on the importance of preserving the palm grove against any attempt to subdivide and, above all, to maintain "for the future" the magnificent panorama of the High Atlas south of the city.

At the same time, Forestier had planned, as for Rabat, new reserves of open space on the periphery of future neighborhoods. At the request of the Bureau des Renseignements, Forestier planned a public garden on the site of one of the former orchards outside the city, Jnan El Harti. While maintaining the original layout and the division into fruit squares, Forestier inserted sport fields of different sizes. Jnan El Harti is the only public garden built by the protectorate outside the ramparts of the medina.

In order to improve the route network used by the military, Forestier sketched pathway profiles along which he proposed walkways to protect from the sun and dust.

Also, Forestier ordered the factories to be moved away from dwellings and palm groves so as not to disfigure the emblematic landscape of the High Atlas to the south. In Marrakech, the role of Prost was to continue the development of the first nucleus created by the military (Guéliz) by connecting it to the medina and realizing, to the south, a new wintering area.

The wintering district was conceived as an ideal garden city, destined for a wealthy foreign clientele, with villas, hotels, a casino and planted walks. Located west of the medina, it was ideally located in the middle of the gardens (Arsa Medina in the east, Jnan El Harti in the center and Menara in the west). The layout of the wintering city was determined by a network of green links (avenues, streets, footpaths) designed to connect the tourist sites near the Grand Hotel de la Mamounia: the Jamâa El-Fna square and the Koutoubia located inside the medina, the intramural gardens (Arsat Moulay Abdeslam and Arsat Mamounia), the El Harti garden and the Menara. In order to preserve the exceptional panorama of the Atlas Mountains to the south, Prost struck the area on the southern edge of the wintering district with servitude not aedificandi. The new city thus followed "a curved path suitable for strolling, taking its views on the ramparts, the Koutoubia, the snowy peaks of the Atlas". Outside the wintering area, Prost merely rehabilitated the old gardens: the Aguedal and Menara gardens as well as the riads and gardens of the old palaces.

3 The City of Casablanca: A Metropolis that Suffocates

Casablanca (white house "in French), locally called Anfa (" hill" in Amazigh), is a city of Morocco, capital of the region of Grand Casablanca and economic capital of the country, located on the Atlantic coast, about 80 km south of Rabat, the administrative capital.

Administratively, its territory (of an area of 386.14 km^2) corresponds to that of the prefecture of Casablanca. Since the return to the principle of unity of the city in 2004, it is composed on the one hand of the municipality of Casablanca (divided into 16 districts distributed in 8 district prefectures), on the other hand, the tiny municipality of Mechouar of Casablanca placed in its center and where sits a royal palace. In the last census of 2004, its population was 2,949,805 inhabitants (2,946,440 for the municipality of Casablanca and 3365 for that of the Mechouar of Casablanca).

Made legendary by the film Casablanca (1942), the city has an important modern architectural heritage, due to the architectural diversity it experienced during the twentieth century, where it was then the "laboratory of modernity" of new generation architects who landed directly benches of the School of Fine Arts in Paris. Casablanca is located on the plain of Chaouia, a historically agricultural region and to this day one of the main centers of agricultural activity in the country. Its position on the Atlantic coast allows it access to maritime resources (mainly related to fishing).

The only forest area bordering the city is Bouskoura, which was planted in the twentieth century and consists mainly of Eucalyptus, Pine and Palm.

3.1 *Urban and Demographic Expansion*

Greater Casablanca is one of the sixteen regions of Morocco. Located in the northwest of the country, the regional capital is Casablanca whose borders are the flap regions to the north and the Atlantic Ocean to the west. The Casablanca region is the largest urbanized region in the Kingdom of Morocco, with a population of 3.6 million inhabitants (2003) which will exceed 4.6 million in 2015 according to estimates. The average population growth was 2.3% per year between 2000 and 2005. 22% of the urban population of Morocco lives in Casablanca.

40% to less than 20 years old (1994). The city spanned only 50 hectares in 1907. Between 1990 and 2000, the area of the city rose from 15,000 to 21,000 ha. In 1996, an administrative reform led to the inclusion of some neighboring municipalities. The region "Grand Casablanca" was created in 1997. It is composed of 8 prefectures on a total area of 869 km^2.

Casablanca is an important economic and financial center and one of the largest ports in Africa. Agglomeration is home to 60% of Moroccan industry (leather, textiles, machinery, food, and chemical industries). This induces a rapid urban growth as shown in the Fig. 1 and that is accompanied by the development of slums ("slums"). New industrial zones are currently being created. A large number of previously rural communes such as Dar Bouazza, Médiouna, Mellil, SidiMaarouf, Lisasfa, AinHarrouda and Bouskoura have been strongly affected by the transformation of agricultural areas.

3.2 *Problems and Confrontations of Casablanca city*

- **The Traffic movement**

The problems of traffic in Casablanca are becoming more and more difficult to overcome. At certain times of the day, many axes of the metropolis are simply blocked because of road infrastructure works in progress, the obvious difficulty of traffic officers to control flows and, above all, indiscipline motorists who engage in

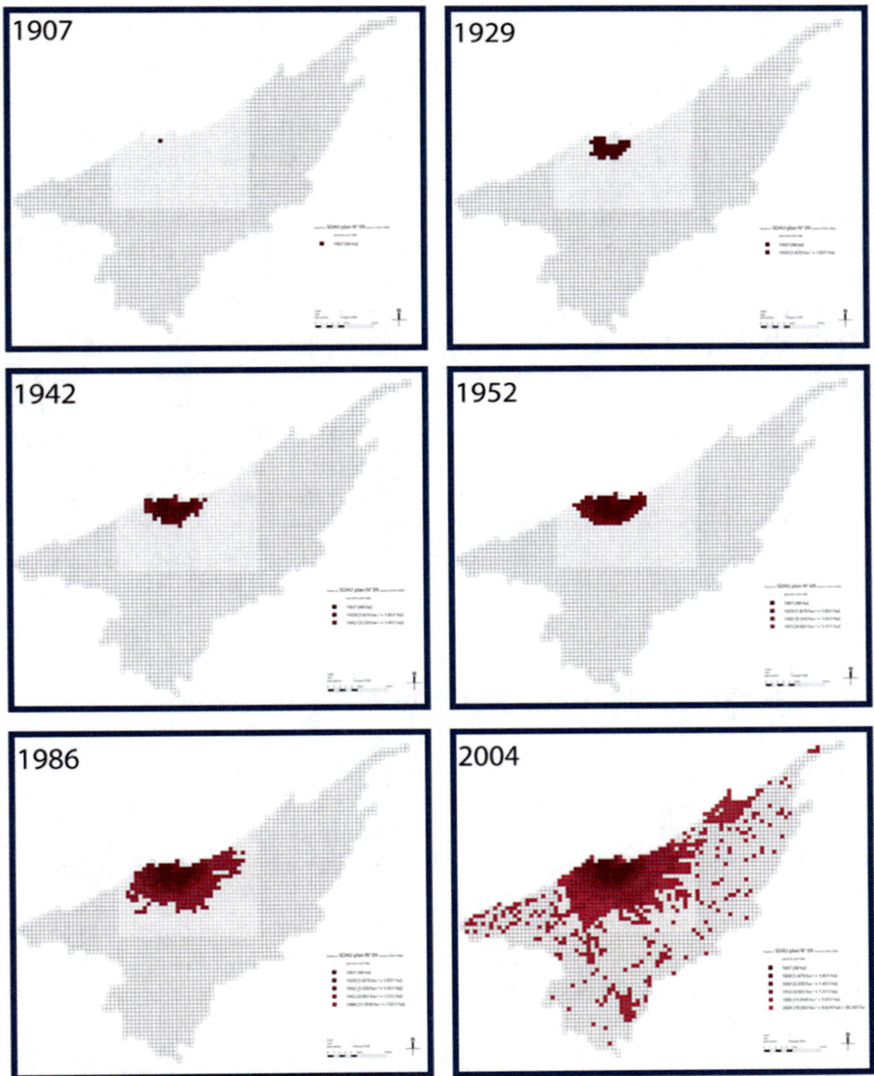

Fig. 1 Urban development of Casablanca between 1907 and 2004. *Source* Urban Agriculture Casablanca-Design as an Integrative Factor of research, p. 10)

a daily war to reach including the city center and neighborhoods. If the numerous studies and diagnoses made in recent years have made it possible to draw the outlines of a global urban transport plan, and a new organization of integrated urban transport around the Casablanca tramway, the transition to the application of these plans on the ground is waiting.

- **The Pollution**

The study of the International Agency for Research on Cancer (IARC), a specialized agency of the World Health Organization (WHO), published recently that the polluted air of Casablanca is carcinogenic. IARC experts believe that there is sufficient evidence to say that exposure to outdoor air pollution causes lung cancer. "We are strongly concerned by this study since the concentration of fine particles is higher in Casablanca than in Paris or Washington," said Mohamed Benjelloun, environmental expert. Same observation from Prof. SaïdKarrouk, climatologist. He believes that the findings of the study in question can be applied to all regions of the planet even though exposure levels may vary from place to place.

Indeed, the metropolis holds the sad national record that sometimes reaches 100 $\mu g/m^3$ while the European standard accepts 40 $\mu g/m^3$ at most. It exceeds 2.5 times the international standards. Moreover, for a good reason, its car fleet and its industrial activities are constantly growing. Indeed, in 2012, the economic capital welcomed 59,731 new cars, against 50,972 a year earlier, an increase of 17.2%. For the current year, 65,000 new registrations have been registered. 80% of these vehicles run on diesel fuel classified in June 2012 in the "certain carcinogenic" category. In total, 1,800 tonnes of exhaust gases are emitted into the air each year in Casablanca, with very serious pathogenic effects.

3.3 *The Pre-Urban Area and Luxury Housing Projects*

Casablanca's urban expansion has contributed to the emergence of new real estate developments. The good prospects for luxury real estate are mechanically attractive and several types of operators have invested or are in the process of positioning themselves on this segment since the "traditional" market is showing signs of slowing down.

Villas around 250,000 Dollars and more, designed as main homes with at least 3 rooms with garden and pool, all just minutes from the downtown, is the lode of the moment for real estate developers. Indeed, there are more developers who offer affordable villas located on the immediate outskirts of Casablanca. In Dar Bouazza and its region, about twenty kilometers from the metropolis, the residences of Dar El Kenz, The Villas of Anfa, Dar El Bahr, Garden Beach, the Villas des Prés or Tamaris City have added these month to the long list of projects that have grown in the region in recent years (Marina Blanca, Tamaris Eden, Puerto Marina, Al Boustane …).The same effervescence is noticeable on the Bouskoura side with, among other projects, CASADIAA, Sun City, Villa Monaco, the Walili Residences, California Eden, Green Residences, Lilas Park …, in addition to the essential mega-projects at the level of forest Bouskoura initiated by Prestigia, CGI and Palmeraie Development.

It is that these formulas of affordable villas in the periphery attract more and more the Casablancais, tired of the urban din. Considered until recently as second homes, these products have gradually conquered a clientele of young well-off couples who settled because they can not access villas in the center. Moreover, the specter of demand has quickly widened to senior executives, retirees or professionals ready to trade an apartment in a posh downtown area for a villa in a more peaceful environment. A whole new market that was created so quickly that it took real estate developers by surprise.

4 The Casa Diaa Project

CASADIAA is a residential complex erected on a surface of 40 ha proposes a new concept composed of 200 villas of master exclusively on one level. Located 15 min from downtown Casablanca, the project is ideally away from urban nuisances. Figure 2 presents the geographic location of CADIAA project.

CASADIAA offers three types of villas built on a plot of 800m^2 including 307 m^2 built area (370 m^2 covered at a price of 400.000 Dollars).The materials used are designed to respect the environment (wastewater recycling for irrigation of green spaces, the use of solar energy …).

CASADIAA will include commercial space for shopping, a fitness center with SPA and indoor swimming pool, a restaurant, a nursery, a youth activity center, multi-sports fields, playgrounds, jogging trails, and many landscaped public gardens of more than 4 hectares of green spaces.

Figure 3 shows the three types of villas. The first villa is Aube Primera which is a construction that has been specially designed for people who have for the first time given in to the temptation to have a villa. On an area of 820 m^2 and more, the built surfaces vary in this case from 280 to 406 m^2. The garden is valued. On the bright side, the architectural lines have been finely designed to have large windows that let light into the house.

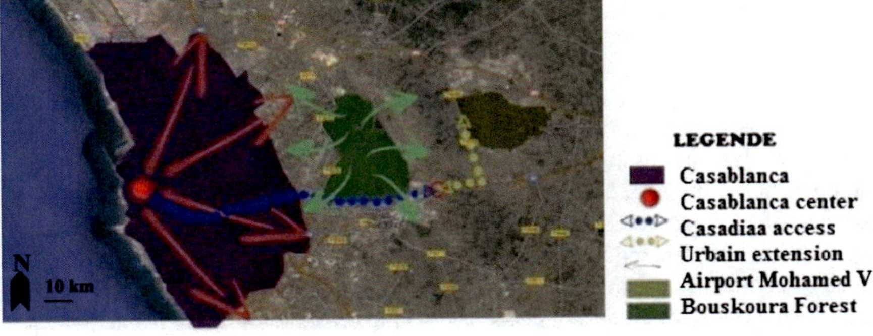

Fig. 2 Geographic location of CADIAA project

Fig. 3 The three types of Casadiaa' villa

The villa Aurore, the second model, signed CASADIAA, is attractive with its modern and innovative style. It differs largely by its reception area ensuring user-friendliness. Already, the built surfaces exceed the level of 500 m^2 over a total area exceeding 1000 m^2. Finally, the villa Halo, classified as the villa of excellence guarantee calm and serenity for its owners. Large in its capacity but also enhanced by a larger pool, this model of villas is 700 m^2 built on a space of 1300 m^2. The three models rare represented in Fig. 3.

The project is located 15 min from downtown Casablanca, on the highway of the Airport and close to the Bouskoura Forest. The village of Bouskoura is located south of Casablanca about twenty kilometers. It is called by this name because it is located near the forest of Bouskoura which is one of the great forests of Morocco and which attracts a large number of Casablancans on weekends and holidays. Most of Bouskoura's trees are eucalyptus.

CASADIAA moves away by about twenty kilometers from the center of Casablanca. It is accessible only using a means of transport by the road boulevard El-Jadida. It can be reached only by personal means of transport or large taxis, the red taxis (Private Taxi) cannot reach it.

The road El Jadida is a road that connects the city of Casablanca with Bouskoura, it is a newly built highway whose circulation is easy and fast, CASADIA presents a system of severe security, and the access to CASADIAA is limited for the inhabitants and their guests. CASADIAA is inhabited by a very high social category and very demanding who have chosen to live in a residential complex of high standing. Since Casablanca is a big city in a very important city, the smothered and polluted city presents a very stressful life and a congestion of the car traffic. Moreover, the green spaces present on Casablanca does not offer activities for all age categories. On the other hand, the lack of security and the bad attendance presents an almost total absence of the high social category. Given these living conditions, there were the appearance of several modern residences inspired by the American and South African cities where the richest chose to move away from the city and create their own places and live in collective cities. very modern, on the outskirts of the city and away from nuisance.

These important social class dwellers are looking for a quiet place away from pollution and traffic jams. They are looking for a cool environment or green area, quiet recreation areas, a place where they can do business safely, Knowing that the inhabitants have family relations or friendships. We will base ourselves in our conception on this relationship to create interactive landscape spaces of relaxation and entertainment where the inhabitants can meet and share good moments in all safety.

5 The Park of the Casa Diaa Project

5.1 Landscape Analysis

The presence of a road/highway south of the project will be a source of nuisance to residents. The nuisance caused by this highway could be reduced by the excessive planting of the south side of the park. The roughness of the CASADIAA project traces the boring and long distance circulation as schematically shown in Fig. 4. In the park, the creation of a different circulation from that of the project could be tried. It must be flexible and having an aspect of discovery.

The repetition of the villa element or grouping of villas can be boring and can contribute to the loss of sense of direction in the project as we could see in the Fig. 5. It can be remedied by the use of a variety of plants as well as materials and equipment.

The park extends throughout the project; it is linear. Access to the park should be easy to access and close to all residents. Their location should be close to the car parking spaces and ensure continuity with the project lanes as shown in the Fig. 6.

The studied site and its environment are a very judicious and strategic choice to create a zone of modern houses. This area would be an attractive place to enjoy a real quality of life in a very quiet and luxurious residence in the wilderness away from the city and its pollution.

The observations and analyzes of the site are summarized in Table 1.

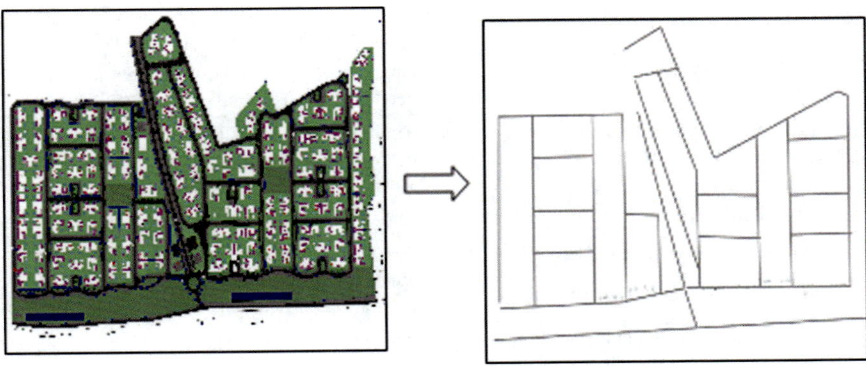

Fig. 4 Rigidity of forms

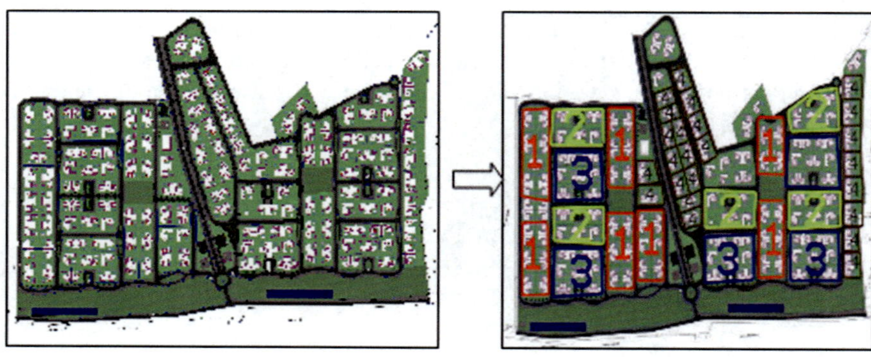

Fig. 5 Repetition of forms

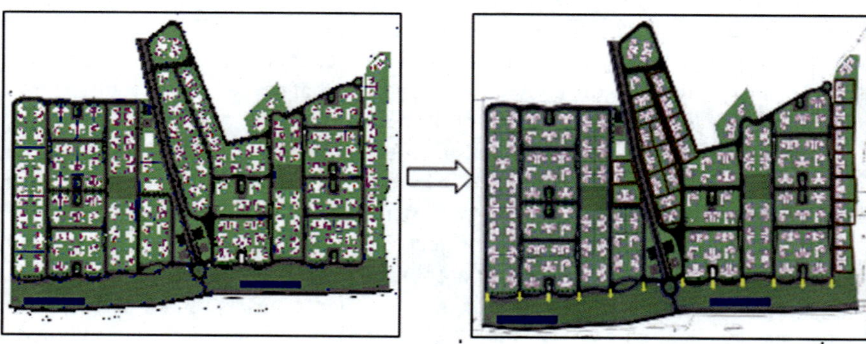

Fig. 6 Continuity between lines and future access

Table 1 Summary of the main findings and necessities of the site

Modern architectures of high standing	Creating space of modern architectural form, with the use of high quality materials to create harmonical outdoor spaces and reflecting the prestigious aspect of existing houses
Inhabited by a very high and too demanding social category	Creation of interactive spaces that meet the needs of security, calm and relaxation
Significant thermal amplitude, presence of frost during winter, and a very hot and dry summer	Avoid the use of plants that fear frost and can not withstand the drought and high temperature of summer
Aclay-limestonesoil	Avoid any acidophilic plant
The site presents an open and clear space	Creating wind breezes to limit the strong winds that will be detrimental to the new vegetation and their aspects
The site is limited on one side by a highway	Try to reduce the nuisance that can be caused by this highway

5.2 Needs and Objectives

CASADIA offers an exceptional lifestyle: high standard villas on one level that provides an escape for its inhabitants while keeping a natural and friendly. It makes them discover a sweetness of life and well-being where the harmony between the different elements of the project is paramount.

Located south of the CASADIAA project, the park is conceived as an inclusive sensitive destination. This is a space that will enrich the project through the contemporary design of the site and the variety of equipment and programs that will be developed. The park is an integration into nature. It reinforces the character of the area by landscaping green spaces that are home to cultural and sports and recreational activities where the man can live enjoying the freedom and developing his skills as well. Physical and intellectual property and this in complete safety. These activities provide a refuge that balances mind and body. Mainly listed areas for relaxation and lively and eventful spaces for entertainment.

This park will be designed keeping the contemporary spirit of the environment while integrating the natural and dynamic aspect. This dynamism will be ensured especially by the selected vegetation (shape, aspects, flowering periods, colors …) as well as by the change of colors of the leaves with the seasons. For this, a plant palette was chosen to adapt to the natural conditions of the site. It includes a good selection of grasses, flowering trees and shrubs and perennials. Thanks to the beauty of the inflorescences and their light, graceful and relaxed general appearance, the grasses form a real moving sea of vegetation, with many changing reflections depending on the wind and the light. They combine very well with all the perennials and roses. They will put each other in value in order to balance the general aspect of the massifs. The use of trees in many aspects and forms, will add on the one hand dynamism thanks to the rustling of the wind in its foliage, and on the other hand increase the dynamism to our park thanks to the flowering of trees and shrubs with flowers, especially during the flowering periods that will be triggered by the seasons. The wind and its action on the vegetation is not the only element capable of giving movement to the park. The water also holds an important place in different aspects. The park will be designed natural, contemporary, diverse and dynamic. The result is a comprehensive concept plan that supports a variety of equipment and activities for all ages and user groups and creates a dynamic focal point for the neighborhood with the following objectives:

Fluid Accessibility: A park accessible to all people and all ages throughout the site. Park features should respond to all universal accessibility standards including path, transfer height requirements, and signage and other directional strategies that will respond to the needs of people with physical and visual impairments.

Sense of place: A strong sense of belonging that reflects the distinct character of the region. Shapes and materials should contribute to strengthening the identity and heritage of the community while preserving existing vegetation in new landscape improvements.

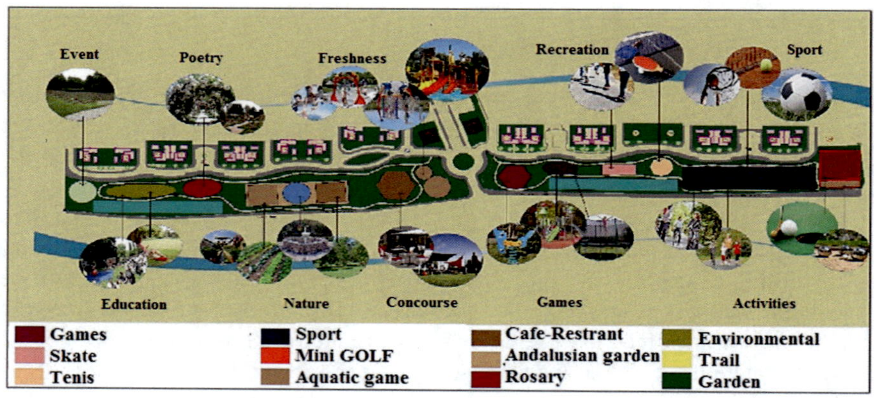

Fig. 7 The concept plan of the park

Secure landscaping: A safe and comfortable public space for all ages and abilities. Safety principles and techniques should be incorporated throughout the design to achieve secure urban spaces.

Sustainability Initiatives: A site that demonstrates a commitment to environmental concerns. Development initiatives must be integrated throughout the site and must include strategies for achieving sustainable development.

Park of Various Activities: A park that can support a variety of educational activities and community events. A variety of "soft" spaces in nature that can be transformed to host events including various activities.

Group for Young People: A safe and welcoming environment that responds to the needs of young people by providing specific facilities that are fun and enjoyable to provide safe and healthy areas for sharing and socializing.

The result is a comprehensive concept plan that supports a variety of equipment and activities for all ages and user groups and creates a dynamic focal point for the neighborhood, presented in the following diagram of Fig. 7.

5.3 *The Design of the Park*

The park offers complimentary entertainment activities that respond to the needs of all age groups. The park would be built in such a way that all residents can safely practice their various activities.

A first part would be dedicated to physical activities and a second part would be educational.

5.3.1 Part I

In addition to gaming machines, this first part offers areas for adults to exercise various sports and play activities which are represented in the Fig. 8.

- **Playground for children**

Movements and games determine the harmonious development of child. They are inseparable from the capacities of discernment, emotional and social behavior. For this purpose, attractive play spaces have been designed, safe and adapted to the needs of children. Four areas defined the playground for children, climbing, swept, sand, and living structure areas.

Moving from one age group to another, children's abilities develop and their needs change, so games must take these criteria into account.

Thus, the playground is designed naturally to develop the child's sensitivity to nature. On the one hand, the playground will be surrounded by an embankment that must be at a slope less than 45 degrees, maintained by a wall of gabion used as benches for rest. This embankment will be planted with grasses that will ensure the dynamism and freedom of this space due to their light, graceful, relaxed appearance. These grasses will be associated with *Buddleia davidii* as a backdrop, which is an upright shrub whose long arched branches bear flowers in long, dark purple spikes in summer. In turn, it will add to the space of dynamism due to the color of its fragrant bloom and its characteristic of attracting butterflies.

Trees and shrub will hide the wall and forget the built-in limit of the project. They will also play a vital role in marking the season by their flowering or changing the colors of the leaves:

- *Jacaranda mimosifolia* with its mauve-blue flowers that appear at the end of spring and fall during the summer to form a beautiful, colorful carpet.
- *Brugmansia* appreciated for its sumptuous trumpet flowers. Often scented, will give an exotic touch to space.
- *Lagerstroemia indica*, is a small tree with a great show, due to its beautiful flowers in bright colors blooming in the heart of summer.

Fig. 8 Design of part I

This selection of tree and shrub will increase children's sensitivity to nature. The children's play area offers different possibilities to develop their flexibility (climbing gear, balancing machines, etc.) on a soft floor that is made of EPDM to ensure their safety against falls. It gives also free rein to their creativity with a sand space and islands of *Pennisetumrebrum*, *Miscanthussinensis*, *Strealitzeaagusta*, and *Phyllostachysbambusoides* which offer to the children a place to enrich their creativity by passing on this island and will activate their senses of touch and which would be a great place for hiding and seek games.

The play area will be planted by shade trees like *Cassia nodosa, Celtisaustralis*, and *Jacaranda mimosifolia* that will provide shade and thus relax flower spaces for parents.

To conclude, the transition between the areas of the playground is designed to easy and flexible. For the turfed part, kikuyu was chosen as a turf species due to: its resistance against settlements, its resistance to heat and extreme conditions of drought and its maintenance is very simple, which will encourage children to practice the games of roulade,... The Fig. 9 presents a 3D graphical representation of the playground.

- **The skate**

The skate game is considered one of the most favorite sports by teens. In order to allow fans of this extreme sport to enjoy it safely and prevent children from exercising their talents on public roads and paths, skateboarding and skating trails have been designed. To strengthen the safety of the practitioners of this sport, clearly visible panels recommend wearing protective gear (helmet, knee pads, elbow pads, wrist guards) must be placed on the edges. Thus a turfed area with a lively and shaded ground by *Brachichitonpopulenus* and *Grevillearobusta* will present a relaxing area for fans of the sport. A profile of this entity is presented in Fig. 10.

- **Table tennis**

Two tennis tables will be placed on a hard surface to prevent missteps as shown on the profile below. Natural soils have a major disadvantage that can form cavities that prevent players from evolving and that fill with water when it rains (Fig. 11).

Fig. 9 3D graphical representation of the Playground for children

Fig. 10 The skate profile

Fig. 11 The table tennis Profile

The passage between these two tables will be provided by slabs placed on the grass. Between these 2 tables, a *Delonixregia* will be planted in isolation that, due to its crown that forms a spectacular parasol, is quite remarkable when its big red flowers pair during the summer.

- **Multiple sports fields**

This part has been designed for multiple sports. There are mainly individual playgrounds and communal playgrounds. The space has been subdivided to include different activities such as mini-football, volleyball and basketball. These sports fields offer people air spaces allowing them to practice their favorite sports in all safety. International standards have been followed to give them the opportunity to be entertained. These lands will be surrounded by trees and shrubs to promote shade, reduce noise, and play a role as a windbreak. They will be placed in series with a north–south exposure. Figure 12 presents a 3D sketch of the sports fields.

The different plants used in the field boundaries are as follows:

- For the football pitch: *Malvaviscusarboreus, Lantana Camara, Grevillearobusta.*
- For the basketball and volleyball court: *Euphorbiacotinifolia, Durantarepens gold, grevillearobusta.*
- For the tennis court:*Viburnum tinus, grevillearobusta.*

- **Mini GOLF and bowling**

The game of mini golf, presented in Fig. 13, will be designed under the standards (courses from 0.7 m to 1.2 m wide and 3.5–15 m long). To this fact, the layout of the

Fig. 12 Multiple sports fields

Fig. 13 The Mini GOLF

courses will be different in terms of plant palette to create a theme specific to each course. For the game of bowling, presented in Fig. 14, we designed 2 sand tracks of dimension 15 m long and 4 m wide shelter by Catalpa bignonioïdes shading trees. These precision games have been designed at the end of the park to allow fans to practice in a quiet and soft space.

- **Trails**

The park trail is one of the most desired facilities; it is designed flexible all along the park to give a natural look and a sense of discovery for the site. It is a circuit

Fig. 14 The Bowling

to run, walk, and use the bike. Also, there is a proposal for some equipment for the physical movements.

The entire trail must be universally accessible in terms of quality. It will consist of a stabilized soil of very good quality in terms of landscape integration and will be delimited by type P2 border. Its use brings a natural connotation to the site as well as a satisfactory comfort for pedestrians and cyclists.

Resting areas are designed along the park's trail edges which are characterized by completely wooden benches or benches constituting gabion. Fitness stations should be provided along park trails creating a "fitness loop" function for the site. Figures 15 and 16 show the resting area and the fitness station.

5.3.2 Part II

To reinforce our concept of "the naturalness, movement and dynamism" of our park we have used in the left part of the park, presented in the Fig. 17, another element of nature that is 'water'. Indeed, water plays an important role in our park; it comes in addition to the action of the wind on the vegetation by water games, streams with jets, and the discreet flow of a fountain.

- **Aquatic playground**

Fig. 15 Resting area

Fig. 16 Fitness station

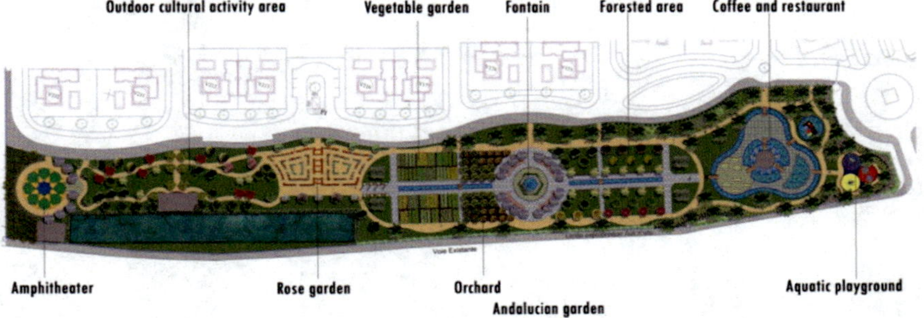

Fig. 17 The second part plan

These aquatic playgrounds, complementary to the playgrounds on the right side of the park, are designed to entertain children and adults alike while refreshing them. They are also places of discovery, meeting and fun that do not require monitoring.

These "dry wading pools", without any water storage, are new points of attraction, that, with play areas adapted to different age groups and people with reduced mobility, are real attractions. These games are likely to significantly increase the reputation of the place and therefore its attendance. The space will be surrounded by *Syagrus romanz offiana* an extraordinary palm tree that has many assets: ringed gray trunk, rapid growth, feathery appearance very exotic by its foliage and good resistance to drought and wind. It will give space a charm beach. At the foot of these palm we plant the *Forium tenax* characterized by its trendy and natural side. We will associate it with grasses, bamboo, and yuccas giving the space a design and contemporary atmosphere.

- **Café- restaurant**

Just next to the water playground we designed a café-restaurant area that we can see in Fig. 18. It is a friendly place: Atmosphere tastefully retro inside, very pleasant terrace on sunny days in a charming place where families can enjoy family or friends moments. Surrounded by a jet stream, this relaxing meeting place allows parents to control their children while playing, safely. This space is equipped with a shaded structure of parasols surrounded by a stream. A large terrace equipped with stretched canvas offering residents a warm atmosphere.

The transition between the different parts of space is ensured by wooden bridges which are presented in Fig. 19.

- **Andalusian garden**

The garden is characterized by a regular and symmetrical style, using the straight line and the right angle. It is in the form of an elongated rectangle. In order to reduce this roughness, soft and undulating paths have been designed around this garden. The entrance to the Andalusian garden is well marked by two olive trees planted in isolation on a lawn that would give the entrance a lush appearance as well as nobility

Fig. 18 The restaurant

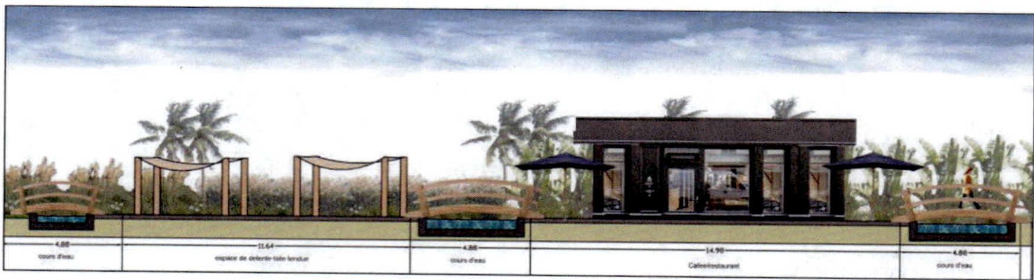

Fig. 19 The restaurant and its wooden bridges

and wealth. This garden reserves a very frank circulation in the sense of a principal axis guided by a watercourse and intersected by 4 secondary vistas which divide it into four parts as shown in Fig. 20.

The first part of the Andalusian garden has been designed as an arboretum that would present an educational space by the presence of a large variety of trees that summarize the species used in our park. On the other hand it would present a place for picnicking and relaxation due to the shade provided by these trees and the picnic tables installed.

Crossing this part, an alignment of *Albizia* embellished the path due to its abundant summer bloom and its spreading, very elegant. At the foot of the Albizia an alignment of agapanthus with its beautiful ornamental foliage and a blooming blue bloom that will add color and freshness in the middle. The harmony between these plants and

Fig. 20 The Andalusian garden

the water will give the space a contemporary and natural touch. Arriving at the center of the garden, an open square circular centered by a fountain creates an atmosphere of freshness and dynamism, and surrounded by Jacaranda that will strengthen the rounded shape of the square. Wooden pergolas will be covered by climbing plants like the wisteria, with spectacular blooms, in the form of hanging clusters with an infinity of elegance and delicately scented. The freshness provided by this space due to the surrounding vegetation, and the discreet song of the fountain would present the relaxing place for inhabitants during the day as well as during the night due to the mood lighting designed for this space. The third part is designed primarily by a lemon and orange orchards that soften the air due to the scent of their flowers and their fruit, and secondly by a kitchen garden and aromatic that will enhance the natural appearance of the park. This garden will, on the one hand, increase the dynamism and strengthen the natural aspect by the diversity of cultures and colors of their leaves, and on the other hand, this space will promote exchanges between neighbors through social, cultural or educational activities.

After having walked through this garden where the profusion of colors and smells of plants reign under bright bursts, white colonnades completely enveloped by climbing roses will be implanted at the edge of the end of aisles will trigger the beginning of romantic space.

- **The rose garden**

The rose garden was designed as a labyrinth where we will plant several species of roses, such as Rosa 'Aloha', ROSA 'DAMASCENA', Rosa 'Snow Fairy', Rosa 'fortunakordes', Rosa' lupokordes' Rosa 'memory kordes', Rosa 'rugosa', Rosa 'rotiliakordes', this space will give the park a natural touch and will increase the dynamism of the park by the diversity of colors of the flowers. Red synonymous with life and gaiety and love; the white, symbol of purity, reflects the light, the yellow recalls the color of the sun, the light and makes think of the joy. This space will invite the inhabitants to activate their senses of touch and feel (Fig. 21).

- **Event space-cultural activity**

Fig. 21 The rose garden

Fig. 22 Event space

This open and green space is designed for informal gatherings, major events of the inhabitants and cultural activities such as reading, music, theater, dance in a natural setting, for which we have designed a center that will welcome the inhabitants Cartier for the practice of these activities. Figure 22 shows the look of this entity.

5.4 *The Park: An Added Value*

As outdoor spaces are particularly rare in the center of the metropolis, it is of course there that they are most sought after where there is a substitute market, for lack of a sufficient supply of townhouses. Energy performance has become a real asset for the valuation and sale of real estate. Housing tells you all you need to know about the emergence of the "green value" which refers to the variation of its value (price or rent) related to its environmental performance. The presence of green spaces and natural sites near the property is also increasingly sought after by buyers. In more than 40% of cases, public green spaces around real estate allow homeowners to set a higher sale price or rent for similar properties, but more away from the greenery.

Note that if this landscaping is carried out professionally, the return rate generated is higher than the other improvement projects that can take place in or around the house: its value of recovery of investment can rise at 100 or even 200%. This means that when reselling his property, the owner can hope to recover all the money that he has invested in his development. In comparison, we can observe a recovery rate much less interesting for a kitchen renovated from 75 to 125%. The rate is even lower for the repair of the bathroom, 20 to 120% and the installation of a pool can recover only 20–50% of the investment.

Also, investing in the outdoor spaces of your garden can also increase its value over time through the growth of young shrubs and new plantations that will continue to mature and beautify the property. Thus, the design and development of the CASA-DIAA project park will surely increase the value of the project and will present an attractive element that will undoubtedly facilitate the sale of villas but also become an ecologically sound model.

6 Conclusions

Today's context, triply marked by economic globalization, socio-technological changes and ideological and political uncertainties, poses new challenges for developing countries. Local governance consists of a set of institutions and mechanisms. Processes that allow citizens and citizens' groups to express their interests and needs, to settle their differences and to exercise their rights and obligations at the local level.

Today, we talk about attractiveness and competitiveness of territories, territorial anchoring of companies, industrial districts, territorial projects, etc.….However, the question is no longer simply to initiate local projects, but to ensure that synergies are created between the actors and accelerate the local dynamics observed in territories that remain to be built. What is the real contribution of these projects to the problem-solving at the local, regional and national levels? How to act effectively and accelerate the dynamics of growth and development? In other words, how to move from localized action to a territorial approach to development problems? These are the questions facing Morocco today regarding local development.

7 Recommendations

1. Create a new label: the "urban and landscape soils" that will recognize the quality of "urban and landscape terroir" in the periurban areas in which, the communes, departments, regions, and the state wish to carry out voluntary, concerted and contractual action. This label would be concretized by:
   - the signing of a charter applicable to a territory;
   - the definition of social, urban and agricultural objectives to respect to manage the space and requalify it when it is degraded;
   - the mobilization of cross-financing State-local authorities.

2. Develop the land policies of local authorities by:

    - -Including an agricultural and landscaping component in city contracts, in order to treat peri-urban space as a heritage and dynamic component of the city;
    - -Reconsidering the question of property assessments, in particular to enable municipalities to purchase l and if they so wish and facilitate the maintenance of agriculture;

3. Create regional environmental public institutions

    These agencies would ensure the protection of peri-urban green and agricultural areas under;
    Allocate them a share of the proceeds of the departmental tax of sensitive natural areas.

References

1. Rodrik D (2013) Unconditional convergence in manufacturing. Q J Econ
2. Kruger J (2008) Productivity and structural change: A review of the literature. J Econ Surv 22(2)
3. Chabbi M (2006) Fonctions et usages des études d'urbanisme dans la production de la ville au Maghreb. In: Villes réelles, villes projetées. Fabrication de la ville au Maghreb
4. Philifert P (2010) La construction d'une politique de développement durable au Maroc: principes, traductions et contradictions. In: Expérimenter la «ville durable» au sud de la Méditerranée
5. Proulous C (2004) L'aménagement du territoire au Maroc: survol des politiques urbanistiques et environnementales, recources naturelles et protection du patrimoine. Université de Montréal.
6. Bennani M (2000) L'identité de l'espace public colonial dans les grandes villes marocaines. Ecole nationale supérieure du paysage
7. Beaujeu-Garnier J, Dkesch J et al (1953) Aspects de la géomorphologie du Maroc. In: L'information géographique, vol 17, n°3
8. Beguin H (1974) L'organisation de l'espace marocain. In: Espace géographique, N°2
9. Refass M (1996) L'organisation urbaine de la péninsule Tingitane. In: Publications de la Faculté des Lettres et des Sciences Humaines, N° 27
10. Naciri M (1999) Territoire: contrôler ou développer, le dilemme du pouvoir depuis un siècle. In: Monde arabe Maghreb Machrek, N°164

The Environment and Sustainable Development in Front of the Artificialisation of the Coastlines: Coasts of Tunisia, Morocco and Algeria

O. Ben Attia, F. Fersi, and H. Rejab

Abstract Throughout the world, the advent of mass tourism and, in particular, seaside tourism has led to significant economic and demographic growth. This growth was most often followed by urban sprawl along the coasts, resulting in dramatic changes in their landscapes. These transformations were essentially an artificialisation of the coasts and the emergence of a barrier between the littoral space and the hinterland. Tunisia, and in particular the Sahelian region, have not escaped this phenomenon, and since the emergence of this economic sector in the early years of independence, there has been a simultaneous coastal and was confined in a narrow ribbon along the beaches thus turning its back to the hinterland. Moreover, in addition to the profound transformations that this urban continuum has produced on the coastal landscapes of this region, it has had other negative repercussions. This is due to its high consumption of spaces, particularly in vulnerable sites, its disturbance in the exploitation of natural resources (water pollution, access to water, reduction of local natural biodiversity, reduction of the coastal agricultural area, reduction of the beach surface, etc.) and its socio-cultural impact through the marginalization of countryside. This impact of coastal artificialisation, notably on the environment and sustainable development is found not only in the Sahelian region of Tunisia but also in several countries of the Mediterranean basin, such as: Morocco, where tourist and resort urbanization has taken place in the form of a ribbon along the beaches and on former coastal agricultural land, and Algeria, where a diffuse urbanization generated in particular by the most widespread form of dwelling which is the form of individual dwelling, has invaded coastlines such as those of the city of Oran.

Keywords Artificialisation · Environment impact · Littoral agriculture · Sustainable development · Urbanization

---

O. Ben Attia (✉) · F. Fersi · H. Rejab
RU «HPE», UR2013 A06 ISA-IRESA-Sousse University, Sousse, Tunisia

M. Abu-hashim et al. (eds.), *Agro-Environmental Sustainability in MENA Regions*, Springer Water, https://doi.org/10.1007/978-3-030-78574-1_9

1 Introduction

The Middle East and North Africa (MENA) are two regions whose major cities are among the most populated areas of the world. This settlement has resulted in a very significant urban expansion, especially with regard to the Mediterranean cities, such as Tunisia, Algeria and Morocco, which suffer from the same principle problem which is the development of a disproportionate urbanization of the coast. This development is mainly due to the tourist infrastructures which is generally settled on arable land often with high potential.

This over-exploitation of the coastline leads to often irreversible consequences such as:

- The deterioration of biodiversity;
- The deterioration of landscapes;
- The deterioration of wetlands;
- The deterioration of the ecosystem …etc.

This development of coastal tourism infrastructure has therefore not been made for the sake of sustainability, given that we have noticed that, in recent decades, tourism has been turning more and more towards new concepts. We are no longer talking about seaside tourism alone (this concept is the main reason for the coastal polarization of tourist infrastructures) but also about other concepts among them cultural tourism, green tourism and agritourism.

Tunisia has important potentials that could be essential in promoting these new concepts. Among these potentials: coastal agriculture. Indeed, the Sahel region is characterized not only by its tourism but also by its agriculture located on the seafront and which is entirely productive. This agriculture could play, in addition to its primary function of food production, other functions that can be assets for the promotion of tourism respectful of the environment and perfectly in agreement with sustainable development of the region. This agriculture is special because of its layout in relation to the sea and the landscapes that it forms in this region. It could become an integral part of the development of this region but not a land reserve. It could be also one of the links in the chain that would lead to environmental protection and sustainable development of this region.

2 Materials and Methods

2.1 *Location of the Study Site*

The Sahel region, the subject of our study, is substantially formed by the hinterland of the old cities of Sousse, Monastir and Mahdia. It is not only a natural region, but it is also and above all a region that owes its originality to its traditional shrub-growing economy and its habitat of villages and densely agglomerated villages. The latters

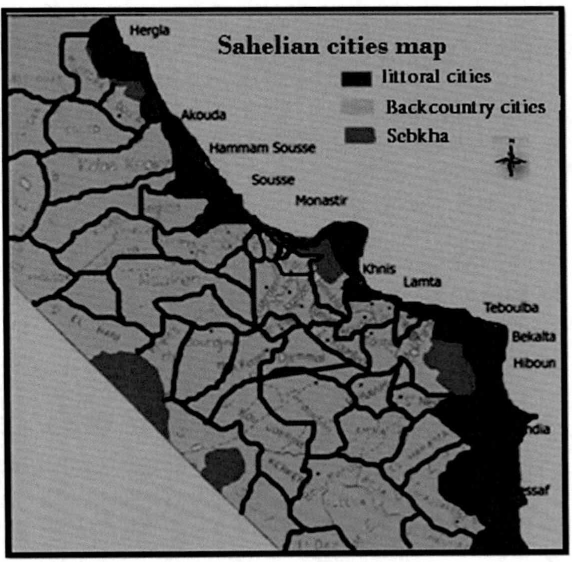

Fig. 1 The Sahelian cities [1]. Looking at this map, we can see that the further we get from the coast, the larger the size of the cities, the coastal towns are represented by the dark greycolor, while the backcountry towns are represented by the light grey color

have, as shown in the following map, a territory which becomes more important as we get further from the coast [1] (Fig. 1).

2.2 *Agri-Landscape Analysis Parameters*

This study work was essentially done in two stages. At first, we began with a bibliographic, iconographic and cartographic analysis that allowed us to understand both the spatiotemporal evolution of the region, to evaluate the degree of spatial transformation that it knew and to estimate the impact of urban growth on the ecology of itsenvironment. Secondly, we carried out an analysis of the bibliographic corpus relating to the evolution of the urban and peri-urban territories in some of the cities of the MENA and, essentially, in the littoral regions of Morocco and Algeria.

3 Results and Discussion

3.1 *The Tunisian Sahel: An Essentially Rural Territory*

Until the early years of Independence, the traditional economy of the Sahel region was mainly based on agriculture and, in particular, olive growing, livestock farming and vegetable crops. This was also the case for all of Tunisia, which had just emerged from decades of colonialism. Thus, at that time, the coastal landscapes of the Sahelian region, in general, and the Sousse region, in particular, were formed, essentially, as the following figures show, by a succession of traditional vegetable crops plots (called *chatts*) and olive groves. The latter were particular and occupied most of the fertile coastal lands of the region [1] (Figs. 2 and 3).

However, this agricultural sector was in decline with an ageingOlivette and a traditional market gardening uncompetitive. The first government of independent Tunisia in 1962 undertook to modernize it in order to combat, in particular, the unemployment that was raging in the region. This modernizationwas accompanied by the setting up of several irrigated public perimeters for the production of mainly

Fig. 2 The agricultural footprint of the ancient surroundings of the *Sidi El Kantaoui marabout*. This old photo shows the ancient surroundings of the *Sidi El Kantaoui marabout*, in the town of Hammam Sousse, under the city of Sousse. The observation of these photos shows us the agricultural footprint in this littoral zone. Indeed, vegetable gardens stretched all along the coastal fringe and were followed by olive groves. Urbanization was almost absent, and agriculture seemed to spread throughout the country [1]

Fig. 3 Section of the sea to the plateau level of the coast of the town of Hammam Sousse. This section, which presents the area of *Sidi El Kantaoui* before the development of the marina, tells us about the land cover at that time. In fact, according to this section, the border dune was followed by traditional vegetable gardening followed by olive groves. This provision was that of most of the coastal agricultural parcels of the Sahelian region [1]

early and late season vegetable crops. The setting up of these perimeters began with the construction of the *Nebhana* dam, located in the Kairouan region, which allows the mobilization of water needed for irrigation. The canal, which carries this water, passes essentially, as the following map shows, by the coastal towns. The choice of this route is that of the fertility of coastal lands, the favorable climate for off-season vegetable crops and the technical skills of farmers in this region. [2] (Fig. 4).

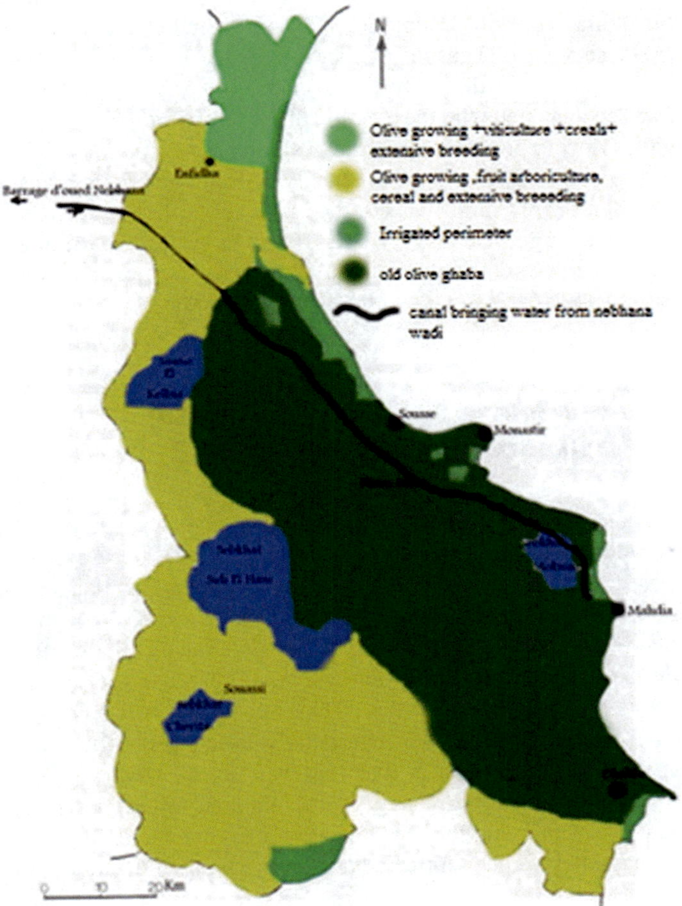

Fig. 4 Map of the land occupation of the Sahel region, 1981 [1]. This map of 1981 shows the predominance of olive plantations, which corresponds to dark green colored spaces, in the Sahelian region. This culture is the most important of the coastal fringe where it is often located behind the irrigated perimeters, the colored spaces in light green, which are maintained mainly on the sea front. The canal carrying the water of the dam of the *WadiNebhana* passes through the littoral zones. Cereal farming, livestock farming and fruit growing are mainly located in the hinterland

The establishment of this perimeter competed with traditional plots that could not compete with their traditional crops based on irrigation from surface wells. From this period, the latter experienced a decline in production competing with intensive crops. On the other hand, this marginalization of traditional cultures hasbeen accentuated by the development of secondary and tertiary sectors and, essentially, the tourism sector which has allowed the creation of new jobs and the revival of others. From the beginning, this area was installed on the waterfront lands in response to a demand for seaside tourism. The development of this sector was preceded by the opening, in 1968, of the Dkhila-Monastir International Airport which would allow the transport of tourists to these regions. Thus this tourist activity has been at the base of coastalurbanization that has often settled on fertile farmland, old orchards and market gardens. This trend was accentuated with the creation of the seaside resort of Port *El Kantaoui*, in 1979, with a view to diversification and development of tourism. This station was built on the site of *Sidi El Kantaoui* in the city of Hammam Sousse, a site that was, as we saw earlier, deeply rural. Indeed, the installation of this station was made on 320 ha of farmland largely private [3]. Thus, from this period, traditional agriculture will increasingly give way to urbanization and we will witness an overexploitation of the coastline, which will not be without consequences.

3.2 The City of Sousse: A Linear Development and Marginalization of the Hinterland

From the second half of the twentieth century, the world has experienced significant economic development that has led to the development of leisure activities and, in particular, mass tourism mainly in its seaside appearance. The main consequence of this development was a significant demographic and economic growth in the coastal towns, which mainly resulted in a strong urban sprawl, which most often took place on agricultural land and therefore on fertile land.. This urban sprawl has led to often dramatic changes in coastal landscapes. This state of affairs has not spared Tunisia whose urban growth has generated, particularly in the case of the Sahelian coast, changes in the coastal landscapes that followed, as we have seen, two essential periods of the development of this region. It is the first followed the modernization of the agricultural sector and the intensification of vegetable crops, and the second, which had the greatest impact, followed the development of the tourism sector in the region.

The study of this region has shown us the spectacular transformations that it has experienced during the last fifty years since it has gone from an agricultural region to one of the most important tourist centers in Tunisia. Indeed, the city of Sousse has always been one of the most important cities of Tunisia and it is considered the capital of the Center-East of the country because of its geographical situation. Its core, like the majority of Tunisian coastal cities, was built, as shown by the following ancient map, around the port which has continuously allowed a significant commercial activity. From this core, the city has developed along its coasts and mainly to the North East but initially, with a certain withdrawal from the sea (Fig. 5).

Indeed, the tourism sector has spawned significant spatial transformations that are taking the landscape of this region, as the following maps show, from predominantly rural landscapes to deeply urban landscapes. Tourist urbanization stretches along a narrow ribbon that forms a screen between the sea and the hinterland (Fig. 6).

This peculiar form of agriculture called *chatts*, or even *ouljas*, which occupied the entire northern coast of Sousse, and which constituted a green and open area of the littoral of the region, is ignored today by the public authorities as it is shown in the next map is enclosed in residual spaces inside the urbanization (Figs. 7 and 8).

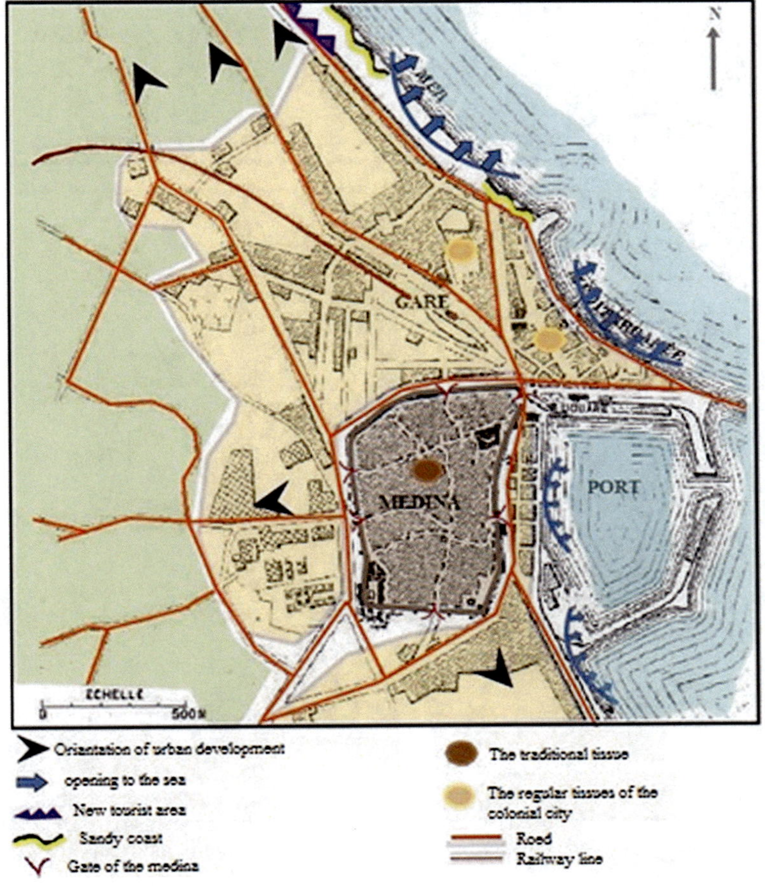

Fig. 5 Map of the city of Sousse from 1928 [1]. As this map shows, the urbanization of the city of Sousse in 1928 revolved around the port and stretched along axes parallel to the coast. Urban development has essentially followed the direction of the north-east coasts of the city on which the new tourist area has been established

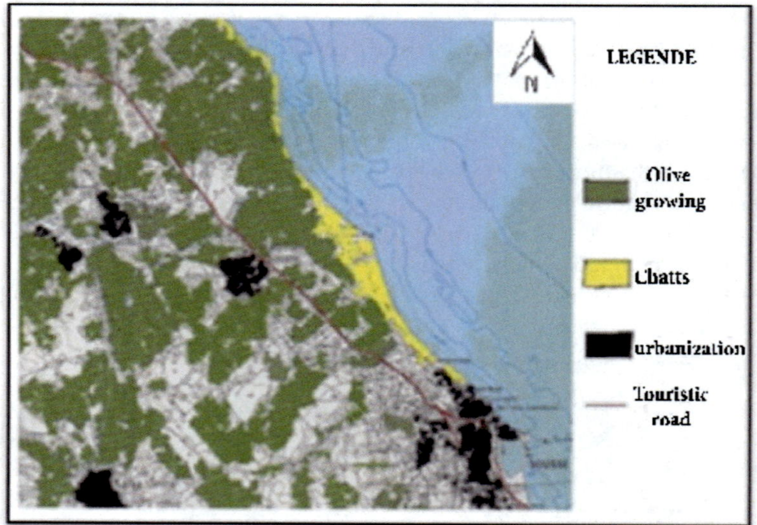

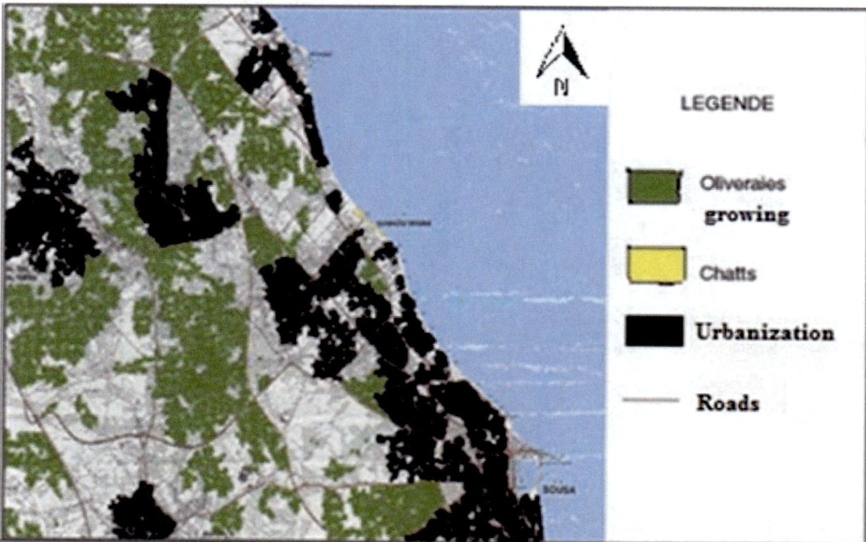

Fig. 6 The city of Hammam Sousse, north-east of Sousse, in 1958 and 1994 [1]. As the first map of 1958 shows, the coastal fringe consisted of *chatts* that covered most of the space. Urbanization was set back from the sea, on the other hand, on the second map of 1994, we see that coastal urbanization has become large and dense, olive groves have become diffuse and in the form of tasks and the *chatts* have almost gone

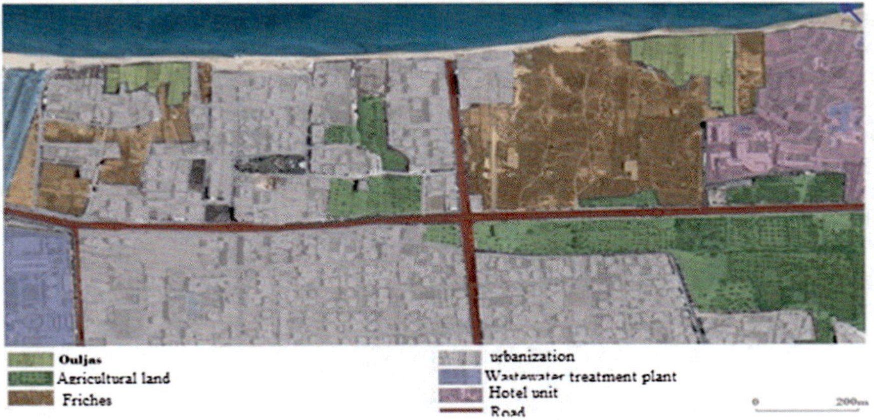

Fig. 7 Regression of *ouljas* in the northeastern region of Sousse (2010) [1]. While until the 1970s, they were along the entire coast of the Sahel region, *ouljas* (or *chatts*) occupy today only a few residual areas enclosed between urbanization

Fig. 8 The Hammam Sousse *Chatt*, Tunisia. [1]. The *chatts*are located at a distance of ten meters from the sea. They are delimited in relation to the latter by a dune edge consolidated by herbaceous plants. This dune edge also plays the role of the windbreak. In these plots, divided into crop rectangles, are planted several kinds of vegetable crops. It is ancient agriculture that often continues to use traditional farming methods

The particularity of these *chatts* was their proximity to the sea, their presence and maintenance within the urbanization itself, their mode of exploitation and the strategies used by their farmers

This form of traditional agriculture is considered, as can be seen on the following map, as a land reserve for tourism and resort town planning [1]. As Ezzeddine Houimli emphasizes in his thesis, "*this agriculture, now in decline as a result of seaside town planning, was mentioned as vegetable gardens on the topographic map of 1958. It is no longer on the topographic map of 1994 while the 1996 aerial photos, combined with the field observation, show the permanence of part of this agriculture currently (in 2006)*" [4] (Fig. 9).

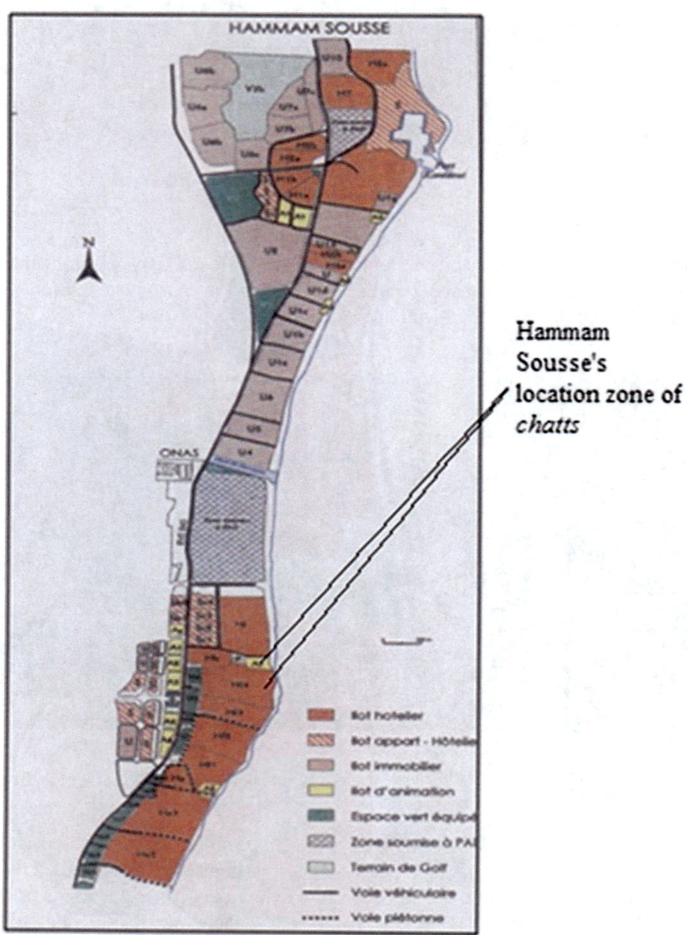

Fig. 9 Extract of the development plan of the city of Hammam Sousse in the late 2000s [1]. This development plan carried out by the Municipality of Hammam Sousse shows us that the municipal authorities consider these traditional waterfront vegetable parcels as a land reserve for tourism. The whole area has been reserved for a tourist destination

3.3 A Tourism Sector that Consumes Large Amounts of Space

Thus, thanks to its geographical situation, its great maritime opening, its long beaches of fine sand, the Sahel was one of the fiefs of the development of the tourism sector in Tunisia which was necessarily based on the seaside practices which resulted in a very important coastal polarization. The Sahel then passed from a country that seemed to turn its back to the sea to a coast that seems to turn its back to the hinterland (as shown in the following figure) leading to dramatic changes in the shoreline (Fig. 10).

Indeed, as we saw earlier, as soon as it was set up in the region of Sousse, tourist activity settled, as shown in the following map, on the coastal fringe, leaving the hinterland for Classic residential areas and agriculture. The majority of the hotels are located on the littoral zone forming a screen between the sea and the rest of the country (Fig. 11).

Also, the construction of the seaside resort of *El Kantaoui* port was at the origin of coastalurbanization which stretches on both sides of this station. This coastal urbanization was established on the *chatts* [1], in addition to their economic, social and landscape functions, played an ecological role in the preservation of the coastline: they contributed "*to the balance of environments particularly sensitive to climate change erosion and pollution*" (PAP/RAC 2005). It has been stretched, as the following map shows, in a linear fashion essentially on both sides of the Hergla-Sousse tourist route and, mainly, in the area between this road and the sea. Thus, "*the tourist-oriented urbanization of the seafront by an almost continuous chain of hotels and sub-hotel facilities has been a driving force in theurbanization processes*

Fig. 10 Urban planning of the coasts of the region of Sousse [1]. This figure assigns a color code to each of the three existing spaces: the first, delimited by a blue line corresponds to the sea, as to the red, it sets the limits of the space invested by the hotel sector and, finally, the grip of the olive groves originates from the green line. The observation of this figure, located north of the Marina *El Kantaoui*, shows us that tourism urbanism is maintained on a linear fringe turning the back to the olive groves

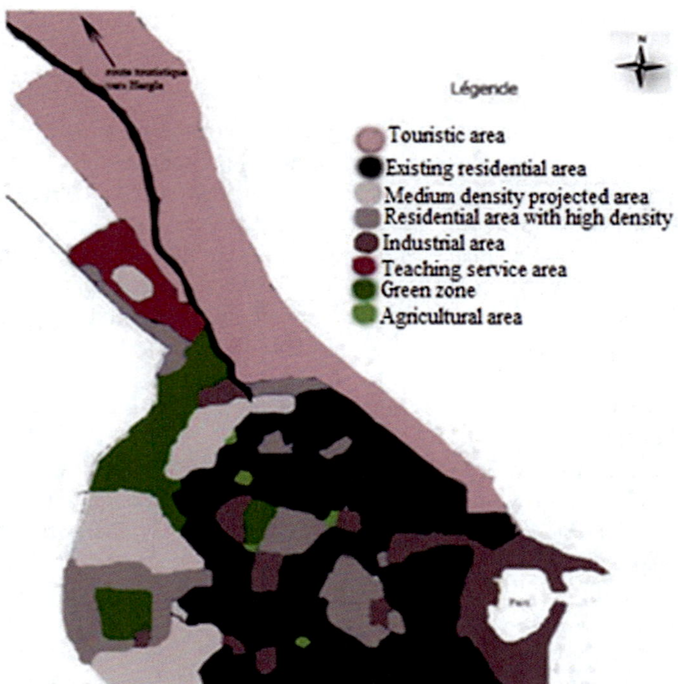

Fig. 11 The development plan of the city of Sousse in the early 1980s [1]. This development plan of the city of Sousse in the early 1980s shows that the coastal fringe is exclusively reserved for the development of the tourist area. The latter is limited by the Hergla Sousse tourist route. Classic residential urbanizationis relegated to the hinterland. It is separated from the sea by the tourist zone

of the northern suburbs of Sousse and the eastern suburbs of Hammam Sousse (El Menchia)" [5] (Fig. 12).

Also, this tourist urbanization has been at the base of the expansion of resort urbanization that has developed, often anarchically, on the coasts of the municipality of Akouda. It is mainly made up of villas and apartment buildings dedicated to coastal resorts and located, without prior planning, according to the availability of land [6]. This holiday resort was set up both on traditional waterfront vegetable parcels and on market gardens that were part of the public irrigated perimeter of Chott Meriem, hence [1]:

- By settling on these parcels of the irrigated public perimeter, this urbanizationwas installed on a project that required significant investments and major implementation work carried out by the State. According to the 1983 Farmland Protection Act, these lands fall into the blackout zone category. However, the spread of urbanization in this area is achieved with the consent of "local officials such as the *Omda*", the latter is the head of the smallest territorial subdivision on the scale of

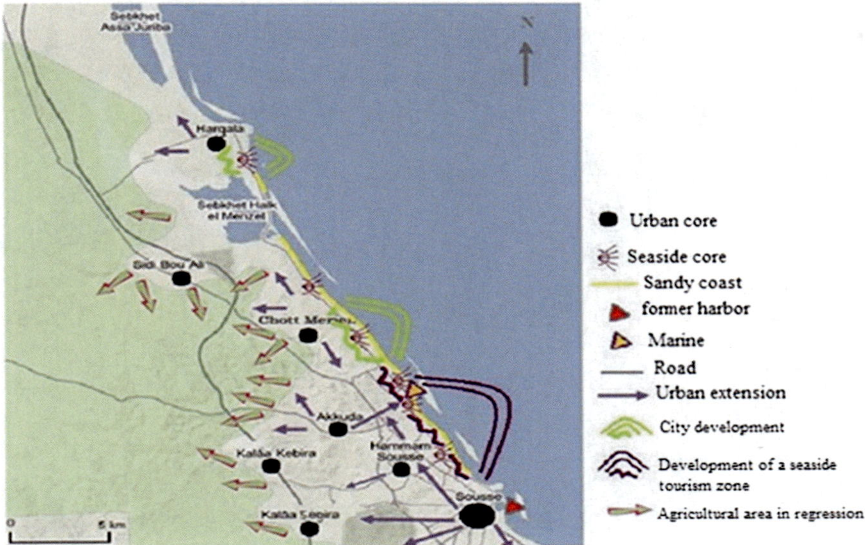

Fig. 12 Summary map of the northeastern region of Sousse [1]. While the urban cores are relatively behind the sea, resort areas and seaside tourism are located on the coastline resulting in a regression of farmland and their retreat to the hinterland

the delegation, which in turn, is an administrative subdivision of the governorate [4];

- In parallel with the irrigated public perimeter, urbanizationwas mainly based on the *chatts* along the coast of Akouda commune. The urbanization of these plots is the consequence of the lack of interest of local authorities in this traditional agriculture. This lack of interest materialized, in particular, by the introduction of these zones into the development plan as a tourist zone. This was also the case of the few *chatts* that persist in the coastal area of Hammam Sousse and that the municipal authorities consider as developed space which was introduced, as we saw earlier, in the development plan of the city of Hammam Sousse as a tourist area.

The result of this coastal urbanization is, as the following maps show, an artificialization of the coastal fringe Sousse Akouda: the *chatts* of this zone have been replaced by a tourist urbanization organized next to an urbanization of resort often anarchic particularly in this which concerns the coastal fringe of the commune of Akouda [1]. Moreover, as we will see in what follows, this coastal polarization has had a lot of negative repercussions on the environmental balance of this area (Fig. 13).

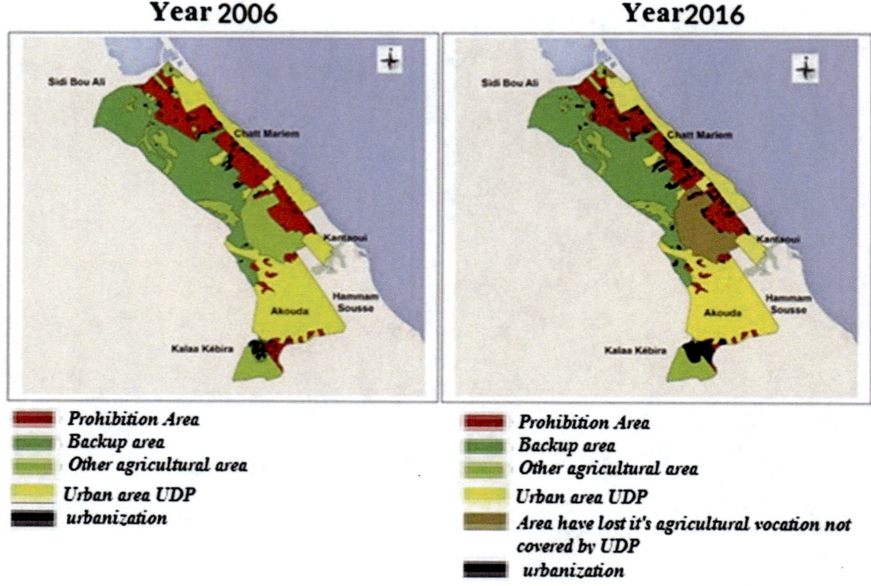

Fig. 13 Comparison between land use maps of the Akouda littoral fringe from 2006 to 2016 [1]. These two maps show us urbanization that continues to invest the agricultural area of the coastal area of Akouda commune

3.4 An Artificialisation of the Coasts and Its Environmental Impact

Indeed, the tourism sector is certainly the source of an indisputable contribution to the national economy but, at the same time, it has negative repercussions on landscape and environmental and socio-cultural plans. In fact, besides the consumption of spaces, particularly in vulnerable sites, such as the coast, the activity in question often causes disturbances in the exploitation of natural resources: water pollution, access to water, reduction of local natural biodiversity, coastal agricultural area, etc. [1]. Thus, from the first years of the implementation of this sector, several consequences were raised [2], which we quote (Figs. 14 and 15):

- A tourist location both coastal and linear to meet the needs of seaside tourism. This location has resulted, as we have seen previously, in the urban sprawl of the city of Sousse along the north-east coast several kilometers long and a few hundred meters wide. This urbanization is dense and compact, as shown in the following figure, and consumes large spaces, especially agricultural;
- A clear improvement of infrastructures and urban equipment: roads, post office, sanitation network, commerce, artistic crafts, travel agencies, car rental agencies, banks, restaurants, …;

Fig. 14 Panoramic view of the coastal fringe of Chott Meriem [1]. The construction of the seaside resort of *El Kantaoui* was at the base of the development of a compact and dense coastal resort on the coastal fringe of Chott Meriem

Fig. 15 The decrease in beach surface south of Port El Kantaoui. [1]. As this picture shows, the holiday resort located south of Port *El Kantaoui* settled directly on the maritime coastal area (DPM), destroying the dune edge and resulting in the reduction of beaches that, in some places, have completely disappeared

- The creation of direct and indirect jobs which has contributed to the reduction of the unemployment that the region knew until the first years of the Independence, the elevation of the standard of living in the region and the reinforcement of the weight of the non-agricultural activities in the economy of the region;
- The acceleration of coastal erosion, as shown in the following figure, due, on the one hand, to the occupations of the seafront by tourist and resort urbanization and, on the other hand, to the destruction of the dune for the construction of the tourist route and the hotel infrastructures;
- The rejection of swimming pool waters degrading spontaneous plantations on the beach, which play an essential role in fixing the latter;
- The privatization of beaches by hotel units that block access to the inhabitants of the region;

- The installation of hotel infrastructures on agricultural land containing a large number of olive trees and traditional vegetable garden plots based on the exploitation of surface wells of the groundwater (*ouljas*);
- The expropriation for public purposes of certain parcels to their owners who did not want to sell their land for tourism;
- The competition with agriculture irrigated perimeters in the water consumption of the dam of the *wadi Nebhana*, etc.

Thus, the tourism sector of the Sahel region was therefore based mainly on seaside practice, which has led, as we have seen previously, to significant coastal urbanization. However, with the new concepts of sustainable tourism, especially those relating to the protection of the environment, new concepts related to the tourism sector have emerged, including agritourism, which is beginning to develop more and more in Tunisia. It is in this sense that, in this region of the Tunisian Sahel, while for several decades, tourism turned its back to the hinterland, today; with the interest that some tourists seem to bring to the agricultural area, it uses the latter for the development of some of its infrastructures. However, due to its peculiarities, Sahelian coastal agriculture could play a major role in diversifying the tourism offer by becoming multifunctional, that is to say, not only productive of food products, but also landscaping and environmental services to tourists and residents by enhancing the landscape of this region. Notwithstanding its potentialities, the Sahelian agricultural area continues to be exploited by the tourism sector, either as a framework, or as support or subject of new activities that it offers to its customers in order to diversify its activities.

However, this artificialisation of the coastlines by the spread of urbanization through the agricultural area is not the only fact of Tunisia. This is also the case in Morocco, where resort urbanization spreads across agricultural land, and from Algeria where urban sprawl is at the expense of agriculture.

3.5 The Sprawl of Moroccan Coasts

The phenomenon observed in the coastal plain of Ksar Sghir, in northern Morocco, was a sprawl of agricultural land. In fact, at the end of the harvest, the farmers rented their plots to the campers to whom they sold well water and fresh vegetables. This phenomenon has evolved with the arrival of urban speculators who encourage farmers to sell their plots. This is how anarchic second homes have emerged at the expense of cultivated land. At the same time, a development plan plans to install a secondary residence station in this region, which has led to a conversion of lands into social wastelands. [7] (Fig. 16).

Also, the coastal area of Oued Laou in northern Morocco has, for its part, been more or less marginalizedin terms of tourism investment. This area is made up of valleys opening on the sea by plains where the inhabitants lived of an irrigated agriculture and national tourism which took importance from year to year with the

Fig. 16 Ksar Sghir (Northern Morocco), a coastalurbanization within former agricultural plots [1]. These two photos show us both a resort that settles on old agricultural parcels and agricultural parcels that are converted into wasteland waiting to be urbanized thus inducing profound changes in the landscapes of this farming village in the North Moroccan

development of anarchic urbanization to the detriment of agricultural land and, in particular, the irrigated area [7] (Fig. 17).

Thus the question of the preservation of agricultural land, located near the urbanization front, has certainly been identified by the public authorities in Morocco, whether national or local, but it is not a primary concern of public action, which, for its part, is more focused on economic development and housing, [8], it is also the case, as we will see, of Algeria.

Fig. 17 Oued-Laou: an important coastal resort located in the heart of an irrigated perimeter [1]. In this photo, we observe coastalurbanization that has settled on land that was irrigated agriculture. This urbanization seems to form a narrow ribbon along the beach. At the rear of this urbanization, land, which seems to be old agricultural parcels, are left undeveloped while waiting to be urbanized. In addition, we can see in this picture two sets of dwellings: the first, on the right, seems anarchic while the second set, located in the center of the photo, seems to us more organized bearing the imprint of the real estate developers but also installing on old agricultural parcels

Fig. 18 The sight of the sky of the city of Oran in Algeria [1]. This photo shows us the importance of the urbanization of the city of Oran, diffuseurbanization, which favors the form of individual housing and is therefore much more space-consuming

3.6 Algeria: Development of Cities on Their Coastal Areas

According to Fafa Rebouha, the city of Oran in northern Algeria is expanding by ignoring its agricultural heritage whose architecture and landscapes remain coveted by urban dwellers in search of individual homes. Thus, the growth of urban units and large urban complexes adjoins the rural area, which becomes the reception area for inhabitants seeking housing and whose agricultural element has become a minority. Indeed, and still according to the same author, the peri-urbanization of the city of Oran was made on the field of individual and collective farms, projects designed for rural areas aim to densify these areas by bringing a population looking for cheaper housing. Also, the ambition of a real estate, offer for more compact housing and subdivision forms less space consuming and more in line with the needs of sustainable development is not yet reached in the urbanization of Algerian cities. Thus, the search for a single-family home has accelerated migration to rural communities: those located on the border of the city of Oran register a significant increase in their population given the proliferation of housing programs that are prioritized by the disappearance of the agricultural sector, since these rural communes bordering the city of Oran benefit from a land use plan that allows this disappearance of agricultural land [9] (Fig. 18).

4 Conclusion

Thus, the coasts of the Sahelian Tunisian region have undergone spectacular transformations throughout the decades that followed the independence of the country.

Indeed, from an open green space, they have become the capital of tourist and resort urbanization. These transformations have not been without consequences including social, landscape and environmental. Indeed, the installation of this urbanization has led to an artificialization of the coast; a barrier seems to have been erected between the coast and the hinterland. This barrier, formed by a narrow cordon of urbanization, has relegated coastal agriculture to the hinterland and has also given rise to a form of social segregation since most of the coastal territory has been invested by the tourist sector which privatizes the beaches.

Also, this artificialisation has led to profound changes in the coastal landscapes of this region, since they have moved from predominantly rural landscapes to deeply urban landscapes in the space of a few decades. All these transformations have not been without environmental consequences. The precarious balance of this portion of the territory has been weakened by the large consumption of space of tourist and resort urbanization and by the acceleration of coastal erosion, mainly due to: the location of the urbanization in front of the sea, the destruction of the coastal dune, the degradation of spontaneous plantations by the discharge of swimming-pool water, etc.

However, until the end of the 1970s, coastal and traditional agriculture occupied most of this territory and, in addition to its social and landscape functions provided an environmental balance for this fragile area. Of this coastal and traditional agriculture (the *chatts* or *ouljas*) there remains, to the present day, only some plots witnesses of formerly rural territory. These plots could play a key role in the development of sustainable, environmentally respectful tourism. Indeed, this particular and rare form of agriculture could be considered as an agricultural heritage and introduced as a territorial component in the development of the city of Hammam Sousse.

However, this state of affairs is not specific to Tunisia, in general, and to its Sahelian region, in particular. Indeed, the urbanization of the coastlines and its social, landscape and environmental impacts are a phenomenon that affects the majority of countries in the Middle East and North Africa and, in particular, Morocco where coastal farmland is giving way to tourist urbanization. And often anarchic resort and, Algeria, where diffuseurbanization, made mainly of the form of individual habitat, spreads on the coastlines of the city of Oran.

5 Recommendations

The tourism development of the Tunisian Sahel region has of course had major repercussions on the economy of the region, which has become one of the most important Tunisian tourist poles, but this has not been without consequences both on the social level and 'ecological. Indeed, this development has led to disruption of the natural environment and a social mix which has triggered conflicts. However, as we have seen previously, this territory is the capital of an urban agriculture which can help to maintain nature in the city: "it thus increases environmental services and therefore constitutes an important vector of resilience of cities" [10].

In addition, today, tourism demand has evolved into other areas of interest. Indeed, new concepts applied to the tourism industry have promoted its sustainability through the development of "sustainable tourism" based, according to the World Tourism Organization (WTM 2005), on respect, preservation and the development of the environmental, economic and socio-cultural aspects of the countries. Tunisian tourism, in general, and Sahelian tourism in particular, must therefore define new directions for its development.

It is therefore a question of halting the increasing densification of tourist areas and alleviating the pressure exerted on several sites, including, in particular, the natural and agricultural sites of the Sahelian coast. The latter are, moreover, the territory of traditional coastal agriculture which should be protected and introduced as an integral part of this new tourist offer, based on the multifunctionality of agriculture and in accordance with the rules of sustainable development. Today, this new tourist offer is in demand "for its ability to share cultural values and diversity, as well as to make known and recognize natural and cultural heritages" [11].

References

1. Ben Attia O (2016) Interactionsbetween the evolution of Tunisian coastal agriculture and the development of urbanization and tourism. The case of peri-urban areas north-east of Sousse and southeast of Monastir. PhD thesis in Landscape, Territory and Heritage. Higher Institute of Agronomy Chott-Mariem, Sousse, Tunisia
2. Jedidi M (1986) Economic growth and urban space in the Tunisian Sahel since Independence. Publication of the University of Tunis. Tome I, Tunis
3. Sethom H, Kassab A (1981) The geographical regions of Tunisia. Publication of the University of Tunis
4. Houimli E (2008) The factors of resistance and fragility of coastal agriculture in the face of urbanization: the case of the region of Sousse Nord in Tunisia. PhD Thesis in Landscape Science and Architecture. Institute of Life Sciences and Industries and Environment (AgroParisTech). Directed by Donadieu, P
5. Lamine R (2000) Cities and townspeople of the Central Sahel. Acts Collection. The gold of time. Tunis
6. Toumi S (2007) Public spaces, places of cultural diversity in the territories of tourism. Comparative approach in the Sahelian territory. Master's in Landscape, Territory and Heritage. Higher Institute of Agronomy Chott-Mariem, Sousse, Tunisia. Co-directed by Donadieu, P. et Vidal, R
7. Berriane M (1996). Development, urbanization of the Mediterranean coast and environment in Morocco. In: Ben Ali D, Di Giulio A, Lasram M, Lavergne M (eds) Urbanization and agriculture in the Mediterranean: conflicts and complementarities. l'Harmattan. Paris
8. Valette E, Chéry PP, Debolini M, Azodjilande J, François M, El Amrani M (2013) Urbanization on the outskirts of Meknes (Morocco) and become agricultural land: the example of the agrarian cooperative Naïji. Agriculture Notebook. Volume XX. N°8
9. Rebouha, F. (2005). Mutation of rural space near urban space: the case of the city of Oran. In international seminar cities and territories: changes and current issues. LaboratoryPUVIT. Sétif
10. Gangneron E, Mayol P (2019) Urban agriculture: a decisive tool for sustainable cities. CESE 15

11. Marcotte P, Bourdeau L, Sarrasin B (2017) Tourism and sustainable development. An exercise of adaptation, integration and reconciliation. TEOROS Tourism Res J

Practices for Sustainable Plant and Soil Production

Sustainable Agriculture in Some Arab Maghreb Countries (Morocco, Algeria, Tunisia)

Messaouda Benabdelkader, R. Saifi, and H. Saifi

Abstract The Arab Maghreb countries (Morocco, Algeria and Tunisia) are located in a fragile zone, in majority semi-arid or Saharan. One of the priorities of the governments of these countries after their independence was the interest of the agricultural sector to ensure their food. They have devoted their efforts to developing laws, plans (PMV, PNDA and PDAR) and strategies to this end. However, policies were not sufficient to achieve self-sufficiency. Land users were not as up to updates of agricultural development. Using intensive agriculture, this has been accelerating in recent years. Farmers have deeply plowed the land, causing the destruction of the physical and chemical structure of the soil. They irrigated at random crops with systems that save not water, resulting in water depletion, organic and mineral matter, soil degradation and salinity in large areas. Through their excessive use of mineral fertilizers and phytosanitary products, they have caused edaphic, atmospheric and aquatic pollution, including contamination of soils, groundwater and surface water by nitrates and phosphates; thus intoxication and cancerous diseases announced by medical reports and university studies; not to mention the deforestation to carry out certain agricultural or industrial activities, by eliminating the biodiversity of the flora and the fauna, causing a climate change and a complete change of the ecosystem. These problems push the countries of the Arab Maghreb out of the path of sustainable agro-socio-economic development, which produces, preserves the environment and natural wealth for future generations. Although there were tests to mitigate the risks of pollution and climate change, deserves to be expanded and deepened as the sub-projects of integration of climate change in the implementation of the Green Morocco Plan (PICCPMV), the adoption of administrative, legislative and economic provisions for the effective and sustainable management of water resources, including the

M. Benabdelkader (✉)
Laboratory of Applied Microbiology, Faculty of Sciences, Ferhat Abbes University, Setif, Algeria

Department of Environment and Agronomy, Faculty of Nature and Life Sciences, Jijel University, Jijel, Algeria

R. Saifi
Department of Agronomical Sciences, Mohamed Khider University, Biskra, Algeria

H. Saifi
Department of Biological Sciences, Tunis El Manar University, Tunis, Tunisia

M. Abu-hashim et al. (eds.), *Agro-Environmental Sustainability in MENA Regions*, Springer Water, https://doi.org/10.1007/978-3-030-78574-1_10

pricing of irrigation water, and the National Irrigation Program for Water Conservation (PNEEI) and the Council Higher Water and Climate (CSEC) in Morocco. Measures considered in the National Planning and Predicted Contribution (INDC) submitted at COP21 in Tunisia, which provides for the agriculture, forestry and land use sector, a mitigation plan that includes the intensification of CO_2 absorption capacities of the forest and arboriculture, thanks to the intensification of reforestation, consolidation and increase of carbon stocks in forest and pastoral environments. An important agricultural heritage peculiar to these countries, it is still cultivated so that the ancestors respects the environment, needs attention and good management to intensify it, as the cultivation of dates, olives, figs, citrus fruits, cocoa and even cereals. Sustainable agriculture requires the organization of these countries: it is necessary to apply on the ground with sincerity the laws of the good management of the water in quantity or quality, either for drinking or for the irrigation of the cultures; cultivate environment-adaptive species in arid zones or climate-sensitive areas; the most important thing is to inform farmers, how to save crop irrigation water by preferably using the drip system, how to limit the use of chemicals (fertilizers and pesticides) by replacing them with organic or natural products, or apply rotations, or select healthy seeds for varieties that are adaptable to the changing environment and resistant to diseases.

Keywords Sustainable agriculture · Environment · Arab Maghreb · Climate change · Pollution · Degradation · Farmer training · Adaptive varieties · Sincere policies

1 Introduction

Agriculture is a human activity that modifies natural ecosystems to produce resources (mainly food) useful to humans. It occupies a large terrestrial surface, with important consequences for the man. The impacts of agriculture on soils and biodiversity have been around for a long time. It was in the 1940s that the first synthetic pesticides were marketed, marking an increase in agricultural yields. Twenty years later, started to Hear harm to people's health and on the environment [1]. Since 1945, the increase in the use of mineral fertilizers, the appearance of pesticides, and the development of irrigation (in the context of the Green Revolution, in particular) and the motorization of agriculture has increased sharply the environmental impacts [2].

Over the past few decades, agriculture has seen a noticeable change and over time has become the main economic sector responsible for multiple environmental impacts. According to Debailleul [3], concentration, specialization and intensification describe this change. Concentration, manifested in a continual increase in the size of farms and a rapid decrease in their number, specialization has led farmers to devote themselves to one or two productions when previously they had a wide range; the third resultant was the increasing use of chemical fertilizers and pesticides. Farmers have used pesticides more extensively to destroy unwanted insects for crops, but unfortunately, this often makes the situation worse as many insects develop resistance to pesticides, and pesticides are only effective for a short period of time [4].

Phytosanitary products are most often applied in the form of liquids sprayed on plants or on the soil. In other cases, they are incorporated, injected or deposited in the soil in the form of granules. During an application, the treatment product was distributed in variable proportion [5]. As soon as they reach the ground or the plant, they degrade or disperse. The active ingredients can volatilize, flow or infiltrate into surface or ground water, absorbed by plants or soil organisms, or remain in the soil. During a season, runoff carries on average 2% of a pesticide applied to the soil, rarely more than 5–10% [6, 7], leaching losses are generally lower [7, 8]. On the other hand, sometimes volatilization losses of 80–90% of the applied product observed a few days after treatment [8, 9].

In general, the environmental impact of a pesticide depends on the degree of exposure (resulting from its spread and concentration in the environment) and its toxic properties [10, 11]. Among of the agreed environmental risk assessment procedures is to measure exposure to pesticides and their effects [12].

Organochlorines are considered a dangerous pesticide family for humans, even if they are in small quantities in water or foodstuffs. Organochlorines are most present in areas where agricultural waterways are located [13, 14]. In these rivers, several pesticides exceeded their limits for the protection of aquatic life [13, 15].

Excessive use of phosphorous in the form of fertilizers and mineral fertilizers leads to saturation of the soil and causes the enrichment of surface water in agricultural ponds [16]. In rainy periods, compost spread brings a much greater amount of phosphorous from runoff to streams [17].

Intensive agriculture exposed Arab Maghreb countries (Morocco, Algeria, Tunisia) to chemical pollution in irrigated areas: contamination and degradation of water ground, watercourses and soil, mainly due to the use of mineral fertilizers and crop protection products [18–23].

Agriculture accounts for about 20% of greenhouse gas emissions in European countries. This is largely due to the emission of nitrous oxide after the application of N fertilizers [24]. Agricultural ecosystems are both sources and sinks of air pollutants and greenhouse gases, and thus play a role in the relationship between climate and air pollution, that is, they contribute to climate change and air pollution [25].

As many countries in this planet exposed to climate change, due to the bad management of this fragile agricultural sector, the reduction of yields in Arab Maghreb countries depends on climate changes [20, 21, 23, 26]. Without forgetting the lower levels of the agriculture users, they had the hard effects on agro systems then on environment [23].

To reduce the effects of random agriculture, many studies have advocated sustainable agricultural development that produces and preserves the environment [27, 28]. Where is the Arab Maghreb countries from this sustainable agricultural development, and what are agricultural sector impacts on environmental compounds in this region?

2 Environmental Impact of Agriculture

In order to increase agricultural production, people add to the soil some products such as fertilizers or pesticides, which leads to an increase in pollution. Many chemicals are used to prevent and control plant pests. Pesticides are phytosanitary products, which are introduced to various parts of the environment by air, water and soil pollution [29]. The presence of phytosanitary products in the natural environment is mainly caused by humans, while fertilizer components, such as nitrogen or phosphorous, are part of the natural biogeochemical cycles. When there is a phytosanitary product in the environment, its equilibrium is disturbed and the possible consequences for human health and natural resources (animal, plant and minerals) [30, 31]. With the recent evolution of treatment products and their consequences on ecosystems [31, 32]: the contamination of air, soil and water by pesticides widely used in agriculture is today more than never central to all debates about environmental protection, it is a chronic and widespread problem [33].

2.1 Soil Degradation

Agriculture is also responsible for soil pollution and degradation. The tillage process improves the physical structure of the soil, which facilitates exchange on the soil surface (air, water and fertilizers) [34]. However, too deep plowing causes the appearance of a hard layer by the action of the natural settlement and the passage of gear for the treatments, the grape harvests, etc. [35, 36]. This compaction phenomenon is accentuated in the case of plowing on wet soil [37].

The soil is completely compacted by removing the weed by chemicals [37], and the harmful effect is very noticeable on the life of the soil microflora and microfauna. The risks of pollution are by the infiltration of molecules of groundwater into the soil and by surface runoff (rain, erosion) are not negligible [38, 39]. Finally, the use of these active molecules is not without danger for the crop (risks of phytotoxicity), the agricultural operators (risks for the health) and the environment in general [40].

Chemical degradations are essentially photo-decompositions, or degradations in contact with organic matter or clays due to a combined effect of acidic conditions and adsorption phenomena [41, 42].

The work of Siimes et al. [43] highlights the transfer of herbicides to agricultural soils (sugar beet cultivation) in Finland. In recent years, models have also been developed to simulate pesticide transfers into the soil based on doses and application dates in order to know and predict associated ecological risks [44–46].

Water erosion is influenced by cropping operations, such as the depth of the soil, the direction in which it is used, the time of plowing, the type of implements and the number of passes. Generally, the less tillage disturbs vegetation or the tailings layer at or near the surface, the less soil tillage results in water erosion. Reduced tillage and no-till are effective ways to limit this type of erosion. Tillage practices in the direction of the slope favor water erosion by providing flow paths for the runoff. Counter-slope tillage and contour plowing techniques oppose the concentration of runoff and limit soil movement [47].

In arid and semi-arid regions, soils of irrigated perimeters can undergo three forms of degradation: salinization (by accumulation of soluble salts, mainly NaCl), waterlogging (by rising water tables), sodization/alkalization (by sodium ion fixation on clay associated with an increase in pH) [48].

2.2 Pollution of Water

Pollution of water by plant protection products creates environmental health problems. These molecules can also be transferred to groundwater and surface water: these two processes, intimately linked in the hydrological cycle, form the basis of the mechanisms that cause surface water pollution [31, 49]. These molecules can also be driven by rainwater and then be the cause of pollution of surface water but also seepage water and soil. The quality of the surface water and infiltration is then highly followed with regard to possible contamination by pesticides [49].

Among the most recent studies, we can mention those of Gao et al. [50] the future of organochlorine pesticides in surface water in China. Similarly, Kalita et al. [51] report pollution of drainage waters by pesticides in maize and soybean crops.

As can be seen from the numerous studies carried out, pollution related to agricultural activities, continued to grow and became one of the primary sources of pollutants that affect the health of rivers and human health. At the origin of the degradation quality of surface water and groundwater, there is the irrational use of fertilizers and pesticides. This deficient management of fertilizers and cultural practices leads to the migration of several contaminants, such as nitrates, microorganisms and pesticides to sources of drinking water. Today, these contaminants are detected in most waters drinking water located in the agricultural region. In some municipalities, contamination has reached such a magnitude that it has been necessary to stop the use of water sources for consumption [52].

2.3 Use of Fresh Water

Agriculture is a sector that consumes a lot of fresh water. A ton of grain requires an average of 1000 tons of water [53].

The importance of water consumption and trade in agricultural products around the world has given rise to the concept of virtual water [54].

The water supply is done in two different ways:

- So-called rainfed agriculture uses only rainwater.
- Irrigated agriculture uses water from rivers, lakes, and reservoirs or groundwater.

In 2000, rain fed agriculture in the world consumed 5000 km^3 of water per year over an area of 1240 million ha. Irrigated agriculture consumed 1500 km^3 of water per year, covering an area of 264 million ha. At the current rate of expansion of the irrigated area, by 2050, it would reach 331 million hectares irrigated; consuming about 500 km^3 per year, more water than today. However, the demand for complimentary water in 2050 is estimated at 4500 km^3 per year because of population growth forecasts. The sole use of irrigation will therefore not be able to meet global needs [55]. Also, about 10% of the water currently used for irrigation comes from non-renewable sources (fossil groundwater) [56].

According to a study from the University of Utrecht, water shortages are expected in many countries, including the three largest grain-producing countries in the world—China, the United States, and India., as well as in countries where the proportion of irrigation water of non-renewable origin is essential: Saudi Arabia, Pakistan, Iran, Mexico, in particular. According to the same study, "the unsustainability of groundwater use for irrigation is a problem for countries that rely heavily on groundwater, but also for the world as a whole since international trade introduces strong correlations between food production in one country and consumption in another [57]".

2.4 Air Pollution

Agriculture also plays a role in the air or atmospheric pollution. Pesticides used in intensive agriculture are inexorably polluting our air (Fig. 1), affecting human health, animal health and biodiversity [58]. Volatilization is one of the main causes of pesticide leakage outside the target area, especially when treatments target the

Fig. 1 Pollution by pesticides used in intensive agriculture [58]

soil or plant surfaces. These losses often exceed those due to chemical degradation, runoff and leaching [8], air transport and deposition are the main contributors to pesticide dispersion on land [59].

Molecules can volatilize resulting in atmospheric pollution. They are also potentially photodegraded by sunlight [49].

The first campaign to measure pesticides in the air starts in France in June 2018. Glyphosate and chlordecone are among the 90 substances that will be sought to better understand the exposure of the population; they have been selected according to their persistence in the air, their degree of use and their danger to health [60].

2.5 *Contribution to Climate Change*

The agricultural sector is a major contributor to the greenhouse effect. FAO publishes detailed statistics on global and country greenhouse gas (methane and nitrous oxide) emissions (1990–2011 average CO_2 equivalents) [61].

Breakdown by continent: Asia 42.6%; Americas 25.3%; Europe 14.1%; Africa 13.9%; Oceania 4.2%.

- Main emitting countries: China 655.5 Mt CO_2eq; India 564.5 Mt; Brazil 366 Mt; United States 354 Mt; Australia 154.6 Mt; Indonesia 138.7 Mt; Russia 117.6 Mt; Argentina 109.3 Mt; Pakistan 104.2; Sudan 93.4 Mt.
- Emissions (carbon dioxide, methane and nitrous oxide) due to energy consumption (burning of fuel and electricity generation in agriculture and fishing): 785.3 Mt CO_2eq in 2010 (growth rate annual average 1990–2010: + 1.6%), of which 54.9% in Asia (+ 2.6% per year), 24% in Europe (– 1.3% per year), 16.5% in the Americas (+ 1.7% per annum), 3.2% in Africa (+ 6.8% per annum) and 1.4% in Oceania; breakdown by fuel: diesel 44.9%, electricity 36.9%, coal 9.5%, natural gas 3.4%, gasoline 2.3%, heavy fuel oil 1.9%, LPG 1.1%.

2.6 *Loss of Biodiversity*

The modification of agricultural practices in the twentieth century has led to an erosion of biodiversity [62] that has led locally to the extinction of many animal species (including butterflies, bees, wasps, beetles, reptiles, amphibians, sticklebacks, larks, etc. very common in the fields or their surroundings until the 1970s). Since the 1990s, biodiversity monitoring experiments [63] have been set up, which have made it possible to quantify the impacts of intensive agriculture and highlight some of the interests of biological agriculture.

In addition to its importance for the conservation of the genetic diversity of ancient varieties, agriculture sometimes plays a vital role for the protection of biological diversity: The European Commission combines three main criteria to measure the interest of an agricultural area in terms of contribution to the preservation of biodiversity. The areas with the highest score are called "high natural value" [64]. 10–30% of agricultural land deserves this title in Europe. In France, 84% of surfaces classified as "high natural value" are in mountains or medium mountains (Alps, Corsica, Franche-Comté, Massif Central, Pyrenees …). These are mostly open-air rearing areas characterized by low stocking density (livestock) per hectare, little or no chemical inputs, and usually greater use of agricultural labor.

3 Reduction of Impacts

3.1 The Sustainable Agriculture

In this issue of sustainable development, agriculture occupies a prominent place. The term "sustainable agriculture" translates the desire to develop an agriculture that contributes to "sustainability" [27, 28, 65, 66]. For the Organization for Economic Co-operation and Development [67], this sustainability is based on four criteria: an economically viable production system; preservation and enhancement of the basic natural resources of the farm; the preservation or enhancement of other ecosystems affected by agricultural activities; the creation of a pleasant natural setting and aesthetic qualities. Thus, sustainable agriculture must preserve the environment while being productive. In this lineage is developed the notion of multifunctionality which attributes to agriculture a landscape function, social and identity in addition to that of production and preservation of the environment. The Ministers of Agriculture of the OECD countries adopted a definition of this concept in 1998: "the act of agricultural production ensures market functions of production, but also other functions, presenting the character of goods public, very rarely paid because they are not marketable" [68]. This definition, however, remains very vague and the non-commercial dimension, in particular, is a source of controversy. It opposes the members of the group of "friends of the multifunctionality" (European Union, Japan, South Korea, Norway and Switzerland) with the Countries of the South and the United States and the Cairns group [69]. Whatever the recognition of the non-market functions of agriculture, the notion of sustainable agriculture gives rise to the emergence of that of "reasoned" agriculture. This is an extension of the concept of "reasoned control" which aims to replace a systematic fight against pests with a fight against the threshold of tolerance of crops. Its application seems to have shown very positive results around the world (see insert). Rational farming is a broader concept: more than the fight against pests, it is the whole work of the farmer that must be reasoned.

3.1.1 The Positive Effects of Sustainable Agriculture (from Brown [53])

In the Netherlands, nearly 550 farmers have reduced their use of pesticides by 30–50% and have totally eliminated the insecticides. Through the modification of practices, they obtained the same yield and the value of the sanitary products decreased. On the other hand, while efficiency gains are a breakthrough, some studies analyzed by Brown [53] have shown that it is possible to replace the use of chemicals with non-polluting methods of improving fertility and controlling insects. Systems of traditional culture, integrating manure and rotations, as well as diversity crops, provide yields equivalent to those of intensive farming. In some regions of China, for example, cropping systems incorporating rotation (rice varieties different) doubled the yield of the farms, while eliminating the use of pesticides.

In Indonesia, the Integrated Pest Management (IPM) method, a kind of near-ecological control of insects, has been tested. This method has no only the cultivation of several species, but also the introduction of natural predators and plants, scattered in the fields, which repel insects. In four years of implementing this in the country, the use of insecticides has been halved, and yields have increased 15%. This IPM is starting to take root in countries as diverse as Kenya, Cuba, Peru, and the state of Iowa in the United States. At the European level, the recognition of multifunctionality comes first and foremost by the reform of the Common Agricultural Policy (CAP) in 1992 [70].

In response to the challenges of multifunctionality and the threats of sustainability of production support (issued during multilateral trade negotiations), this new CAP establishes a decoupling between the level of support to farmers and the quantities produced. Modifications are also made in the structures policy. A set of three so-called "accompanying measures", financed by the European Agricultural Guidance and Guarantee Fund (EAGGF), comes directly under the new agricultural policy: agri-environment, afforestation of agricultural land, early cessation of activity. Following the Cork [71] declaration confirmed in 1999 by the Berlin Agreement [72], the Rural Development Regulation (RDR) [73] becomes the second pillar of the CAP, taking into account of environmental challenges, the diversification of economic activities and the preservation of rural heritage. The Regulation establishes the framework for Community support for sustainable rural development from 1 January 2000.

4 Contribution of Some Maghreb Countries (Morocco, Algeria and Tunisia) to Sustainable Agriculture

4.1 Morocco

4.1.1 The Importance of the Agricultural Sector

The agricultural sector in Morocco generates 14% of GDP. The country's growth rate is strongly correlated with that of agricultural production. Agriculture remains the country's largest provider of jobs. More than 40% of the populations live in this sector. The agricultural area is estimated at 8,700,000 ha (12.25% of the total area of the country). The issue of water is crucial for the development of agriculture in the country. The main vegetable products of the country are cereals (wheat, barley), citrus fruits (oranges, clementines), olives, fruit rosaceae (almonds, apples, apricots …), sugar beets, legumes, vegetable crops including potatoes and tomatoes. Livestock (sheep, goats, cattle, camels, and poultry) is also an essential component of the agricultural sector [74].

4.1.2 The Socio-agro-environmental Impacts on the Agricultural Sector

Water Deficiency

The scarcity of water resources is a significant constraint on a global scale. For the Mediterranean, climatic constraints, particularly the scarcity of water resources have always been structural characteristics. Water insecurity, a threshold that concerns countries whose water availability per year and inhabitant is less than 1700 m^3, is already affecting 10 Mediterranean States (Libya, Israel, Egypt, Jordan, Morocco, Algeria, Tunisia, Malta, and Palestine, Syria). As a result, food security will become increasingly dependent on the efficient management of water resources and the optimization of irrigation systems. Indeed, Morocco is a Mediterranean country with essentially semi-arid and arid climate. The water resources are limited by its geographical situation and its exposure to climatic hazards [75].

The Moroccan agricultural sector, a pillar of the country's food security, is highly dependent on climatic hazards, particularly drought (Fig. 2) and the scarcity of water resources. The lack of water and its scarcity are first and foremost a problem for agriculture, and the solutions to deal with them are imperative for agriculture. According

Fig. 2 The drought causing by global warming in Morocco [76]

to the report of El Aissi in 2016: Morocco is also challenged by the impact of climate change on its agricultural production, with very irregular water supplies from one year to another. In this unfavorable climate context and to support its development [75].

Water Quality and Agricultural Impact

The use of intensive agriculture has accelerated over the last 20 years with the development of irrigation. This was done with a gradual but ultimately significant recourse to chemical fertilizers. The level of agricultural pollution by phosphates and nitrates was estimated at 10,000 tons/year in 1994 [77]. For many groundwater nitrate levels have reached thresholds exceeding 50 mg/L which are set by the OMS for drinking water quality. Levels ranging from 50 to 70 mg/L have been observed in the vegetable growing areas of many irrigated areas (Tadla, Gharb, Moulouya, etc.). As agricultural intensification is likely to increase, especially for the valorization of agricultural water, the risks of increased chemical pollution by fertilizers are certain if measures of good practice are not implemented. The other problem that potentially threatens the quality of water is the increasing use of pesticides with what it generates as contamination by toxic organic products. About 1 million hectares are treated annually with pesticides, 60% of which are irrigated [78]. The quantities of pesticides used by farmers are considered low and estimated at around 9400 tons. The data currently available in Morocco indicate that little is known about the active substances, formulations and application conditions of pesticides. Levels and types of water and soil pollution are also poorly known; it is, however, estimated that 0.5–1% of these products are found in watercourses [79]. Estimates of contamination of groundwater by pesticides are now indeterminate.

4.1.3 Measures of Impact Mitigation and the Sustainability of Agriculture in Morocco

Strategies to reduce agricultural drought risk fall into three categories [80]:

- In irrigated agriculture, water can be saved by minimizing losses and improving water use efficiency.
- In pasture and forest areas, evaporated water can be used by developing pasture and fruit tree ecosystems.
- In rain fed areas, dry farming techniques consisting of improved water harvest, storage and use at farm and plot levels can increase productivity.

Significant progress has been made for Morocco in terms of techniques and practices for saving irrigation water, its collection and mobilization. The strategy of irrigation and water economy is the basis of the country's agricultural policy. Since its independence in 1956, Morocco has put water at the center of its development equation, has oriented its water management policy towards the mobilization, rationalization and development of conventional and unconventional water resources. Concretely, the answers to these issues are reflected in the agricultural development strategy called "Green Morocco Plan" launched in 2008 based on programs of mobilization and responsible management of irrigation water [75]:

- National Program for Water Economy in Irrigation (PNEEI): which aims to convert from traditional irrigation systems to modern and water-saving systems, saving a volume of water of more than 1.4 Md m^3 in 2016, or 2546 m^3/ha.
- Irrigation Extension Program (PEI): which aims to upgrade 1.5 billion cubic meters of water by hydro-agricultural development covering an area of 160,000 ha by 2020.
- Public–Private Partnership Program in Irrigation (PPP): the objective is to ensure the efficiency of water management and to ensure a better valorization in the existing perimeters and the new perimeters as well as the safeguarding of irrigation in areas with high agricultural production potential, particularly through desalination projects for seawater.

Planning for Water Management

From independence until the 1970s, water planning was done in Morocco on a sectoral or project basis. During the 1980s, there was a need for planning at the level of the Water Management. Indeed, the scarcity of water and the emergence of competition in its use have made it imperative to put in place integrated planning taking into account the resources and needs in the relationship between development and the preservation of the water sector and the socio-economic economic [81].

The pricing of irrigation water has been launched. In the 7 ORMVAs which priced irrigation water (Moulouya, Loukkos, Gharb, Doukkala, Tadla, Haouz, and Souss-Massa) a water royalty adjustment plan was adopted in 1997. This plan was aimed to promoting efficient, economical and productive use of irrigation water [82]. This planning of water management has enabled Morocco to accomplish during these last decades, the following objectives [81]:

- Irrigation of one million hectares in 2000 and the total irrigable land potential in 2020.

- Ensure a balance between needs and available resources.
- To allow balanced access to the water of all the regions of the kingdom.
- Adoption of administrative, legislative and economic provisions allowing efficient management moreover, sustainable water resources.

Legislative and Regulatory Aspects

As part of the overhaul of national legislation in the field of water a single law on water adopted by the House of Representatives on July 15, 1995. The law on the water today constitutes the legal basis of the water policy in the Kingdom. It is based on a number of principles that derive from several objectives, namely:

- The regulation of activities likely to pollute the water resources: in particular, the forecast sanctions and the creation of water police to repress any illegal exploitation of water or any act likely to alter its quality, the introduction of the principles "sampler-payer" and "polluter pays".
- The search for the greater agricultural value of water by improving the conditions for improving moreover, use of agricultural water.

This law introduced the notion of participatory, concerted and decentralized water management through the higher Council for Water and Climate (CSEC), the creation of basin agencies and the development of contracting [83].

4.1.4 The Climate Change Integration Project in the Implementation of the Green Morocco Plan (PICCPMV)

Like all development projects. The World Bank coordinating this project requires that a study is conducted to evaluate the positive or negative impacts of the project on the environment and society, as well as to define mitigation measures for the most important risks.

- The ESIA is performed in accordance with:
  Operational policies for the environmental and social safeguarding of the World Bank.
- Provisions of all relevant laws and regulations, including Moroccan laws and regulations.
- The spirit of the national charter of the environment and sustainable development (framework law in preparation).

World Bank Safeguard Policies

Programs and development projects. They include:

- Environmental Assessment (OP 4.01/PB 4.01).
- Pest control (OP 4.09).

- Forests (OP 4.36/PB 4.36).
- Natural habitats (OP 4.04/PB 4.04).
- Dam safety (OP 4.37/PB 4.37).
- Projects related to international waterways (OP 7.50).

Description of the Pillar II Project

The Climate Change Integration Project in the implementation of the Green Morocco Plan (PICCPMV), financed by the Global Environment Fund (GEF) and coordinated by the World Bank (WB), It is grafted on a group of projects Pillar II, have been identified in five targeted regions of Morocco: Chaouia-Ouardigha, Rabat-Sale-Zemmour-Zaer, Gharb-Chrarda-Beni Hssen, Tadla-Azilal and Doukkala-Abda. The success of these PICCPMV Sub-Projects will provide evidence of the competitive advantage of climate change vulnerability reduction strategies, encouraging farmers in neighboring regions to adopt similar measures.

The present Pillar II project "Extension and enhancement of fig production in the region of Doukkala-Abda" has been selected to benefit from adaptation measures to climate change.

The Doukkala-Abda region has been identified as vulnerable to climate change and has significant agricultural potential. The fig tree has been identified as a sector that will become a priority in this region with regard to the extension program provided for in the Regional Agricultural Plan. It thus requires adaptation measures to climate change.

The area covered by the PICCPMV Subproject is 165 ha. Beneficiaries and intervention sites will be identified by the DRA Doukkala-Abda.

Proposed Climate Change Adaptation Measures

The measure that can have a significant impact on the adaptation of this sector is the compost based on earthworm (lombri-compost) which improves the retention capacity of rainwater or irrigation and improve the fertility of soils. Two applications of lombri-compost are required, the first as a substantive amendment and the second as a cover amendment to be made each year on the soil surface from the second year. The amendments are made as follows:

- 5 kg of lombri-compost per fig tree plant should be brought as a substantive amendment at one time during the first year when the new plants are planted.
- 2 kg of lombri-compost per fig tree plant are to be provided, as a cover amendment, for each of the last 3 years (6 kg over the last 3 years).

Reference Situation

The baseline describes the current state of the environment and society before the implementation of the PICCPMV. It serves as a basis of comparison for measuring the environmental and social impact of the PICCPMV Sub-Project. This involves collecting and analyzing existing information on the intervention area of the PICCPMV and therefore on the corresponding Pillar II project.

General State of Water and Soil Degradation in Morocco

Degradation of Water Quality

In Morocco, between 1950 and 1980, water quality was average to excellent irrespective of water availability [84], then deteriorated significantly due to the main sources of pollution, which are: domestic pollution, industrial and agricultural. For example, during the year 1998–1999, the water quality was judged degraded by the Directorate General of Hydraulics, in 37% of the sampled stations and good in 53% and 1999–2000 it was judged degraded in more than 50% of the stations. Critical pollution levels are observed in several stretches of the watercourse. Groundwater is of better quality with the exception of some important aquifers on the Atlantic coast. The deterioration of the quality of water resources is observed in all the watersheds and more particularly in the Sebou basin which has reached alarming levels of pollution downstream of large discharges of domestic and industrial wastewater [22].

The main causes of water quality degradation in Morocco are:

- The intrusion of marine waters generated by excessive pumping in coastal areas.
- The significant delay in the field of sanitation and the purification of domestic wastewater.
- The delay in the field of industrial and domestic depollution.
- The unsustainable use of fertilizers especially nitrates in irrigated areas: Tadla, Berrechid, Triffa, Doukkala, etc. The distribution of fertilizer consumption by crop shows that about 32% of fertilizers are used by citrus fruits, sugar crops and vegetable crops which occupy only five percent of the cultivated area and are mainly irrigated (58% of the tonnage). As for cereal crops, which occupy 63% of the cultivated area, they mobilize only about 43% of global tonnages [19]. The level of agricultural pollution by phosphates and nitrates was estimated at 10,000 tons/year in 1994. For several groundwater nitrate levels reached thresholds exceeding 50 mg/L which are set by the OMS for potability waters.
- The unsustainable use of pesticides in agriculture. About 1 million ha are treated annually with pesticides, 60% of which are in irrigated areas, while the semi-arid and semi-arid zones are poorly treated due to the weakness and irregularity of the rains. The quantities of pesticides used by the 14 farmers are considered low and estimated at nearly 9400 tons. However, the levels and types of pollution of water and soil by pesticides are also poorly known, and it is estimated that

Fig. 3 Water erosion with disappearance of biodiversity in Morocco [85]

0.5–1% of these products are found in watercourses. These causes are aggravated by the context of scarcity and irregular rainfall and the consequent low flows, the concentration of socio-economic activities and agricultural intensification.

Soil Degradation

In Morocco, soils are being eroded at rates that far exceed international standards. The average specific degradation varies between 212 and more than 2000 t/km^2/year. These rates are excessive especially for the basins of the North and North-West where they exceed 2000 t/km^2/year. According to Badraoui [26], the main forms of soil degradation in Morocco are: water erosion (Fig. 3), wind erosion, secondary salinization, alkalization, loss of organic matter, crusting, drainage, compaction and urbanization.

Cultural Practices

The total usable area of the municipality of Ouled Frej is 8600 ha, of which 300 ha are uncultivated, 400 ha irrigated by private pumping (PMH). The number of exploitation is estimated at 2000.

The rural commune of Ouled Sidi Ali Ben Youssef has a total estimated area of 11,500 ha, of which 600 ha irrigated by private pumping (PMH), the number of holdings is 3000 Cereals are the main crops in both communes.

Expected Environmental Situation

The PICCPMV Sub-Project will have a positive impact on soil quality through the improvement of the general soil fertility through water conservation, the amelioration of the structure through the incorporation of organic matter, and the betterment of soil biology. It will also have a positive impact on the water balance thanks to the upturn of the water supply in the soil due to the improved infiltration of rainwater,

Fig. 4 Fig tree culture adaptable to climate warming [86]

the reduction of soil evaporation by the plants of fig tree, and the conservation of water by organic matter. The adoption of the climate change adaptation measures proposed by this project will have direct impacts on the increase of yield compared to old fig trees, the improvement of the resistance to drought, and the possibility of valorization of the fig tree by the biological conduct.

Expected Social Situation

Climate Awareness Level for Climate Change Adaptation Measures

Farmers are aware of the impacts of aridification of the climate and wish moreover develops the fig tree culture (Fig. 4).

The project through the diffusion of the technique of Lombri-compost intervenes several levels: better mastering techniques of valorization of the water, better control of the tools and techniques of treatment of the plantations in conditions of aridity. Awareness sessions will enable beneficiaries to identify better areas suitable for fig plantations and those that can receive cactus plantations.

The PICCPMV Sub-Project will have a positive environmental impact on soil quality, water balance and crops. It will also have a positive social impact on the level of awareness of climate change adaptation measures.The World Bank's safeguard policies, analysis of the expected environmental situation, shows that the impacts are positive. The analysis of the expected situation at the social level shows that the impacts are positive on all social components. As a result, the expected impacts of the PICCPMV Sub-Project are generally neutral or positive on the environment (soil quality) or society (between beneficiaries).

4.2 Algeria

4.2.1 The Importance of the Agricultural Sector

Since independence, various policies have been put in place for this sector, particularly since the year 2000 (PNDA, PDAR) (MADR 2012 cited in Si-Tayeb [23]). In 2005 according to the survey carried out by the MADR, agricultural employment was evaluated at 2,237,867 workers on 1,052,602 agricultural holdings of which 90,471 were above ground. In 2014 agricultural employment accounted for 10.8% of the active population, which had reached 11,500,000 people [23].

In the early 1960s, agriculture played a relatively important role in the Algerian economy. By 1966, its value added in the gross domestic product had fallen and then changed little [87].

The agricultural sector in Algeria remains significant, as it contributes on average to about 10% of the annual GDP and a quarter of the active population depends on this sector. It also helps reduce food dependency on other countries [23].

4.2.2 The Socio-agro-environmental Impacts on the Agricultural Sector

Agriculture in Algeria faces many constraints among which we quote:

- Fertilizer supply Algerian farmers are not only struggling to source fertilizer, but they are also struggling to benefit from subsidies for it.
- Renewable concession titles. The difficulty for many farmers working on concession lands to obtain a definitive concession or ownership titles.
- Weaknesses of agricultural warning stations Weak public warning structures for farmers such as agricultural warning stations in the event of crop or animal problems such as the appearance of diseases, or in the case of parasitic attacks or particular climatic events.
- Low technical level of farmers, many farmers have a very low technical level which is a problem for agricultural development, and empirical research on the issue of farmers' adoption of innovations is often fragmented or differentiated [88–90] (Figs. 5 and 6).
- Agricultural production dependent on climate and landforms Algerian agriculture is still dependent on rainfall, particularly in mountainous regions, so agricultural yields in Algeria depend on climatic hazards [23]. Landforms are also a constraint for Algerian agriculture, especially when they are associated with the climate,

Fig. 5 Random use of pesticides by untrained peasants. Author source 2017

Fig. 6 Irrigation by a canal system that does not save water by unformed farmer in Algeria [91]

because the majority of plains useful for agricultural activity are characterized by an arid or semi-arid climate and the majority of wetlands are mountainous [18].

- A threatened agricultural area In addition to the small agricultural area that focuses on the coast, especially compared to the total area of Algeria, it is threatened by desertification in the south and by urbanization in the north because of population growth. The useful agricultural area is also threatened by erosion and soil degradation, in this regard Bessaoud and Montaigne [18] note: "the per capita UAA is in continuous reduction, from 0.75 ha in 1962 at 0.25 ha today. This drop is mainly due to population growth, but also to the loss of agricultural land, erosion and soil degradation"; Added to this is the unorganization of agricultural markets, the weakness of material means, the fragmentation of Algerian parcels, the problems of ownership of land [92].

The agricultural sector in Algeria remains weak. Nevertheless, it remains a strategic sector, because it plays an important socio-economic and political role. On the socio-economic level, it contributes an average of 10% of GDP and employs almost a quarter of the active population, politically it contributes to the improvement of food security, thus reduced food dependency. However, its importance in the Algerian economy presents a good and a bad, a good because in addition to what we quoted previously, it contributes to partially reduce the scale of the recurrent economic crises which affects the Algerian economy. It is a misfortune because of its weakness and its inability to achieve food self-sufficiency are a significant burden for the Algerian economy through food imports is vital and increasing with the increase in the Algerian population.

4.2.3 Measures of Impact Mitigation and the Sustainability of Agriculture in Algeria

Standard Algerian agriculture suffers from a lack of sustainable competitiveness and weak integration into external markets. Traditional policies and successive agricultural development plans have produced only modest results in terms of the country's potential and needs. Faced with such a finding, organic farming can prove to be an interesting alternative for developing local resources. Nature is conducive to the cultivation of a whole range of products perfectly adapted to the Algerian climate and soils (Figs. 7 and 8) (certain fruits and vegetables, dates, figs, olive trees, certain medicinal plants, etc.), offering our farmers the opportunity to diversify their offers and work seamlessly during the four seasons. Exporters who are well trained in the export trade will not have great difficulty in selling these products to foreign markets, especially those in Europe willing to pay a heavy price [93, 94].

Fig. 7 An oasis of palm trees passing through a dry area—Biskra—in Algeria. Author source 2017

Fig. 8 Adaptable agriculture to global warming in Algeria [95]

4.3 *Tunisia*

4.3.1 The Importance of the Agricultural Sector

The agricultural sector occupies an important place in the Tunisian economy by contributing to the creation of employment and the balance of payments balance through exports, in addition to its role in the guarantee of the food security of the country [96]. It is estimated that today, one in six Tunisians works in agriculture. Sector revenues account for more than 10% of gross domestic product, and food exports account for 11% of goods exports. Tunisian agrosystems combine rainfed and irrigated crops, livestock and oasis crops [97].

Agriculture is practiced on a total agricultural area of about 10 million ha, with a plowable area equal to 5 million ha and an area equipped for irrigation of about 4% of agricultural land and 8.2% arable land. Agriculture is provided through 516,000 farms with an average size of 10 ha with about 75% of the farmers having an area of less than 10 ha. Livestock constitutes one-third of agricultural GDP and employs 39% of the agricultural labor force [97]. The main agricultural products of the country are cereals, olives, dates and citrus for the vegetable sector [96].

4.3.2 The Socio-agro-environmental Impacts on the Agricultural Sector

According to OTEDD's report [20, 21], most of the agriculture is practiced under the most fragile natural conditions. It, therefore, has a relatively high environmental cost. Under these conditions, the economic unsustainability could pose the problem of the socio-political feasibility of openness and growth choices. This dimension is one of the key issues in the search for sustainability of agriculture, whether economic or ecological.

The irrigated sector, which occupies 425,000 ha during the year, consumes about 81% of water resources and contributes 37% of agricultural GDP. The reduction of water resources combined with the impact of climate change poses a threat to this irrigated sector and therefore to the country's economic equilibrium.

Also, the agricultural sector faces other environmental challenges that are primarily related to:

- The intensive use of non-renewable resources in the South of the country for the needs of oases.

- The use of high salinity waters and unsuitable soils. It is estimated that 60% of the public irrigated land in Tunisia is moderate to highly sensitive to secondary salinization following irrigation. This rate reaches 86% in the private perimeters.
- The limited extension of urban planning.
- Decreased fertility of irrigated soils is a dominant form of degradation.
- Chemical pollution in irrigated areas mainly due to improper or improper use of mineral fertilizers and crop protection products.
- The reduction of yields due to the impacts of climate change.
- Increasing the frequency and intensity of extreme drought and flood events and disrupting the seasonal distribution of precipitation.
- Also, the assessment of coastal vulnerability to accelerated sea-level rise by 2050 suggests an average sea-level rise of + 30 to + 50 cm, inducing an annual rate of retreat ranges from 20 to 135 cm, depending on the littoral and the regions.

The direct and indirect effects of climate change (decreasing rainfall, raising the temperature that would increase the water needs of the various socio-economic sectors and the accelerated rise in sea level) will strongly impact the water resources and especially groundwater that is heavily over-exploited. Thus, the situation will worsen in the coming years as a result of climate change, with a decline in conventional water resources estimated at 28% by 2030 and a decrease in surface water of about 5%.

These climate projections also indicate that under the effect of the drought the areas of cereals would decrease by 20% and those olive-growing by 40% by 2020. The effect of the drought would be accompanied by a fall in the production of cereals. Cereals, in rainfed, at around 40% in 2030, dry olive production would fall by 52% for the same horizon. Animal production would decrease by 36–50%, respectively, for sheep and goats.

4.3.3 Measures of Impact Mitigation and the Sustainability of Agriculture in Tunisia

Tunisia has already realized the dangers of climate change. In the Mediterranean, the country plays a pioneering role in the field of adaptation to climate change and also in the agricultural sector (Fig. 9). For example, an adaptation strategy aims to strengthen the resilience of this sector to the impacts of change [98].

Several institutional, technical and economic measures are advanced. In the area of water management, the Water Code and the regulation for the reduction of water consumption by large consumers must be applied more rigorously and reinforced in risk areas. A desalination strategy and improved wastewater treatment will increase the availability of water [98].

In relation to the management of agrosystems, the Agriculture sector (agricultural production, forests and land-use change) has mobilized itself in the field of greenhouse gas mitigation, even though it is not an emitting sector (the absorptions are of the order of 3 million t CO_2eq) [98].

Fig. 9 Olive farm adapted to the rise of global warming [99]

The reconversion of certain activities that are economically unprofitable and climatically unsustainable, the revision and adaptation of the agricultural map to the impact of climate change or the adoption of new technologies of sustainable production and consumption constitute measures of adaptation to integrate into national planning. Several of these measures have been considered in the National Planned, and Determined Contribution (INDC) submitted at COP21 in August 2015 which provides for the agriculture, forestry and land use sector, a mitigation plan that includes the intensification of CO_2 absorption capacities of the forest and arboriculture, thanks to the intensification of reforestation, consolidation and increase of carbon stocks in the forest and pastoral environments [97].

It is also planned to reduce the carbon footprint of agriculture by using less emission-intensive practices such as organic farming or conservation as well as the energy recovery of animal waste or the optimization of their diets food. These mitigation measures are also adaptation measures given the resulting environmental benefits: dam protection, an increase of soil fertility, stabilization of production in relation to climate variability, improvement of the forage balance for the region. Livestock feeding, job creation and income-generating activities in rural areas as well as the reduction of rural exodus. Of the US$18 billion needed to mitigate greenhouse gases projected for Tunisia's INDC, $1533 billion is earmarked for agriculture; forestry and land use [97].

The 2016–2020 development plan provides for an agricultural growth rate of 5% and foresees a budget of around 2409 billion dinars for the protection of the environment, sustainable development and the green economy of which 12% for the fight against pollution, 40% for sanitation, 19% for waste management, 23% for coastal protection and only 1.3% for the protection of natural resources. In the absence of the integration of climate volatility in the agricultural and economic policy of the country, this investment scheme alone will save the agricultural sector in Tunisia? [100].

5 Conclusions

The Arab Maghreb countries (Morocco, Algeria and Tunisia) are among the growing countries whose economy is based on natural resources, agriculture and industry. Agriculture is practiced in a traditional and modern way, with large areas, irrigated and non-irrigated, which have absorbed a lot of work to ensure the availability of food. Since the independence of these countries have changed the policies, plans and programs for the recovery of this sector, but so far have not achieved the desired objectives, which are demonstrated by the large volume of the bill of the importation of basic foodstuffs.

The geographical situation of the Arab Maghreb countries in the South of the Mediterranean sea, which is characterized by a semi and arid climate, lack rainfall generally, in addition to the random and excess use of water in the agricultural system, and even in drinking make these countries exposed to water shortage crises several times in their history, and this has had a negative impact on their environment in general: on soil structure, on vegetable biodiversity, and even on climate change; and these changes have also had a negative impact on the agro-socio-economic system.

The impact studies available in these countries, though rare, have been shown to be intensive agriculture using many chemicals (fertilizers and pesticides) plus the human factor (unskilled) leads to soil pollution, water and air, as well as human exposure to health problems.

Sustainable agricultural production and conservation of the environment is a dream for these countries comes after the dream of food insurance for lack of qualified human potential and material possibilities.

Recommendations

The water is a natural resource; where there is there is a life, balance and sustainability. It is important to these Maghreb countries to study how to exploit this dear gold to their life, eventually for their food security:

Storage of rainwater in dams and distribute tightly in agriculture and drinking. Establish strict laws prohibiting water loss. It is best to irrigate cultures by drip system. Cultivate varieties in appropriate places according to their environmental needs (soil, water and temperature). Uses desalinate water of the sea or the ocean.

It is necessary to fertilize the poor soil, and to control the plant diseases with bio products to avoid negative impacts of chemical products.

Climate change, always followed by special attention to the soil and the cultivation of plant varieties adapted to the new climate.

Most important for preserving safety agriculture without impact on the environment; It is the human factor that practices agriculture; it needs to be structured to understand what it is for sustainable agriculture. As well as coordination between all actors in the agricultural sector.

References

1. Carson RL (1962) Silent spring. Riverside Press, Cambridge
2. Ministère de l'Agriculture et Ademe, Interactions entre agriculture et environnement; Quels outils de diagnostic? Actes du colloque du 2 avril 1997, Paris
3. Debailleul G (1998) Le processus d'intensification de l'agriculture québécoise et ses impacts environnementaux: une rétrospective à méditer. Vecteur Environ 31(2):49–54
4. Poirié M, Pasteur N (1991) La résistance des insectes aux insecticides. La Recherche: l'actualité des Sciences 22(234):874–881
5. Pimentel D, Levitan L (1986) Pesticides: amounts applied and amounts reaching pests. Bioscience 36:86–91
6. Leonard RA (1990) Movement of pesticides into surface waters. In: Pesticides in the soil environment. Soil Science Society of America Book Series, n° 2, Madison, WI, USA, pp 303–349
7. Schiavon M, Perrin-ganier C, Portal JM (1995) La pollution de l'eau par les produits phytosanitaires: état et origine. Agronomie 15:157–170
8. Taylor AW, Spencer WF (1990) Volatilization and vapor transport processes. In: Pesticides in the soil environment. Soil science society of America book series, n°2, Madison, WI, USA, pp 213–269
9. Glotfelty DE, Taylor AW, Turner BC, Zoller WH (1984) Volatilization of surface-applied pesticides from fallow soil. J Agric Food Chem 32:638–643
10. Emans HJB, Beek MA, Linders JBHJ (1992) Evaluation system for pesticides (ESPE) 1. Agricultural pesticides. National Institute of Public Health and Environmental Protection (RIVM), report n°679101004, Bilthoven, Pays-Bas
11. Severn DJ, Ballard G (1990) Risk/benefit and regulations. In: Pesticides in the soil environment: processes, impacts and modeling, pp 467–491
12. Klein AW, Goedicke J, Klein W, Herrchen M, Kördel W (1993) Environmental assessment of pesticides directive 91/414/EEC. Chemosphere 26:979–1001
13. Giroux I, Morin C (1992) Contamination du milieu aquatique et des eaux souterraines par les pesticides au Québec. Revue des différentes activités d'échantillonnage réalisées de 1980 à 1991. Ministère de l'Environnement et de la Faune, Direction du milieu agricole et du contrôle des pesticides, Québec, 74p
14. Rondeau B (1996) Pesticides dans les tributaires du fleuve Saint-Laurent 1989–1991. Environnement Canada, Région du Québec, Conservation de l'environnement, Centre Saint-Laurent, Rapport scientifique et technique ST-62, 58p
15. Berryman D, Giroux I (1994) La contamination des cours d'eau par les pesticides dans les régions de culture intensive de maïs au Québec. Campagnes d'échantillonnage de 1992 et 1993. Ministère de l'Environnement et de la Faune, Direction des écosystèmes aquatiques, Québec, 134p
16. Simard RR, Cluis D, Gangbazo G, Beauchemin S (1995) Phosphorus status of forest and agricultural soils from a watershed of high animal density. J Environ Qual 24:1010–1017
17. Gangbazo G, Couillard D, Pesant AR, Cluis D (1993) Effets du lisier de porc sur la charge d'azote et de phosphore dans l'eau de ruissellement sous des pluies simulée. Can Agric Eng 35:97–103
18. Bessaoud O, Montaigne E (2009) «Quelles réponses au mal-développement agricole? Analyse des politiques agricoles et rurales passées et présentes». In: Abis S, Blanc P, Lerin F, Mezouaghi M (coord) Perspectives des politiques agricoles en Afrique du Nord. CIHEAM, Paris, pp 51–91 (Option s Méditerranéennes: Série B. Etudes et Recherches; n°64)
19. FAO (2006) Utilisation des engrais par culture au Maroc. Organisation des Nations Unies pour L'alimentation et L'agriculture, Rome. ftp://ftp.fao.org/agl/agll/docs/fertusemaroc.pdf
20. OTEDD (office tunisienne d'environnement et de développement durable) (2013) Rapport général sur la durabilité de l'agriculture en Tunisie

21. OTEDD (2016) Rapport national sur l'état de l'environnement et du développement durable de l'année 2015. In: Lerin F, Mezouaghi M (coord) Perspectives des politiques agricoles en Afrique du Nord. CIHEAM, Paris, pp 51–91 (Option s Méditerranéennes: Série B. Etudes et Recherches; n°64)
22. SEE (2009) État de la qualité des ressources en eaux au Maroc. Secrétariat d'État auprès du Ministère de l'Énergie et des Mines, de l'Eau et de l'Environnement chargé de l'Eau et de l'Environnement. Département de l'Eau. Maroc, 135p
23. Si-Tayeb H (2015) «Les transformations de l'agriculture Algériennes dans la perspective d'adhésion à l'OMC». thèse de doctorat en science agronomique, option économie rurale, université Mouloud Mammeri de Tizi Ouzou, p 282
24. Cellier P, Bethenod O, Castell JF, Germon JC (2008) Contribution de l'agriculture à l'effet de serre. Importance de l'azote et interactions avec l'ozone. Environnement. OCL 15(5):317–323
25. Cellier P, Genermont S (2016) L'agriculture entre pollution atmosphérique et changement climatique. Pollution atmosphérique. Agriculture (Numéro spécial):64–75
26. Badraoui M (2008) Connaissance et utilisation des ressources en sol au Maroc. http://www.rdh50.ma/fr/pdf/contributions/GT8-3.pdf
27. Campbell D (2003) Intra-and intersectoral effects in environmental disclosures: evidence for legitimacy theory? Bus Strateg Environ 12(6):357–371
28. Moyano E, Garrigo F (1998) Acteurs sociaux et politique agri-environnementale dans l'Union Européenne. Le Courrier de l'environnement de l'INRA 33(33):106–114
29. Calvet R (2005) Les pesticides dans le sol. Conséquences agronomiques et environnementales. Edition Agricole, 563p
30. Chabrier C, Dore M (1996) Impact des pesticides sur l'environnement: étude de la contamination des eaux de ruissellement. Fort-de-France, CIRAD, 6p
31. Ramade F (1992) Précis d'écotoxicologie. Ed. Masson, Paris, 299p
32. Moreau JP (1991) La protection des cultures, les pesticides et l'environnement. Le courrier de l'Environnement n°14:5
33. Louchart X (1999) Transfert de pesticides dans les eaux de surface aux échelles de la parcelle et d'un bassin versant viticole. Étude expérimentale et éléments de modélisation. Thèse de doctorat. École Nationale Supérieure Agronomique de Montpellier, 263p
34. Coulouma G, Boizard H, Trotoux G, Lagacherie P, Richard G (2006) Effect of deep tillage for vineyard establishment on soil structure: a case study in Southern France. Soil Tillage Res 88:1321–1343
35. Boizard H, Richard G, Roger-Estrade J, Dürr C, Boiffin J (2002) Cumulative effects of cropping systems on the structure of the tilled layer in northern France. Soil Tillage Res 64:149–164
36. Richard G, Boizard H, Roger-Estrade J, Boiffin J, Guérif J (1999) Field study of soil compaction due to traffic in northern France: pore space and morphological analysis of the compacted zones. Soil Tillage Res 51:151–160
37. Lagacherie P, Coulouma G, Ariagno P, Virat P, Boizard H, Richard G (2006) Spatial variability of soil compaction over a vineyard region in relation with soils and cultivation operations. Geoderma 134:2072–2116
38. Landry D, Dousset S, Fournier JC, Andreux F (2005) Leaching of glyphosate and AMPA under two soil management practices in Burgundy vineyards (Vosne Romanée, 21-France). Environ Pollut 138:191–200
39. Zhang M, Geng S, Ustin SL, Tanji KK (1997) Pesticide occurrence in groundwater in Tulare county, California. Environ Monit Assess 45:101–127
40. Magne C, Saladin G, Clément C (2006) Transient effect of the herbicide flazasulfuron on carbohydrate physiology in Vitis vinifera L. Chemosphere 62: 650–657
41. Armstrong DE, Chesters G, Harris HF (1987) Atrazine hydrolysis in soil. Soil Sci Soc Am Proc 31:61–66
42. Yaron B (1989) General principle of pesticide movement to groundwater. Agric Ecosyst Environ 26:275–297

43. Siimes K, Rämö S, Welling L, Nikunen U, Laitinen P (2006) Comparison of the behaviour of three herbicides in a field experiment under bare soil conditions. Agric Water Manag 84:53–64
44. Holman IP, Dubus IG, Hollis JM, Brown CD (2004) Using a linked soil model emulator and unsaturated zone leaching model to account for preferential flow when assessing the spatially distributed risk of pesticide leaching to groundwater in England and Wales. Sci Total Environ 318:738–748
45. Renaud FG, Brown CD (2008) Simulating pesticides in ditches to assess ecological risk (SPIDER): II. Benchmarking for the drainage model. Sci Total Environ 394:124–133
46. Whelan MJ, Davenport EJ, Smith BG (2007) A globally applicable location-specific screening model for assessing the relative risk of pesticide leaching. Sci Total Environ 377:192–206
47. MAAARO (Ministère de l'Agriculture, de l'Alimentation et des Affaires rurales d'Ontario) (2012) Fiche technique sur l'érosion du sol- Causes et effets
48. Bertrand R, Keita B, Kabiroue NM (1993) Cah Agric 2:316–329
49. Colin F (2000) Approche spatiale de la pollution chronique des eaux de surface par les produits phytosanitaires. Cas de l'atrazine dans le bassin versant du Sousson (Gers, France). Thèse de 'Sciences de l'eau' l'ENGREF, 255p
50. Gao J, Liu L, Liu X, Lu J, Zhou H, Huang S et al (2008) Occurrence and distribution of organochlorine pesticides—lindane, p, p′-DDT, and heptachlor epoxide—in surface water of China. Environ Int 8:10971–11103
51. Kalita PK, Algoazany AS, Mitchell JK, Cooke RAC, Hirschi MC (2006) Subsurface water quality from a flat tile-drained watershed in Illinois, USA. Agr Ecosyst Environ 115:1831–1893
52. Rousseau H (1995) Les contaminants chimiques de l'eau brute ou issus des canalisations du réseau de distribution. In: Air intérieur et eau potable, sous la direction de Pierre Lajoie et Patrick Levallois. Les Presses de l'Université Laval, Sainte-Foy, pp 221–243
53. Brown LR (2001) Éco-économie, une autre croissance est possible, écologique et durable, Seuil, p 76
54. Blanchon D (2010) Les échanges d'eau virtuelle via les produits agricoles, 1997–2001. Cartes, Documentation Française
55. Varet J (2005) L'eau souterraine. Revue géosciences. BRGM Edition. N° 2, p 2
56. www.agr.gc.ca/...irrigation/...irrigation/contenu-archive-durabilite-de-l-irrigation-activ
57. Wada Y, Van B, Marc LB (2012) Non-sustainable grown water sustaining irrigation: a global assessment. Département de géographie physique, l'Université d'Utrecht (Pays-Bas)
58. http://www.ompe.org/quand-lagriculture-pollue-latmosphere/
59. Atlas EL, Schauffler S (1990) Concentration and variation of trace organic compounds in the north pacific atmosphere. In: Kurtz DA (ed) Long range transport of pesticides. Lewis Publishers, Chelsea, Michigan, USA, pp 161–183
60. Mandard S (2018) La première campagne de mesure des pesticides dans l'air démarre en France. Lemonde https://www.lemonde.fr/pollution/article/2018
61. FAOStat (2014) Émissions - Agriculture
62. INRA (Institut Nationale de Recherche Agricole) (2008) Agriculture et biodiversité: chap 1, Les effets de l'agriculture sur la biodiversité. Rapport d'expertise à la demande de MAP et MEEDDAT
63. Clergué B, Amiaud B, Plantureux S (2004) Évaluation de la biodiversité par des indicateurs agri-environnementaux à l'échelle d'un territoire agricole, séminaire 2004 de l'École doctorale RP2E Ingénierie des ressources, procédés, produits et environnement, Nancy, 15 janvier 2004. ISBN 978-2-9518564-2-4
64. AEE (Agence Européenne de l'Environnement) (2004) Rapport sur «High nature value farmland». n°1/2004
65. Young MD (1991) Towards sustainable agricultural development. OCDE, Belhaven Press, London, 346 p
66. Bonny S (1994) Les possibilités d'un modèle de développement durable en agriculture, le cas de la France. Le courrier de l'Environnement, n°23: 11

67. OCDE (1993). L'intégration des politiques de l'agriculture et de l'environnement. Paris, 114 p
68. Communiqué de la réunion du Comité de l'agriculture de l'OCDE au niveau ministériel, les 5 et 6 mars 1998
69. Constitué en août 1986, à Cairns, en Australie: Argentine, Australie, Bolivie, Brésil, Canada, Chili, Colombie, Costa Rica, Guatemala, Indonésie, Malaisie, Nouvelle-Zélande, Paraguay, Philippines, Afrique du Sud, Thaïlande et Uruguay
70. Bazin G, Kroll JC (2002) La multifonctionnalité dans la Politique Agricole Commune: projet ou alibi? Colloque de la Société Française d'Economie Rurale "La multifonctionnalité de l'activité agricole et sa reconnaissance par les politiques publiques", Paris Institut National Agronomique, 21–22 mars 2002, 15p
71. Conférence européenne sur le développement rural réunie à Cork, Irlande, du 7 au 9 novembre 1996
72. RDR, [CE] 1257/99 du 17 mai 1999
73. Sommet de Berlin, mars 1999
74. AgriMaroc (2016) Le secteur Agricole au Maroc. www.agrimaroc.ma/secteur-agricole-au-maroc/
75. Sadiki M (2017) Cas du Secteur Agricole au Maroc. Séminaire de haut-niveau, « rareté de l'eau : défis et opportunités » Rome, Italie, 17 Novembre 2017
76. Infomediaire (2017) Lutte contre le réchauffement climatique: Le Maroc sur le podium mondial. https://www.infomediaire.net/lutte-contre-le-rechauffement-climatique-le-maroc-sur-le-podium-mondial/
77. CSEC (Conseil supérieur de l'eau et du climat) (1994)
78. Erraki L, Ezzahiri B, Bouhache M, Mihi M (2004) «Élaboration d'une base de données des pesticides à usage agricole utilisés au Maroc». In: Proceedings du 5eme congrès de l'AMPP, Rabat, 30–31 mars 2004
79. MAUHE (Ministère de l'aménagement du territoire, de l'urbanisme, de l'habitat et de l'environnement) (2001) «État de l'environnement du Maroc»
80. Balaghi R, Jlibene M, Tychon B, Mrabet R (2007) Gestion du risque de sécheresse agricole au Maroc. Sécheresse 18(3):169–176
81. Jellali M (1995) «Développement des ressources en eau au Maroc». Jellali M. – Direction générale de l'hydraulique – Ministère des travaux publics
82. CSEC (2001) « L'Économie d'Eau, une opportunité et un défi pour le secteur de l'irrigation », 9e Session du Conseil Supérieur de l'Eau et du Climat, Agadir, mars 2001
83. Balaferj R (2000) «Une bonne lecture de la loi de l'eau pour une véritable gestion de la ressource». Académie du Royaume du Maroc – session de novembre 2000
84. Agoumi A, Debbarh A (2008) Ressources en eau et bassins versants du Maroc – 50 ans de développement 1955–2005. http://www.rdh50.ma/fr/pdf/contributions/GT8-1.pdf
85. IRES (Institut royal des études stratégiques) (2017) Rapport stratégique intitulé «Panorama du Maroc dans le monde: les enjeux planétaires de la biosphère». LEVERT.MA (journal électronique marocain)
86. AgriMaroc (2017) La filière du figuier: Entre pari de valorisation et défi de commercialisation. http://www.agrimaroc.ma/figuier-valorisation-commercialisation/
87. Bedrani S, Bensouiah R, Djenane AM (2000) «Développement rural en méditerranée: Algérie» In: Abaab A, Olivereira Baptista F, Bedrani S, Bessaoud O, Campagne P, Ceña Delgado F, Elloumi M, Goussios D (eds) Agricultures familiales et développement rural en Méditerranée. Edition Karthala, Paris (France), p 692
88. Derderi A, Daoudi A, Colin JP (2015) Les jeunes agriculteurs itinérants et le développement de la culture de la pomme de terre en Algérie. L'émergence d'une économie réticulaire. Cah Agric 24(6):387–395. https://doi.org/10.1684/agr.2015.0784
89. Naouri M, Kuper M, Hartani T (2020) The power of translation: innovation dialogues in the context of farmer-led innovation in the Algerian Sahara. Agric Syst 180:102793. https://doi.org/10.1016/j.agsy.2020.102793

90. Ould Rebai A, Hartani T, Chabaca MN, Kuper M (2017) Une innovation incrémentielle: la conception et la diffusion d'un pivot d'irrigation artisanal dans le Souf (Sahara algérien). Cah Agric 26(3):35005. https://doi.org/10.1051/cagri/2017024
91. El-Moudjahid (2016) Agriculture: Encourager l'investissement des jeunes. http://www.elmoudjahid.com/fr/actualites/97314
92. Benyoucef B (2015) «Le rôle de l'agriculture dans le développement économique et social Qu'en est-il de l'Algérie?». Communication au séminaire sur l'agriculture Organisé par l'Université Mohamed Boudiaf de M'sila (Algérie), p 30
93. Grim N (2017) Intensive ou bio: Quelle agriculture pour l'Algérie? Algérie Eco newsletter, 02 novembre 2017
94. Hadjou L, Cheriet F, Djenane AM (2013) Agriculture biologique en Algérie: potentiel et perspectives de développement. Cahiers du CREA (105–106):113–132. https://prodinra.inra.fr/record/274041
95. LNA (2017) Développement de l'agriculture dans le Sud du pays: l'Algérie vise à irriguer 600.000 hectares. La nation
96. Bakari S (2017) The impact of citrus exports on economic growth: empirical analysis. MPRA Paper No. 82414, 23p. https://mpra.ub.unimuenchen.de/82414/1/MPRA_paper_82414.pdf
97. Gafrej R (2016) Avec le changement climatique, quel avenir de l'agriculture en Tunisie? CIHEIM. Watch Letter n°37 - Septembre 2016, 7p
98. Hoernlein L (2014) Rapport sur le secteur agricole tunisien face au changement climatique. Deutsche Gesellschaft für Internationale Zusammenarbeit (GIZ) GmbH en coopération avec Ministère de l'Agriculture des Ressources Hydrauliques et de la Pêche de la Tunisie, 2p
99. Webdo (2018) Tunisie: Quatre millions d'oliviers plantés en 2017. http://www.webdo.tn/2018/01/25/tunisie
100. MF (Ministère des finances) (2016) Programme d'Analyse de Performance du secteur agricole. http://www.gbo.tn

Possibilities of Mineral Fertilizer Substitution Via Bio and Organic Fertilizers for Decreasing Environmental Pollution and Improving of Sesame (*Sesamum indicum* L.) Vegetative Growth

Mohamed Said Abbas, Heba Ahmed Labib, Mohamed Hamza, and Sayed A. Fayed

Abstract Possibilities of partial or complete substitution of mineral fertilizer via bio and organic fertilizers were assessed for improving the characteristics of sesame vegetative growth and contributing to decrease environmental pollution. The two experiments were implemented in two summer seasons of 2015 and 2016 at the Agricultural Experimental and Research Station, Faculty of Agriculture, Cairo University. The experimental design was based on a randomized complete block design with seven treatments. Sesame was tested under three types of fertilizers i.e., 33, 50, and 100% of mineral, compost, and 100% of bioformulations. Results revealed that the treatment of 50% of mineral in interaction with 50% of compost gave the highest values of the dry weight of whole plant (g), the dry weight of shoot (g), absolute growth rate (mg day^{-1}), and unit shoot rate (mg mg^{-1} day^{-1}) in the first season. Also, the treatment of 50% of mineral in interaction with 50% of compost attained maximum values of absolute growth rate (mg day^{-1}) and unit shoot rate (mg mg^{-1} day^{-1}) in the second season and in both seasons. It can be concluded that the partial substitution of mineral fertilizer via 50% of mineral in interaction with 50% of compost for improving the characteristics of sesame vegetative growth and contributing to decrease environmental pollution.

Keywords Sesame · Bio-fertilizers · Compost · NPK · Total chlorophyll content · The dry weight of whole plant · The dry weight of shoot · The dry weight of root · The dry weight ratio of shoot/root · Leaf area ratio · Absolute growth rate · Unit shoot rate · Vegetative measurements

---

M. S. Abbas (✉) · H. A. Labib
Department of Natural Resources, Faculty of African Postgraduate Studies, Cairo University, Cairo, Egypt

M. Hamza
Department of Agronomy, Faculty of Agriculture, Cairo University, Cairo, Egypt

S. A. Fayed
Department of Agricultural Biochemistry, Faculty of Agriculture, Cairo University, Cairo, Egypt

M. Abu-hashim et al. (eds.), *Agro-Environmental Sustainability in MENA Regions*, Springer Water, https://doi.org/10.1007/978-3-030-78574-1_11

1 Introduction

Sesame plant is an important oilseed crop and it belongs to the family *Pedaliaceae* [1–4]. It has a medical and pharmaceutical uses such as mildly laxative, emollient, demulcent, anti-inflammatory, antibacterial, healing wounds, and antioxidant [1]. Its seed oil consist of 39.0% of monounsaturated fatty acids, 46.0% of polyunsaturated fatty acids, and 14.0% of saturated fatty acids [1]. Also, its seeds contain some of vitamins and minerals such as copper, calcium, a high content of phosphorus, iron, magnesium, manganese, zinc, and vitamin B1 [1].

Nitrogen is an essential element for other vital processes like carbohydrates, proteins, metabolism, and (reinforcing both stages of cell division and cell enlargement) [5, 6]. Potassium is playing a vital role to stimulate the enzymes and fight biotic and abiotic stresses such as cold, diseases, water logging, other opposite conditions [7], pests, drought, salinity, and frost [8]. Siddik et al. [9] pointed out that the addition of 60 or 120 kg N ha^{-1} recorded a maximum and significant value of the dry weight of shoot, but the addition of 60 kg N ha^{-1} gave a maximum and significant value of the dry weight of root in a sandy loam soil.

The two strains viz., *Azospirillum lipoferum* and *Azospirillum brasilense* bacteria as a biofertilizer can increase the plant resistance against some diseases, invasion of insects, wind, and (reinforce the growth of shoot and development of root) [10]. Also, it can transform Ca_3 $(PO_4)_2$ from the unavailable form to available form in the soil [10]. Additionally, *Azospirillum* bacteria can transform the atmospheric nitrogen gas to ammonia for providing 50% N of the plant needs via the roots [10]. Abdel-Sabour and Abo El-Seoud [11] found that the combination of the two organic wastes of compost viz., biosolids + water hyacinth positively impacted and recorded a maximum value of total chlorophyll content $plant^{-1}$ in a sandy soil. While, Kumar et al. [12] noticed that the inoculation of *Azotobacter* and phosphate solubilizing bacteria together or the individual inoculation with Azo resulted in an increase in the dry weight of shoot.

However, Eifediyi et al. [13] found that the addition of 200 kg NPK ha^{-1} or 4 ton neem seed cake ha^{-1} or their interaction significantly increased leaf area. Whereas, Dubey et al. [14] indicated that the inoculated seeds by *Azotobacter chroococcum* (AZO2) in combination with 50% of NPKS i.e., 60 kg urea:15 kg diammonium phosphate:15 kg muriate of potash:30 kg gypsum ha^{-1} gave higher the dry weight of shoot and root than the addition of 100% of NPKS in a sandy loam soil. While, Abdullahi et al. [15] concluded that the interaction of *Azospirillum brasilense* with *Glomus mosseae* × 5 ton poultry manure ha^{-1} recorded the highest value of leaf area when compared with the control treatment in a sandy loam soil. However, Ghosh et al. [16] concluded that the interaction of the inoculated seeds by 1.5 kg *Azospirillum* ha^{-1} × 40 kg urea:20 kg single superphosphate:20 kg muriate of potash ha^{-1} × 50% N via FYM or vermicompost recorded a maximum value of the dry weight $plant^{-1}$ in a sandy loam soil.

Mineral fertilizers are expensive, not easily available, and increasing its doses about plant needs lead to environmental pollution. In Egypt, the production of the

Table 1 Average values of physical and chemical properties of compost in 2015 and 2016 seasons

Bulk density (kg m^{-3})	538
Moisture content (%)	33
pH	7.44
Electrical conductivity EC (dS m^{-1})	4.44
Total nitrogen (%)	1.19
Ammoniacal nitrogen—NH_4^+ (ppm)	383
Nitrate nitrogen—NO_3^-	Nil
Organic matter (%)	39.65
Organic carbon (%)	22.99
Ash (%)	60.35
C:N ratio	1:19.32
Total phosphorus (P_2O_5) (%)	1.09
Total potassium (%)	0.53
Weed seeds	Nil
Nematodes	Nil

local edible oil is inadequate [17] and it represent 10% of the consumption [3]. Therefore, the objectives of this study were for improving the characteristics of sesame vegetative growth, also contributing to decrease environmental pollution by assessing the possibilities of partial or complete substitution of mineral fertilizer via bio and organic fertilizers.

2 Materials and Methods

2.1 Experimental Site

The two experiments were implemented in the two summer seasons of 2015 and 2016 at the Agricultural Experimental and Research Station, Faculty of Agriculture, Cairo University, Egypt. The Egyptian cultivar Shandweel-3 was utilized. Physical and chemical properties of compost are shown in Table 1. Whereas, physical and chemical properties of soil before the planting day are shown in Table 2.

2.2 Physical and Chemical Analysis Methods of Soil and Compost

2.2.1 Moisture content: organic material samples were oven dried at 70 °C, but soil samples were oven dried at 105 °C to a constant weight. The difference

Table 2 Average values of physical and chemical properties of soil sample before the planting day in 2015 and 2016 seasons

Physical properties			Chemical properties	
Particle size distribution	Coarse sand (%)	6.48	pH	7.85
	Fine sand (%)	12.01	Electrical conductivity EC (dS m^{-1})	1.34
	Silt (%)	39.00	Total nitrogen (%)	0.05
	Clay (%)	42.51	Total phosphorus (%)	0.32
Textural class		Clay	Total potassium (%)	0.18
Bulk density (g cm^{-3})		1.42	Available nitrogen (mg kg^{-1})	80.50
Total porosity (%)		46.41	Available phosphorus (mg kg^{-1})	26.80
Hydraulic conductivity (cm h^{-1})		0.80	Available potassium (mg kg^{-1})	169.0
Field capacity (%)		27.12	Available iron (mg kg^{-1})	13.27
Wilting point (%)		13.04	Available manganese (mg kg^{-1})	8.39
Available water (%)		14.08	Available zinc (mg kg^{-1})	7.58
			Available copper (mg kg^{-1})	3.15
			Organic matter content (%)	1.32

between the fresh and corresponding the dry weight equals the moisture content, which was calculated as a percentage for each material according to APHA [18].

2.2.2 The pH value was directly measured in the (1:5) soil:water suspensions and in (1:10) compost:water mixture using a pH glass electrode of Orion Expandable ion analyzer EA 920.

2.2.3 Electrical conductivity measurements were run in a suspension of (1:5) soil:water or (1:10) compost:water determined by Richards [19] using EC meter, ICM model 71150.

2.2.4 Organic matter content of compost was determined by glowing the dried samples at 550 °C to a constant weight as recommended in APHA [18]. From the values of organic matter, the quantity of organic carbon percentage was calculated by multiplying by 0.58. While, organic matter content of soil was determined by the determination of total carbon (C %) according to the method of Hesse [20] and multiplying the result by 1.724.

2.2.5 Total nitrogen in dry compost or soil samples were determined by micro-Kjeldahl method as recommended by Jackson [21].

2.2.6 Total phosphorus: acid solutions of the digested soil or compost materials were used for determination of total phosphorus contents as described by APHA [18].

2.2.7 Total potassium: digested solutions of soil or compost materials were used for the determination of total potassium using flame-photometric method according to APHA [18].

2.2.8 Available nitrogen (ammoniacal and nitrate-N): nitrogen forms in compost materials were determined according to the methods outline by Page et al. [22] as follows: soluble nitrogen forms viz., ammonia and nitrate were extracted from the soil or compost samples in KCl solutions and distilled in the presence of MgO for extracting NH_4, meanwhile NO_3 were extracted by running the distillation process another time in the presence of MgO and Devarda alloy. In both steps, the evolved ammonia was collected in 4% H_3BO_3 using a mixed indicator of methyl red and bromocresol green and titrated against 0.05 NH_2SO_4 solution.

2.2.9 Available phosphorus: soil available P was obtained by the method introduced by Olsen et al. [23]. About 10 g of soil samples or 5 g of compost were extracted with 100 ml 0.5 M $NaHCO_3$ (pH 8.5) and 5 ml aliquot of the extract was analyzed for soluble phosphorus. The concentration of phosphorus in compost materials or soil were determined colourimetrically according to Troug and Mayer [24] at a wave length of 660 nm and calculated as ppm.

2.2.10 Available potassium: compost materials or soil were extracted with a normal solution of ammonium acetate (pH 7) in ratio of 1:2.5 (soil/solution, W/V) and determined by Chapman and Pratt [25]. Soluble potassium concentration in the filtrate was estimated using a flame photometer (Corning 410).

2.2.11 Available iron (mg kg^{-1}), available manganese (mg kg^{-1}), available zinc (mg kg^{-1}), and available copper (mg kg^{-1}) in soil were determined according to Jackson [26].

2.2.12 Physical analysis methods of soil:

Particle size distribution was determined by the method of Day [27]. Bulk density (g cm^{-3}) was determined by using paraffin wax method according to Dewis and Freitas [28]. Total porosity (%) was determined by Vomocil [29] method. Hydraulic conductivity (cm h^{-1}) was carried out according to Klute [30]. Field capacity (%), wilting point (%), and available water (%) were determined by Peters [31].

2.3 *Treatments*

The treatments were $T_1 = 100\%$ of mineral (control treatment) (M), $T_2 = 100\%$ of compost (C), $T_3 = 100\%$ of bioformulations (B), $T_4 = 50\%$ of mineral in interaction with 50% of compost, $T_5 = 100\%$ of bioformulations in combination with 50% of mineral, $T_6 = 100\%$ of bioformulations mixed with 50% of compost, and $T_7 = 100\%$ of bioformulations combined with 33% of compost and 33% of mineral.

2.4 Agricultural Practices

Sesame seeds were obtained from Oil Crops Research Department, Field Crops Research Institute (FCRI), Agricultural Research Center (ARC) and sown in hills by hand on the 17th July in both seasons. The distance between each hill was 20 cm. Approximately 164.7 g of the seeds were soaked with bioformulations i.e., 200 ml of biofertile bioformulation viz., *Azospirillum brasilense*, *Azotobacter chroococcum*, *Bacillus polymyxa*, *Enterobacter agglomerans*, and *Pseudomonas putida* in combination with 200 ml of biocontrol bioformulation viz., *Bacillus polymyxa*, *Bacillus macerans*, *Bacillus circulans*, and *Enterobacter agglomerans* for an hour before the planting. Additionally, bioformulations were provided from Environmental Studies and Research Unit (ESRU), Department of Agricultural Microbiology, Faculty of Agriculture, Cairo University, Egypt. While, compost rates of green manure source i.e., 3.17, 4.8, and 9.6 ton ha^{-1} were obtained from Beni Suef compost company and added on the same day of the planting. Sesame seedlings were also sprayed with 1 L of biofertile bioformulation:7 L of the irrigation water and 1 L of biocontrol bioformulation:7 L of the irrigation water as a foliar treatment at 8–10 A.M. on the 22th day of the planting. In the early bloom phase on the 35th day of the planting the mineral fertilizers viz., ammonium nitrate (33.5% NH_4 NO_3) i.e., 94.56:143.28:286.56 kg ha^{-1}, single superphosphate (15.5% P_2O_5) i.e., 153.29:232.27:464.52 kg ha^{-1}, and potassium sulfate (48% K_2O) i.e., 79.2:120:240 kg ha^{-1} were applied on the seedlings. Also, the seedlings were thinned to two seedlings per hill to avoid the competition between the plants directly after the mineral fertilization on the same day. The two experiments were irrigated by flooding irrigation on the 18th, 35th, 51th, and 84th day of the planting. Weeds were controlled by hand and hoeing on the 14th, 33th, and 44th day of the planting. Finally, the crop was harvested on the 113th day of the planting when 80–90% of the leaves and capsules turned to yellow.

2.5 Studied Traits

2.5.1 Vegetative Measurements

Three plants were randomly taken from the two central ridges of each plot through different growth stages to determine the following attributes:

Total chlorophyll content (%) was measured on the 45th, 66th, and 78th day using a hand-held dual-wavelength chlorophyll meter (SPAD 502; Minolta) [32]. Whereas, the dry weight of whole plant (g), the dry weight of shoot (g), the dry weight of root (g), the dry weight ratio of shoot/root, leaf area ratio (cm^2 mg^{-1}), absolute growth rate (mg day^{-1}), and unit shoot rate (mg mg^{-1} day^{-1}). For these attributes the different parts of sesame plant were collected on the 46th, 63th, and 79th day, as well as dried in the oven at 65 °C for three days to a constant weight.

2.6 *Experimental Design*

A randomized complete block design (RCBD) with three replications was utilized. The experiment consisted of 21 plots for each season. Each plot consisted of 4 ridges, 0.60 m apart, and 3 m long (7.2 m^2).

3 Results and Discussion

All treatments did significant impact on vegetative measurements, as well as the treatment of 50% of mineral in interaction with 50% of compost attained the best results, while the treatment of 100% of bioformulations attained the worst results in both seasons.

3.1 *Total Chlorophyll Content (%)*

3.1.1 In the First Age

Results in Fig. 1 illustrated that the treatment of 100% of mineral (control treatment) (T_1) recorded the highest significant value (24.62%), but the treatment of 100% of bioformulations (T_3) recorded the least value (13.53%) in the first season. Results in Fig. 2 illustrated that the treatment of 100% of mineral (control treatment) (T_1) recorded the highest value (19.11%), but the treatment of 100% of bioformulations (T_3) recorded the least value (12.79%) with significant difference between them in the second season. Results in Fig. 3 illustrated that the treatment of 100% of mineral (control treatment) (T_1) recorded the highest significant value (21.87%), but the

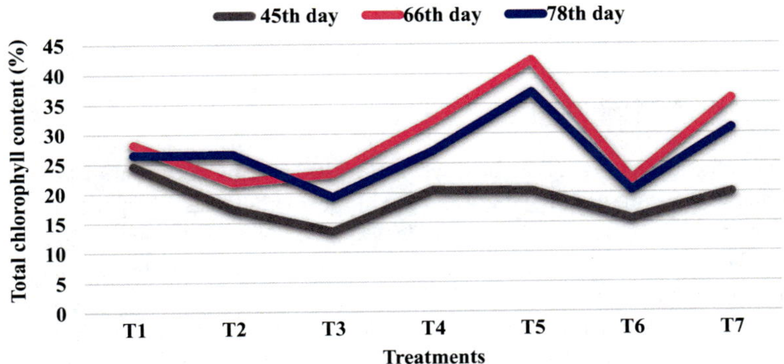

Fig. 1 Effect of bio, organic, and mineral fertilizers on total chlorophyll content % in 2015 season

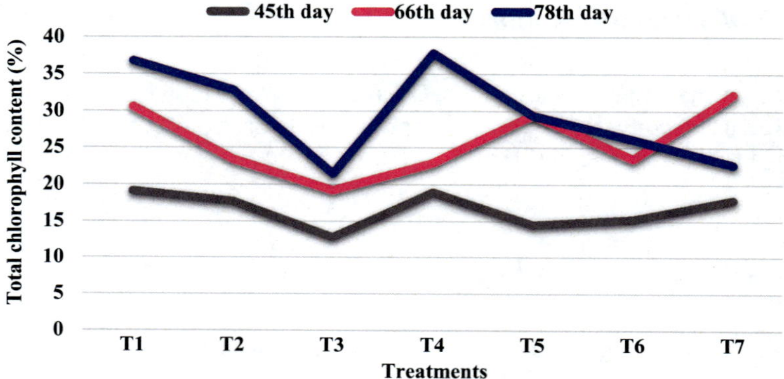

Fig. 2 Effect of bio, organic, and mineral fertilizers on total chlorophyll content % in 2016 season

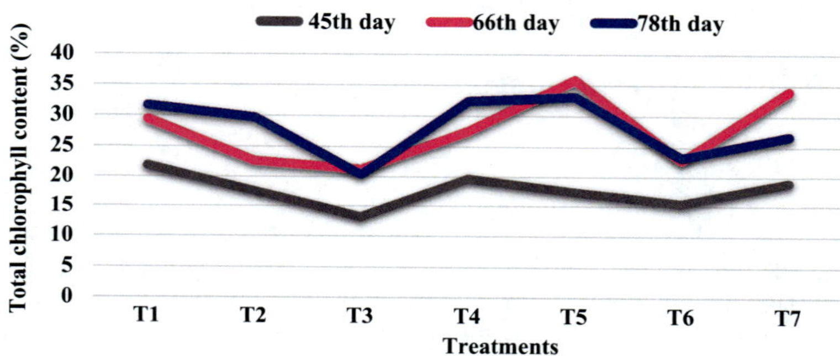

Fig. 3 Effect of bio, organic, and mineral fertilizers on total chlorophyll content % in 2015 and 2016 seasons

treatment of 100% of bioformulations (T_3) recorded the least value (13.16%) in both seasons.

3.1.2 In the Second Age

Results in Fig. 1 illustrated that the treatment of 100% of bioformulations in combination with 50% of mineral (T_5) recorded the highest value (42.22%), but the treatment of 100% of compost (T_2) recorded the least value (21.86%) with significant difference between them in the first season. On the other hand, Abdel-Sabour and Abo El-Seoud [11] found that the combination of the two organic wastes of compost viz., biosolids + water hyacinth positively impacted and recorded a maximum value of total chlorophyll content plant^{-1} in a sandy soil. Results in Fig. 2 illustrated that the treatment of 100% of bioformulations combined with 33% of compost and 33% of mineral (T_7) recorded the highest value (32.23%), but the treatment of 100% of bioformulations (T_3) recorded the least value (19.20%) with significant difference between them in the second season. Results in Fig. 3 illustrated that the treatment of 100% of bioformulations in combination with 50% of mineral (T_5) recorded the highest value (35.84%), but the treatment of 100% of bioformulations (T_3) recorded the least value (21.25%) with significant difference between them in both seasons.

3.1.3 In the Third Age

Results in Fig. 1 illustrated that the treatment of 100% of bioformulations in combination with 50% of mineral (T_5) recorded the highest significant value (36.90%), but the treatment of 100% of bioformulations (T_3) recorded the least value (19.33%) in the first season. Results in Fig. 2 illustrated that the treatment of 50% of mineral in interaction with 50% of compost (T_4) recorded the highest value (37.80%), but the treatment of 100% of bioformulations (T_3) recorded the least value (21.45%) with significant difference between them in the second season. Results in Fig. 3 illustrated that the treatment of 100% of bioformulations in combination with 50% of mineral (T_5) recorded the highest value (33.14%), but the treatment of 100% of bioformulations (T_3) recorded the least value (20.39%) with significant difference between them in both seasons.

3.2 *The Dry Weight of Whole Plant (g)*

3.2.1 In the First Age

Results in Fig. 4 indicated that the treatment of 100% of mineral (control treatment)

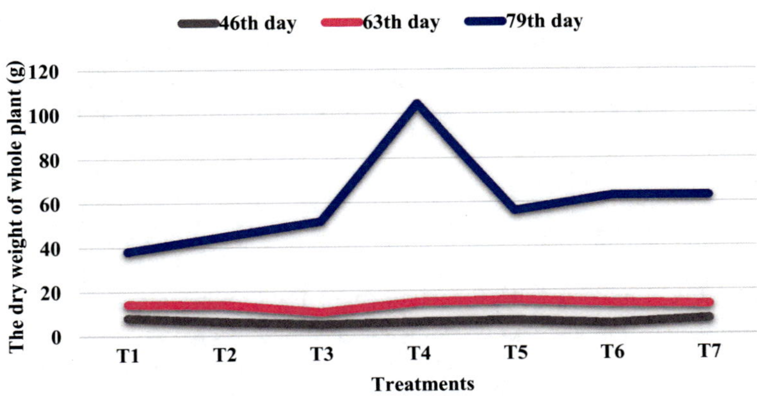

Fig. 4 Effect of bio, organic, and mineral fertilizers on the dry weight of whole plant (g) in 2015 season

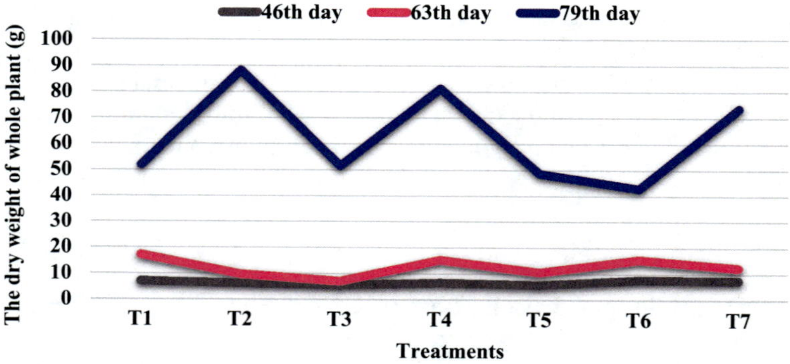

Fig. 5 Effect of bio, organic, and mineral fertilizers on the dry weight of whole plant (g) in 2016 season

(T_1) attained a maximum value (8.17 g), but the treatment of 100% of bioformulations (T_3) attained a minimum value (4.89 g) with significant difference between them in the first season. This result deviated from those obtained by Kumar et al. [33] who found that the inoculated seeds by PGPR *Pseudomonas aeruginosa* (LES4) significantly increased the dry weight of shoot and root. Results in Fig. 5 indicated that the treatment of 100% of bioformulations mixed with 50% of compost (T_6) attained a maximum value (7.73 g), but the treatment of 100% of bioformulations (T_3) attained a minimum value (5.64 g) without significant difference between them and the other treatments in the second season. On the other hand, Sabannavar and Lakshman [34], Sabannavar and Lakshman [35] noticed that the interaction of the two varieties viz., DS 1 and E 8 × the inoculation of *Glomus fasciculatum* (*Thax. sensu Gerd.*) *Gerd.* (Gf), *Azotobacter chroococcum* (Ac), and *Pseudomonas fluorescens* (Pf) or the interaction of TSES 1 variety × the inoculation of Gf, Ac, and Pf or the interaction of TSES 4 variety × the inoculation of *Acaulospora laevis*, Ac, and Pf significantly increased the dry weight of shoot and root in a sandy loam soil. Results in Fig. 6 indicated that the treatment of 100% of mineral (control treatment) (T_1) attained a maximum value (7.60 g), but the treatment of 100% of bioformulations (T_3) attained a minimum value (5.27 g) with significant difference between them in both seasons.

3.2.2 In the Second Age

Results in Fig. 4 indicated that the treatment of 100% of bioformulations in combination with 50% of mineral (T_5) attained a maximum value (15.54 g), but the treatment of 100% of bioformulations (T_3) attained a minimum value (10.48 g) without significant difference between them and the other treatments in the first season. This result is in accordance with those obtained by Kumar et al. [36] who noticed that the combination of the inoculated seeds by *Pseudomonas aeruginosa* (LPT5)

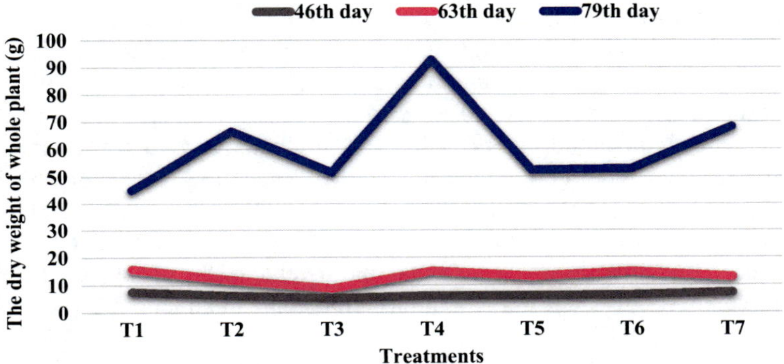

Fig. 6 Effect of bio, organic, and mineral fertilizers on the dry weight of whole plant (g) in 2015 and 2016 seasons

+ 50% of NPKS i.e., 60 kg urea:15 kg diammonium phosphate:15 kg muriate of potash:30 kg gypsum ha^{-1} gave the highest values of the dry weight of shoot and root in a sandy loam soil. Moreover, Dubey et al. [14] indicated that the inoculated seeds by *Azotobacter chroococcum* (AZO2) in combination with 50% of NPKS i.e., 60 kg urea:15 kg diammonium phosphate:15 kg muriate of potash:30 kg gypsum ha^{-1} gave the highest values of the dry weight of shoot and root in a sandy loam soil. This result deviated from those obtained by Maheshwari et al. [37] who noticed that the inoculated seeds via *Azotobacter chroococcum* (TRA2) significantly increased the dry weight of shoot and root in a sandy loam soil. Results in Fig. 5 indicated that the treatment of 100% of mineral (control treatment) (T_1) attained a maximum value (17.19 g), but the treatment of 100% of bioformulations (T_3) attained a minimum value (7.10 g) with significant difference between them in the second season. On the other hand, Ahmed et al. [10] found that the inoculation of 350 kg bio-NP fertilizer viz., *B. polymyxa*, *Azotobacter* spp., 3 g/100 kg boron active, 200 g/100 kg P_2O_5, 4000 g/100 kg calcium, and bio-stimulator fed^{-1} significantly increased the total dry weight $plant^{-1}$. Results in Fig. 6 indicated that the treatment of 100% of mineral (control treatment) (T_1) attained a maximum value (15.76 g), but the treatment of 100% of bioformulations (T_3) attained a minimum value (8.79 g) with significant difference between them in both seasons.

3.2.3 In the Third Age

Results in Fig. 4 indicated that the treatment of 50% of mineral in interaction with 50% of compost (T_4) attained a maximum and significant value (104.47 g), but the treatment of 100% of mineral (control treatment) (T_1) attained a minimum value (38.26 g) in the first season. This result is in parallel with those obtained by Yadav et al. [38] who noticed that the interaction of the two organic manures i.e., 2.5 ton farmyard manure ha^{-1} with 0.5 ton need cake ha^{-1} × 40 kg N:20 kg P:20 kg K:30 kg

S:20 kg zinc sulfate:25 kg ferrous sulfate ha^{-1} significantly increased the dry weight $plant^{-1}$ in a sandy clay soil. Results in Fig. 5 indicated that the treatment of 100% of compost (T_2) attained a maximum value (88.22 g), but the treatment of 100% of bioformulations mixed with 50% of compost (T_6) attained a minimum value (42.89 g) with significant difference between them in the second season. It could be due to the high content of ammoniacal nitrogen—$NH4^+$ (383 ppm), organic matter (39.65%), and organic carbon (22.99%) in compost. This result is in harmony with those obtained by Nurhayati et al. [39] who mentioned that the addition of 30 ton cow manure ha^{-1} on white sesame gave a maximum value of the dry weight of plant in a coastal sandy soil. Results in Fig. 6 indicated that the treatment of 50% of mineral in interaction with 50% of compost (T_4) attained a maximum and significant value (92.87 g), but the treatment of 100% of mineral (control treatment) (T_1) attained a minimum value (45.08 g) in both seasons. This result is in accordance with those obtained by Haruna [40], Haruna et al. [41] who found that the combination of 15 ton poultry manure ha^{-1} and 60 kg urea ha^{-1} recorded a maximum value of the total dry weight $plant^{-1}$.

3.3 *The Dry Weight of Shoot (g)*

3.3.1 In the First Age

Results in Fig. 7 showed that the treatment of 100% of mineral (control treatment) (T_1) gave the highest value (5.78 g), but the treatment of 100% of bioformulations (T_3) gave the least value (3.71 g) with significant difference between them in the first season. This result is in parallel with that obtained by Abou-Taleb [42] who stated that the addition of 40 kg urea fed^{-1} gave the highest value of the dry weight of shoot in the second season under clay loam soil conditions. This result deviated

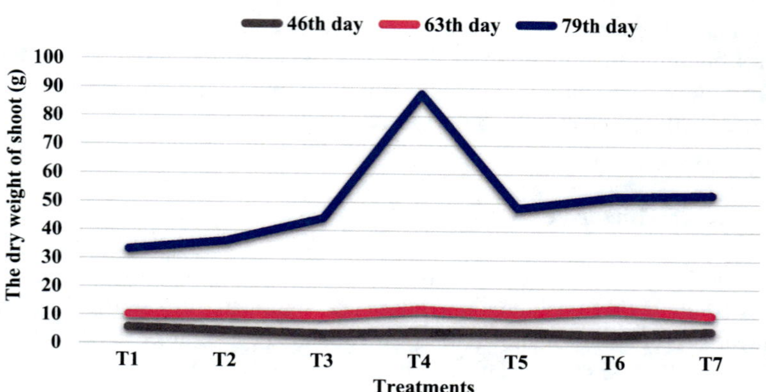

Fig. 7 Effect of bio, organic, and mineral fertilizers on the dry weight of shoot (g) in 2015 season

from those obtained by Kumar et al. [33] who found that the inoculated seeds by PGPR *Pseudomonas aeruginosa* (LES4) significantly increased the dry weight of shoot in a sandy loam soil. Results in Fig. 8 showed that the treatment of 100% of bioformulations combined with 33% of compost and 33% of mineral (T_7) gave the highest value (5.79 g), but the treatment of 50% of mineral in interaction with 50% of compost (T_4) gave the least value (3.62 g) with significant difference between them in the second season. Results in Fig. 9 showed that the treatment of 100% of bioformulations combined with 33% of compost and 33% of mineral (T_7) gave the highest value (5.49 g), but the treatment of 100% of bioformulations (T_3) gave the least value (3.87 g) with significant difference between them in both seasons. On the other hand, Sabannavar and Lakshman [34, 35] noticed that the interaction of the two varieties viz., DS 1 and E 8 × the inoculation of *Glomus fasciculatum* (*Thax. sensu Gerd.*) *Gerd.* (Gf), *Azotobacter chroococcum* (Ac), and *Pseudomonas fluorescens*

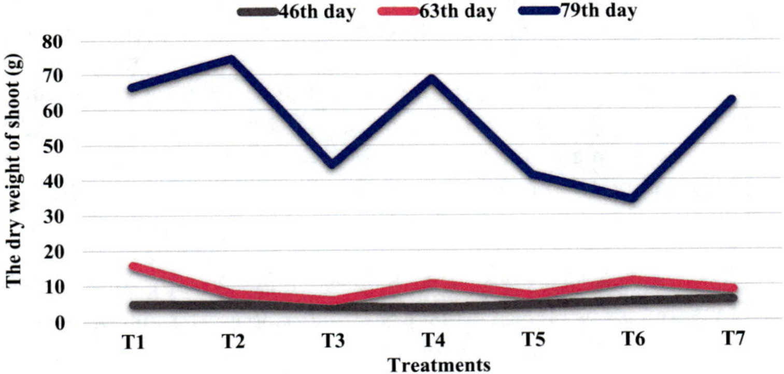

Fig. 8 Effect of bio, organic, and mineral fertilizers on the dry weight of shoot (g) in 2016 season

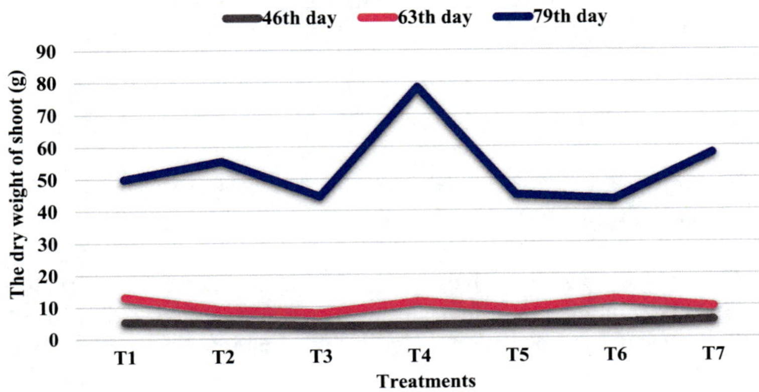

Fig. 9 Effect of bio, organic, and mineral fertilizers on the dry weight of shoot (g) in 2015 and 2016 seasons

(Pf) or the interaction of TSES 1 variety × the inoculation of Gf, Ac, and Pf or the interaction of TSES 4 variety × the inoculation of *Acaulospora laevis*, Ac, and Pf significantly increased the dry weight of shoot in a sandy loam soil.

3.3.2 In the Second Age

Results in Fig. 7 showed that the treatment of 100% of bioformulations mixed with 50% of compost (T_6) gave the highest value (12.81 g), but the treatment of 100% of bioformulations (T_3) gave the least value (10.02 g) without significant difference between them and the other treatments in the first season. This result deviated from that obtained by Abou-Taleb [42] who stated that the inoculation of 400 g *Azospirillum brasilense* fed^{-1} resulted in an increase in the dry weight of shoot in the first season under clay loam soil conditions. Results in Fig. 8 showed that the treatment of 100% of mineral (control treatment) (T_1) gave the highest significant value (15.87 g), but the treatment of 100% of bioformulations (T_3) gave the least value (5.71 g) in the second season. This result is in harmony with those obtained by Boghdady et al. [2] who found that increasing N-levels from 50 to 100 and 200 kg ammonium sulfate fed^{-1} with increasing P-levels from 50 to 100 and 200 kg calcium superphosphate fed^{-1} significantly increased the dry weight of shoot plant^{-1} of Shandaweel-3 cv. in a clay loam soil. On the other hand, Maheshwari et al. [37] noticed that the inoculated seeds via *Azotobacter chroococcum* (TRA2) significantly increased the dry weight of shoot in a sandy loam soil. Results in Fig. 9 showed that the treatment of 100% of mineral (control treatment) (T_1) gave the highest value (13.07 g), but the treatment of 100% of bioformulations (T_3) gave the least value (7.86 g) with significant difference between them in both seasons. This result is in accordance with those obtained by Siddik et al. [9] who pointed out that the addition of 60 or 120 kg N ha^{-1} recorded a maximum and significant value of the dry weight of shoot in a sandy loam soil. This result deviated from those obtained by Kumar et al. [12] who noticed that the inoculation of *Azotobacter* and phosphate solubilizing bacteria together or the individual inoculation with Azo resulted in an increase in the dry weight of shoot.

3.3.3 In the Third Age

Results in Fig. 7 showed that the treatment of 50% of mineral in interaction with 50% of compost (T_4) gave the highest significant value (87.90 g), but the treatment of 100% of mineral (control treatment) (T_1) gave the least value (33.35 g) in the first season. On the other hand, Boghdady et al. [2] found that increasing N-levels from 50 to 100 and 200 kg ammonium sulfate fed^{-1} with increasing P-levels from 50 to 100 and 200 kg calcium superphosphate fed^{-1} significantly increased the dry weight of shoot plant^{-1} of Shandaweel-3 cv. in a clay loam soil. Furthermore, Siddik et al. [9] pointed out that the addition of 60 or 120 kg N ha^{-1} recorded a maximum and significant value of the dry weight of shoot in a sandy loam soil. Results in Fig. 8 showed that the treatment of 100% of compost (T_2) gave the highest value (74.70 g), but the treatment of 100% of bioformulations mixed with 50% of compost (T_6) gave the least value (34.23 g) with significant difference between them in the second season. It could be due to the high content of ammoniacal nitrogen—$NH4^+$ (383 ppm), organic matter (39.65%), and organic carbon (22.99%) in compost. Results in Fig. 9 showed that the treatment of 50% of mineral in interaction with 50% of compost (T_4) gave the highest significant value (78.32 g), but the treatment of 100% of bioformulations mixed with 50% of compost (T_6) gave the least value (43.32 g) in both seasons.

3.4 The Dry Weight of Root (g)

3.4.1 In the First Age

Results in Fig. 10 displayed that the treatment of 100% of bioformulations combined with 33% of compost and 33% of mineral (T_7) gave the highest value (2.12 g), but the treatment of 100% of bioformulations (T_3) gave the least value (1.26 g) with significant difference between them in the first season. This result deviated from those obtained by Kumar et al. [33] who found that the inoculated seeds by PGPR *Pseudomonas aeruginosa* (LES4) significantly increased the dry weight of root in a sandy loam soil. Results in Fig. 11 displayed that the treatment of 100% of mineral (control treatment) (T_1) gave the highest value (2.26 g), but the treatment of 100% of bioformulations in combination with 50% of mineral (T_5) gave the least value (1.39 g) with significant difference between them in the second season. This result is in parallel with those obtained by Siddik et al. [9] who pointed out that the addition of 60 kg N ha^{-1} gave a maximum and significant value of the dry weight of root in a sandy loam soil. On the other hand, Kumar et al. [36] noticed that the combination of the inoculated seeds by *Pseudomonas aeruginosa* (LPT5) + 50% of NPKS i.e., 60 kg urea:15 kg diammonium phosphate:15 kg muriate of potash:30 kg gypsum ha^{-1} gave the highest value of the dry weight of root in a sandy loam soil.

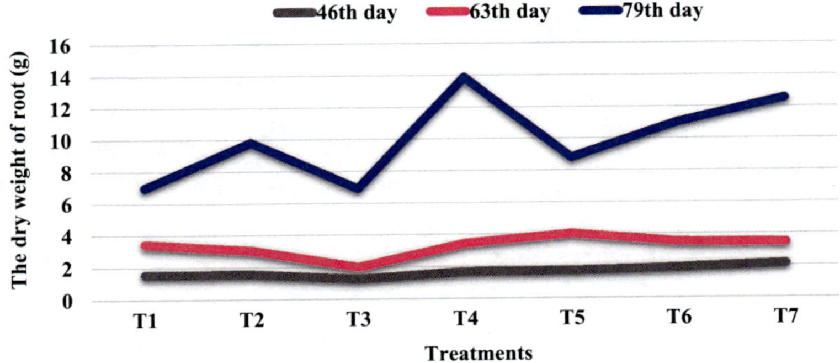

Fig. 10 Effect of bio, organic, and mineral fertilizers on the dry weight of root (g) in 2015 season

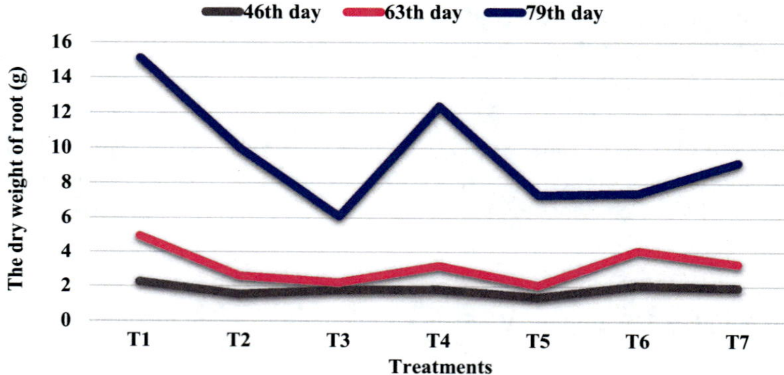

Fig. 11 Effect of bio, organic, and mineral fertilizers on the dry weight of root (g) in 2016 season

Results in Fig. 12 displayed that the treatment of 100% of bioformulations combined with 33% of compost and 33% of mineral (T_7) gave the highest value (2.03 g), but the treatment of 100% of bioformulations (T_3) gave the least value (1.52 g) with significant difference between them in both seasons. This result deviated from those obtained by Sabannavar and Lakshman [34], Sabannavar and Lakshman [35] who noticed that the interaction of the two varieties viz., DS 1 and E 8 × the inoculation of *Glomus fasciculatum* (*Thax. sensu Gerd.*) *Gerd.* (Gf), *Azotobacter chroococcum* (Ac), and *Pseudomonas fluorescens* (Pf) or the interaction of TSES 1 variety × the inoculation of Gf, Ac, and Pf or the interaction of TSES 4 variety × the inoculation of *Acaulospora laevis*, Ac, and Pf significantly increased the dry weight of root in a sandy loam soil.

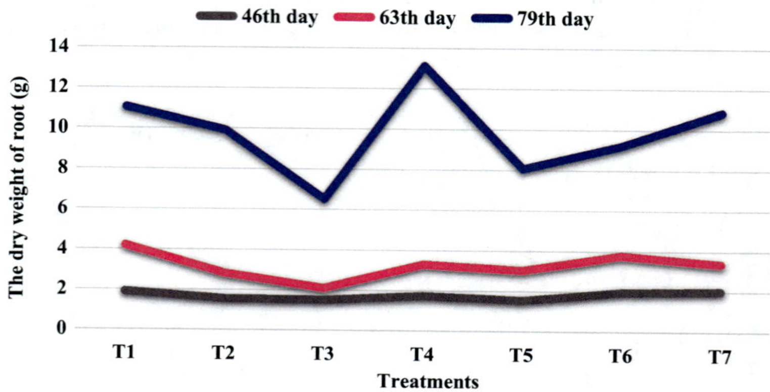

Fig. 12 Effect of bio, organic, and mineral fertilizers on the dry weight of root (g) in 2015 and 2016 seasons

3.4.2 In the Second Age

Results in Fig. 10 displayed that the treatment of 100% of bioformulations in combination with 50% of mineral (T_5) gave the highest value (4.00 g), but the treatment of 100% of bioformulations (T_3) gave the least value (1.98 g) with significant difference between them in the first season. This result is in harmony with those obtained by Kumar et al. [36] who noticed that the combination of the inoculated seeds by *Pseudomonas aeruginosa* (LPT5) + 50% of NPKS i.e., 60 kg urea:15 kg diammonium phosphate:15 kg muriate of potash:30 kg gypsum ha^{-1} gave the highest value of the dry weight of root in a sandy loam soil. Additionally, Dubey et al. [14] indicated that the inoculated seeds by *Azotobacter chroococcum* (AZO2) in combination with 50% of NPKS i.e., 60 kg urea:15 kg diammonium phosphate:15 kg muriate of potash:30 kg gypsum ha^{-1} gave the highest value of the dry weight of root in a sandy loam soil. On the other hand, Maheshwari et al. [37] noticed that the inoculated seeds via *Azotobacter chroococcum* (TRA2) significantly increased the dry weight of root in a sandy loam soil. Results in Fig. 11 displayed that the treatment of 100% of mineral (control treatment) (T_1) gave the highest value (4.94 g), but the treatment of 100% of bioformulations in combination with 50% of mineral (T_5) gave the least value (2.06 g) with significant difference between them in the second season. This result deviated from those obtained by Sabannavar and Lakshman [43] who noticed that the combination of TSES 4 variety + the inoculation by *Acaulospora laevis* with *Pseudomonas striata* + 30 kg dose of rock phosphate ha^{-1} significantly affected the dry weight of root in a sandy loam soil. Moreover, Dubey et al. [14] indicated that the inoculated seeds by *Azotobacter chroococcum* (AZO2) in combination with 50% of NPKS i.e., 60 kg urea:15 kg diammonium phosphate:15 kg muriate of potash:30 kg gypsum ha^{-1} gave the highest value of the dry weight of root in a sandy loam soil. Results in Fig. 12 displayed that the treatment of 100% of mineral (control treatment) (T_1) gave the highest value (4.19 g), but the treatment of 100% of bioformulations (T_3) gave the least value (2.10 g) with significant difference between them in both seasons.

3.4.3 In the Third Age

Results in Fig. 10 displayed that the treatment of 50% of mineral in interaction with 50% of compost (T_4) gave the highest value (13.88 g), but the treatment of 100% of bioformulations (T_3) gave the least value (6.91 g) with significant difference between them in the first season. Results in Fig. 11 displayed that the treatment of 100% of mineral (control treatment) (T_1) gave the highest value (15.11 g), but the treatment of 100% of bioformulations (T_3) gave the least value (6.10 g) with significant difference between them in the second season. Results in Fig. 12 displayed that the treatment of 50% of mineral in interaction with 50% of compost (T_4) gave the highest value (13.13 g), but the treatment of 100% of bioformulations (T_3) gave the least value (6.51 g) without significant difference between them and the other treatments in both seasons.

3.5 *The Dry Weight Ratio of Shoot/Root*

3.5.1 In the First Age

Results in Fig. 13 illustrated that the treatment of 100% of mineral (control treatment) (T_1) recorded a maximum value (5.04), but the treatment of 100% of bioformulations mixed with 50% of compost (T_6) recorded a minimum value (2.13) with significant difference between them in the first season. Results in Fig. 14 illustrated that the treatment of 100% of bioformulations in combination with 50% of mineral (T_5) recorded a maximum value (3.10), but the treatment of 50% of mineral in interaction with 50% of compost (T_4) recorded a minimum value (2.04) with significant difference between them in the second season. Results in Fig. 15 illustrated that the treatment

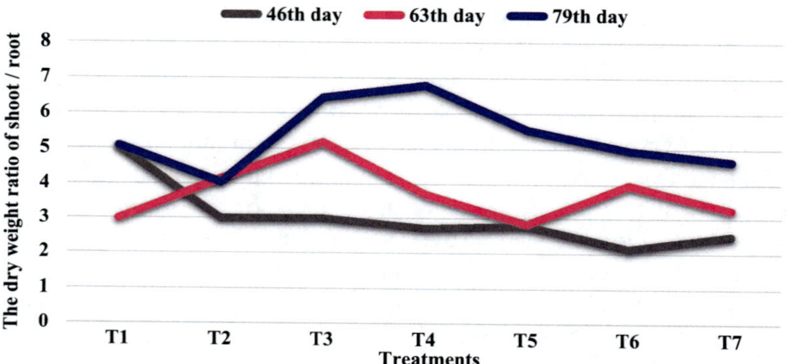

Fig. 13 Effect of bio, organic, and mineral fertilizers on the dry weight ratio of shoot/root in 2015 season

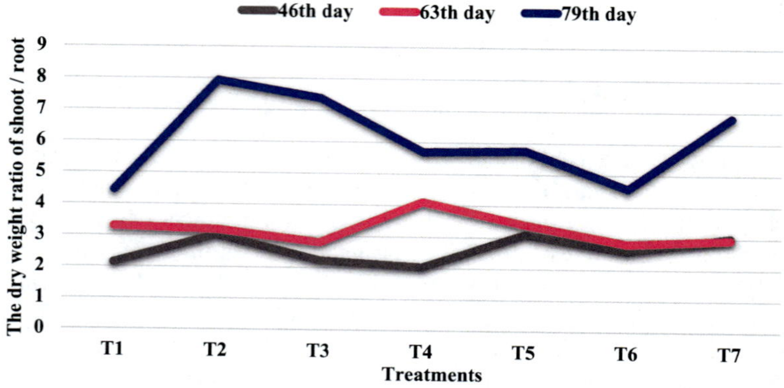

Fig. 14 Effect of bio, organic, and mineral fertilizers on the dry weight ratio of shoot/root in 2016 season

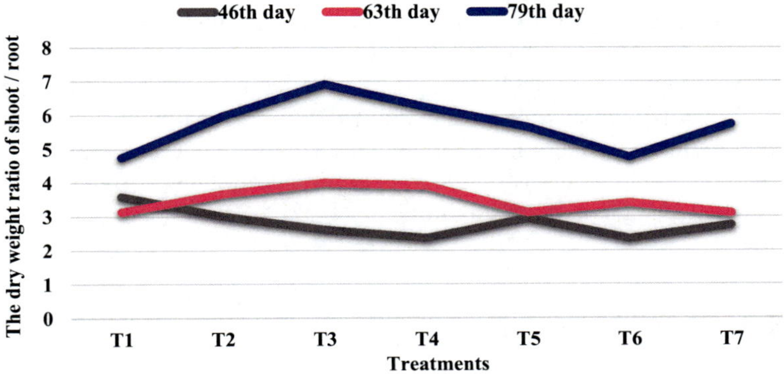

Fig. 15 Effect of bio, organic, and mineral fertilizers on the dry weight ratio of shoot/root in 2015 and 2016 seasons

of 100% of mineral (control treatment) (T_1) recorded a maximum value (3.59), but the treatment of 100% of bioformulations mixed with 50% of compost (T_6) recorded a minimum value (2.35) without significant difference between them and the other treatments in both seasons.

3.5.2 In the Second Age

Results in Fig. 13 illustrated that the treatment of 100% of bioformulations (T_3) recorded a maximum value (5.18), but the treatment of 100% of bioformulations in combination with 50% of mineral (T_5) recorded a minimum value (2.83) with significant difference between them in the first season. Results in Fig. 14 illustrated that the treatment of 50% of mineral in interaction with 50% of compost (T_4) recorded a maximum value (4.11), but the treatment of 100% of bioformulations (T_3) recorded a minimum value (2.81) without significant difference between them and the other treatments in the second season. Results in Fig. 15 illustrated that the treatment of 100% of bioformulations (T_3) recorded a maximum value (4.00), but the treatment of 100% of bioformulations combined with 33% of compost and 33% of mineral (T_7) recorded a minimum value (3.10) without significant difference between them and the other treatments in both seasons.

3.5.3 In the Third Age

Results in Fig. 13 illustrated that the treatment of 50% of mineral in interaction with 50% of compost (T_4) recorded a maximum value (6.79), but the treatment of 100% of compost (T_2) recorded a minimum value (4.04) with significant difference between them in the first season. Results in Fig. 14 illustrated that the treatment of 100% of compost (T_2) recorded a maximum value (7.93), but the treatment of 100% of mineral (control treatment) (T_1) recorded a minimum value (4.45) with significant difference between them in the second season. It could be due to the high content of ammoniacal nitrogen—$NH4^+$ (383 ppm), organic matter (39.65%), and organic carbon (22.99%) in compost. Results in Fig. 15 illustrated that the treatment of 100% of bioformulations (T_3) recorded a maximum value (6.91), but the treatment of 100% of mineral (control treatment) (T_1) recorded a minimum value (4.76) with significant difference between them in both seasons.

3.6 *Leaf Area Ratio* (cm^2 mg^{-1})

3.6.1 The Relationship Between First and Second Age

Results in Fig. 16 indicated that the treatment of 100% of bioformulations mixed with 50% of compost (T_6) achieved the highest value (0.0026 cm^2 mg^{-1}), but the treatment of 100% of bioformulations (T_3) achieved the least value (0.0014 cm^2 mg^{-1}) with significant difference between them in the first season. Results in Fig. 17 indicated that the treatment of 50% of mineral in interaction with 50% of compost (T_4) achieved the highest value (0.0022 cm^2 mg^{-1}), but the treatment of 100% of bioformulations mixed with 50% of compost (T_6) achieved the least value (0.0016 cm^2 mg^{-1}) without significant difference between them and the other treatments in the second season. Results in Fig. 18 indicated that the treatment of 100% of bioformulations mixed with 50% of compost (T_6) achieved the highest value (0.0021 cm^2 mg^{-1}), but the treatment of 100% of bioformulations combined with 33% of compost and 33% of mineral (T_7) achieved the least value (0.0017 cm^2 mg^{-1}) without significant difference between them and the other treatments in both seasons.

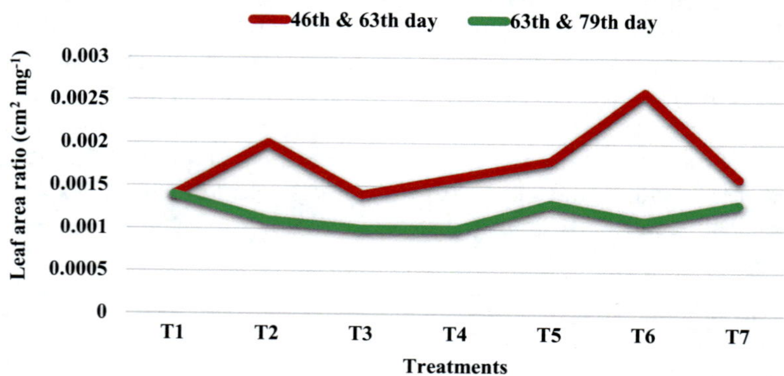

Fig. 16 Effect of bio, organic, and mineral fertilizers on leaf area ratio (cm^2 mg^{-1}) in 2015 season

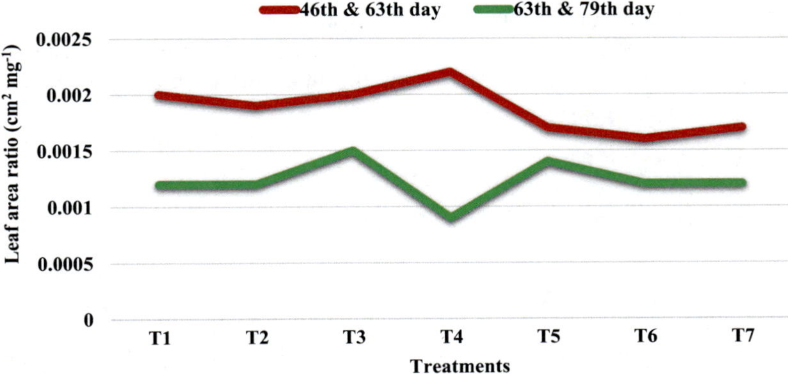

Fig. 17 Effect of bio, organic, and mineral fertilizers on leaf area ratio ($cm^2\ mg^{-1}$) in 2016 season

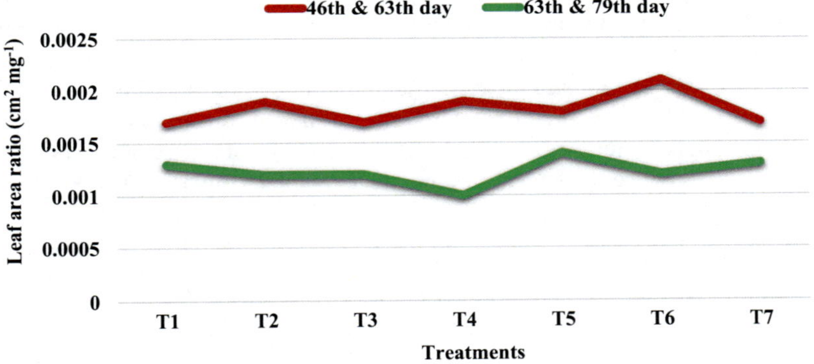

Fig. 18 Effect of bio, organic, and mineral fertilizers on leaf area ratio ($cm^2\ mg^{-1}$) in 2015 and 2016 seasons

3.6.2 The Relationship Between Second and Third Age

Results in Fig. 16 indicated that the treatment of 100% of mineral (control treatment) (T_1) achieved the highest value (0.0014 $cm^2\ mg^{-1}$), but the treatment of 100% of bioformulations (T_3) achieved the least value (0.0010 $cm^2\ mg^{-1}$) without significant difference between them and the other treatments in the first season. This result is in accordance with those obtained by Kalegore et al. [44] who reported that the addition of 30 kg P_2O_5 ha^{-1} attained the highest significant value of leaf area $plant^{-1}$ in a clay soil. Results in Fig. 17 indicated that the treatment of 100% of bioformulations (T_3) achieved the highest value (0.0015 $cm^2\ mg^{-1}$), but the treatment of 50% of mineral in interaction with 50% of compost (T_4) achieved the least value (0.0009 cm^2 mg^{-1}) with significant difference between them in the second season. Results in Fig. 18 indicated that the treatment of 100% of bioformulations in combination with 50% of mineral (T_5) achieved the highest value (0.0014 cm^2 mg^{-1}), but the treatment of 50% of mineral in interaction with 50% of compost (T_4) achieved the least value (0.0010 cm^2 mg^{-1}) without significant difference between them and the other treatments in both seasons.

3.7 Absolute Growth Rate (mg day^{-1})

3.7.1 The Relationship Between First and Second Age

Results in Fig. 19 showed that the treatment of 100% of bioformulations in combination with 50% of mineral (T_5) attained a maximum value (15,167.32 mg day^{-1}), but the treatment of 100% of bioformulations (T_3) attained a minimum value (10,190.20 mg day^{-1}) without significant difference between them and the other treatments in the first season. Results in Fig. 20 showed that the treatment of 100% of mineral (control treatment) (T_1) attained a maximum value (16,775.16 mg day^{-1}), but the treatment of 100% of bioformulations (T_3) attained a minimum value (6768.10 mg day^{-1}) with significant difference between them in the second season. Results in Fig. 21 showed that the treatment of 100% of mineral (control treatment) (T_1) attained a maximum value (15,314.05 mg day^{-1}), but the treatment of 100% of bioformulations (T_3) attained a minimum value (8479.15 mg day^{-1}) with significant difference between them in both seasons.

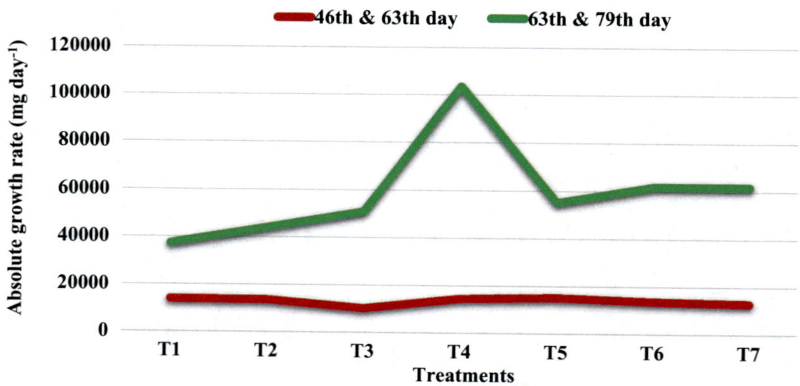

Fig. 19 Effect of bio, organic, and mineral fertilizers on absolute growth rate (mg day^{-1}) in 2015 season

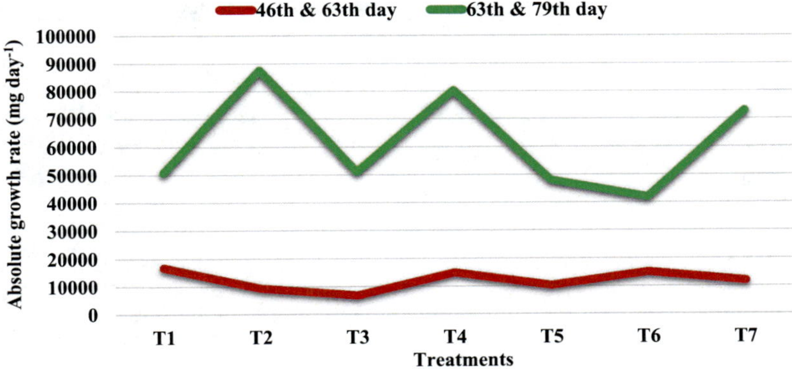

Fig. 20 Effect of bio, organic, and mineral fertilizers on absolute growth rate (mg day^{-1}) in 2016 season

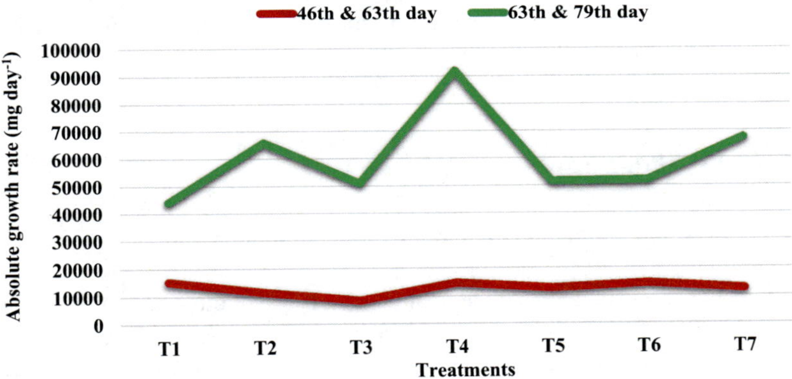

Fig. 21 Effect of bio, organic, and mineral fertilizers on absolute growth rate (mg day^{-1}) in 2015 and 2016 seasons

3.7.2 The Relationship Between Second and Third Age

Results in Fig. 19 showed that the treatment of 50% of mineral in interaction with 50% of compost (T_4) attained a maximum and significant value (103,542.57 mg day^{-1}), but the treatment of 100% of mineral (control treatment) (T_1) attained a minimum value (37,359.17 mg day^{-1}) in the first season. Results in Fig. 20 showed that the treatment of 100% of compost (T_2) attained a maximum value (87,610.21 mg day^{-1}), but the treatment of 100% of bioformulations mixed with 50% of compost (T_6) attained a minimum value (41,919.31 mg day^{-1}) with significant difference between them in the second season. It could be due to the high content of ammoniacal nitrogen—$NH4^+$ (383 ppm), organic matter (39.65%), and organic carbon (22.99%) in compost. Results in Fig. 21 showed that the treatment of 50% of mineral in interaction with 50%

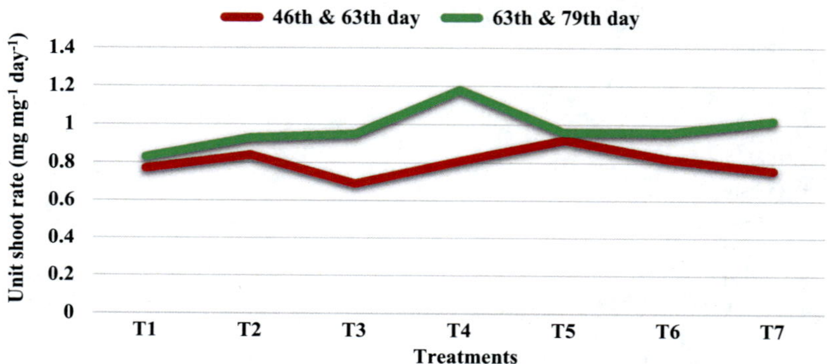

Fig. 22 Effect of bio, organic, and mineral fertilizers on unit shoot rate (mg mg^{-1} day^{-1}) in 2015 season

of compost (T_4) attained a maximum and significant value (91,927.43 mg day^{-1}), but the treatment of 100% of mineral (control treatment) (T_1) attained a minimum value (44,092.43 mg day^{-1}) in both seasons.

3.8 *Unite Shoot Rate (mg mg^{-1} day^{-1})*

3.8.1 The Relationship Between the First and Second Age

Results in Fig. 22 displayed that the treatment of 100% of bioformulations in combination with 50% of mineral (T_5) recorded the highest value (0.92 mg mg^{-1} day^{-1}), but the treatment of 100% of bioformulations (T_3) recorded the least value (0.69 mg mg^{-1} day^{-1}) without significant difference between them and the other treatments in the first season. Results in Fig. 23 displayed that the treatment of 50% of mineral in interaction with 50% of compost (T_4) recorded the highest value (1.00 mg mg^{-1} day^{-1}), but the treatment of 100% of bioformulations (T_3) recorded the least value (0.62 mg mg^{-1} day^{-1}) with significant difference between them in the second season. Results in Fig. 24 displayed that the treatment of 50% of mineral in interaction with 50% of compost (T_4) recorded the highest value (0.90 mg mg^{-1} day^{-1}), but the treatment of 100% of bioformulations (T_3) recorded the least value (0.66 mg mg^{-1} day^{-1}) with significant difference between them in both seasons.

3.8.2 The Relationship Between Second and Third Age

Results in Fig. 22 displayed that the treatment of 50% of mineral in interaction with 50% of compost (T_4) recorded the highest significant value (1.18 mg mg^{-1} day^{-1}), but the treatment of 100% of mineral (control treatment) (T_1) recorded the least

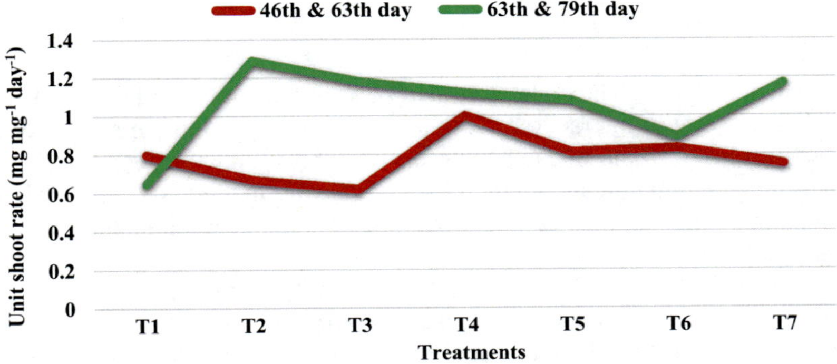

Fig. 23 Effect of bio, organic, and mineral fertilizers on unit shoot rate (mg mg^{-1} day^{-1}) in 2016 season

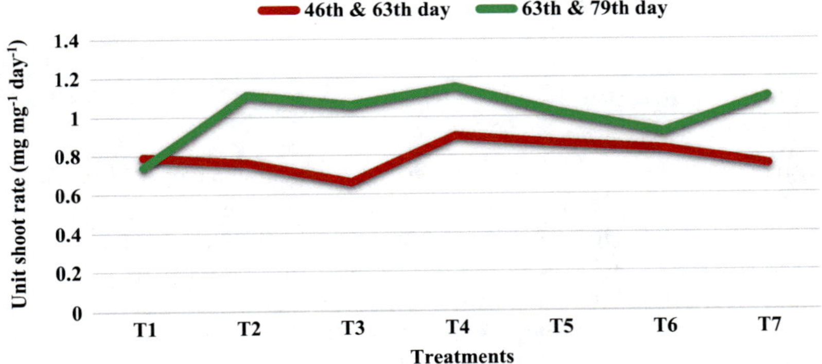

Fig. 24 Effect of bio, organic, and mineral fertilizers on unit shoot rate (mg mg^{-1} day^{-1}) in 2015 and 2016 seasons

value (0.83 mg mg^{-1} day^{-1}) in the first season. Results in Fig. 23 displayed that the treatment of 100% of compost (T_2) recorded the highest value (1.29 mg mg^{-1} day^{-1}), but the treatment of 100% of mineral (control treatment) (T_1) recorded the least value (0.65 mg mg^{-1} day^{-1}) with significant difference between them in the second season. It could be due to the high content of ammoniacal nitrogen—$NH4^+$ (383 ppm), organic matter (39.65%), and organic carbon (22.99%) in compost. Results in Fig. 24 displayed that the treatment of 50% of mineral in interaction with 50% of compost (T_4) recorded the highest value (1.15 mg mg^{-1} day^{-1}), but the treatment of 100% of mineral (control treatment) (T_1) recorded the least value (0.74 mg mg^{-1} day^{-1}) with significant difference between them in both seasons.

4 Conclusion

It can be concluded that the mineral fertilizer was partially replaced by the treatment of 50% of mineral in interaction with 50% of compost gave the highest values of the dry weight of whole plant (g), the dry weight of shoot (g), absolute growth rate (mg day^{-1}), and unit shoot rate (mg mg^{-1} day^{-1}) in the first season. Also, the treatment of 50% of mineral in interaction with 50% of compost attained maximum values of absolute growth rate (mg day^{-1}) and unit shoot rate (mg mg^{-1} day^{-1}) in the second season and in both seasons. Treatment of 100% of bioformulations recorded the highest value of the dry weight ratio of shoot/root in the first season and in both seasons. Treatment of 100% of bioformulations in combination with 50% of mineral achieved a maximum value of total chlorophyll content (%) in the first season and in both seasons. Treatment of 100% of bioformulations combined with 33% of compost and 33% of mineral gave a maximum value of the dry weight of root (g) in the first season and in both seasons.

5 Recommendations

It can be recommended that the treatment of the treatment of 50% of mineral in interaction with 50% of compost was the best treatment for the dry weight of whole plant (g), the dry weight of shoot (g), absolute growth rate (mg day^{-1}), and unit shoot rate (mg mg^{-1} day^{-1}) in the first season. Also, the treatment of 50% of mineral in interaction with 50% of compost was the best treatment for absolute growth rate (mg day^{-1}) and unit shoot rate (mg mg^{-1} day^{-1}) in the second season and in both seasons.

References

1. Anilakumar JR, Pal A, Khanum F, Bawa AS (2010) Nutritional, medicinal and industrial uses of sesame (*Sesamum indicum* L.) seeds—an overview. Agric Conspectus Sci 75(4):159–168
2. Boghdady MS, Nassar RMA, Ahmed FA (2012) Response of sesame plant (*Sesamum orientale* L.) to treatment with mineral and bio-fertilizers. Res J Agric Biol Sci 8(2):127–137
3. Noorka IR, Hafiz SI, El-Bramawy MAS (2011) Response of sesame to population densities and nitrogen fertilization on newly reclaimed sandy soils. Pak J Bot 43(4):1953–1958
4. Shaban KA, Abd El-Kader MG, Khalil ZM (2012) Effect of soil amendments on soil fertility and sesame crop productivity under newly reclaimed soil conditions. J Appl Sci Res 8(3):1568–1575
5. Mahrous NM, Abu-Hagaza NM, Abotaleb HH, Fakhry SMK (2015) Enhancement of growth and yield productivity of sesame plants by application of some biological treatments. Am-Eurasian J Agric Environ Sci 15(5):903–912

6. Shehu HE, Kwari JD, Sandabe MK (2010) Effects of N, P and K fertilizers on yield, content and uptake of N, P and K by sesame (*Sesamum indicum*). Int J Agric Biol 12(6):845–850
7. Jat R, Naga SR, Choudhary R, Jat S, Rolaniya MK (2017) Effect of potassium and sulphur on growth, yield attributes and yield of sesame (*Sesamum indicum* L.). Chem Sci Rev Lett 6(21):184–186
8. Wang M, Zheng Q, Shen Q, Guo S (2013) The critical role of potassium in plant stress response. Int J Mol Sci 14(4):7370–7390
9. Siddik A, Shirazy BJ, Islam MM, Hoque A, Haque M (2016) Combined effect of nitrogen and NAA on morphological characters of sesame (*Sesamum indicum* L.). Int J Biomater Sci Eng 3(1):7–14
10. Ahmed AG, Hassanein MS, Zaki NM, Khalifa RKM, Badr EA (2015) Effect of bio-NP fertilizer on yield, yield components and some biochemical components of two sesame varieties. Middle East J Appl Sci 5(3):630–635
11. Abdel-Sabour MF, Abo El-Seoud MA (1996) Effects of organic-waste compost addition on sesame growth, yield and chemical composition. Agric Ecosyst Environ 60:157–164
12. Kumar N, Kumar A, Shukla A, Ram A, Bahadur R, Chaturvedi OP (2018) Effect of application of bio-inoculants on growth and yield of *Arachis hypogaea* L. and *Sesamum indicum* L. Int J Curr Microbiol Appl Sci 7(1):2869–2875
13. Eifediyi EK, Ahamefule HE, Remison SU, Aliyu TH, Akanbi N (2017) Effects of neem seed cake and NPK fertilizer on the growth and yield of sesame (*Sesamum indicum* L.). Cercetări Agronomice în Moldova 2:57–72
14. Dubey RC, Maheshwari DK, Kumar V, Pandey RR (2012) Growth enhancement of *Sesamum indicum* L. by rhizosphere-competent *Azotobacter chroococcum* AZO2 and its antagonistic activity against *Macrophomina phaseolina*. Arch Phytopathol Plant Prot 45(4):437–454
15. Abdullahi R, Sheriff HH, Lihan S (2013) Combine effect of bio-fertilizer and poultry manure on growth, nutrients uptake and microbial population associated with sesame (*Sesamum indicum* L.) in North-Eastern Nigeria. J Environ Sci Toxicol Food Technol 5(5):60–65
16. Ghosh AK, Duary B, Ghosh DC (2013) Nutrient management in summer sesame (*Sesamum indicum* L.) and its residual effect on black gram (*Vigna mungo* L.). Int J Bio-resour Stress Manage 4(4):541–546
17. Hamza M (2010) Response of some promising safflower genotypes to different nitrogen levels under modern irrigation systems in newly reclaimed soils. Ph.D. thesis, Faculty of Agriculture, Cairo University, Egypt, p 1
18. APHA (1989) Standard methods for the examination of water and wastewater, 17th ed. American Public Health Association, Washington, D.C., p 2672
19. Richards LA (1954) Diagnosis and improvement of saline and alkali soil. The pollination system of *Melilotus* species. Ecol Plant 12:383–394
20. Hesse PR (1971) In: Murry J (ed) A text book of soil chemical analysis. Ltd, 50 Albermarle Street, London, p 520
21. Jackson ML (1973) Soil chemical analysis, 2nd ed. Prentice-Hall, Englewood Cliffs, p 930
22. Page AL, Miller RH, Keeney DR (1982) Methods of soil analysis part 2. Soil Science Society of America, Madison, p 310
23. Olsen SR, Cole OV, Watanaba FS, Deen AL (1954) Estimation of available phosphorus in soils by extraction with sodium bicarbonate. U.S. Department Agriculture Circulation No., p 939
24. Troug E, Mayer AH (1949) Improvements in the deniges colorimetric method for phosphorus and arsenic. Ind Eng Chem Anal 1:136–139
25. Chapman HD, Pratt FP (1961) Methods of analysis for soils, plants and water. University of California, Division of Agriculture Science, Riverside, p 309
26. Jackson ML (1973) Iron, manganese, copper, zinc, molybdenum, and cobalt determinations. In: Soil chemical analysis, Chap 15. Prentice-Hall of India, Private Limited, New Delhi, pp 388–415

27. Day PR (1965) Particle fraction and particle-size analysis. In: Black CA, Evans DD, White JL, Ensminger LE, Clark FE (eds) Methods of soil analysis. Part 1, Physical and mineralogical properties, including statistics of measurement and sampling, Chap 43. American Society of Agronomy Inc., Madison, Wisconsin, p 552
28. Dewis J, Freitas F (1970) Physical and chemical methods of soil and water analysis. Soil Bull 10 (FAO, Rome)
29. Vomocil JA (1965) Porosity. In: Black CA, Evans DD, White JL, Ensminger LE, Clark FE (eds) Methods of soil analysis. Part 1, Physical and mineralogical properties, including statistics of measurement and sampling, Chap 21. American Society of Agronomy Inc., Madison, p 300
30. Klute A (1965) Laboratory measurements of hydraulic conductivity of saturated soil. In: Black CA, Evans DD, White JL, Ensminger LE, Clark FE (eds). In: Methods of soil analysis. Part 1, Physical and mineralogical properties, including statistics of measurement and sampling, Chap 13. American Society of Agronomy Inc., Madison, pp 210–220
31. Peters DB (1965) Water availability. In: Black CA, Evans DD, White JL, Ensminger LE, Clark FE (eds) Methods of soil analysis. Part 1, Physical and mineralogical properties, including statistics of measurement and sampling, Chap 19. American Society of Agronomy Inc., Madison, pp 279–285
32. Dehnavi MM, Misagh M, Yadavi A, Merajipoor M (2017) Physiological responses of sesame (*Sesamum indicum* L.) to foliar application of boron and zincunder drought stress. J Plant Process Funct 6(20):27–35
33. Kumar S, Pandey P, Maheshwari DK (2009) Reduction in dose of chemical fertilizers and growth enhancement of sesame (*Sesamum indicum* L.) with application of rhizospheric competent *Pseudomonas aeruginosa* LES4. Eur J Soil Biol 45:334–340
34. Sabannavar SJ, Lakshman HC (2008) Interactions between *Azotobacter*, *Pseudomonas* and arbuscular mycorrhizal fungi on two varieties of *Sesamum indicum* L. J Agron Crop Sci 194(6):470–478
35. Sabannavar SJ, Lakshman HC (2011) Synergistic interactions among *Azotobacter*, *Pseudomonas*, and arbuscular mycorrhizal fungi on two varieties of *Sesamum indicum* L. Commun Soil Sci Plant Anal 42(17):2122–2133
36. Kumar T, Wahla V, Pandey P, Dubey RC, Maheshwari DK (2009) Rhizosphere competent *Pseudomonas aeruginosa* in the management of *Heterodera cajani* on sesame. World J Microbiol Biotechnol 25:277–285
37. Maheshwari DK, Dubey RC, Aeron A, Kumar B, Kumar S, Tewari S, Arora NK (2012) Integrated approach for disease management and growth enhancement of *Sesamum indicum* L. utilizing *Azotobacter chroococcum* TRA2 and chemical fertilizer. World J Microbiol Biotechnol 28(10):3015–3024
38. Yadav RA, Tripathi AK, Yadav AK (2009) Effect of micronutrients in combinations with organic manures on production and net returns of sesame (*Sesamum indicum*) in Bundelkhand tract of Uttar Pradesh. Ann Agric Res New Ser 30(1&2):53–58
39. Nurhayati DR, Yudono P, Taryono, Hanudin E (2016) The application of manure on sesame (*Sesamum indicum* L.) under coastal sandy land area in Yogyakarta, Indonesia. Int Res J Eng Technol 3(4):2047–2051
40. Haruna IM (2011) Dry matter partitioning and grain yield potential in sesame (*Sesamum indicum* L.) as influenced by poultry manure, nitrogen and phosphorus at Samaru, Nigeria. J Agric Technol 7(6):1571–1577
41. Haruna IM, Aliyu L, Olufajo OO, Odion EC (2011) Growth of sesame (*Sesamum indicum* L.) as influenced by poultry manure, nitrogen and phosphorus in Samaru, Nigeria. Am-Eurasian J Agric Environ Sci 10(4):561–568
42. Abou-Taleb SM (2012) Response of sesame crop to the biofertilizer cerealine with or without mineral nitrogen fertilization. Res J Agric Biol Sci 8(1):45–54

43. Sabannavar SJ, Lakshman HC (2009) Effect of rock phosphate solubilization using mycorrhizal fungi and phosphobacteria on two high yielding varieties of *Sesamum indicum* L. World J Agric Sci 5(4):470–479
44. Kalegore NK, Kirde GD, Bhusari SA, Kasle SV, Shelke RI (2018) Effect of different level of phosphorus and sulphur on growth and yield attributes of sesame. Int J Econ Plants 5(4):163–166

Animal and Rangeland Resources in Shalatin–Abou Ramad—Halaib Triangle Region, Red Sea Governorate, Egypt: An Overview

Hassan M. El Shaer

Abstract Shalatin–Abou Ramad—Halaib Triangle Region (SAHTR) has vital and strategic importance to Egypt. It is located at the southeastern corner of Red Sea governorate. It is characterized by broad biodiversity of both plant and animal resources. Income of most of the inhabitants in the triangle depends basically on raising, trading, and marketing of animals (mainly camels, goats and sheep). Raising such animal species is the most important economic activity for the majority of local Bedouins though coal manufacture and medicinal plants trading are practiced by few of the Bashaari people. Livestock production systems mainly depend on natural vegetation of rangelands in SAHTR. Seasonal fluctuations cause a periodical restriction in feed quality and quantity. Sheep, goats, and camels are main ruminants species in the region. The majority of offspring are born in late winter and spring seasons, while they are weaned towards the warm and dry season with insufficiently high-quality forage available as the fodder quality decreases and the available feed does not fully meet the nutrient requirements for early growth. There are numerous constraints that affect animal performance, trading and marketing. Therefore, the article focuses on classification and identification of both animal and ranges resources in the region that are critically needed to raise the social life and income of local inhabitants. General current features of animal and rangeland resources, main factors affecting animal and range production resources, constraints of animal resource production in the SAHTR and the suggested policies and strategies are also stressed.

Keywords SAHTR · Red sea governorate · Inhabitants · Livestock · Rangelands · Forage · Biodiversity · Egypt · Shalatin-Halaib

1 Introduction

Red Sea Governorate is one of the most promising regions in Egypt which covers an area of about 204,000 km^2 representing around 20% of Egypt's territory. It is located between Nile Valley in the west and the Red Sea in the east and stretching

H. M. El Shaer (✉)
Animal and Poultry Production Division, Desert Research Center, Cairo, Egypt

M. Abu-hashim et al. (eds.), *Agro-Environmental Sustainability in MENA Regions*, Springer Water, https://doi.org/10.1007/978-3-030-78574-1_12

the Sudanese borders in the south. It is characterized, in general, by irregular scarcity rainfall, with an annual average of 80 mm. It is, also, characterized by eroded valleys system (so-called Wadies). The wadies are rendered as involving dynamic ecosystems and are comprised of a highly interconnected series of watershed complexes that are vital water resources as well as flora biodiversity and refuge for different mammals species. The north of the Red Sea Governorate (RSG) is well developed due to oil and tourism activities. The south part is Shalatin–Abou Ramad—Halaib Triangle Region (SAHTR) which starts from South of Mersa Alam up to Hederba valley and Elba mountain, nearby the Sudanese borders and represents more than two-thirds of the RSG. The SAHTR has the highest incidence of poverty in Egypt which is further compounded by serious problems of unemployment, lack of health services, and education facilities [1]. Raising sheep, camel, and goats is the most important economic activity of the local Bedouins. The region is also an important trading centre for camels from Sudan sold to Egyptian merchants. Coal manufacture and trading in medicinal plants are practiced by few Bashaari people. Bedouin women make rugs and blankets from hair and wool [2].

Rainfall in Shalatin–Abou Ramad—Halaib Triangle Region (SAHTR) is the limiting factor that affects all agriculture practices. It is highly variable in time and place with no specific pattern, but most rains occur in spring and autumn. Therefore, the farming system is defined as a drylands farming system where limited areas of grain sorghum and millet are cultivated on wadi runoff for food and feed, especially in wadi Sufera (Shalatin) and Deieb (Abo Ramad) and in small mountain catchments [1, 3].

Attention has been recently directed the Government of Egypt towards improving and 'developing the animal resources in the Triangle region' [4–7]. It is aimed at improving the economic and social circumstances of the existing population in the Red Sea governorate as it moves forward towards achieving its millennium development goals including those related to hunger and poverty.

This chapter aims to throw lights on animal and natural range resources in Shalatin–Abou Ramad—Halaib Triangle Region (SAHTR) to identify the potentialities of animal and rangeland resources; main constraints and problems that would help in planning specific strategies for developing animal production in the region and enhancing local Bedouins welfare.

2 Features of Shalatin–Abou Ramad—Halaib Region (SAHTR)

The Eastern Desert of Egypt is approximately 275,000 km^2 which represents about one-quarter of the total territorial area of Egypt. It is located between the Red Sea in the east and the Nile River in the west. It is the eastern gate of Egypt. It has coasts extend for more than 1080 km on the Red Sea. It represents one-eighth of the whole area of Egypt. The RSG is bordered on the North by the Suez Governorate. Its East boundary is the Red Sea. The governorates of Aswan, Qena, Sohag, Assiut, al-Minia and Beni Suef are its West boundary. The SAHTR is bordered by the Red Sea State of Sudan from the South.

Shalatin–Abou Ramad—Halaib Triangle Region (SAHTR), belongs to Red Sea Governorate, is located in the south eastern corner of the Eastern desert which occupies approximately 18,000 km^2. The region has vital and strategic importance to Egypt. It looks like a triangle with a bottom side of about 300 km parallel to 22° latitude (Egyptian-Sudanese borders). The top point of the triangle is located at the Red Sea nearby the Shalatain well. It across of Jeddah in Saudi Arabia; thus Africa and Asia could be connected at this point. The Triangle region is a mountainous desert with several valleys dissecting mountains. Total surface area of the coastal plain representing the deltas of several wadis draining the region is about 6000 km^2. Rainfall sometimes starts from October up to March; it is erratic and unpredictable.

Shalatin, Abu Ramad and Halaib cities, mainly are located at the Red Sea coast and are considered the most important cities of the region. The majority of the population is semi-nomadic which composed of two main tribes of the Beja: Ababda in the northern section and Besharia in the south. Each tribe consists of clans, which trace their descent from a single ancestor. These people are characterized by very low income, low level of education, if any, frequent health problems and high infant mortality rates [1]. They lack the basic skills necessary to pursue alternative livelihoods and have become increasingly impoverished. Due to harsh climatic conditions, low and erratic rainfall precipitation and lack of knowledge, there is no any traditional agriculture activities [8]. The major income and food source have been traditionally through nomadic animal husbandry and trading. A part from livestock, other traditional sources of income are charcoal, wild honey, medicinal herbs, and seeds collection. A combination of severing long-lasting drought and loss of access to traditional grazing area has dramatically undetermined the continued viability of their nomadic pastoralist way of life. Most of them loose their animals and become extremely poor and food insecure, with high levels of malnutrition amongst women and children. Several preliminary surveys, conducted by the researchers of the Desert Research Center, Egypt [1, 9] indicated that there are, too some extents, potential fertile soils and ground water storage which are available in some areas in the region, particularly in Qusair and Safaga areas. However, these waters need more exploration and proper exploitation.

Concerning the location and geomorphology of Shalatin–Abou Ramad—Halaib Triangle Region (SAHTR), the land next to the Red Sea in Egypt is generally mountainous, feactured on the western side by the range of coastal mountains (1705–2187 m amsl). Between the shoreline and the mountains extends a gently sloping sand plain varies in width and is dissected by series of wadis running eastward to drain their waters into the sea. The climatic features of the region, generally, those affect the Red Sea basin show uniformity throughout the year which may be divided into two main seasons: cool and hot seasons. Winter usually extends from mid-October to mid-April and summer from mid-April to mid-September. The mean daily maximum temperature in January (winter) ranges from 20 °C in the far north to about 29 °C in the far south. The corresponding July (summer) figures being 35 °C and 40 °C, respectively. In winter, lower night temperature resulting from the prevailing cool wind provides relief from the heat of the day. Rainfall is deficient throughout the Red Sea coast. The mean annual rainfall of the Red Sea shoreline in Egypt generally ranges between 3 mm up to 30 mm. Precipitation in Gebel Elba range may reach 60–100 mm/year. The erratic rainfall, which characterizes the arid areas, is obvious along the Red Sea coast. It is highly variable in time and place with no specific pattern, but most rains occur in spring and autumn. No accurate rainfall records are available, but since the region represents an extension of the Sudanese savanna, rainfall is expected to be abundant in some years especially to the south [10]. The only source of surface water is the thunderstorms on the Red Sea Mountains leading to the accumulation of strong wadi flows every 3–4 years. Mean annual minimum and maximum temperature are respectively 10 and 34 °C. During winter (October–April) temperature varies from 19 to 28.8 °C which is suitable for new vegetation growth with high biomass production.

3 Animal Production Resources in the SAHTR

Total numbers of animals (cows, buffaloes, sheep, goats and camels) in the desert governorates in Egypt are presented in Table 1 during years 2010, 2012 and 2017.

It seems that there some fluctuation in animal species numbers from year to year and between governorates as well. Such fluctuations may be due, mainly, to some environmental factors in particular rate, distribution and duration of rainfalls which affect biomass and nutritional values of natural rangelands.

The number of cows and buffaloes in the RSG are not numerous as compared to other governorates in Egypt according to the census of FAOSTAT [11, 12]. In Red Sea governorate, the total numbers of cows, buffaloes, sheep and goats were decreased from year 2012 to 2017 while camels numbers were increased (Table 1). It is clear that the main animal species in RSG are sheep followed by camels then goats in year 2017 (according to *Statistics*, FAOSTAT [11, 12]).

During 2012 the total number of camels was increased significantly and approached 35,637 heads then reached to 53,713 heads in 2017. The majority (at least 90%) of sheep, goats, and camels are raised in Shalatin–Abou Ramad—Halaib Triangle Region.

There are several factors which affect the animal production in the region as reported by El Shaer et al. [10], Nassar [13], and El Shaer and El-Khouly [8]. These factors are number of each animal species, flock size and movement, grazing pressure and systems all round year, the reproductive efficiency of animal and health status and marketing systems. Sheep seem to be the dominant species in the region (52%) followed by camels and goats (26 and 22%, respectively). The total number of camels imported from Sudan was approximately 100 thousand yearly during the last five years according to records derived from the Veterinary Department in Shalatain city [14]. The total animal figures could be far behind the real ones due to continuous moving and travel from area to area and interaction between tribes in Egypt and

Table 1 Animal species populations in desert governorates in Egypt (Animal, Poultry and Fish Resources Statistics, FAOSTAT [11, 12])

Animal species	North Sinai Governorate	South Sinai Governorate	New Valley Governorate	Red Sea Governorate
Cows				
2010	2966	257	151,714	789
2012	457	1481	104,526	559
2017	1606	486	35,630	359
Buffalos				
2010	161	207	577	814
2012	189	556	905	647
2017	191	69	497	255
Camels				
2010	2593	2066	1687	1404
2012	2409	1035	2115	35,637
2017	2718	1282	1712	53,713
Goats				
2010	119,766	24,919	97,255	58,731
2012	88,087	41,153	123,498	65,545
2017	38,470	26,560	87,225	44,213
Sheep				
2010	93,976	14,694	81,497	128,779
2012	95,953	62,672	89,264	98,107
2017	46,863	33,563	112,125	73,105

Sudan. Also, no accurate surveys were conducted due to topographical, climatic and social difficulties. Thus, an accurate survey should be conducted to assess any program for animal production development in the region.

The main animal species could be described in brief (as reported in El Shaer [1, 9], Mohamed [15] and Askr et al. [16]) as follows:

1. ***Sheep***

There are several sheep breeds in the SAHTR region. El Kanzy, Abou Deleak, and El Manaeet breeds are considered the most dominant breeds all over the SAHTR region. They are well adapted to suit the local environmen. Generally, these breeds are tall; have no horns, and their bodies are covered with hair and sometimes with wool which has different colors ranging from white to dark brown.

El Kanzy breed is characterized by moderate twinning rate, and their coat goes from dark brown to black. Abou Deleak breed is very common in the region and is sometimes called El Daboul. It is characterized by a curvey nose; an extra skin underneath the neck; high twinning and milk production. El Manaeat breed has a small size with hairy coat.

2. ***Camels***

Camels play an important role in the social and economic life of migratory pastoralists and nomads in Shalatin–Abou Ramad—Halaib Triangle region. Approximately 100 thousands of camels were transported out of the region last year into the other Egyptian markets [14]. About 75% of these animals were brought through the Egyptian–Sudanese borders whereas the rest were locally raised in the region, according to information derived by the Veterinary Department in Shalatain. During their long, harsh trip, up to the local camel market in Shalatain, they severely suffer due to many constraints which lead to heavy losses for camel's producers and traders. Unfortunately, so far information on such constraints are not available. El Zabaida (so called El Rashaida) and El-Ashab breeds are most distinguished camel breeds in the region. El Zabaida breed is a heavy bodyweight animal, and the color of his coat ranges from light brown to light red. It is characterized by high milk and meat production. El Ashab breed is used mainly in transportation as well as racing since it is characterized by a light bodyweight and a white coat. It is more expensive as a race animal. Camels are mainly of the Zubeidi breed which is a fairly large animal (300 kg) and the Ashab breed which is a small racing camel [15, 17].

3. ***Goats***

Goats in the SAHTR look like the Desert black goat. They are small in size; the black hair is dominant. Although the number of goats is less than that of sheep, goats have economic importance in the triangle since they are the principal source of meat and milk for local consumption. However, goats are variable but are generally of small size body covered with thick hair of different colours. So far, there are scarce information concerning feed and water resources availability; nutritional and health problems; traditional management practices; and production characteristics of animal species [13, 15]. All these information should be classified and identified for applying proper approaches to improving animal production resources in the region and consequently to achieve optimum return and high income for pastoralists and nomads.

Animal Housing

Simple animal housing systems are commonly practiced under restricted grazing systems in the region which fit the environmental conditions. It depends mainly on locally available raw materials. The non-ceiled fences systems are widely adopted where fences are made of wood or stones under Acacia trees as a natural shed [9]. However, such housing systems are not applied under open grazing practices where animals in a continuous movement and are allowed to graze freely on the rangelands.

4 Natural Vegetation and Grazing Resources of the SAHTR

It is known that natural rangelands include desert and forests and all natural grass lands. Grazing is defined as the consumption of existing forage i.e. edible grass and forbs by livestock or wildlife, whereas browsing is the consumption of edible leaves and twigs of the woody plants i.e. trees and shrubs by the animal [8, 18, 19]. The total rangelands area in SAHTR is estimated approximately 600,000 ha occupying the south east corner of Egypt extending for about 300 km from Ras-Binas north to the borders with Sudan (Latitude 22° N), and bounded to the Red Sea to the east and the Red Sea mountains to the west [19]. The general trend for the rangelands in the region is decrease of areas and reduce in mass and quality of forage production [10, 20]. Negative changes were recorded in the rangeland plant community, i.e. more annuals, less palatable and more unpalatable and harmful species in most areas of the range. The joint effects of overgrazing, uproot woody plants and expansion of rain-fed agriculture exacerbated the decrease in production of native forage and the degradation of the genetic repository of good native forage plants and hence the biodiversity. Also, Pastures in this region are also constrained by drought stress, low productivity and water scarcity. Hence, utilization is mostly opportunistic depending on the sporadic rainfall that induces plant growth and provides surface water. As rainfall is not only sparse, but also unpredictable, so, grazing resources vary widely between different range types and between different sites within the same range type. There is also a great variation from year to year and between summer and winter seasons (particularly in wadis range type) depending on rainfall and amount of run off water) as pointed out by El Shaer [20, 21], El-Khouly [22], Askr et al. [3]. They, also, reported that the permanent vegetation in the region includes trees, shrubs, and semi shrubs. Annual flora might be grow with great frequency in favorable regions after winter rainfall. The valleys systems existing a numeral of habitats differing in soil salinity, moisture, and texture. At wadi ends, *Avicennia marina* is repeated. The salt marshes in the area close to the sea are dominated by species tolerant to salinity i.e. *Arthrocnemum glaucum* and *Suaeda monoica.* Additional inland in the dry salt marshes *Salsola baryosma, Suaeda monoica* and *Anabasis* group are dominant. In the sandy plains along the raised wadi beds, *Salsola baryosma* is present, accompanied by *Anabasis setifera, Heliotropium,* and *Acacia tortilis.* At the foothills, *Crotalaria aegyptiaca* is dominant with *Panicum turgidum.* Wadi Hederbah in Halaib is characterized by richest variety resources and the highest potential for development of all wadis [20]. *Panicum turgidum* and *Artistida mutablis* are two main plant species in this wadi giving good natural ranges for camels as well as small-ruminants in winter and summer, alongside *Lythium shawii* and *Artemisia* which are valuable for camel grazing.

Generally, it is observed that herbaceous plant communities in wadis of Shalatain basin are dominated by the unpalatable species of *Sasola baryosma* and *Francoeria crispa.* Meanwhile, there are many palatable species dominated by *Panicum turgidum* in wadis of Halaib basin which is a good forage grass [17–19, 22]. Likewise, Wadi Hederba in Halayeb Basin is characterized by the richness of grazing resources and the highest potential for improvement compared to other wadis. The forage species *Panicum turgidum, Aristida mutabilis, Artemisa judaica* and *Lycium shawii* are the most important in Wadi Hederba that could provide good valuable grazing resources for small ruminants and camels in winter and summer. Some of these shrubs are very palatable to all animal species and often overgrazed. It is believed that *Panicum turgidum* is the main dominant forage crops in Wadi Hederba where most animals rely on such plant all over the year [20]. *Panicum turgidum* a desert grass, growing in dense bushes, up to 1 m. high or more; culm woody, densely branched, with clusters of brown empty leaf-sheaths at the swollen nodes. Rootlets felty, often covered with adherent sand, thus of a corky appearance, abundant in sandy and stony deserts. Along the Red Sea, it often covers the whole coastal plain. The most common plant species in the SAHTR area and their palatability index to sheep, goats, and camels are shown in Table 2 as cited from El Shaer et al. [17].

It is indicated that some of these shrubs are very palatable to all animal species and were often overgrazed. The most palatable forage species for all animal species and always overgrazed are:

- *Maerua crassifolia*
- *Farsetia aegyptiaca*
- *Acacia tortilis*
- *Indigafera spinosa*
- *Taverniera aegyptiaca*
- *Lycium shawii*
- *Indigafera argente*
- *Convolvulus hvstrix*
- *Panicum turgidum*

Though, palatability of such plants differs from season to season and from grazing area to another in the region based on some factors, i.e. environmental factors, plant relatives and populations and animal species [17, 18, 22]. *Lycium shawii*, *Taverniera aegyptiaca*, *Maerua crassifolia* and *Farsetia aegyptiaca* are the greatest palatable species all round year to sheep, goats, and camels in several locations. It should be noted that these shrubs must be protected from overgrazing, increasing their reproduction, and managing re-seeding. There are three main grazing districts in the SAHTR:

- Coastal salt marshes district
- Coastal desert plain district
- Wadies dissecting mountainous district

Most grazing practices are concentrated in these districts all over the year depending on the season of the year. These grazing districts briefly could be described by El Shaer [1, 9] as follows:

1. ***The coastal salt marshes district***

It represents the area adjacent to the Red Sea coast from the north to the south. Grazing activities are practiced most likely in summer and autumn seasons where salt marsh shrubs are grown intensively. The most common halophytes identified in this area according to their coverage and intensity are:

- *Arthrocnemum glaucum*
- *Suaeda monaica*

Table 2 Common natural ranges, its Arabic names and their palatability index for sheep, goats, and camels in the SAHTR

Plant species	Arabic name	Animal species		
		Goat	Camel	Sheep
Acacia albida	Arg	PP	HP	PP
Acacia etbaica	Qrtbi	PP	HP	PP
Acacia mellifera	Al Khashshab	FP	HP	PP
Acacia reddiana	El Saial	PP	HP	FP
Acacia tortilis	El Samraa	FP	HP	NP
Arocnemom glaucum	El Shanan	–	HP	NP
Astragalus eremophilus	Om Elkrin	HP	HP	HP
Avicennia marina	El Shora	–	HP	HP
Blepharis ciliaris	Shule Eldeb	PP	HP	HP
Cadaba farinosa	Cally	HP	PP	NP
Cadaba oblonifia	Al Qrti	HP	HP	HP
Calligonum comosum	Shbrq	PP	HP	HP
Convolvulus hystrix	El Graba	HP	HP	HP
Halopeplis prefaliala	Haloubils	HP	HP	FP
Heliotropium luteum	Al Ghbira	–	PP	NP
Indigofera argente	Toukhit	PP	HP	PP
Indigofera spinosa	Sunkid	HP	FP	PP
Leptadenia pyrotechnica	Al Mrkh	HP	HP	HP
Lycium shawii	Awssag	HP	HP	HP
Maerua crassifolia	Kamoub	HP	HP	HP
Ochradenus baccatus	Al Qrdi	HP	HP	FP
Panicum turgidum	Thommaam	HP	HP	HP
Pergularia tomentosa	Om Ellbin	HP	HP	FP
Plantago ciliata	Houlakit	FP	HP	FP
Salsola baryosma	Al Khrit	FP	PP	–
Salsola tetrandra	Al Dmran	–	HP	PP
Stipagrostis ciliata	Hmorate	PP	HP	HP
Suaeda monaica	Al Sowaida	HP	FP	–
Taverniera aegyptiaca	Dhasir	PP	HP	HP
Trichodesma ehrenbergii	Khroua ElGamel	HP	HP	PP
Zygophyllum coccineum		PP	FP	PP

HP highly palatable; *FP* fairly palatable; *PP* poorly palatable; *NP* non palatable
Source El Shaer et al. [17] and Mohamed [15]

Table 3 Average values of chemical constituents (%, on DM basis) of most common halophytes in the coastal salt mashes district [17]

Plant species	DM	CP	CF	EE	Ash	NFE
Avicennia marina	33.2	10.5	13.9	3.30	25.9	46.40
Arthrocnemum glaucum	39.1	6.33	17.3	1.98	39.7	34.69
Suaeda monoica	40.2	9.40	16.7	3.00	30.3	40.6

DM dry matter; *CP* crude protein; *CF* crude fiber; *NFE* nitrogen free extracts

- *Helopeltis perfoliata*
- *Limonium axillare*
- *Avicennia marina*

The chemical composition of the dominant halophytic species which were grazed particularly by camlets is presented in Table 3. *Avicennia marina* is occasionally overgrazed by camels in particular in summer and autumn seasons when palatable shrubs are not presented.

These halophytes are characterized by higher ash content, particularly *A. glaucum* (approximately 40%) in addition to reasonable levels of fiber content (around 15.9%). *A. marina* and *S. monaica* comprise moderate levels of CP (10.5 and 9.4%, respectively) which appear to be sufficient to cover the protein necessities for animals [23]. These figures are close to those obtained on similar halophytes grown in Sinai [10]. Constraints of palatability such as specific anti-nutritional factors (tannins, saponins, nitrate, etc.) should be tested. Then proper approaches should be practiced through certain processing treatments to reduce the effect of such nutrients and improve palatability and nutritive value of these forages species [24, 25].

2. ***Coastal desert plain district***

It extends between the coastal salt marshes district and wadis dissecting mountainous chains district. It includes some valleys for example Houdean, Sheab, Ebeb and Hederbah which contain excessive varieties of annual and perennial shrub plants. *Acacia tortilis* and *Salsola baryosma* community types represent the most dominant plant communities and associate with numerous plant species. *Lycium shawii*, *Maerua crassifolia*, *Acacia tortilis* and *Acacia albida*, are greatest important palatable range fodders to all animal species over the year.

Table 4 show mean values of the chemical composition of dominant plant species in Hederbah Valley. *Acaia albida* and *Lycium shawii* followed by *Maerua crassifolia* and *Acacia tortilis* contains reasonable levels of crude protein that covers the nutritional needs of camel, sheep, and goats [23]. Such plant species should be protected from intensive overgrazed, besides Bedouins tend to cut and collect wood from these trees to make charcoal.

3. ***Wadies dissecting mountainous chains district***

It represents the main grazing district during all seasons of the year in the region. It has great potentialities in terms of numerous plant species and varieties: annuals,

Table 4 Average values of chemical composition (%, on DM basis) for dominant plant species in Hederbah Valley [17]

Plant species	DM	CP	EE	CF	Ash	NFE
Acacia albida	51.3	13.2	2.66	27.9	9.59	*46.65*
Acacia tortilis	71.3	10.3	2.11	30.2	6.60	50.79
Blepharis ciliaris	56.9	8.72	1.53	23.3	3.6	52.85
Lycium shawii	39.8	12.5	1.17	34.9	8.65	42.78
Maerua crassifolia	60.8	11.5	1.98	30.7	10.2	45.62
Panicum turgidum	57.2	5.99	4.56	34.1	6.25	49.1

DM dry matter; *CP* crude protein; *CF* crude fiber; *NFE* nitrogen free extracts

perennials, shrubs, semi-trees, and trees. Several plant community types are recognized and identified in this area such as (1) *Acacia tortils* community type, (2) *Acacia raddiana* community type, (3) *Panicum turgidum* community type. Each one associates numerous annual and perennial plant species that vary in their palatability. The plant communities and associations were identified in three main wadies: (1) Metab valley, (2) Auglahoug valley and (3) Sarmatay valley.

Table 5 show chemical constituents of the dominant plant species in Metab valley as cited from El Shaer et al. [17]. It might be clear to mention that *Acacia tortilis, Capparis decidua, Farsetia aegyptiaca, Lycium shawii, Maerua crassifolia* and *Panicum turgidum* forms the main forage resources in the wadis dissecting mountainous region judging from their palatability to all animal species all round year, in addition to their coverage, intensity, and dry yield.

It is reported that protein content varied from 5.43% (*Panicum turgidum*) to 13.8% (*Farsetia aegyptiaca*) whereas fiber content ranged from 21.1% (*Farsetia aegyptiaca*) to 35.6% (*Lycium shawii*). *Maerua crassifolia, L. shawii* and *F. aegptiaca* appear to be highly palatable and nutritious for all animal species. *Panicum turgidum* represents the most important natural grass in this grazing district and provides a good feed resource during dry seasons (summer and autumn).

Table 5 Average values of chemical constituents (%, on DM basis) for dominant plant species in Metab Valley [17]

Plant species	DM	CP
Acacia tortilis	63.70	10.2
Capparis decidua	23.20	7.22
Farsetia aegyptiaca	35.80	13.8
Lycium shawii	40.80	11.8
Maerua crassifolia	59.60	11.1
Panicum turgidum	54.30	5.43

5 General Current Features of Animal and Rangeland Resources

From the previous details on the description of the existed animal and rangeland resources in the region, the general characteristics and features could be summarized as reported in several studies [8, 16, 19, 21, 22] in the following points:

1. Livestock productivity
   - High parental mortality of lambs,
   - High barrenness,
   - Low conception and fertility rates,
   - Low prolificacy and low birth and weaning weights,
   - Slow growth rate of growing animals, and
   - Reduced animal productivity (meat, milk, etc.)
2. Livestock health and disease control
   - Ecto- and endo-parasites can inflict great economic losses
   - Fertility disorders, reduced body weight gain, deterioration of animal products
   - Viral and/or bacterial diseases and toxic plants could have major impact on animal health
   - Frequency of coccidia infection is high
   - The incidence of parasitic infection among camels is very high, particularly in the SAHTR areas
   - The toxic plants affect animal production on the rangelands and make lots of troubles to livestock holders
3. Range resources
   - The indigenous natural vegetation is the principal feed resources
   - Fluctuate in biomass production, palatability, and nutritive value from season to season and from area to area
   - The imbalance between livestock feed requirements and supplies from the native ranges is common
   - *Atriplex* spp., *Salsola* spp., *Medicago* spp., *Stipagrostis* spp., *Nitraria retusa*, *Zygophylum* spp., *Acacia* spp., *Hamada scoparia* and *Plantago* sp., are the most common plants
   - Signs of overgrazing and degradation are evident in several sites
   - Misuse of rangelands is common resulting in:
     i. The decrease of desirable and palatable plant species,
     ii. The increase of less palatable and toxic plants,
     iii. The disappearance of the genetic resources of forage plants.
4. Water shortage:
   It is a common problem particularly in the region in terms of water quantity and quality where water points are not enough available on the rangelands.

6 Constraints of Animal/Range Resource Production in the SAHTR

Problems and constraints of animal production development in the region could be summarized in the following points:

1. Erratic and short duration of rain precipitation leads to long drought period (from April up to October). To causes shortage of feed resources in that period which has a negative impact on animal production activities and social life.
2. Unavailability of feed supplements, particularly traditional feed concentrates which brought from the Nile Valley and extremely expensive.
3. Unavailability of drinking water points, in particular during the dry seasons (Summer and Autumn).
4. Transportation of camels and sheep through the southern Egyptian borders to the local markets could transfer several serious diseases.
5. Slaughter houses are not available in the region, that cause hygienic problems.
6. Veterinary services and extension are limited in the coastal desert plain zone, moreover, mot available at all in the wadis dissecting mountainous chains zone where thousands of animals are raised.
7. Precise survey on animal production, productivity, movement, flock size, ownership, etc. are not available for many reasons.
8. Animal production is specifically constrained by the following:
    - The relatively poor production of rangelands
    - Lack of an effective plan for range development
    - Poor livestock/range extension and veterinary services
    - The lack of integration between crop and range production
    - The lack of cultivated forages.

7 Factors Affecting Animal and Range Production Resources

The animal and rangeland resources and their productivity and production systems in the region are generally affected at various degrees by main four limiting factors, e.g. environmental, range resources, animal nutritional, socio-economic stresses. Such factors will be summarized as follows:

1. Environmental factors
   The most critical environmental factors affecting livestock productivity are:
    - Erratic and short duration of rain precipitation leads to long drought periods
    - Salinity stress
    - Erosion stress
    - Durrinal variation in ambient temperature

- Heat and wind stress.

2. Range resources factors
   There are many factors which can drastically affect rangelands resources and its utilization as animal feeds. The main limiting factors are:
   - Rangelands deterioration and overgrazing
   - Shortage of fodder crops integrated with ranges
   - Unavailability of common feed supplements on ranges
   - The relatively poor production of rangelands
   - Lack of an effective plan for range development
   - Poor livestock/range extension and veterinary services
   - The lack of integration between crop and range production
   - The lack of cultivated forages.
3. Nutritional factors
   - Shortage of feed resources (quantitative and qualitative)
   - Seasonal starvation of animals
   - Unavailability of drinking water
   - The high salinity of the available feed and drinking water
   - Improper economic inter-relationship between the productivity of grazing animals and potential utilization of indigenous ranges.
4. Socio-economic stress
   - Veterinary workers and services and extension are limited
   - Lack of knowledge and communication
   - Trans-boundary system
   - Poor interaction between the researchers, farmers, extension workers and policy makers
   - Lack of the economic incentives caused by unfavorable price policies, poor marketing and infrastructure and lack of financial resource.

8 Suggested Polices and Strategies

New polices, and strategies could be suggested for coping with desertification effects and livestock development:

- Integration of livestock development, agriculture an agriculture industries.
- Modernization of the marketing by the realization the slaughterhouse and the improvement of facilities of rural markets.
- The installation of private veterinaries and the adoption of a regulation or legal executive.
- Development of milk production and dairy products by the incitement to the creation of centers for milk collection and processing particularly in rural areas in the Nile Valley and Delta.

- Setting up pastoral associations for the better management of grazing resources.
- Establishing an effective early warning system and it must be accompanied by an infrastructure for the implementation of drought relief measures.
- Conserve and improve management of natural resources utilization (water, land, plant, animal, etc.).
- Support and develop coordination programmas between concerned institutions at the national, regional and international levels in the fields of development of livestock production, rangelands rehabilitation and control of desertification and drought.

9 Conclusions and Recommendations

Shalatin–Abou Ramad—Halaib Triangle Region (SAHTR) has vital and strategic importance to Egypt; it is characterized by broad biodiversity of both plant and animal resources. Livestock production on the native ranges of Shalatain–Abou Ramad—Halaib Triangle Region, Red Sea Governorate is facing several major constraints mainly: drought, feed and water shortage, diseases, and environmental stresses. All efforts should be directed towards setting up a sustainable national strategy to improve both rangelands and livestock production which has the following tasks:

- Optimum utilization of the available local feed and drinking water resources
- Applying proper approaches for native range management and conservation
- Development of watering points combined with proper control of grazing
- Utilizing the available non-conventional feed resources efficiently
- Genetically constitution improvement, particularly through the selection
- An effective program for the improvement of feeding, disease control, and management should precede any attempt at genetic improvement
- Providing proper assistance and advice to local Bedouins or investors to best utilize available resources
- Strengthening the cooperation between scientists, specialists, policy makers and Bedouins
- Reader who are interested in sustainable development of the areas could consult [5].

References

1. El Shaer HM (2001) Project "Development of animal resources (sheep, goat, and camel in Halaib/Shalatain region, Red Sea Governorate" phase 2. Final report, Academy for Scientific Research and Technology, Egypt, 1998–2001
2. CAPMAS (2013) The Central Agency for Public Mobilization and Statistics (CAPMAS), Cairo, June 2003, The Statistical Year Book

3. Askr AR, Salama R, El Shaer HM, Raef O (2019) Effect of animal species and supplementary feeding on digestion and energy utilization by sheep and goats grazing the arid rangelands. In: Proceedings of the joint meeting of the FAO-CIHEAM networks on sheep and goats and on Mediterranean forages and fodder crops, Meknes, Morocco, 23–25 Oct 2019
4. Badawy NS (2016) Effect of supplementary feeding on sheep skin and coat characteristics in Halaieb, Shalateen and Abou Ramad Triangle. Egypt J Nutr Feeds 19(3):473–483
5. Elewa HH, Nosair AM, Ramadan EM (2020) Sustainable development of mega drainage basins of the eastern desert of Egypt. Halaib–Shalatin as a case study area. In: Negm A (eds) Flash floods in Egypt. Advances in science, technology & innovation (IEREK Interdisciplinary series for sustainable development). Springer, Cham. https://doi.org/10.1007/978-3-030-29635-3_9
6. Mohamed AH et al (2019) Estimating grazing capacity for desert rangelands of Wadi Hederbah in Southeastern Egypt. Adv Environ Biol 13(10):22–31. https://doi.org/10.22587/aeb.2019.13.10.3
7. Nassar MS (2020) Effect of grazing time on productive performance of goat in Halaib–Shalateen region. Egypt J Nutr Feeds 23(1):37–54. https://doi.org/10.21608/EJNF.2020.95805
8. El Shaer HM, El-Khouly AA (2016) Natural resources in saline habitats in South East of Egypt. In: Natural resources in saline habitats in South East of Egypt. Academy of Scientific Research and Technology, 168p
9. El Shaer HM (1998) Project "Development of animal resources (sheep, goat, and camel) in Halaib/Shalatain region, Red Sea Governorate", Egypt, 1995–1998. Final report, Academy of Scientific Research and Technology, ASRT, Egypt
10. El Shaer HM, Kandil HM, Khamis HS, Abou El Nasr HM (1996) Feeding desert goats on some ensiled fodder shrubs supplemented with ground date seed in Sinai. In: Proceedings of regional workshop on native and exotic fodder shrubs in arid and semi-arid zones, Hammamat, Tunis, 27 Oct–4 Nov 1996
11. FAOSTAT (2013) Animal, poultry and fish resources statistics. Economic Sector, The Central Agency for Agricultural Economics, Ministry of Agriculture and Land Reclamation, vol 8, May 2013
12. FAOSTAT (2017) FAOSTAT: food and agriculture data. FAOSTAT, Rome. Accessed 04 Jan 2017
13. Nassar MSM (2008) Nutritional aspects and supplementary feeding for local goats on ranges of Halaib and Shalatain region. Ph.D. thesis, Faculty of Agriculture, Cairo University, Egypt
14. MALR (2017) Agriculture department in Red Sea Governorate. Ministry of Agriculture & Land Reclamation, Cairo, Egypt
15. Mohamed ORH (2012) Nutritive evaluation of natural ranges in the south eastern corner of Egypt. M.Sc. thesis, Faculty of Agriculture, Al-Azhar University, Cairo, Egypt
16. Askr ARR, Salama R, El Shaer HM, Safwat MA, Poraei M, Nassar MS, Badway HS, Raef O (2014) Evaluation of the use of arid-areas rangelands by grazing sheep: effect of season and supplementary feeding. Small Rumin Res 121:262–270
17. El Shaer HM, Kandil HM, Abou El-Naser HM, Khamis HS (1997) Features and constraints of animal resources development in Shalaten-Halaib region. Egypt J Nutr Feeds 1:121–128
18. El-Khouly AA, Khedr AA (2007) Zonation pattern of *Avicennia marina* and *Rhizophora mucronata* along the Red Sea Coast, Egypt. World Appl Sci J 2(4):283–288
19. Khalifa ES (2016) Potential of mangrove rehabilitation using different silvicultural treatments at Southeastern Coast of Egypt. J Biodivers Environ Sci 8(2):298–305
20. El Shaer HM (2002) Range resources inventory and assessment of grazing potential in the South Eastern region of Egypt, "A case studies". In: Proceedings of the regional workshop for the Africa and North Africa/Middle East regions, 21–24 Apr 2002, Sultanate of Oman
21. El Shaer HM (1999) Conservation and improvement of natural rangelands as animal feed resources in the north western coast of Egypt. In: The third conference on desertification and environmental studies beyond the year 2000. Center for Desert studies, King Saud University, Kingdom of Saudi Arabia, 30 Nov–4 Dec 1999

22. El-Khouly AA (2006) Influence of water stress and salinity on germination and seedling growth of three medicinal desert plant species. AL-Azhar Bull Sci 17(1):103–129
23. Kearl LC (1982) Nutrient requirements of ruminants in developing countries. International Feedstuff Institute, Utah
24. Bayoumi MT, El Shaer HM (1992) Impacts for halophytes on animal nutrition health. In: Proceedings of the international workshop on "Halophytes for rehabilitation of saline wastelands and a resource for livestock", 22–27 Nov, Nairobi, Kenya
25. Elshaer H, Gihad EA (1992) Halophytes as animal feeds in Egyptian deserts. In: Proceedings of international workshop on "Halophytes for rehabilitation of saline wastelands and a resources for livestock", 22–27 Nov, Nairobi, Kenya

Industrial, Landscape, Touristic and Political Approaches for Agro-environment Sustainability

Sustainable Mining Site Remediation Under (Semi) Arid Climates in the Middle East and in Northern Africa: The Djebel Ressas Mine in Tunisia as an Example of the Orphaned Mines Issue

C. Dietsche

Abstract Humans have mined a large number of sites for minerals, ores and other natural resources over the past centuries. When looking at a practical benefit of applying quality targets to the brownfield site remediation of an Orphaned Mine, the costs directly correlate with the planned target levels. The main problem seems to be the flue dust and mining waste deposits (Alayet et al. in J Appl Remote Sens 11(1), 2017 [1]), so the research follows the three pillars of sustainable development:

1. In order to remove heavy metals contents (Pb, Cd, Zn, Cu, and As—see Table 4 for an overview) safely from a former mining village (cf. Achour et al. in Assessment heavy metals (Pb, Zn, Cd) and metalloids (Sb, As) mobility in carbonated soils contaminated by mining wastes (Northern Tunisia). The poster can be downloaded from: https://www.researchgate.net/publication/335608443_Assessment_Heavy_Metals_Pb_Zn_Cd_and_Metalloids_Sb_As_mobility_in_Carbonated_Soils_Contaminated_by_Mining_Wastes_Northern_Tunisia, (2019) [2] on a regional scale), measures have to be taken for protecting local water reservoirs (E, environmental).
2. The reuse of the deposits as material for road construction needs to be studied and decided in terms of costs versus benefits (E, economic).
3. Once the topsoil is exchanged with less contaminated soil, after a proper site remediation, the use of the former deposit area for gardening and agriculture should follow health and safety criteria (cf. Béjaoui et al. in Health risk assessment in calcareous agricultural soils contaminated by metallic mining activity under mediterranean climate. CSAWAC CLEAN: Soil Air Water 44(10):1261–1427, 2016 [3, Ghorbel in Contamination métallique issue des déchets de l'ancien site minier de Jebel Ressas: modélisation des mécanismes de transfert et conception de cartes d'aléa post-mine dans un contexte carbonaté et

C. Dietsche (✉)
Interdisciplinary Degree Programme: Environmental Sciences "Infernum", University of Hagen, Hagen, Germany
e-mail: carsten.dietsche@studium.fernuni-hagen.de

Present Address:
Am Berliner Platz 2, 63329 Egelsbach, Germany

M. Abu-hashim et al. (eds.), *Agro-Environmental Sustainability in MENA Regions*, Springer Water, https://doi.org/10.1007/978-3-030-78574-1_13

sous un climat semi-aride. Evaluation du risque pour la santé humaine. Thesis, University of Toulouse, France, 4]): Some of the inhabitants grow deep-rooting vegetables on site (S, social) that accumulate heavy metals.

Proposed Solutions:

A. Control the spread of heavy metals into the towns and villages of an affected area, and into agricultural land, and monitor it—for example via biomarkers.
B. Exchange the topsoil in areas where people grow vegetables, especially deep rooting species. At best, utilize a local workforce to create more jobs.
C. Additionally, extract heavy metals and stabilize the affected soil by adapted plants. Their leaves and stems accumulate metals that the roots extract. The biomass can be harvested (as a kind of mining) to prevent recontamination of the soil on site.
D. Define and try to meet quality targets, e.g. to re-settle certain species in the area.

This chapter gives a clear benefit of applying realistic quality targets for orphaned mines' brownfields. For its site remediation, the main limiting factors are the costs, as "Contra" arguments. So the targets should be re-adopted during the site remediation. For example, practical solutions allow stronger releases of heavy metals after the remediation, compared with the original targets.

Keywords Mine tailings · Ecotoxicity · Orphaned mines · Brown site remediation · Environmental quality targets · Phytoremediation · Phytomining · Bioleaching · Biomarkers · Biosorption · Phytoextraction · Heavy metals · Phytostabilization · Agromining · Hyperaccumulation · Field experiments · Bioavailability · Conservation biology

1 Importance of Sustainable Soil and Land Management

The necessity for a sustainable land and soil management is derived from.

- global scarcity and shrinking availability of undisturbed, ecologically well functioning and even altered soils that can still produce biomass, e.g.
  (a) either in a natural state, including some hazardous natural states
  (b) or as cultivated soil in an anthropocentric point of view [5] source of nutrition, source of raw materials, foundation for buildings, etc.
- the needs to avoid the degradation of soils in the context of
  (a) a natural surroundings, for example influences of desertification, erosion
  (b) an agricultural landscape as a part of a cultural landscape
  (c) an industrialized landscape, including mining [6]

One of the possible aims is to

Table 1 Examples of soil degradation—acidification and salinization

Causes	Effects	Possible consequences
Bad management practice Industrial pollution Natural causes	Acidification: (a) Acid precursors: Nitrogen oxides, NO_x (b) Photochemical oxidants: HO, HO_2, H_2O_2, Ozone O_3 (c) Acids: sulfuric acid, H_2SO_4, Nitric Acid HNO_3	Damage of plants: Increases phytotoxicity Changes bioavailability in the soil: change of pH values may release heavy metals from Fe–Mn complexes, etc.
Wrong irrigation practices, also application of poultry (e.g. chicken) manure	Salinization: (a) Salt concentration in the soil is rising	Increase phytotoxicity; decrease soil health, decrease number of adopted food crops for humans

- set quality targets for a remediation [7] of degraded soils according to the needs to achieve realistic aims for a sustainable land management
  (a) in order to specify if a remediation should return the soil to its natural state (100% reversion)
  (b) or to accept a less degraded state as the target
- reflect on overly anthropocentric views of soils [5, 8] and to see both soils and human beings, animals and plants in a holistic point of view

Table 1 shows some aspects of how soils can become degraded and disturbed:

Table 1 shows that many soil management practices, industrial pollution, wrong irrigation practices, but also natural causes may have detrimental effects on soils: The pH values can shift or the salt concentration can rise, thus releasing heavy metals from the soil or damaging plants.

Looking at the effects and consequences, there are both qualitative and quantitative ones. Likewise, the remedy could be divided into qualitative and quantitative environmental, social and economic quality targets, according to the definition of a Sustainable Development [9]. To cite former mining areas such as Djebel Ressas in Tunisia as an example (for others, see [2]), even the undisturbed, natural landscape had its very limited release of hazardous heavy metals, with some plants being in place that are well be adopted to this environment.

Applying the same structure to the orphaned mines 'issue where the disturbed soil is not rehabilitated, there are similarities with other soil degradations.

Table 2 shows that many soil management practices, industrial pollution, wrong irrigation practices, but also natural causes may have detrimental effects on soils: The shift in the soil's pH values could release heavy metals, phosphorus, as well as certain radioactive elements. The heavy metals negatively affects plants that are not adopted to them.

Also, a rise or fall of groundwater levels always has an effect on the flora, but also often on building foundations and other structures, too. Then, the water scarcity of

Table 2 Examples of mining soil degradation—chemicals and water issues

Causes	Effects	Possible consequences
Bad land management practice Industrial pollution Natural causes	Release of heavy metals ($\rho >$ 5.0 g/cm^3) from soil, e.g. Pb, Hg, Cr, Mo, Ni, Co, Zn (cf. [10]) Release of P and S compounds from the soil Release of radioactive elements from the rocks or from the soil, e.g. U, Pu, Rn	Damage of plants: Increases phytotoxicity Changes bioavailability in the soil: change of pH values and soil disturbance may release heavy metals from Fe–Mn complexes, etc. cf. Norme tunisienne/Tunisian Standard NT 106-02 [11]
Wrong water management: Pumping of mine waters out of the mine may cause desastrous side effects	Groundwater: (a) levels are declining (b) is being released into surface (fresh) waters	(a) Damage in deep-rooting plants, of building walls etc. (b) Increase of toxicity levels both in ground and in surface waters due to release of heavy metals (c) Decreased soil health

the MENA regions under a semi arid climate is an important factor for agriculture, as the next chapter describes.

2 Targets for Sustainable Soil and Land Management

For a remediation, it is often difficult to reconstruct the original groundwater levels and their underground flow for one specific site.

Before applying quality targets for developing a site, conserving a site, or for a site remediation, the targets should take into account samples of the undisturbed, undegradated soil on-site in and around the village of Djebel Ressas or on locations nearby [10, p. 76, 78], such as the Jurassic Limestone (carbonate sedimentary rock) of the Djebel Ressas mountain [12, 13]. For the mobilization of heavy metals, the pH value of the soils and rocks in the study area as well as of the rain water throughout the year need to be determined. The assumption is that acid rain caused by SO_2 and NO_x ambient air pollution from urban conglomerates can mobilize heavy metals in rocks and soils. On the one hand, limestone is acid sensitive and can be dissolved by acid rain (low pH). On the other hand, limestone can increase the alkalinity of water and buffer weak acids.

As a starting point, both quantitative and qualitative targets [7] can be derived from laboratory test results. This is beneficial in terms of cost saving to adopt the targets to the local situation, and to allow a Plan-Do-Check-Act cycle before, during and after the implementation, for a cost control [8, 13].

The broad environmental, social, and economic targets, as explained above, can then be fine-tuned into more detailed sub-targets, for example:

Both anthropocentric and non-anthropocentric site remediation [14] can suffer from unrealistic targets, for example: In geography and conservation biology, some experts sometimes implicitly use the pre-industrial, idealized Graeco-Roman "arcadian landscape" as a quality target to which to restore an agricultural, or even industrial present state [5, pp. 379–383, 386–388].

If such culturally influenced target is only implied and not expressed in public, target setting may imply failure since the targets cannot win the support of the local population. So the targets should be explained in detail, citing the pro and the contra aspects, especially if the project is a cross-cultural one.

3 Problem: The Orphaned Mine at Djebel Ressas in Tunisia as an Unsolved Issue of Sustainable Land and Soil Management

For local populations, orphaned mines after coal, metals, oil shale, and other extracting, are a universal problem all over the world, in different cultures [14, p. 76].

As primary effects, if the underground or open cast mining is stopped, abandoned mines can cause unstable underground structures or a heightened exposure of minerals releasing chemicals, for example.

As secondary effects, deforestation, erosion and groundwater contamination from isolating metals can follow, among others. As a tertiary effect, areas near open-cast mines can become inhabitable.

For example, as mining often affects the groundwater levels via pumping, the stopping of the mining activities does not only restore the former groundwater levels. It may also create an artificial hydrogeological situation where the underground aquifers are even destroyed. Even if the site remediation addresses the groundwater levels' restoration, the water quality could be deteriorated by heavy metals and other pollutants such as radioactive substances.

For agriculture and human habitats, the availability of fresh, unpolluted water is a key issue. The semi arid climate limits the water availability in the MENA area, so water issues will be among the top priorities for future challenges.

In Tunisia, approx. 30 km SE of Tunis and circa 10 km SE of Mornag City (36° 36′ 21″ N, 10° 20′ 10″ E), the Djebel (or Jebel) Ressas Mine جبلالرصاص is located.

4 Definition of Quality Targets: Qualitative and Quantitative Approach

Specific to the site and its soil are the following chemical elements [15] that can be quantified in the flue dust and in the soil: Pb, Cd [16], and Zn, Cu [17], as well as As [15, p. 2981]. Another challenge is the contamination of drinking water [16, 18] and of water reservoirs used for agricultural irrigation, for example in case of the one southwest of Djebel Ressas [18].

Many of the contemporary studies show a concern for human health and plant health via heavy metals contaminated soil [19], and via the spread of flue dust by the locally strong wind but do not mention fauna health except two examples: The earthworm *Eisenia andrei* is used as a sentinel animal for soil contamination and is tested for genotoxic responses to heavy metals exposure in case of Pb, Cd, Zn, and Cu [17, 20]. Then, chicken raised around Djebel Ressas accumulate heavy metals, which then enter the human food chain (cf. [21]). These chicken were tested for heavy metals, too [22].

As a recent Polish study shows, wasps which also live in Tunisia such as *Polistes nimpha* (Paper Wasp) can be used as a bio marker for bio monitoring, too (Mielczarek and Wojciechowicz-Żytko [23], who cite it as *Polistes nimphus*).

The actual measured air concentrations of PM 10 (Particulate Matter $\leq 10\ \mu m$), for Pb and Cd exceed maximum WHO levels [16] and pose a serious health thread for the local population [17, 24]. The major exposure paths can be inhalation and ingestion. The latter one is mainly defined by eating deep-rooting vegetables for example.

The air quality around Djebel Ressas had been bio monitored using the *Opuntia ficus-indica*, Figuier de Barbarie, a cactus [25].

Caused by the contaminated water and the spread of flue dust by the wind, the media water, air, and soil are affected. Most studies about Djebel Ressas focus on the human exposure to heavy metals, mainly on the exposure paths. Since results from local residents' blood samples are still missing, these would be a desirable outcome of future medical studies (cf. [26]), for a more fine-tuned targeting (see Table 3).

Looking at the effects and consequences, there are both qualitative and quantitative ones. So in order to remedy the consequences, the targets that can be set may also either encompass quality or quantity. They could be divided into environmental, social and economic quality targets:

So one possible quantitative approach for a quality target is to limit the human exposure to heavy metals to levels well below WHO limits (cf. [26]). This would also benefit plant and animal health. There is still an anthropocentric focus in most studies, however, which is only slightly broadened by this study's proposed indicator species. For a proposed qualitative approach, the author proposes a bundle of measures to limit the spread of heavy metals into the soil by wind and affluent surface waters. For this purpose, the results from a literature research are summarized in Table 3: In order to formulate quality targets for orphaned mines under a sustainable land and soil management, would it be a good starting point to determine the quality of the undisturbed soil at the Djebel Ressas site "before the mining era"?

Table 3 Examples of environmental, social, and economic quality targets and indicators

Quality target (objective)	Quality standard	Quality indicator
Environmental and social: limit the exposure to Pb, Cd, Zn, Cu, and As for humans, flora, and fauna e.g. by phytoremediation, living fences, and drainage system in the former mining areas	Observe WHO limits and/or Tunisian limits for agricultural soils and products, cf. Norme tunisienne/Tunisian Standard NT 106.02 [11]	Let adopted fauna grow on land after site remediation, and quantitatively measure PM 10 concentrations; create subjective human well being of local population (cf. [26])
Environmental: (Re)-Introduce monkeys to the Djebel Ressas mountain after site remediation, as an indicator species for mountain habitat quality	Observe national laws for nature [7, 8, 27] reserves that offer sanctuaries for wildlife	*Macaca sylvanus*, Barbary macaque/ape, or magot), *Polistes nimpha* (Paper Wasp) [23] can also be used for bio monitoring (cf. [28, 29]), besides earthworms [17, 20]
Environmental and economic: Plant live fences to trap flue dust spread by the wind, adopted fauna for phytoremediation, steepen terraced slopes of the abandoned mine inwards to control the flow of rain water, apply animal manure to terraces, mining waste dumps	Apply best practices for site remediation [24, pp. 131–134, 30, 31, pp. 1–5, 32, pp. 41–51], affecting primarily the abandoned mountain top to the north that is no longer mined As a second step, to remedy the cement plant's site to the south could be a future option, but only after the end of all local excavation activites	For example: Can scientists identify plants adopted to the specific site and the specified purpose? [33, p. 312, 34, pp. 75–77] Do the altered terraces release less heavy metals than before?
Social and economic: Conduct a site remediation around the village so that inhabitants can safely grow even deep-rooting vegetables	WHO limits for heavy metals, and international standards for earth worm lab tests: ISO/DIS 23611-1:2016 [35]: Soil quality—sampling of soil invertebrates—Part 1: Hand-sorting and extraction of earthworms; ISO 11268-1:2012 [36], ISO 11268-2:2012 [37], ISO 11268-3:2014 [38], ISO/DIS 23611-1:2016 [35]—cf. VDI [39]; as well as ISO 17512-1:2008 [40]	*Eisenia andrei* and deep-rooting flora can be used as objects for laboratory tests: Boughattas et al. [17, 20] ISO 11268-1:2012 [36] ISO 11268-2:2012 [37] ISO 11268-3:2014 [38] ISO 17512-1:2008 [40] ISO/DIS 23611-1:2016 [35] cf. [39] for a parallel water monitoring, species such as Caenorhabditis elegans (Nematoda) can be suitable, too [41], also see ISO/DIS 23611-1:2016 [35]
Economic and environmental: use of mining residues as material for road construction (except for areas with drinking water wells or irrigation water reservoirs, for example)	cf. Tunisian Standard NT 106-02 [11] and German Industrial Standard DIN 4301:2009-06 "Ferrous and non-ferrous slag for civil engineering"	For marker organisms and further standards for bio monitoring: see text above

(continued)

Table 3 (continued)

Quality target (objective)	Quality standard	Quality indicator
Economic: allow growth of extractive activities south of the village, also promote growth of eco tourism (e.g. via *Macaca sylvanus* as an attraction)	Local regulations, national laws	For example, does the remediation lead to an economic growth by increase in income per time? Does it lead to an increase in local enterprise taxes, among other economic factors such as a growth in tourism?

5 Scope for Design: Can There Be a Practical Benefit of Applying Quality Targets to the Brownfield Site Remediation of an Orphaned Mine?

If the main problem seems to be the flue dust and mining waste deposits (cf. [1]), do their heavy metals contents allow them, for example.

- to be removed safely from the former mining village, while protecting local water reservoirs (environmental sub-targets)?
- and to be re-used as material for cement or road construction (economic sub-target)?
- that once the topsoil is exchanged with less contaminated soil, after a proper site remediation, that their area can safely be used for gardening, using even deep-rooting vegetables (social sub-target)?

If all of these questions can be answered with "Yes" as "Pro" arguments at best, there can be a clear benefit of applying realistic quality targets for orphaned mines' brownfield site. For its remediation, the main limiting factors are the costs, as "Contra" arguments. So the targets should be re-adopted during the site remediation. For example, a practical solution can be to allow a stronger release of heavy metals after the remediation compared with the original planning targets. One of the main questions to be answered in this respect is:

Can some of the mining waste as dumped next to the village be re-used in road construction by applying state-of-the-art environmental impact assessment methods? Management both used for spatial planning and road construction should look for areas that are not near aquifers, rivers or wells for drinking water, to avoid the further spread of heavy metals.

Another suitable alternative can be to centralize the mining waste dumps near the village of Djebel Ressas from three main ones to just one. This can well be achieved by planting a living fence with trees, bushes, and by installing dead fences made of cut wood from treed and bushes around the village first [42].

Another top priority is to move the mining waste to one central spot. This method can be combined without the re-use of some of the dumped waste for road construction, or with any re-use as allowed by the Tunisian laws and standards.

Iteb Boughattas, one of the researchers who had researched on Djebel Ressas [17], has remarked to the author: In comparing Tunisia and Germany, the centralizing of the mining waste dumps from both underground and open-cast copper mining in the *Mansfelder Land* area in Eastern Germany was the key remediation measure applied there. The three centralized dumps had been secured then, approximately within ten years after the mining activities had been stopped, to avoid the further spread of contaminated dust by the wind. This can be verified visually by looking at map data from the southeast of Hettstedt, Germany, at the geographic location of 51° 37′ 25.87″ N, 11° 32′ 1.35″ E (coordinates system: WGS84), available online [last accessed on 2020-07-18] under https://www.openstreetmap.org/?mlat=51.623852&mlon=11.533708&zoom=11#map=14/51.6239/11.5337.

Here is one peculiar situation typical for Eastern Germany. The speedy site remediation within ten years was one of the national and regional priority measures after Germany's reunification in 1990 and cannot be generalized as a common method for other German sites. Many of these in Western Germany still remain in their orphaned state as abandomed around the same time as Djebel Ressas. So commonly said, many sites all over Germany resembles Djebel Ressas very much.

6 Solution: What We Can Do to Reach All Three Dimensions of Sustainable Development via the Use of Quality Targets?

If there are economically and socially beneficial effects of a site clearing and a soil remediation, e.g. by employing a local workforce to reduce health-threatening local contamination on site, we can set targets to reach all three dimensions of a sustainable development: the social, economic, and environmental ones:

The study identifies some top priority sites in and around the village, and in some area in the shape of a box, marked dark green, on the mountain top between the green and the yellow line in Fig. 2. The green areas show mining waste dumps that have not been centralized or covered yet. Incidentially, they are right next to the houses in the village (cf. Fig. 1).

The quality objectives Obj. 1 to Obj. 5 sum up the remediation measures and roughly correlate with social, economic, and ecological dimensions of a possible site remediation project. A dam on the bottom left hand site shows how delicately close both former and present mining areas are to the dam reservoir "W". There is a recent extractive plant in the light brown area south of the peak (with the summit at 750 m) that excavates rock for construction purposes. So any site remediation may have a direct impact on water that is being retained in the reservoir "Barrage El hamma", apparently used for agricultural purposes.

See Table 4 for an overview of the chemical elements. In brief, the project should immobilize Pb, Cd, Zn, Cu, and As by the following means:

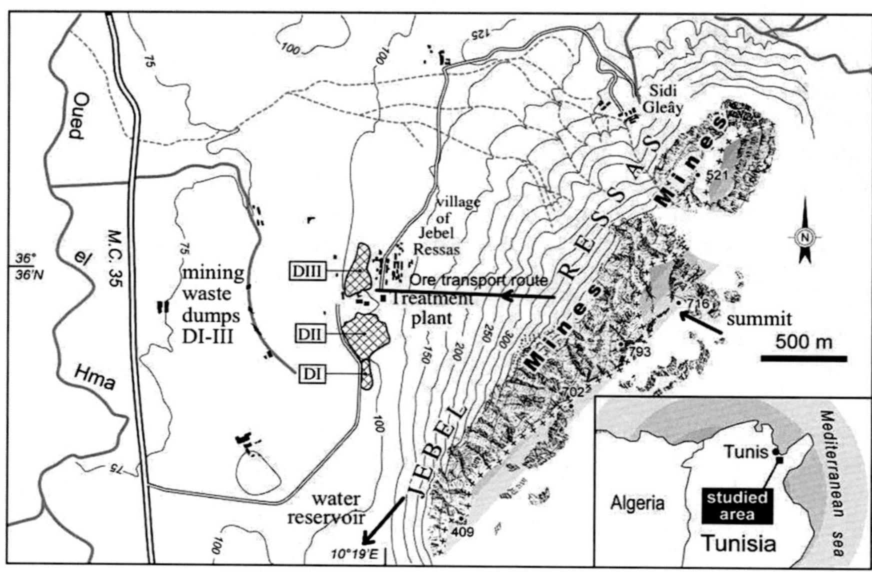

Fig. 1 Geographic localization of Djebel Ressas Village, modified, originally taken from [15]. Copyright © Editions I'Harmattan, 2009. **Legend**: The environmental target to centralize or cover up at the three major mining dumps DI, DII, and DIII (see [15], Illustration 1) could be a first step. The mines, now orphaned, date back to Roman times. Throughout the nineteenth/early twentieth century until the 1950s [43, pp. 1–2] during the French colonial rule [44], Pb–Zn–Cd had been mined (cf. [6]) near the mountain peak (summit in), as one of the first two mines of their kind in Tunisia [45]. A village related to the orphaned mine still exists. Gravel and rock extraction for building and road construction, continues southeast of the peak. A Tunisian-Turkish-Danish consortium was formed in 2010 to set up and run a cement plant with a production capacity of 5800 tons/day clinker [46] next to the water reservoir. Its location southwest of the summit can be seen in . Ghorbel et al. [16] described the main problem caused by flue dust spread from the remaining mine dumps and from the former mines: "Pb and Cd concentrations exceeded WHO guidelines", cf. [12, 15]. Boughattas et al. [17, 20] observed high Zn–Cu concentrations in earth worms, as marker organisms for potential human health risks of the village people [3, 4, 12, 15, 16] and occupational hazards for the plant workers. So if the degraded soil's situation is compared with other degrading factors, could there be a possible remedy (cf. [33, p. 310])?

- Objective 1: Plant living fences (green line) with bushes and trees. Control the spread into the village of Djebel Ressas, and into a larger settlement towards the mountain's Northeast, and monitor it—p.ex. via biomarkers. Apply phytoextraction, phytostabilization of the affected soil by adapted plants: For that purpose, lucerne/alfalfa, *Medicago sativa,* could be used [30]
- Phytoremediation: "Roots take up metals from contaminated soils and transport the metals to stems and leaves. Leaves accumulate metals and are harvested to prevent (a) soil recontamination" [58]
- Objective 2: Inward sloping terraces should be installed to control the surface water during the rainy months of the year.

Fig. 2 Quality objectives for site remediation at Djebel Ressas—the left half of the satellite photo roughly covers Fig. 1—scaled photo (taken by ESA 2019 [47] on 2019-Apr-29). **Legend**: The village is shown northeast of the water reservoir "W" and east of the summit (750 m). The grey area around the village is identical with the major mining dumps DI, DII, and DIIIas in Fig. 1. This is where the remediation could start. The green line suggests the areas where living fences can be planted. The grey areas show where mining dumps can be centralized, or where the disturbed surface can be sealed next to the ancient mines. The yellow line suggests some alteration of the slopes so that the rain water does not carry away easily the polluted soil or small rock particles.

- Objective 3: The aim is to stabilize the deposit area and avoid the spread of flue dust and contaminated soil by surface water [59, pp. 3019–3021, 3026]. In some instances, for example if manure is added, again, lucerne/alfalfa plants can help stabilize the mining dump deposits.
- Objective 4: In order to avoid the spread of heavy metals into the water reservoir "W" southwest of the mountain range, a drainage system should be connected with the slopes from Objective 2.
- Objective 5: Exchange the top soil in areas where people grow vegetables. One key measure is to centralize, adjacent to the former mining area, the three mining waste *"dump and deposit areas"* DI, DII and DIII as shown in Fig. 1 to just one dump site. Most likely, this would be site DI or also some area in the southern part of DII on the village's border, as this is the site far away from most of the inhabitants' houses. Top priority would be the complete remediation of site DIII west of the village, as it borders the inhabited area directly [16]
- Objective 6: To employ the local residents for the remediation work whenever possible would ideally create jobs in and around Djebel Ressas.

Table 4 Well adopted plants from Northern Africa, Tunisia for site remediation

Metal	Species as relevant for Djebel Ressas in Tunisia (Source)
Pb, Lead	*Opuntia ficus-indica*, Figuier de Barbarie [25], can mainly be used for biomonitoring *Coronilla juncea* and *Globularia alypum* [48] *Rosmarinus officinalis/Salvia rosmarinus*, Rosemary [49] *Chrysopogon zizanioides*, Vetiver grass [43, 50] *Atriplex halimus*, Mediterranean saltbush [51] *Brassica juncea*, Brown Mustard [52, p. 898, 53] *Vallisneria americana*, Tape Grass [53, p. 898], not found on land, suitability for fresh water reservoirs to be confirmed *Medicago sativa*, lucerne or alfalfa [30]; *Helianthus annuus*, Sunflower [50]
Cd, Cadmium	*Coronilla juncea* and *Globularia alypum* [48] *Rosmarinus officinalis* [49] *Atriplex halimus*, Mediterranean saltbush [51] *Silene colorata*, Cloven-petalled campion [43] *Vallisneria spiralis*, Eel grass, not found on land, suitability for fresh water reservoirs to be confirmed [53, p. 891] *Bidens pilosa*, a daisy species [54, p. 2] *Solanum nigrum*, black nightshade or blackberry nightshade [54, pp. 1–2], which can be poisonous both for humans and for macaques, however
Zn, Zinc	*Coronilla juncea* and *Globularia alypum* [48] *Helianthus giganteus*, giant sunflower, and *Medicago sativa*, alfalfa [55] *Rosmarinus officinalis* [49] *Atriplex halimus*, Mediterranean saltbush [51] *Rumex bucephalophorus*, also known as horned, red, or ruby dock [43]
Cu, Copper	*Coronilla juncea* and *Globularia alypum* [48] *Helianthus annuus*, Sunflower [53, p. 898]
As, Arsenic	*Coronilla juncea* and *Globularia alypum* [48] *Helianthus giganteus*, giant sunflower, and *Medicago sativa*, alfalfa [55] *Pteris vittata*, Chinese brake/Chinese ladder brake/ladder brake (neither typical nor endemic in Tunisia), cf. Hakeem et al. [56], see in detail: Hasanuzzaman et al. [57]

7 Conclusions

For a post-mining site remediation [14, 31, 32] finding and application of environmental quality targets [17] is feasible and can be modelled after projects like "Saumon 2020" or its predecessor "Saumon 2000" [27], for example:

Resettle an almost extinct species in an area where it had been living in the past [28, 29]. For some of the areas in Tunisia, *Macaca sylvanus*, Barbary macaque/ape, or magot, can be a native candidate. It is suitable for bio monitoring [27], to track the heavy metals contents. Biomass samples can be taken via fur samples or via biopsies (cf. [17] for earth worms).

Redevelop the current nature reserve around the Djebel Ressas mountain top into an area in which *Macaca sylvanus* is to be resettled into a habitat suitable for that species [29]. That can be implemented if the growing cement works still allow such a measure. As the summit of Djebel Ressas, see Fig. 1, is right below 800 m, the minimum height of a habitable area is barely met. The Barbary macaque mostly lives in areas above 700 m [29]. For phytoextraction, phytostabilization of the affected soil by adapted plants [58], lucerne/alfalfa could be used ([30], also see Table 4). However, as the Barbary macaque needs trees such as in "high cedar forests *(Cedrus atlantica)*, cedar/holm oak *(Quercus ilex)* mixtures, pure holm oak forests and cliffs and gorges dominated by scrub vegetation.", the fauna must be adopted [29]: The dietary habits of the ape species need to be considered: "Its diet is primarily composed of cedar (*Cedrus atlantica*) and the oak (*Quercus* sp.), which make up over 50% of its total intake. It eats fruits (33% of its intake), tree leaves (16%), and other plant parts (24%)" [29].

The local population should be fenced, in a pleasurable and natural way by trees and bushes [42], from the mining deposits' flue dusts (compare [15–17, 20, 30]) so that the exposure to heavy metals via inhalation and ingestion is closer to international WHO levels than under the current conditions [15, 16]. Also, under a site remediation scheme, jobs in local projects can be created to ensure high acceptance levels among the inhabitants of Djebel Ressas village and the surrounding area.

8 Recommendations

A site remediation project around Djebel Ressas village should ideally satisfy all three aspects of sustainable development as defined by the Rio AGENDA 21 [9]:

1. **Economic factors**:
   Local jobs can be created by tearing down the remains of the old mining facilities and by centralising the mining waste deposits from three dumps to just one, as immediate, short-term measures. By planting a living fence around the village and the mountain can employ even more people on a smaller, long term scale. The sales of polluted soil for a controlled use in road construction, see NT 106-02(1989), cf. DIN 4301:2009-06 "Ferrous and non-ferrous slag for civil engineering" (cf. [29]), outside areas with aquifer and groundwater wells could generate profit for the project. All in all, one must ensure a high level of acceptance among the locals.
2. **Social factors**:
   By diminishing direct exposure paths, that is inhalation of heavy metals flue dust unsettled by the wind, and the eating of polluted vegetables and fruits, the health hazards for the local population can be reduced. The former mining facilities must be demolished and the mining waste deposits be removed, via an exchange of the top soil with less polluted one so people can safely grow vegetables that are not deep rooting. The site remediation can thereby create short and mid-term employment and has a long-term benefit for the village populace.

3. **Environmental factors**:
By planting lucerne/alfalfa plants on site of the former mining waste dumps, by creating a drainage system that stops polluted water from entering the water reservoir southwest of the village, the spread of heavy metals can be slowed down. The reintroduction of the Barbary macaque could be the long-term environmental quality target. For this purpose, heavy metals-tolerant, well adopted flora (cf. [34, 60]) and moreover trees that are suitable for feeding the macaques can be planted.

9 Buzz Words

Table 5 summarize a number of terms that often occur in the context of phytoremediation.

Table 5 Buzz Words within site remediation using plants

Buzz word	Quick explanation
Bioleaching	"Biohydrometallurgy uses the metabolism of microorganisms to dissolve minerals and bring the metals into aqueous solution in the course of metal extraction or to remove metals from water in the course of water purification or metal extraction, e.g. by mineral formation or by means of complexing." [61, cf. 62]
Phytoremediation	"Roots take up metals from contaminated soils and transport the metals to stems and leaves. Leaves accumulate metals and are harvested to prevent (a) soil recontamination" [58, p. 9]. Phytoremediation can be a framework term for phytoextraction and phytoharvesting in the context of brownfield site remediation [33], p. 310)
Phytoextraction	Special well adopted flora can • either extract heavy metals that are found in the soil and that are to be removed for site remediation purposes • or extract metals and other substances that can be useful for further processing, for example for industrial purposes. Some sources [24, 34, 58] describe this method as some sort of mining in a broader sense Regardless its purpose, phytoextraction describes the use of well-adopted flora in order to obtain solid matter out of other solid matter. It is also called Phytomining if used in a more industrial scale style, or agromining, by using special plants that hyperaccumulate [34] heavy metals in their bio mass [33, 60] Also, metals can be extracted by bioleaching
Phytostabilization	When conducting a brownfield site remediation, a stabilization of the affected soil can be reached by using certain suitably adapted plants, for example in areas where there is a pollution with heavy metals. These would be stabilized by the roots and by the incorporation into the plants' biomass. The phytoextraction by harvesting these plants can lead to a successful phytoremediation

(continued)

Table 5 (continued)

Buzz word	Quick explanation
Well adopted flora	This describes a group of heavy metals tolerant plants that grow in areas with a high level of heavy metals in the soil, either because of natural reasons or man-made disturbances of the ground (cf. [19]) For example, in some German mining areas, metallophytes ([34], p. 76, 83) may such as • the botanical class of *Violetum calaminariae*, dominated by *Viola calaminaria* • metallicolous vegetation (*Armerietum halleri*), mainly formed by *Silene vulgaris* var. Humilis, *Minuartiaverna* ssp. Hercynica, and *Armeria maritime* ssp. Halleri Further studies are needed to confirm the existence of similar plant societies in Tunisia [43], or of other species that can be used for phytoremediation [58]

Acknowledgements I thank God for giving me wisdom and an exploring mind, and for the help of fellow human beings to achieve things together, through good and bad times. The research was made possible with the generous funding of the German Academic Exchange Service (DAAD).—Thank you very much, 2016 DAAD Study Group from Tunisia and Germany, for the inspiration to publish this text. Dr. Borhane Mahjoub, Higher Institute of Agricultural Sciences of Chott-Mariem, University of Sousse, Tunisia, has provided the systematics to fill in the tables. Dr. Evelyn Dietsche, Senior Business Lead, swisspeace (Swiss Peace Foundation) Basel, Switzerland, has provided me with some challenging approaches, pointing at fundamental literature to research on post-mining site remediation, for example (EPA, 2012); (ICMM, 2011). I thank her very much. Also, I remember Dr. Bruno Tauché and his excursions to the German area "Mansfelder Land" and its post-mining landscapes around Hettstedt quite well.

Last not least, his colleague Prof. Dr. Peter Knauer, formerly of Martin Luther University of Halle-Wittenberg, who had suddenly died on 1996-Aug-15 during his academic studies, is gratefully remembered for establishing the concept of *environmental quality targets* for a sustainable development.

References[1]

1. Alayet F, Mezned N, Sebei A, Abdeljaouad S (2017) Continuum removed band depth analysis for carbonate mining waste quantification using X-ray diffraction and hyperspectral spectroscopy in the north of Tunisia. J Appl Remote Sens 11:1. The report can be downloaded from: https://doi.org/10.1117/1.JRS.11.016021
2. Achour Y, Souissi R, Tlil H, Motelica Heino M, Souissi F (2019) Assessment heavy metals (Pb, Zn, Cd) and metalloids (Sb, As) mobility in carbonated soils contaminated by mining wastes (Northern Tunisia). The poster can be downloaded from: https://www.researchgate.net/publication/335608443_Assessment_Heavy_Metals_Pb_Zn_Cd_and_Metalloids_Sb_As_mobility_in_Carbonated_Soils_Contaminated_by_Mining_Wastes_Northern_Tunisia
3. Béjaoui I, Kolsi-Benzina N, Sappin-Didier V, Munoz M (2016) Health risk assessment in calcareous agricultural soils contaminated by metallic mining activity under mediterranean

[1] For further articles not cited, also see http://www.utm.rnu.tn/visirech/Fr/utm/fst/lrme/publication/P3/.

climate. CSAWAC CLEAN: Soil Air Water 44(10):1261–1427. The report can be accessed from https://doi.org/10.1002/clen.201500512
4. Ghorbel M (2012) Contamination métallique issue des déchets de l'ancien site minier de Jebel Ressas: modélisation des mécanismes de transfert et conception de cartes d'aléa post-mine dans un contexte carbonaté et sous un climat semi-aride. Evaluation du risque pour la santé humaine. Thesis, University of Toulouse, France. The thesis can be downloaded from: https://tel.archives-ouvertes.fr/tel-00760685/
5. Kuester H (2015) Symbole der Mittelmeervegetation und ihre Nachahmung im Gebiet nördlich der Alpen' (Symbols of Mediterranean vegetation and their re-modelling in areas north of the Alps). Braunschweiger Geobotanische Arbeiten 11:379–389. The reports can be downloaded from: http://www.digibib.tu-bs.de/?docid=00059137
6. Celamin Holdings (2018) ASX Announcement 31 October 2018. Djebba Zinc-Lead Project, Tunisia—Historical Resource Estimate. Celamin Holdings Ltd. South Melbourne. The report can be downloaded from: https://www.asx.com.au/asxpdf/20181031/pdf/43ztt7bqp0d7ym.pdf
7. Knauer P (1997) Leitbilder, Qualitätsziele, nachhaltige Entwicklung. Diskussion über Umweltqualitätsziele (Discussion about Statement of purpose. Environmental Quality Targets, and Sustainable Development). In: Rauschelbach B, Klecker P M (eds) Regionale Leitbilder—Vermarktung oder Ressourcensicherung. ISBN 978-3-923623-18-1. Kuron, Bonn
8. Helbron H (2008) Strategic environmental assessment in regional land use planning (PhD Thesis). © Brandenburg University of Technology. The report can be downloaded from: https://opus4.kobv.de/opus4-btu/files/452/Doktorarbeit_Hendrike_Helbron_Druckfassung_17_11_2008.pdf
9. UNCED, United Nations Conference on Environment & Development (1992) AGENDA 21, Rio de Janeiro, Brazil, 3 to 14 June 1992. © UNCED 1992. The report can be downloaded from: https://sustainabledevelopment.un.org/content/documents/Agenda21.pdf
10. Jemmali N, Souissi F, Vennemann TW, Carranza EJM (2011) Genesis of the Jurassic Carbonate-Hosted Pb–Zn Deposits of Jebel Ressas (North–Eastern Tunisia). Evidence from Mineralogy, Petrography and Trace Metal Contents and Isotope (O, C, S, Pb) Geochemistry. Resour Geol 61(4):367–383. The report can be downloaded from: https://doi.org/10.1111/j.1751-3928.2011.00173.x
11. Norme tunisienne/Tunisian Standard NT 106-02(1989) Protection de L'environnement - Rejets d'effluents dans le milieu hydrique. INNORPI (Tunisian Standards Organization). The standard can be downloaded from: http://193.95.93.125:8080/web/guest/normes?p_p_id=EXT_2&p_p_action=0&p_p_state=maximized&p_p_mode=view&p_p_col_id=&p_p_col_pos=0&p_p_col_count=0&_EXT_2_struts_action=%2Fext%2Fcatalogue_normes%2Fview&_EXT_2_tabs1=normes-tunisiennes&_EXT_2_group=&group=&nor
12. Ghorbel M, Munoz M, Courjault-Radé P, Destrigneville C, Parseval P, Souissi R, Souissi F, Ben Mammou A, Abedeljaoued S (2010) Health risk assessment for human exposer by direct ingestion of Pb, Cd, Zn bearing dust in the former miners village of Jebel Ressas (NE Tunisia). Eur J Mineral 22:639–649. The report can be downloaded from: https://doi.org/10.1127/0935-1221/2010/0022-2037
13. Herrick JE (2000) Soil quality: an indicator of sustainable land management?. Appl Soil Ecol 15:75–83. The report can be downloaded from: https://doi.org/10.1016/S0929-1393(00)00073-1
14. García A, Álvarez E (2008) Soil Remediation in Mining Polluted Areas. © Macla. Revista de la sociedad española de mineralogía 10:76–84. The report can be downloaded from: http://www.ehu.eus/sem/macla_pdf/macla10/Macla10_76.pdf
15. Ghorbel M, Courjault-Radé P, Munoz M, Maire E, Destrigneville C, Souissi R, Souissi F (2009) Un risque d'origine anthropique: la contamination chronique par les métaux lourds à proximité d'anciens sites miniers. Le cas de la mine (plomb, zinc, cadmium) de Jebel Ressas (Tunisie nord-orientale). Risques et environnements : recherches internationales sur la vulnérabilité des sociétés. © L'Harmattan 2009:271–284 <halshs-01057339>. The report can be downloaded from: https://halshs.archives-ouvertes.fr/halshs-01057339

16. Ghorbel M, Munoz M, Solmon F (2014) Health hazard prospecting by modeling wind transfer of metal-bearing dust from mining waste dumps: application to Jebel Ressas Pb–Zn–Cd abandoned mining site (Tunisia). F Environ Geochem Health 36:935. https://doi.org/10.1007/s10653-014-9610-y. The report can be downloaded from: https://www.ncbi.nlm.nih.gov/pubmed/24760620
17. Boughattas I, Hattab S, Boussetta H, Sappin-Didier V, Viarengo A, Banni M, Sforzini S (2016) Biomarker responses of *Eisenia andrei* to a polymetallic gradient near a lead mining site in North Tunisia. Environ Pollut 218:530–541. The report can be downloaded from: https://doi.org/10.1016/j.envpol.2016.07.033
18. Yousfi R, El Ouear Z, Dahri N, Ouddane B, Rigane H (2019) Evaluating the heavy metals-associated ecological risks in soil and sediments of a decommissioned Tunisian Mine. Pol J Environ Stud 28(4):2981–2993. The report can be downloaded from: https://doi.org/10.15244/pjoes/86198
19. Daldoul G, Souissi R, Tlil H, Elbahri D, El Hamiani O, Chebbi N, Boularbah A, Souissi F (2019) Assessment of heavy metal toxicity in soils contaminated by a former Pb–Zn mine and tailings management using flotation process, Jebel Ghozlane, Northern Tunisia. Environ Earth Sci 78:703. The report can be downloaded from: https://doi.org/10.1007/s12665-019-8720-3
20. Boughattas I, Hattab S, Mkhinini M, Rochdi H, Banni M (2019) Dynamic of organic matter and available nutrients in heavy metal contaminated soil under the effect of Eisenia Andrei Earthworms. Int J Environ Sci Nat Res. 22(1):13–20. 556077. The report can be downloaded from: https://juniperpublishers.com/ijesnr/IJESNR.MS.ID.556077.php
21. Ali H, Khan E (2018) Trophic transfer, bioaccumulation, and biomagnification of non-essential hazardous heavy metals and metalloids in food chains/webs—concepts and implications for wildlife and human health. Hum Ecol Risk Assess Int J. The report can be downloaded from: https://doi.org/10.1080/10807039.2018.1469398
22. El Kribi-Boukhris S, Boughattas I, Zitouni N, Helaoui S, Sappin-Didier V, Coriou C, Bussiere S, Banni M (2020) Ecotoxicity of trace elements to chicken *GALLUS gallus domesticus* exposed to a gradient of polymetallic-polluted sites. Environ Pollut 265, Part A 114831. Available online 20 May 2020. The report can be downloaded from: https://doi.org/10.1016/j.envpol.2020.114831
23. Mielczarek A, Wojciechowicz-Żytko A (2020) Bioaccumulation of Heavy Metals (Zn, Pb, Cd) in *Polistes nimphus* (Christ, 1791) (insertion: should read: *Polistes nimpha—the author*) (Hymenoptera, Vespidae) Living on Contaminated Sites. Pol J Environ Stud 29(6):1–8. https://doi.org/10.15244/pjoes/118746. The report can be downloaded from: http://www.pjoes.com/pdf-118746-53046?filename=Bioaccumulation%20of%20Heavy.pdf
24. Elouear Z, Bouhamed F, Boujelben N et al. (2016b) Assessment of toxic metals dispersed from improperly disposed tailing, Jebel Ressas mine, NE Tunisia. Environ Earth Sci 75:254. The report can be downloaded from: https://doi.org/10.1007/s12665-015-5035-x
25. El Hayek E, El Samrani A, Lartiges B, Kazpard V, Benoit M, Munoz M (2015) Air pollution biomonitoring using successive cladodes of *Opuntia ficus-indica*: A lead isotopic study. hal-02182871. The report can be downloaded from: https://hal.archives-ouvertes.fr/hal-02182871/document
26. Bouhouch RR, El-Fadeli S, Andersson M, Aboussad A, Chabaa L, Zeder C, Kippler M, Baumgartner J, Sedki A, Zimmermann M B (2016) Effects of wheat-flour biscuits fortified with iron and EDTA, alone and in combination, on blood lead concentration, iron status, and cognition in children: a double-blind randomized controlled trial. Am J Clin Nutr 104:1318–1326. https://doi.org/10.3945/ajcn.115.129346. The report can be downliaded from: https://academic.oup.com/ajcn/article-abstract/104/5/1318/4668558
27. CIPR, Commission Internationale pour la Protection du Rhin (2016) Saumon 2000/Saumon 2020. Ce programme avait pour but de restaurer l'écosystème du Rhin afin que le saumon et d'autres poissons migrateurs puissent se réimplanter dans le fleuve d'ici l'an 2000. Saumon 2020 s'inscrit dans le prolongement des actions positives réalisées pour les poissons migrateurs dans le cadre du programme Saumon 2000. The informatiuon can be accessed online: https://www.iksr.org/fr/cipr/rhin-2020/saumon-2020/saumon-2020

28. Affenberg Salem (2020) At Affenberg Salem (Germany) nearly 200 Barbary macaques roam freely in a 20 ha forest. The information can be accessed online: https://www.affenberg-salem.de/en/barbary-macaques/
29. Butynski TM, Cortes J, Waters S, Fa J, Hobbelink ME, van Lavieren E, Belbachir F, Cuzin F, de Smet K, Mouna M, de Iongh H, Menard N, Camperio-Ciani A (2008) *Macaca sylvanus.* The IUCN Red List of Threatened Species 2008: e.T12561A3359140. The report can be downloaded from: https://doi.org/10.2305/IUCN.UK.2008.RLTS.T12561A3359140.en
30. Elouear Z, Bouhamed F, Boujelben N, Bouzid J (2016a) Application of sheep manure and potassium fertilizer to contaminated soil and its effect on zinc, cadmium and lead accumulation by alfalfa plants, Sustainable Environment Research 26:131–135. The report can be downloaded from: https://doi.org/10.1016/j.serj.2016.04.004
31. EPA, Environmental Protection Agency, United States Government (2012) Green Remediation Best Management Practices: Mining Sites, EPA 542-F-12-028. © EPA, Office of Solid Waste and Emergency Response (5102G), Washington. The report can be downloaded from: https://www.epa.gov/remedytech/green-remediation-best-management-practices-mining-sites/
32. ICMM, International Council on Mining & Metals (2011) Planning for integrated mine closure: Toolkit. © ICMM, London. The report can be downloaded from: http://www.icmm.com/document/310
33. Morel JL, Echevarria G, van der Ent A, Baker AJM (2018) Conclusions and outlook for agromining. In: van der Ent A et al. (eds) Agromining: farming for metals, mineral resource reviews, pp 309–312. The book chapter can be downloaded from: https://doi.org/10.1007/978-3-319-61899-9_20
34. Reeves RD, van der Ent A, Baker AJM (2018) Global distribution and ecology of hyperaccumulator plants. van der Ent A et al (eds) Agromining: farming for metals, mineral resource reviews, pp 75–92. The book chapter can be downloaded from: https://doi.org/10.1007/978-3-319-61899-9_5
35. ISO/DIS 23611-1:2016: Soil quality—sampling of soil invertebrates—Part 1: Hand-sorting and extraction of earthworms); German and English version prEN ISO 23611-1:2016
36. ISO 11268-1:2012 Soil quality—effects of pollutants on earth worms—Part 1: Determination of acute toxicity to Eisenia fetida/Eisenia andrei
37. ISO 11268-2:2012 Soil quality—effects of pollutants on earthworms—Part 2: Determination of effects on reproduction of Eisenia fetida/Eisenia andrei
38. ISO 11268-3:2014 Soil quality—effects of pollutants on earthworms—Part 3: Guidance on the determination of effects in field situations
39. VDI, Verein Deutscher Ingenieure, The Association of German Engineers (2008) VDI 4230—Part 2 German Engineering Guidance—Biological procedures to determine effects of air pollutants (bioindication)—Biomonitoring with earthworms (announced for revised publication after 2020). © VDI. The standard can be ordered from: https://www.beuth.de/de/technische-regel/vdi-4230-blatt-2/102511225
40. ISO 17512-1:2008: Soil quality—avoidance test for determining the quality of soils and effects of chemicals on behaviour—Part 1: Test with earthworms (Eisenia fetida and Eisenia andrei)
41. Dietsche C (2010) Vogelschutz auf Golfplätzen? – ein deutsch-japanischer Vergleich (Wild Bird Protection on Golf Courses – a German-Japanese Comparison). ISSN 0030–5723. Ornithologische Mitteilungen – Monatsschrift für Vogelbeobachtung und Feldornithologie (German Monthly Field Ornithology and Bird Watching Journal) 62:226–232. The magazine can be overviewed from: http://www.ornithologische-mitteilungen.de/2010
42. Benjes H (1997) Die Vernetzung von Lebensräumen mit Benjeshecken (German). English translation: Creating networks of habitats with Benjes hedges. München/Munich: Natur & Umwelt, 9th edition, ISBN 978-3-924749-15-6
43. Chaabani S, Abdelmalek-Babbou C, Ben Ahmed H, Chaabani A, Sebei A (2017) Phytoremediation assessment of native plants growing on Pb–Zn mine site in Northern Tunisia. Environ Earth Sci 76. Article number 585. The report can be downloaded from: https://doi.org/10.1007/s12665-017-6894-0

44. Entreprises coloniales (2019) Société des mines du Djebel-Ressas (Tunisie). Mise en ligne: 1er juin 2015. Dernière modification: 10 avril 2019. The report can be downloaded from: http://entreprises-coloniales.fr/afrique-du-nord/Djebel_Ressas_Miniere.pdf
45. Office National des Mines (2020) Industrie minérale extractive en Tunisie. The homepage can be accessed via http://www.onm.nat.tn/fr/index.php?p=indminier
46. PROKON (2020) Djebel Ressas Cement Plant (5800 Tons/Day Capacity). The report can be downloaded from: http://www.prokon.com.tr/en/project/djebel-ressas-cement-plant-5800-tons-day-capacity/
47. ESA, European Space Agency (2019) Modified GALILEO Copernicus Sentinel-2 L1C Data, Satellite Photo of Djebel Ressas taken on 2019-Apr-29, used under General Public Licence. The satellite data as well as the reader can be downloaded for free from: https://scihub.copernicus.eu/dhus/#/home
48. Heckenroth A, Rabier J, Dutoit T, Torre F, Prudent P, Laffont-Schwob I (2016) Selection of native plants with phytoremediation potential for highly contaminated Mediterranean soil restoration: tools for a non-destructive and integrative approach. J Environ Manag 183(3):850–863. The report can be downloaded from: https://doi.org/10.1016/j.jenvman.2016.09.029
49. Majeri Gaïda M, Rassaâ Landoulsi N, Rejeb M N, Smiti S (2013) Growth and photosynthesis responses of *Rosmarinus officinalis* L. to heavy metals at Bougrine mine. Afr J Biotechnol 12(2):150–161. https://doi.org/10.5897/AJB11.2802. Available online at: http://www.academicjournals.org/AJB
50. Sahithya G, Asheeka A, Sakthi Prasanth K (2020) Bioremediation of Lead from Contaminated Soil and Groundwater. Studia Rosenthaliana (J Study Res) XII(I):40–47. ISSN NO: 0039-334. The report can be downloaded from: https://doi.org/10.13140/RG.2.2.27473.07529
51. Acosta JA, Abbaspour A, Martínez GR, Martínez-Martínez S, Zornoza R, Gabarrón M, Faz A (2018) Phytoremediation of mine tailings with *Atriplex halimus* and organic/inorganic amendments: A five-year field case study. Chemosphere 204:71–78. The report can be downloaded from: https://doi.org/10.1016/j.chemosphere.2018.04.027
52. Ma Y, Rajkumar M, Rocha, Oliveira RS, Freitas H (2015) Serpentine bacteria influence metal translocation and bioconcentration of *Brassica juncea* and *Ricinus communis* grown in multi-metal polluted soils. Front Plant Science|Plant Biotechnol 5:1–13. Article 757. The report can be downloaded from: http://dx.doi.org/https://doi.org/10.3389/fpls.2014.00757
53. McCutcheon SC, Schnoor JL (2003) Phytoremediation: transformation and control of contaminants. Print ISBN: 9780471394358|Online ISBN:9780471273042. https://doi.org/10.1002/047127304X. Wiley, New Jersey
54. Maqbool A, Ali S, Rizwan M, Saleem AM, Yasmeen T, Riaz M, Hussain A, Noreen S, Abdel-Daim MM, Saad A (2020) N-Fertilizer (Urea) enhances the phytoextraction of cadmium through *Solanum nigrum* L. Int J Environ Res 17(11):3850. The report can be downloaded from: https://doi.org/10.3390/ijerph17113850
55. Zargan J, Fakharyfar M (2016) Phytoremediation of barium, copper, zinc and arsenic contaminated soils by sunflower and alfalfa. J Biodivers Environ Sci (JBES). ISSN: 2220-6663 (Print) 2222-3045 (Online). J Bio Env Sci 8(6):171–180. The report can be downloaded from: https://innspub.net/download/?target=wp-content/uploads/2016/06/JBES-Vol8No6-p171-180.pdf_15404
56. Hakeem K, Sabir M, Munir Ozturk M, Mermut A (2015) Soil remediation and plants. Prospects and challenges. ISBN: 978-0-12-799937-1. Elsevier, London
57. Hasanuzzaman M, Nahar K, Hakeem K R, Öztürk M, Fujita M (2015) Arsenic toxicity in plants and possible remediation (Chap. 16). In: Hakeem et al (eds) Soil remediation and plants, pp 433–501
58. Becker H (2000) Phytoremediation—using plants to clean up soils. Agricult Res 06:4–9. The report can be downloaded from: https://agresearchmag.ars.usda.gov/AR/archive/2000/Jun/soil0600.pdf
59. Aydi A (2015) Assessment of heavy metal contamination risk in soils of landfill of Bizerte (Tunisia) with a focus on application of pollution indicators. Environ Earth Sci 74:3019–3027. The information can be accessed online: https://doi.org/10.1007/s12665-015-4332-8. See discussions, stats, and author profiles for this publication at: https://www.researchgate

60. Rosenkranz T, Hipfinger C, Ridard C, Puschenreiter M (2019) A nickel phytomining field trial using *Odontarrhena chalcidica* and *Noccaea goesingensis* on an Austrian serpentine soil. J Environ Manag 242:522–528. The report can be downloaded from: https://doi.org/10.1016/j.jenvman.2019.04.073
61. TU Bergakademie Freiberg (2020a) Mikrobielle Laugung - Bioleaching. The homepage can be accessed from: https://tu-freiberg.de/fakultaet2/bio/environmental-microbiology/forschungsprojekte/mikrobielle-laugung-bioleaching
62. TU Bergakademie Freiberg (2020b) Publications of Prof. Dr. Sabrina Hedrich. The homepage can be accessed from: https://tu-freiberg.de/fakultaet2/bio/arbeitsgruppen/biohydrometallurgie-mikrobiologie/jun-prof-dr-sabrina-hedrich

The Agricultural Landscape of Sahel Bizertin: A Heritage in Peril

Sondès Zaier and Saida Hammami

Abstract The present work focuses on illustrating the environmental consequences of the development of urbanization on the agrarian landscape of the Tunisian coast. It will be initially to show how this landscape is singular and evaluate in a second time the impact of rapid urbanization on the transformation of coastal agricultural space. Until they become a territory coveted by vacationers. The small towns of the sahel Bizertin have evolved in a very slow way because they are enclosed between the sea on the one hand and hills or mountain of one other. That developed an original form of appropriation of farmlands in the form of small plots of land dedicated to the cultivation of cereals, the arboriculture, the truck farming and the small breeding. The originality of this system comes on the one hand, to its vicinity with the mouth Oued Medjerda, the country's most important waterway but with a peasant genius that h as developed a natural environment whose formation and evolution are a unique case in Tunisia. This is necessary to know, a culture that has largely contributed to shaping the landscape of this region, which, in view of its practices, will have given this region a character and an identity. These singular landscapes today suffer from the new pressure exerted on the land by the vacationers who chose this area as a new seaside place. The resulting pressure of urbanization will only aggravate the deterioration of this agrarian system and therefore cause degradation of the coastal landscape in the region. We are now entitled to ask about its future. It will then be necessary, inevitably, to de-encrypt this mixed landscape to identify the values and to determine the management strategies and its conservation and to find the means of adapting it to contexts in permanent evolution.

Keywords Agrarian system · Singular landscapes · Identity · Patrimonial · Gattayas · Urbanization

S. Zaier (✉) · S. Hammami
UR 13 AGR07/Higher Institute of Agronomy—Chott Meriem, University of Sousse, Sousse, Tunisia

M. Abu-hashim et al. (eds.), *Agro-Environmental Sustainability in MENA Regions*, Springer Water, https://doi.org/10.1007/978-3-030-78574-1_14

1 Introduction

The knowledge of landscapes as a value has become very distinctive and topical in the last decades. The landscape with harmonious and consistent relations between human activity and preserved nature is an essential competitive advantage. A recognisable outstanding landscape presents above all the identity of a certain area at different levels. The treatment of the landscape also depends on the attitude of society towards it, thus influencing the state and image of landscapes and the quality of our life [1].

The recognition of the agrarian landscapes of the Bizerte Sahel region must be reflected in its characteristic climatic conditions, its ecosystems and its typical vegetation. The climate and the terrain influence the diversity of the coastal landscapes. Geological diversity and diversity of landforms create a wide variety of land use, spatial orientation, and hydrology. In the Bizerte Sahel agriculture has largely contributed to shaping the landscape of Ghar El Melh [2]. Through its practices, it has given this region, known for its agrarian landscapes, a character and an identity. This region, today, its farmland is declining, especially on the banks of the lagoon. This regression leads to a decrease in agricultural surfaces and an irreversible change in its exceptional landscape [3].

The chapter aims to show the singularity of agrarian landscapes of the Sahel of Bizerte and to evaluate the landscape and environmental consequences of the development of the resort in the countryside.

2 The Sahel Bizertin: A New Contemporary Resort Area

2.1 Presentation of the Site

The Sahel of Bizerte is located about 60 km north of the capital Tunis.

It is a coastline stretching over 36 km, ranging from Ghar el Melh to Cap Zebib and passing through RAF-RAF, Renin, Ras Jebel and Chott Memi (Bani Ata) and Metline (Fig. 1).

The region lies at the end of the Atlas mountain range, offering a variety of landscapes: Lagoon Coast to Ghar El Melh followed by fine sand beaches and rocky shores to the south, all bordering a narrow strip limited by a mountainous terrain to the north [4]. The coastline of the Sahel of Bizerte is a unit of remarkable landscapes scarred by the know-how of peasants of Andalusian origin. The landscape is marked by a rocky spur and especially by the oued Medjerda, who created a Lake enclosed by an arrow natural [4]. The artificial boom, drawn by the peasants, delimits a lagoon which each acre of land is cultivated uniquely. This natural site and rural landscapes, reflecting the Andalusian traditions are currently threatened by the phenomenon of development of the secondary residence and that of the orbiting of the cities of the Northeast small and medium at 30–40 and 60–70 km from Tunis [5]. This chapter

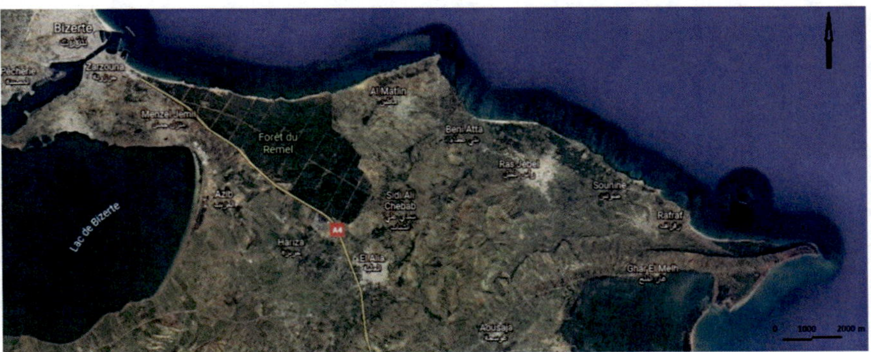

Fig. 1 Map of the region of Tunis and Bizerte, the various localities of the Bizerte Sahel

presents how the territory of the Sahel of Bizerte is unique, and provides a heritage to protect and then exposes the pressure exerted on this territory.

2.2 *The Seaside Boom and Urban Development of the Bizerte Sahel*

The saturation of the beaches of the Gulf of Tunis and the degradation of the tunisois coast caused, during the seventies, the emergence of new categories of landowners. Tunis elite buy land in near coastal communities of Tunis seeking agricultural profits, speculative game but especially looking for new places of the resort away from the saturated Tonisian beaches.

The transformation of the resorts on the outskirts of Tunis in places of permanent habitat had so pushed the demand for second homes to new sites, which sometimes escape any intervention by the public authorities [6]. The first form of the attraction of the Sahel of Bizerte is related to the presence of a sandy coast still a virgin [2]. All historical centres were established at least 1 km from the coast according to a traditional Tunisian cities organization, which always takes place in the little fertile and sometimes elevated sectors [7, 8]. The shore was reserved for agriculture because the land is very fertile [9]. The image of a peaceful region with wide beaches began to spread at the beginning of the century last thanks to the multiple stories [10] describing this area and representations that have been made especially by Roubtzoff paintings dating from the beginning of the century.

The first cottagers were those who, fleeing beaches saturated Tunis, have discovered that the Sahel of Bizerte site ideal relay through the diversity of its landscapes, the quality of its beaches and the sea. This first advantage linked to the natural elements, was facilitated by a second linked to the development of the road network. Geographic implantation of different bathing areas followed the nature of the busy coast [11, 12]. In other words, a simple observation of an aerial photograph of the

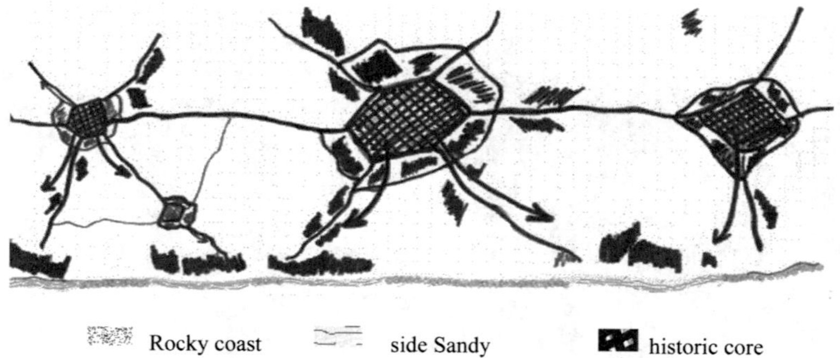

Fig. 2 Schematic representation of the urban development of the four localities of the Sahel bizertin

region shows that homes appear—feel there where there is a sandy coast and go where there appears a rocky coast (Fig. 2).

In the Middle, the town of Ras Jebel, growing in the direction of the sea to the Chott Memi Beach, left the city of Raf–Raf with development of two seaside districts: one at Raf–Raf Beach and the other at Ain Martin, who joined the seaside neighborhood of Sounine. Finally on the right, the city of Metline who also sees the birth of a seaside neighborhood through a sandy coast. Seaside urbanization expands to the time near the beaches avoiding the cliffs and rocky coasts.

If this phenomenon of the resort is spontaneous, that the reorganization of the entire region of Bizerte in metropolization of greater Tunis area is programmed. As the master plan for development of the National territory: "The central objective is firstly to succeed the metropolization" and the master plan for and use planning of the great Tunis (SDAT), which says "metropolisation would defer on peripheral cities some of the growth and public investment".

3 The Traditional Agricultural Landscapes of the Sahel Bizertin (Identity)

The Sahel of Bizerte is rich in heritage and landscape identity reflecting the evolution of the human occupation of the territory. The heritage value of the landscape is mainly due to the time some. It gives an account of the mode of development of that time which based on various elements: the fragmented agricultural rural, village, the port and the fortresses kernels Street frame. Land of the permanent agricultural zone on the edge of the lagoon and the terraces watching, even today, the Andalusian cutting originated at the beginning of the seventeenth century [13]. Farmland, the ranks and upgrades make a frame adapted to the morphology and geography ground on which is based the development of smaller village centers. Gradually, small communities formed leading to the creation of municipalities.

To study the singularity of this agricultural system and the landscapes that result, we choose two localities based on the administrative page discovered:

- The community of Ras **Jebal**: In this community, the choosing place the Raf–Raf city.
- The community of **Ghar El Melh** who the city of Ghar El Melh will be a support of analyse.

3.1 The Gattays of Ghar El Melh

Ghar El Melh, formerly known as Porto Farina, is a coastal north of the Tunisia, living the fishing, agriculture and known for its historic site and its beach village. It is located at the bottom of the lagoon bearing his name [3]. The lagoon is a body of water of 30 km^2. It consists of the basin of Ghar El Melh, Sebkhet El Ouafi and Sebkhet Sidi Ali El Mekki isolated by a coastal boom and relayed by marshy Islands. Powered by the flooding of the oued Medjerda, the large body of water is dominated, on the side North by the jbel Ennadhour. Two arrows bound the central part of the lagoon: an artificial to the West and the other natural East. Between the Lake and the mountains spread agricultural land. These small parcels receive different cultures: grain, arboriculture, gardening and small livestock. Enclosed between the sea on the one hand and the mountain on the other hand, the city extends about 500 m wide. These two natural limits are that the land has a very high value. What has also developed an original form of appropriation of farmland and worked in different ways by farmers.

Thanks to the patient work of amendment of the salty soils and freshwater runoff collection (Fig. 3). The slopes are landscaped terraces of arboreal culture. The edges of the lagoon grown in gardens market gardeners, while on the arrow artificial, formed by a daily intake of baskets of Earth, residents, Andalusian origin, won a polder protected hedgerows of Palms where the crops are protected by media carefully aligned reeds. This old complementarity between agricultural and fishing association shows a remarkable ability to organize the space.

The originality of this system is on the one hand, its proximity to the mouth of the Wadi Medjerda, the largest rivers of the country and on the other hand, the peasant genius who was able to highlight a natural mid-rise whose training and evolution are a case unique in Tunisia [14].

Installation of the Andalusians in the region at the beginning of the seventeenth century and later decline and shutdown of the race are the origin of a very intensive earned both local operation on the mountain and the sea. “The Andalusian science of gardening” more enriched in the nineteenth century in contact with Maltese immigrants who developed the construction of terraces of cultures and continued to expand arable land at the expense of the lagoon of Ghar El Melh and the old arrow Edhrea and El Gattaya [15]. Very low, the banks of the lagoon of Ghar El Melh are almost all bordered by marshy spaces sometimes very extensive. It is especially the north-eastern sector of the lagoon that attracts the most attention with the gardening system

Fig. 3 Carefully worked agricultural parcels, today threatened with disappearance at the urging of the cottagers on land to build

that there is practiced: a multitude of tiny farms, gardens very carefully soul swum called the Gattayas (Fig. 4).

The Gattayas show today a unique knowledge to the ecosystem of Ghar El Melh. These spaces of culture attest to an old tradition, which operates a deep knowledge of the behaviour of the freshwater in the coastal and marine environment for agricultural purposes. The surface of farms is conquered with contributions of land spread and regularized to a carefully calculated altitude. The obtained soil must have a thickness such that the roots of the plants that are growing can escape salt water but benefit

Fig. 4 The singularity of the place: the lagoon in gardens by fishermen-peasants, March 2016

Fig. 5 Aerial view of the organization of the Gattayas

from the movement of fresh water from the water table. Located a few Decimeters of depth, this tablecloth is subjected to a vertical swinging in connection with the variations in the level of the Lake as a result of a semi-diurnal tide and 20 cm of amplitude. Cultivated plants benefit of moisture from the ground or even naturally irrigated, down, twice a day.

This "meticulous gardening", on the edge of the lagoon, is at the origin of a unique landscape of gardens, tiny, very neat and sometimes lost in the middle of the marshes [10]. Farmers cultivate all types of vegetables on nearly 8885 hectares in total. Also, the borders of Lake and coastal belts was filled and covered sand desalinise and fertilising elements including organic fertilisers; These new lands gained at the expense of the sea have been subsequently used as gardens (Fig. 5). V. Guerin, who visited the city in 1860, maintains us "these rich and laughing orchardsextending on the edge of the Lake and at the foot of the mountain, and even the coastal belts, gardens planted of olive trees, fig trees, almond trees and several other trees" [16].

3.2 *The Planted Terraces of Raf–Raf and the Mixed Cultures of Sounine*

The site occupied by the city of Raf–Raf is characterized by a topography that is diverse, including hills and narrow Plains. The historic core of the town occupies a slope of a hill overlooking the sea unlike the other localities of bizertin Sahel that turns its back to the sea. The traditional Center of the city is connected to Headquarters Raf–Raf Beach by an old road that goes down to the sea and following a steep slope. This topography is, according to Mourad Ben Jaloun, one of the factors that don't have much favored the development of human installation on the seafront, despite the existence of a former group of habitation in the area called Dhar-Ayed [3]. The reason also according to Pierre Signoles abreast than Tunisian ancient cities are often

Fig. 6 The sides worked in terraces to Raf Raf to Sahel Bizertin

little fertile land. In this way and since the slopes leading down to the sea were very fertile, they were reserved for agricultural activities.

Raf–Raf is characterized by precarious of terroirs specializing in viticulture. She is known for its Muscat grapes produced by of extensive vineyards planted in terraces (Fig. 6). These vineyards, which have Mediterranean fame, formed a Crown around the historical core of the city extending to the shore of the sea [17]. This culture on the terrace was probably developed by Andalusian immigrants [18].

Between the Lake and the mountains stretch—transversely from the ground—dependent agricultural land in the city. Plots are parallel to the contour lines structures and receive diversified cultures: gardening, arboriculture, grain and small livestock. The action continues to man since the eighteenth century, to extend surfaces cultivated at the expense of the mountain, has shaped the landscape around the town. Thus, the land was torn off in the mountains, by his side terraced landscaping, creating cultures in floor performed by built-in stone retaining walls. These cultures extend up to a level high enough on the side and are mostly arboreal plantations.

The Bay of Raf Raf is made of reliefs low and scenery to the fields sometimes open, sometimes closed, but always carefully and intensely cultivated. Agriculture benefits from the presence of a shallow water table and several hillside dams. Gardens all around each town are trussed intensive cultures both dry (olive trees, almond trees, fig trees, …) and irrigated (gardening). However, each region natural conditions allowed a certain specialization of terroirs: potatoes in Ghar el Melh, olive trees in El Alia, citrus Sounine and vines of the table to Raf raf (Figs. 7 and 8).

Fig. 7 The Bay of Raf Raf carefully and intensely cultivated

Fig. 8 The Sahel bizertin gardens: the cutting geometry of small plots cultivated with care. *Picture* The Atlas of Tunisia

4 Deterioration of the Agrarian System and Degradation of the Coastal Landscape Generated

This specific agricultural territory becomes part of the area of Metropolitan solidarity of Tunis where the farmer controls less and less agricultural activity and where the nibbling of the agricultural space is achieved through the diversion of land to the urban speculation more remunerative [19]. Among the problems of the villagers in the Sahel of Bizerte, the estate subdivision of parcels is a challenge illustrated by the report of the size of the population to the agricultural land, in Ghar El Melh, for example, each has only 0.5 ha. If this little island of land has still to be divided between the heirs, it is clear that the new generation leaves the land. Urban expansion is especially carried out at the expense of farmland by permanent snacking. Also, development of land speculation resulted in the expansion of abandoned land. Small farmers who hold the microparcelles, especially in close to the beach, to abandon their activities pending a lucrative sale.

Today several agricultural parcels are missing and have been replaced by homes. Cultures in terraces leading down to a long fringe of the beach are coming out say despite the situation of this space outside the municipal scope. These very steep inaccessible terraces to the classic mechanization were long worked by hand. Today, there are few farmers who maintain these terraces manually. The spatial expansion of the resort, which now continues at a high pace, generates more and more environmental problems (Ministry of Equipment, Housing and Territorial Development/DGAT/Urbaconsult-URAM-BRAMMAH [20], study of the Master Plan of Greater Tunis, final report, January (the source is in French)). This growth has touched all side that has experienced an accelerated development concentrated on the coastal strip, which led to profound changes in the landscape and a nibble of farmland. This damage is amplified by the advance of second homes. The extension is done on fertile farmland to urbanization. One of the major problems in this region also concerns the excessive fragmentation of properties related to the dominance of private ownership. Although agricultural land often remains the property of the same family, the excessive land fragmentation causes many problems in agricultural activity in the region of the regulatory construction.

5 Discussion

Farmers in the Sahel of Biezrte suffer today the new pressure exerted on their land by beachgoers who chose this location as a new seaside place. The pressure of urbanization resulting will worsen the situation of the deterioration of the agrarian system and therefore cause the degradation of the coastal landscape in the region. Today, we are entitled to wonder about its future. Urban sprawl throughout the Sahel of Bizerte especially carried out at the expense of farmland by permanent snacking. Also, development of land speculation resulted in the expansion of abandoned land. Small farmers who hold the can, especially in close to the beach, to abandon their activities pending a lucrative sale. The locality of Ghar El Melh knows in the summer an increase considered-able in the number of its inhabitants more than 30,000 people flowing daily to the seaside neighborhood according to local estimates. The important extension of 1′ built-up area is 1′ origin of the pollution of the environment and including the beach and the forest. Also, groundwater from shallow waters suffers from over-exploitation of renewable resources [6]. The I-safe have been adopted in the planning regulations, which are added to older protections on the coast, to those of the so-called national law "law coastline" as well as the protection of a large part of Sidi Ali El Mekki in respect of sensitive areas. Despite these measures, the construction of secondary residence continues and local authorities resign themselves to let. The extension of the urbanization in Ghar El Melh and regression of coastal farmland translates a urban duplication and a linear stretch of beachfront buildings. This phenomenon tends to the standardization of coastal landscapes and causes an intensification of coastal space that is gaining more and more the hinterland. Indeed, the proliferation of homes does not grow only Ghar El Melh. In nearby locations, the phenomenon began since the years eighty. A Raf–Raf, the precarious balance that characterized the operating system has been broken following the demise of the terroirs specializing in viticulture. The extension of the resorts is still spontaneously resulting in serious degradation of the quality of public space. Thus, urban expansion on the coast occurred also at the expense of farmland, which pushed many farmers, who hold the microparcelles to abandon their activities pending a lucrative sale. The consequences of no action to preserve the sustainability of natural resources and sustainability of regional agriculture would be serious as well at the local level, regional and national levels. The projection of the current trend shows that the degradation of soils, of vegetation cover in hydro infrastructure and availability reduction of irrigation water would be such as production systems can no longer continue to exist and farms can no longer withstand the regression of the ac-agricultural activities and because of the decline in their income. Thus, the situation in the medium and long term would be characterized by a worsening of the risk of non-sustainability of farming with inevitable repercussions at economic and social level.

6 Conclusions

Throughout this article, we have shown that the interest of the agricultural landscapes of the Sahel of Bizerte is not limited to the fact that they offer a remarkable aesthetic beauty. We have also been able to show that the key to the preservation of an agro-biodiversity an overall meaning and present a valuable cultural heritage, but they also provide sustainable multiple solutions related to work, to food and to the well-being of farmers.

Although in most countries of the world, modernity has been characterized by process of cultural homogenization and economic, in many rural areas specific cultural groups are still associated with a geographical context and social in which there are special forms of agriculture. Dynamic conservation of these sites and their cultural identity may be the starting point for territorial development and social and cultural renewal.

Agriculture in small towns can participate today in the construction of a sustainable city if it ensures "good, healthy, sufficient and sustainable food" [21] and it participates in improving the environment. Life as well as adaptation to climate change through the practice of ecological principles and can thus become a technical solution that reduces the effect of "heat islands" [22]. The generalization of the concept of sustainable city can thus generate an update of the city-agriculture links which must imperatively involve the local communities [23].

Recommendations

It is essential today to involve the local community in local government. With the new direction taken by the Tunisian government in 2018, promoting decentralization, it is necessary to involve the population for a participatory approach.

The local government to support by the participation of the population can prove the solution for the future and the safeguarding of the agricultural lands of Ghar El Melh.

References

1. Nicolas S and al (2014) Heritage to the rescue of family farmers? Mediterranean lighting revue tiers monde 2014/4 (n ° 220), 137–158. https://doi.org/10.3917/rtm.220.0139
2. Oueslati A (2010) Beaches and urbanization in Tunisia: avatars of the twentieth century experience to the uncertainties of the future. Mediterranean Geogr Rev Mediterranean Countries 115:103–116 (the source is in French)
3. Gafsi SA (2008) In: El Melh G (ed) Heritage development and cultural promotion agency, Tunis (the source is in French)
4. Jallel A (2008) The landscape atlas of Tunisia (the source is in French)
5. Hbib D (2010), "Interface Logic and Globalization: Levers of the" Emerging Metropolitan Region "of Tunis. In: Dlala H (ed) Globalization and urban change. CPU, Tunis, pp 7–36 (the source is in French)
6. Oueslati A (2017) Urbanization, tourism development and erosion of beaches in the southern shores of the Mediterranean. In: Proceedings of the 3rd symposium space of action, space in action: The Mediterranean at the invitation of geography: http://www.wadi.unifi.it/Oueslati_2017.pdf (the source is in French)
7. Oueslati A, Labidi O, Elamri TH (2015) Atlas of the vulnerability of the Tunisian coast to sea level rise. APAL- UNDP, 67 p
8. Oueslati A, Ouadii El Aroui O, Sahtout N (2015) On the great vulnerability of the lido of the lagoon complex of Ghar El Melh and its wetlands (northern Tunisia): erosion, risk of maritimization and threats to the original Ramli terroir. Rev Médit
9. Amor M (1994) Tunisian coastal towns and the peripheral agricultural area. In: Review of the association of Tunisian geographers, n° 12 and 13, pp 28–30 (the source is in French)
10. Victor G (1862) Archaeological trip in the Regency of Tunis. H. Plon, Paris, T. 2, and François Bournand, Tunisia and Tunisians, by Paris: A. Taffin-Lefort, 1893 (the source is in French)
11. APAL-UNDP (2012) Study of the vulnerability map of the Tunisian coast due to climate change. Final report Phase1, 402 p (the source is in French)
12. APAL-UNDP (2012) Study of the vulnerability map of the Tunisian coast due to climate change. Final Report Phase II, 189 p. Encyclopedia Universalis Dictionary (the source is in French)
13. Paul S (1998) Tunis: history of a city, Paris: L'Harmattan (the source is in French)
14. Saadaoui A, Djelloul N. Ghar el Melh: a Tunisian port city of the seventeenth century. Africa XV. 200 (the source is in French)
15. Ameur O, Charfi F, Baccar F (2006) The lower valley of Oued Majerda and the Ghar El Melh lagoon. In: International Meeting INCO-CT-2005–015226. Tunis; December 6–7–8–9 (the source is in French)

16. Mourad BJ (1996) Raf-Raf: new spatial dynamics of a small town in the Bizerte Sahel. Tunisian J Geogr 30 (the source is in French)
17. Angles S (1996) Les aspects récents de la viticulture tunisienne. In: Le Gars C, Roudié P, de Lemps AH (eds) Des vignobles et des vins à travers le monde, colloque tenu à Bordeaux les 1, 2 et 3 octobre 1992. Presses Universitaire de Bordeaux, p 596
18. Hbib D (2007) Metropolisation and Territorial Recomposition of Northeastern Tunisia. Cybergeo 410 (the source is in French)
19. Hbib D (1999) New industrial littorality in Tunisia, globalization and spatial planning. Geogr Space 28(1):49–58 (the source is in French)
20. Ministry of Equipment, Housing and Territorial Development/DGAT/Urbaconsult-URAM-BRAMMAH (2010) Study of the Master Plan of Greater Tunis, final report, January (the source is in French)
21. Giuseppe S et al. (2015) Nourishing the planet, energy for life. Official catalog. Editions Electa, Expo Milan 2015
22. Marry S (under the direction) (2018) Sustainable territories from research to design, parenthesis editions, 223 p
23. Françoise J (2018) Public farmland and urban development: the case of public lands in Lausanne, Rural studies [Online], 201|2018, posted on January 01, 2020. http://journals.openedition.org/etudesrurales/12336. https://doi.org/10.4000/etudesrurales.12336

The Urban Tree: A Key Element for the Sustainable Development of Tunisian Cities

Ikram Saïdane

Abstract In Tunisia, until the nineteenth century, the tree had a purely nourishing role and was mainly found in the orchards outside the medina walls. In the large Tunisian cities, the urban tree was born with the extension of the European city (during the French protectorate) in the alignments of trees of the main avenues but also in the public gardens and parks of the time. The ornamental tree was thus introduced and symbolized urban aesthetics, hygienism and modernity, concepts in vogue at that time. It was only thirty years after the independence of Tunisia (1956) that a "green policy" based on new concepts emerged: the environment and the sustainable development. The tree is at the heart of this environmental motivation and green spaces have been created to protect flora and fauna, especially urban and peri-urban forests, to combat pollution and to increase the ratio of green spaces. The Tunisian State has sought to stimulate a dynamic of sustainable development by implementing a series of institutional measures and specific programs encouraging the creation of green spaces such as the National Urban Parks Program or the Ville- garden. The present work will therefore aim to demonstrate the way in which the Tunisian Government has implemented a green policy through the study of the method of programming green spaces but also through the major projects. These projects carried out, to make the most of the place of nature in the urban environment in order to envisage a sustainable development of the Tunisian cities.

Keywords Green spaces · Urban tree · Sustainable development

1 Introduction

In recent decades, Tunisian cities have undergone a major metamorphosis in the areas of urban expansion and population growth. The relevant indicators showed that nearly 70% of the total population currently lives in cities [1], in addition to the great incessant development of urbanized perimeters to the detriment of agricultural and

I. Saïdane (✉)
Higher Agronomic Institute of Chott Mariem, Sousse, Tunisia

Landscape, Horticulture and Environment Research Unit, Sousse, Tunisia

M. Abu-hashim et al. (eds.), *Agro-Environmental Sustainability in MENA Regions*, Springer Water, https://doi.org/10.1007/978-3-030-78574-1_15

silvicultural zones and ecosystems. It should be noted that these changes in demography, urbanization and lifestyles in general had a direct impact on the management of urban areas in general and on the quality of life in particular. A number of programs and projects aiming to improve the quality of life in urban and rural areas, have been developed by the Tunisian Ministry of the Environment and implemented. We can cite the program of the promotion of urban aesthetics (green spaces, urban parks, environmental boulevards, boulevards of the earth, strategic routes and entrances of cities …), besides the program of assistance to the preparation and implementation of Agenda 21 and urban development strategies.

Green spaces are an essential element for the aesthetics, setting and quality of life of a city. They help to ventilate the cities and must be considered as the lungs of the city. They are places of relaxation, walk, rest, games for the youngest, sports fields for all ages. They also have a significant impact on health, as many studies have shown: they create "oases of better health around them," according to published research in the Journal of Epidemiology and Community Health [2]. These green spaces are actively a specific part of Tunisian environmental protection programs.

This chapter will focus on demonstrating the role of the tree, a living component of green spaces in the construction of sustainable cities. To do this, it was necessary to return to the role of the tree in the city, a role that merges with the three pillars of sustainable development because of its benefits for the territory, for man and society. Then, it was about to see the short history of the introduction of the tree in urban areas in Tunisia and especially in the city of Tunis. The tree that was having a food role only, is having now an ecological, an ornamental role and even a symbolic one. We will see that this interest for the tree in an urban environment manifested itself during the period of the French protectorate parallel to the expansion of the city according to an imported model. Since then, and with the increasing spatial influence of Tunisian cities, there has been a real awareness of public authorities of the ecological and social importance of green spaces. This concern resulted in the establishment of a series of programs for the promotion and realization of green spaces. We will focus specifically on the National Program of Urban Parks (NPUP), a program still in progress and which claimed the creation of 100 parks throughout the Tunisian territory. We will draw conclusions as to their presence in sufficient numbers and their judicious distribution in the city.

2 Tree and Sustainable Development

The benefits of the presence of the plant in the city is no longer to be proven. They can be divided under the three pillars of sustainable development. For man, they concern health and well-being, social link and identity. For the environment and natural balances, it is about biodiversity, thermal regulation, air quality, water flow and soil protection. Finally, for the economy, it conributes to building valuation, valorisation of plant products, urban agriculture and attractiveness of the territory [3].

2.1 Benefits for Man

For men, access to green spaces and nature in the city contributes directly to the health of residents and wellness [4]. Regardless of why trees provide so many benefits by reducing stress, promoting physical activity, improving the living environment and the state of health felt particularly for the sensitive populations that are children and elderly people from popular backgrounds [5]. This is nevertheless highly dependent on the quality of the site, its level of maintenance and its connection to the rest of the city [6]. This is the reason why some associations work for the creation or the improvement of therapeutic gardens in medico-social establishments [7]. Through their frequentation and the activities that take place there, green spaces reinforce social cohesion locally. Public green spaces create opportunities for contact between people from diverse social and ethnic backgrounds. These interactions are all ways to participate in the life of the community and develop a sense of friendliness. Community commitment is decisive for the involvement of the population in development choices, in dialogue and consultation [8]. Several authors speak of the role of green spaces in social cohesion and some rely on field surveys. Those who do, recognize that green spaces potentially offer more opportunities for social interaction than any other space, thanks to their ease of access and their characteristics. Community attachment seems reinforced by the presence of quality green spaces near dense residential areas [9]

2.2 Benefits to the Environment

For natural balances, especially for biodiversity, green islands, urban parks, connected within a multifunctional green network, have an essential role in the conservation of biodiversity. Recent scientific work carried out in France as part of the Urban greenway study, has shown the importance for urban biodiversity to reconnect parks with each other to create multifunctional frames that respond to environmental and social issues [10]. The second consequence on the environment is thermal regulation since it is proven, that the presence of vegetation in the city, reduces the urban effect and contributes to better energy efficiency of buildings. The presence of trees around a building increases the overall roughness of the surface, reducing wind speed and his penetration force. This effect reduces hot air in buildings in summer and cold air in winter, and allows for increased energy efficiency [11]. One of the strongest arguments of nature in town today is certainly the effect of vegetation on the local climate. Trees can reduce the temperature of a street by 2° and in a context of climate change, this role takes an obvious interest and creates an urban microclimate. The quality of urban air is largely influenced by the presence of plants in the city [12]. Air quality is a major concern in urban areas in order to protect public health and the environment. Several studies show that vegetation filters atmospheric particles (including PM2.5) and absorbs pollutants (proven in particular for NO_2 and

SO_2). Parks, green roofs, urban woods, etc. represent as many permeable surfaces, offering points of temporary retention, slowing of the flow, and even infiltration of rainwater reducing the environmental risks, in particular the floods and the erosion of grounds [13]. An urban area dedicated to green spaces and vegetation is a form of guarantee of soil preservation against its artificialization, the loss of its physical properties, and its ecosystem functions or services (hydraulic functions, purification, support of life). The presence of a vegetative cover protects it from erosion and settlement caused by the impact of the precipitations and their flow. The root system of vegetation in the city creates a real protection architecture in the soil, which makes it possible to structure and prevent it even more against erosion.

2.3 Benefits to the Economy

From the economic point of view, the public and private urban green spaces favored by the users lead to real estate gains for the housing located nearby as would the view on a pleasant landscape or a waterfront. The proximity of a green space increases the price of land [14]. The intensity of this phenomenon is very variable from one city to another or even from one park to another. The quality of the living environment and more generally of the urban environment is likely to contribute to the social and economic development of the territory and increase its attractiveness. Parks and gardens are frequented by residents, but also by locals and visitors. Sustainable urban tourism is growing, and is seen as a "constant opportunity for the conservation of biological and social diversity, job creation and the improvement of the quality of life". This form of cultural tourism combines the attraction for gardens and nature spaces with the discovery of the city like architecture or culture. All the plant developments contribute to the image of the city, to the general atmosphere, to its beauty, to its attractiveness. A prestigious park can be a reason to visit [15].

3 Introduction of the Urbain Tree in Tunisian Cities

3.1 The Tree in the Medinas

It was from the introduction of the French protectorate at the end of the nineteenth century that the urban tree took a significant place in the city. This does not mean that the tree was not present in the medina, but it was rare to see it in the narrow streets of it. At the time of colonization, this urban model would be confronted, with the introduction of the protectorate, to a new imaginary and new practices related to aesthetic and economic values that were not those of the medina. The tree was inside the patios of the Arab houses and the essences corresponded to species belonging to the sacred because quoted in the Koran or utility because one wanted to join the useful to the pleasant one [16]. However, the tree was not as rare as one might think in the life of the Tunisians of old. It was widely used in swani (large orchards and vegetable gardens on the outskirts of the city) in which some of the Tunisians went regularly. Indeed, the agricultural fields around Tunis, consisting of an olive tree and other fruit trees, surrounded the city. There were also trees in the gardens of aristocratic residences and holiday homes in the Beyli, outside the medina: in the palaces of Bardo and Manouba and the summer residences of Marsa [17].

3.2 The Tree in the European City

Indeed, it was during the colonial era that the first public green spaces in Tunisia were created and which accompanied the construction of the European city of Tunis established outside the ramparts of the medina [18]. They had to convey the image of a modern city that had nothing to envy to that of French cities. The most convincing examples would undoubtedly be the belvedere park created in 1899 and the Habib Bourguiba avenue promenade built in 1881. This walkway (at the time called 'Avenue de la Marine' in 1881 and 'Jules Ferry' in 1900), hosted the first alignment trees (Ficus nitida) to contribute to the construction of the city center at the end of nineteenth century. This quadruple alignment of the Ficus with an architectural size curtain was arranged to bring the perspective from the gate of France (today Bab Bhar) to the port. Ficus plantations have played an attractive role, shading certainly but also amenity. These plantations have led to the affirmation of the notion of public space, visual perspective and monumentality (Fig. 1).

In the European city of Tunis, alignment trees have helped to mark avenues and streets by creating urban spaces [19]. Several squares appeared such as the square of the station located in the present place of Barcelona, the square of the Qasba, currently Place of the government. However, the tree in the urban park appeared first at the Belvedere Park first made in the late nineteenth century. This park was a place of attraction, rich in a wide variety of ornamental species. The Belvedere Park was

Fig. 1 Avenue Jules Ferry 1920. *La tunisie illustrée, Revue illustrée de vulgarisation tunisienne, éd. Imprimerie Weber, 1921*

created as a recreational and attractive space for a city that was lacking. This park was a place for walking and meeting but also in a hygienist concern, providing shade and freshness for a city known for its hot climate. It was designed in the English style of landscape gardening with woods, streams and clearings and the plant species that inhabited it were exotic and passed through the test garden where the different plants from different colonies. This park is now in the heart of the capital and has undergone several changes and the landscape concern has given way to environmental concerns, health and education and is home to several unique species that exist nowhere else in Tunisia [20].

Tunisia, and particularly Tunis, inherited, at the end of the period of the French protectorate (1881–1956), a new urbanistic landscape model according to the occidental model in opposition to that of the medina and its suburbs. Since then, and with the increasing spatial influence of Tunisian cities, there has been a real awareness of the Tunisian public authorities of the ecological and social importance of green spaces, their presence in sufficient numbers and their judicious distribution in the city as evidenced by the green policy implemented by the Tunisian State.

4 The Tunisian Green Policy

4.1 The Legislative and Institutional Framework

Since its creation, the ministry in charge of the environment reflects the commitment of the country on the path of sustainable progress to initiate and establish an environmental culture. It aims to propose the general policy of the State in the fields of the protection of the environment, safeguarding nature, promoting the quality of life and establishing the foundations of sustainable development in the State's general and sectoral policies. It is responsible for developing a national strategy for sustainable development, defining measures to adapt the methods of planning and management of the structures and institutions of the State to the foundations of the mode of sustainable development. It is also the Ministry of the Environment that conducts prospective studies on the environment and its relationship with economic and social development, in order to help guide general and sectoral policies.

The protection of the environment in Tunisia is ensured by a rather significant legal arsenal which reflects on the one hand, a political will mindful of the problems related to the management of natural resources and confirms, on the other hand the commitment of the country to use rationally and durably the heritage of future generations.

Since the independence, several codes and laws relating to the protection of certain elements of the environment have emerged, for example the Forestry Code (1966, then recast in 1988), the Water Code (1975), the urban planning code (1979 reorganized in 1994), the law for the protection of agricultural land.

The rhythm of legislative and regulatory measures relating to the protection of the environment has been strengthened since 1988, when the first public institution for the protection of the environment was created, the National Agency for the Protection of the Environment (ANPE).

During the last two decades, several public institutions acting in the field of the environment were successively set up, such as the Agency for protection and development of the coast (APAL) created by the law no. 95-72 of 24 July 1995, the International Center for Environmental Technologies of Tunis (CITET) created by law no. 96-25 of 25 March 1996. It should also be noted that, in parallel with these projects and programs, the Ministry has sought to involve and guide stakeholders in integrating the environment into the various development programs and projects (Portail de l'environnement: Le développement durable, www.environnement.gov.tn/ [21]).

4.2 *Specific Programs*

The notion of sustainable development appeared in Tunisian political discourses and real environmental policy was born, this is visible through the multiplication of green space creation programs [22]. In terms of programming, green spaces are projected according to two approaches: an urban planning approach that takes shape with the different urbanism documents, and a project approach with the specific programs devoted to the development of green spaces [23]. The urban planning documents in Tunisia are presented at different scales: the National Spatial Planning Scheme (SDAT) sets the development guidelines on a national scale whereas the Master Plan of Development (SDA) is elaborated at the Scale of governorates. Urban planning is done at the scale of the town and gives rise to urban development plans (PAU). Finally, subdivision plans are developed at the neighborhood scale and must conform to plan specifications. According to the Territorial Planning and Town Planning Code (CATU), the open spaces or wooded areas as well as the urban or natural landscapes to be maintained or created must appear on the SDA cartographic documents.

According to the same document, the development of green areas, public parks and natural areas should be included in the program plan, which constitutes the achievement of the SDA, as well as all programs related to basic infrastructure, major equipment and services and the development and enhancement of archaeological and historical sites. The PAUs, documents of urban planning and regulation of the use of soils made by the local communities, are the principal regulatory framework in which the zones with the vocation of green space are delimited. Any zone can not have another type of occupation. The change of the regulatory role of green space can only take place through a presidential decree as stipulated in the CATU (1994). In practice, all existing green spaces in the city are either programmed and realized from the PAU, or integrated into these documents during their revision, for those programmed and realized after the approval of the PAU or in the framework of operations subdivisions.

The PAUs are the main documents from which it is possible to determine the location of all green spaces in the city. They thus make it possible to define the system of green spaces, in the absence of specific planning.

Green space programming is also done through specific programs generated by the Ministry of the Environment such as the National Urban Parks Program (PNPU) and the Garden Cities Program. The PNPU was launched in the early 1990s to form a "green policy" that aims to achieve a hundred parks throughout the Tunisian territory.

It has a twofold objective: to provide major cities in the country with urban parks, equipment intended to improve the living environment of city dwellers, but also to safeguard peri-urban forests. The National Project Cleanliness and Environmental Esthetics (PNPEE), is a project carried out with the support of the National Agency for Environmental Protection (ANPE). Among other actions, he attributes to cities that fulfila number of criteria, the label city-garden. It can thus be considered as generating green spaces because it encourages the municipalities to build public gardens on their territory. To qualify for the garden-city label, the candidate city must, among other criteria, have a minimum of 14 m^2 of green space per inhabitant, have at least five major green spaces including an urban park and a boulevard de la environment and at least three well-developed main well-developed arteries. As part of the promotion of the urban aesthetics of Tunisian cities, two other programs have been launched, namely the national program of the Boulevards de l'Environnement in each municipality and the national program for the creation of Boulevards de la Terre in each municipality seat of the governorate. The green space promotion program is added to the list listed above in order to increase the green space ratio, which is still considered too low. Other urban and environmental planning tools are added to this list and we can cite the green plan that aims to implement a policy of preservation and enhancement of "natural spaces" and to reinforce the presence of nature in the city.

4.3 The National Urban Parks Program

I will, however, focus on the National Urban Parks Program because it is still ongoing today. This program aimed to create 100 urban parks with the goal of preserving nature and natural resources. Since its launch, this program has been placed under the responsibility of the Ministry of the Environment and Spatial Planning. It led to the creation of 39 parks scattered throughout the Tunisian territory. It is also the program that most expresses the voluntarist policy of the Tunisian State to preserve the green heritage and particularly the urban and peri-urban forests that constitute socio-cultural and economic issues.

These natural spaces present opportunities to improve the ecological environment of the city, to safeguard the natural wealth and to offer city dwellers recreational, rest, relaxation, education, culture, approaching nature. Their existence, their evolution through the power relations between city/nature and man/nature defining three types of ecological, socio-cultural and economic issues, having an impact:

- At the level of the environmental sector (ecological, the ratio of green spaces, environmental education of the citizen and its approximation of nature)
- At the level of other sectors (integration of space-forest/city, urban development, employment, tourism and recreation, health, education.
- At the level of the microsite itself (improvement—redeployment of the site, promotion-attractiveness and viability once developed)

The ultimate objective of this program should be the strict protection of urban and peri-urban forests against any kind of degradation, the stopping of any urban invasion and waste dumping inside these areas, the improvement of aesthetics but also the contribution to the establishment of sustainable cities.

5 A Case to Study: The Park of Sidi Bou Saïd

The urban parks, concretized by the National Urban Parks Program (PNPU), have taken on an environmental vocation and are managed summarily while keeping the forest vocation of the site to the maximum and consequently this limits the plant biodiversity. Nevertheless the Sidi Bou Saïd Park, also called "El Abrachia" and carried out in 2004 according to the same program stands out and is a good case study because it obeys the main objectives and this in view its location, the vegetation it contains and also the social practices related.

5.1 A Strategic Location in the Heart of the Village of Sidi Bou Saïd

The park is located in Sidi Bou Said, a small coastal town in the North Suburbs of Tunis. Sidi Bou Saïd is a village in Tunisia steeped in history and culture. It was the place of passage or retreat of artists and painters, Sidi Bou Saïd implanted on a hill, overlooking the Mediterranean Sea, the Gulf of Tunis, the ancient city of Carthage and its archaeological remains. The village originated in the retreat of the eponymous Sufi saint who was in search of meditation. The white and blue colors of its architecture have been adopted by the Tunisian population under the influence of Baron Erlanger, musicologist and painter who also settled in the village. Powerfully publicized, these two colors have become emblematic of the village marking a well-established landscape identity (Fig. 2).

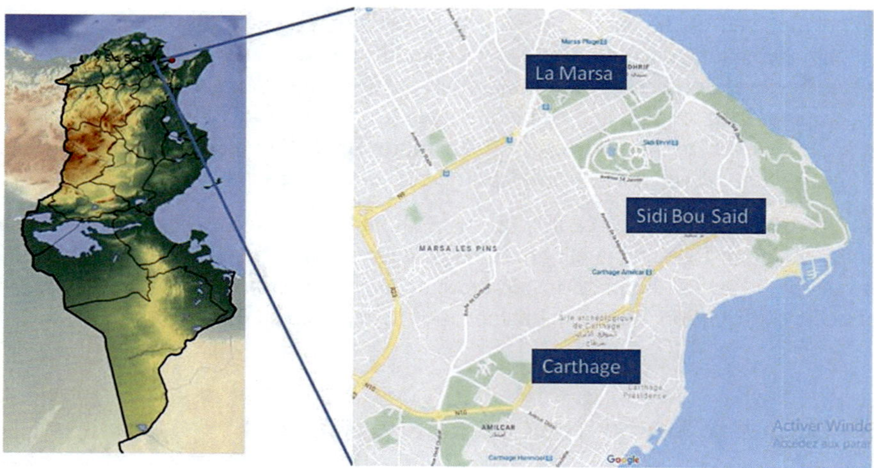

Fig. 2 Geographic situation of Sidi Bou Said's park

5.2 *An Ecological Continuity with the Surrounding Spaces*

The Sidi Bou Saïd Park, is a green infrastructure within an urban area strongly marked by the presence of natural spaces, including forests and green spaces of different natures but richly planted. Indeed, near the park of Sidi Bou Saïd, we find the park of Carthage also created in 2004, the American cemetery open to the public side of Carthage (see Fig. 3). The Essaada Park at Marsa is also nearby. It was realized in 2002 on the garden of an old Beylical palace become the seat of the municipality. These "pieces of nature" are linked together by axes planted so that contact with "nature" is never lost. This constitutes the very foundations of a well-constituted urban greenway.

5.3 *Integration with the Surrounding Urban Area*

The park of Sidi Bou Saïd, is in a low-density urban fabric made up of residential districts, University Institutes like the Preparatory Institute of Scientific and Technical Studies and the National School of Architecture and Urbanism as well as a primary school. The park is easily accessible thanks to a hierarchical road network since it overlooks the Avenue du 14 janvier, on the road to the Archbishop's Palace and a secondary road on the side of the primary school on the east side. It should also be noted that the park has three entrances, which proves its connection to the surrounding urban area (Fig. 4).

Fig. 3 Continuaty of the park with the green spaces

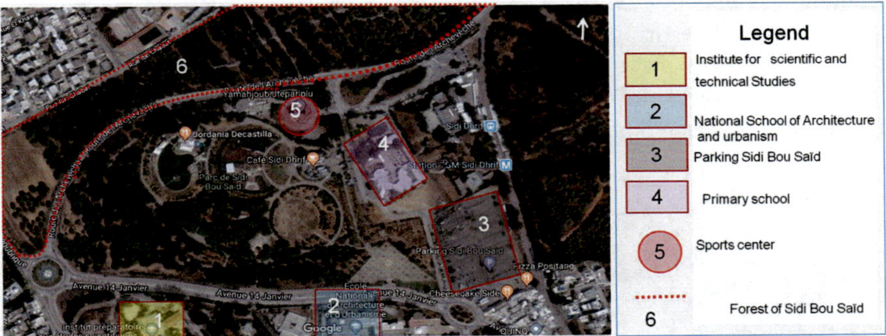

Fig. 4 Sidi Bou Said Park's environment

5.4 A Modern Landscape Reflection

Unlike other parks designed as part of the PNPU that have benefited from a summary development, the Sidi Bou Said park was the subject of a development study entrusted to the Tunisian landscape designer Jalel Abdelkefi. Landscape thinking is contemporary and in line with international landscape practices. The style of landscape gardening is mixed and oscillates between regular forms (at the level of the plots at the level of the entrances) and circular shapes entangled in the rest of the park (Fig. 5).

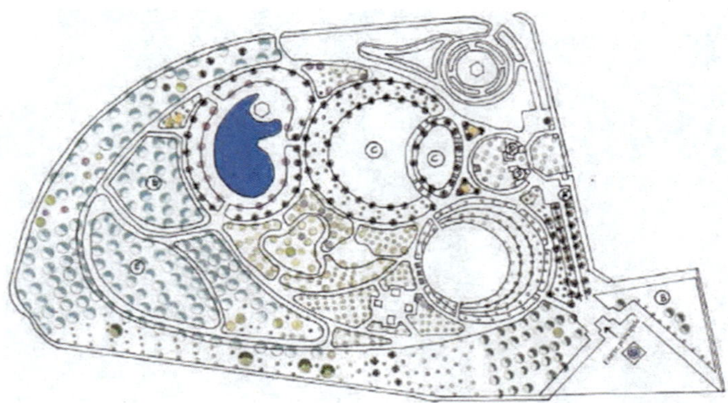

Fig. 5 Ground plan of Sidi Bou Said's park. Turki and Zaafrane Zhioua [23]. *Analyse de la répartition spatiale de l´aménagement des espaces verts programmés par les documents d´urbanisme dans le Grand Tunis*

The latter generate curved perspectives inviting the progressive discovery of the park. Note, however, the existence of a straight perspective in the long driveway to Andalusian coffee. The park is spread over 31 ha in a pine forest, including 15 ha designed to accommodate visitors. It consists of small squares, a gazebo, an arboretum, a playground, a fitness trail, a green theater, a lake and a picnic area. Vegetation biodiversity shows a rich Mediterranean palette, it is possible to find coherence between the distribution of vegetation and the choice of species with the theme of sub-spaces: for example, a plot of bitter orange trees for the Andalusian coffee park. Sidi Bou Saïd, a pine forest for the health trail of the same park (Fig. 6).

Fig. 6 **a**, **b** Planted pathways in Sidi Bou Said's park

5.5 Social Interaction Opportunities

The park of Sidi Bou Said is a green space connected to the city, open to the public and offering aesthetic and functional spaces (although it has experienced the last five years a relative deterioration of the materials of its equipment). It is a space welcoming all social categories and all ages and has a steady attendance. Strong benefits are associated with access to the Sidi Bou Saïd Park and the massive presence of plants and areas suitable for relaxation or the practice of various recreational and sports activities. The organization of events, educational and/or participatory actions in the Sidi Bou Said park, testifies to a certain level of satisfaction of the users and represents as many opportunities of meetings and exchanges, also contributing to the culture and environmental education.

6 Discussion

The urban parks built under the (PNPU), have taken an environmental vocation. The developments were carried out in peri-urban forests and therefore outside the city and are summarily arranged as if one wanted to keep the original aspect of the site, as natural as possible. The works are generally carried out in two phases: a portico of entry, often monumental as well as the beginning of a fence are first edified, then a second phase which consists of arranging the interior of the park. In most cases, only a small proportion of the site is developed, the rest being left in the natural state, that is to say a treed wilderness. These spaces, which have been created by the legislation of some specific programs, even if they are only summarily and particularly adapted, are counted, over their entire area as a green space, and thus contribute to achieving the objectives quickly fixed in terms of square meters of green space per inhabitant. It can be said that Tunisian green policy sometimes tends to operate more quantitatively than qualitatively. Some developed sites are peri-urban forests and do not constitute local spaces to improve the living environment and the introduction of plants in the city to form green lungs. As for the periurban forests, we find mainly the original vegetation of the site, that is to say, most often, forest species planted before or after the creation of the park which generates relatively low biodiversity for these spaces. The most common species in these parks are: acacia, oleander, pine or olive tree. This has led to spaces with poor plant biodiversity that contrast with the ecological and environmental vocation displayed by these parks. However, this heritage remains fragile; it is not only a question of revealing it, of putting it in value but also and especially of making it last what is not easy. However, it should be noted that some public parks are sometimes places of incivility or increased concern for users because of the presence of groups of marginal populations. Support through consultation, appropriate mediation and awareness-raising among residents often facilitate acceptance and social cohesion.

7 Conclusions

Indeed, in Tunisia and in the largest Tunisian cities, green spaces were born with the extension of the European city outside the medinas where this notion is totally illusory. Tunisia has also become, over the years, an attractive tourist destination, since then, the public authorities have tried to promote the creation of green spaces by multiplying the national programs of the creation of urban parks (PNPU), development of city entrances, creating boulevards of the environment in all Tunisian cities. The creation of green spaces in urbanized areas has become a recurring message in messages conveying an environmental policy, especially after adhering to the precepts of "sustainable development". Green spaces have been rationalized as urban equipment as well as cultural, sports or health facilities. They have been standardized in the form of m^2 per capita (the public authorities have set a minimum of 10 m^2 green space/inhabitant). This minimum standard was required for any new development project. Tunisian green policy should be more in the direction of a green infrastructure aimed at restoring connections between existing natural spaces and improving the ecological quality of the environment as a whole. For this, it may include several natural components (protected areas, wetlands, forests, etc.) as well as multifunctional areas in which priority is given to land uses that maintain or restore healthy and diverse ecosystems. This could support the construction of "sustainable city" while taking into account the preservation of biodiversity in current urban planning. The notion of green grid appears as a way to take into account in the urban planning the heterogeneity of the urban space by giving a sense of landscape and ecological to the city. The objective of this study is to provide all the stakeholders concerned with the keys to reflection and the arguments needed to restore the plant to its place in the urban fabric and integrate the benefits that are linked to it in planning. The proposed evaluation consists of three steps: knowledge of the territory, through the collection of information; the promotion and enrichment of the dialogue with the cotoyens, via the shared diagnosis; decision-making, through the identification of assets and levers of action. The scientific literature shows that the role of the plant in the city can be evaluated in many different ways. The main determinants of the benefits of the plant can be organized into five transversal axes of study: Functional and aesthetic amenities, the accessibility of public green spaces, the quantity of vegetated surfaces, the environmental regulation capacities and the ecological balances and continuities.

8 Recommendations

In Tunisia, development policies have for a long time, neglected environmental and living environment issues. For the last two decades, the Tunisian state tried to improve comfort and attractiveness but also to warn about the vulnerability of the territory.

To answer this problem, we must be convinced that the very condition of sustainable urban development comes from the reconciliation between nature and the city.

Unrecognized, the tree heritage is an active component of the city offering many benefits and services: air conditioning of the city, participation in water management, air and soil remediation, benefits on the psyche of urban … Faced with these realities, we must be aware of this ally and promote the place of nature to consider sustainable development of the city and of our lives. In this purpose, we need a planning and a management tool able to gather all the actors of the territory whose professions, actions or sensitivity are interfacing with the question of green spaces. This could generate a support for shared knowledge and principles to build upon to improve and harmonize practices ensuring sustainable and universally accepted protection. To achieve this, we should not forget the civil society which, within the framework of a participative democracy, has a crucial role to play in ensuring the sustainability of the green spaces and their tree heritage making up the landscapes of our cities.

References

1. Food and Agriculture Organization of the United Nations (FAO) (2016) Guidelines on urban and peri-urban forestry. In: Salbitano F, Borelli S, Conigliaro M, Chen Y (eds) FAO Forestry Paper 178: Rome: Food and Agriculture Organization of the United Nations
2. Maas J, Verheij RA, Groenewegen PP, al, (2006) Green space, urbanity, and health: how strong is the relation? J Epidemiol Community Health 60:587–592
3. Laille P, Proven D, Colson F, Salanie J (2013) The benefits of the plant in the city: study of scientific works and method of analysis. Plante & Cité, Angers, p 31
4. Donovan GH (2017) Including public-health benefits of trees in urban forestry decision making. Urban For Urban Greening 22:120–123. https://doi.org/10.1016/j.ufug.2017.02.010
5. Turner-Skoff JB, Cavender N (2019) The benefits of trees for livable and sustainable communities, plants, people. Planet 1(4). https://doi.org/10.1002/ppp3.39
6. Kaczynski AT, Henderson KA (2007) Environmental correlates of physical activity : a review of evidence about parks and recreation. Leisure Sci 29(4):315–354
7. Summers JK, Vivian DN (2018). Ecotherap—A forgotten ecosytem service: a review. Front Psychol 9:1389. https://doi.org/10.3389/fpsyg.2018.01389
8. Coley R, Kuo FE, Sullivan WC (1997) Where does community grows? The social context created by nature in urban public housing. Environ Behav 29(4):468–494
9. Arnberger A, Renate E (2012) The influence of green space on community attachment of urban and suburban residents. Urban For Urban Greening 11(1):41–49
10. Clergeau P (dir.), Blanc N (dir.) (2013) Trames vertes urbaines: de la recherche scientifique au projet urbain. Editions du Moniteur, Paris, pp 31–32
11. Bowler DE, Buyung-Ali L, Knight TM, Pullin AS (2010) Urban greening to cool towns and cities: a systematic review of the empirical evidence. Landsc Urban Plan 97(3):147–155
12. Hirons AD, Thomas PA (2018) Applied tree biology. Wiley, Oxford, UK
13. Endreny TA (2018) Strategically growing the urban forest will improve our world. Nat Commun 9(1):1160. https://doi.org/10.1038/s41467-018-03622-0
14. Donovan GH, Butry DT (2010) Trees in the city: valuing street trees in Portland, Oregon. Landsc Urban Plann 94(2):77–83. https://doi.org/10.1016/j.andurbplan.2009.07.019
15. Nesbitt L, Hotte N, Barron S, Cowan J, Sheppard S (2017) The social and economic value of cultural ecosystem services provided by urban forests in North America: A review and suggestions for future research. Urban For Urban Green 25:103–111

16. Bennour-Azooz M, Donadieu P, Bettaîeb T (2012) L'arbre à Tunis: hypothèses pour une histoire de l'espace public. Projets de paysage. http://www.projetsdepaysage.fr/fr/l_arbre_tunis_hypoth_ses_pour_une_histoire_de_l_espace_public
17. Zaïer S (2004) «Le temps des jardins husseinites: le cas du jardin Ksar Essaâda à la Marsa», mémoire de master dirigé par Pierre Donadieu. ISA de Chott Mariem/université de Sousse, p 26
18. Saloua F (2018) Les places publiques à Tunis sous le protectorat. Naissance, essor et prémisses de disparition, centre de publication universitaire, Tunis
19. Nabli N (2004) mémoire de master dirigé par Pierre Donadieu. ISA de Chott Mariem/université de Sousse
20. Imène Z (1998) Le parc du belvédère de Tunis, mémoire de DEA. Ecole d'architecture de Paris La Villette, p 38
21. Portail de l'environnement: Le développement durable, www.environnement.gov.tn/
22. Belfekhih S, Allouche-Khebour F, Saidane I, Mechmech Y, Bettaieb T (2018) State of the art of greenway concept application in Tunisian green policy: a case study of an urban landscape in Sousse city. Int J Environ Geoinf 5(1):36–50
23. Turki SY, Zaafrane I (2006) Analyse de la répartition spatiale de l´aménagement des espaces verts programmés par les documents d´urbanisme dans le Grand Tunis, Actes du séminaire étape de recherche en paysages, n°8, Ecole nationale supérieure du paysage à Versailles, p 72

Interchange Between Agriculture and Tourism in Hergla (Tunisia) in the Context of Sustainability

Mohamed Hellal

Abstract Hergla is an old farming village, located on the eastern coast of Tunisia, in the governorate of Sousse. The natural, agricultural and patrimonial assets at Hergla would allow this typical village in the Sahel to develop a new type of tourism up-to-date with the new international requirements in terms of sustainable tourism. Although the integrated resorts project is suspended, the current tourism development which is on the way seems to be soft. It is set up jointly with the development of the residential and leisure sectors under the initiative of private investors. This urban development is at the expense of agricultural land. Current urban control and planning on the part of the public authorities seem meager. Thus, the integrated resorts project which is now on the agenda would be carried out at the expense of the coastal forest, farmland and an already fragile coastal environment. Here, the state-led governance to regularize stakeholders and role-players is contrary to the principles of sustainable development.

Keywords Sustainable tourism · Agriculture · Environment · Heritage · Urban

1 Introduction

Hergla is an old farming village, located on the eastern coast. It is located to the north of the integrated El Kantaoui Tourist Resorts, in the governorate of Sousse. The success of the first integrated tourism resorts in Tunisia in El Kantaoui, set up at the end of the 1970s, motivated the Tunisian state to invest more in tourism development up north of the governorate of Sousse, in particular in Hergla. Faced with the natural and cultural specificities of the site, the authorities are compelled to adjust tourism spaces otherwise in response to new international standards in terms of sustainable tourism. "The UN 2030 Agenda for Sustainable Development sets a series of sustainable development goals (SDGs) "to end poverty, protect the planet and ensure prosperity for all" by 2030. The Agenda influences tourism policy even

M. Hellal (✉)
Faculty of Letters and Human Sciences of Soussa, University of Soussa, Research Laboratory: Governance and Territorial Development, Tunis, Tunisia

M. Abu-hashim et al. (eds.), *Agro-Environmental Sustainability in MENA Regions*, Springer Water, https://doi.org/10.1007/978-3-030-78574-1_16

though the Agenda resolution only mentions tourism three times. A “heterogeneous constructionism” approach is adopted to examine the managerial ecology of tourism and the SDGs. Managerial ecology involves the instrumental application of science and economic utilitarian approaches and in the service of resource utilisation and economic development” [1].

To achieve the principles of sustainable development, the Agendat initiate the creation of new partnership relationships between the main actors involved in this activity, in particular, the territorial actors, to forge the hope of developing more responsible tourism in regard with our common heritage. Moreover, tourism development is part of a global land management, which integrates local actors.

Originally, Hergla is a typical village of the Tunisian Sahel. It is distinguished by its ruins, its typical urbanism, and its rural landscape. Although these assets provide an opportunity to develop sustainable tourism, with the total absence and withdrawal of the state, role-players, particularly private stakeholders, are turning it into a lucrative business. We wonder here how the mechanisms of divergent interests and the withdrawal of the state attempt once again, despite the previous experiences, to destroy the ecology and the heritage of an area at the expense of the local populations and for the sole benefit of a purely market valuation of investors on a short-term basis.

Let’s suppose the following hypothesis: the state governmentl to regularize the stakeholders with divergent interests in this project would be contrary to the principles of sustainable development. Through our field studies, particularly through our direct observations and interviews, we would be able to confirm or disprove this hypothesis.

2 Potentials and Vulnerabilities of Hergla’s Resources

2.1 The Situation and Heritage Potential of Hergla

The town of Hergla has 7423 inhabitants according to the population census of 2014. It constitutes a Sahelian village, a “buffer” which marks the northern limit of the social Sahel, where its village agglomerations are inhabited by populations that resemble in their habits the townspeople of the region, and the southern limit of the nomadic territory [2].

Hergla is located in the middle of a tombolo and in a coastal sandspit, about 6 km long, carved in the cliff by the sea (Fig. 1). The nucleus of the city developed on the site of the ancient Horrea Coelia. This was a Phoenician port which was prosperous during the Roman period, as testified by the “Horrea”, which are warehouses of cereals meant for export. The ruins scattered between the buildings also testify to the Byzantine presence in the region.

During the Muslim period, Hergla became a small religious center organized around the tomb of the “Sidi Bou Mendil” Shrine or Mausoleum. The main mosque

Fig. 1 Geographic localization of the agglomeration of Hergla in relation to its natural environment. *Source* Google Earth, 2020

in the city bears its name. Here, the population is somewhat conservative and its character is closer to that of a village than of a city.

Although Hergla is today a small town, it still ranks as rural. In 2017, in the delegation of Hergla, exist find 8572 ha arable, 4305 no arable and 1277 ha in course. In addition, there are no irrigated agricultural perimeters. The rural landscape is marked by the presence of olive trees and palms. If the olive trees are productive, then the palm trees are rather a decoration, especially in the coastal zones.

In terms of production, in 2017, 1300 tons of cereals and 2700 tons of olive oil were produced. Moreover, in Hergla there are 3 oil mills with a transformation capacity of 1000 ton/24 h.

On the urban level, the present agglomeration has a radio-centric structure. The urban network of Hergla, which is stuck between the fields of olive groves and the sea, still preserves the old core despite the effects of modernity. In the past, the traditional structures of the "Dar Aarbi" type, which did not exceed the heights of an olive tree, were in harmony with the rural landscape. These houses are always built with the same materials: limestone for the external structure, stoneware for the walls of the doors and the vaults as well as lime to bleach the walls and the roofs. The plans of the houses do not reflect the origin and the social class of the owner but rather respond to the needs of the rural and urban life of the inhabitants of Hergla. The *skifa*, the reception space, and the patio, which are main components of the typical house, are useful for domestic activities that are related to the rural life of this population that works in agriculture and handicrafts.

The situation of the agglomeration in relation to the water spots gives it the character of a Hergla peninsula; The sea to the east, the sebkha of Aassa Djriba to the north, the sebkha Halk Elmejel to the west and the south. This natural situation is partly responsible for its relative distance from major roads. It spared Hergla and its coastline from the great urban and tourist changes that characterized the great urban poles of the Sahel. However, the rural landscape, the remarkable architectural stamp

Fig. 2 Marabout in Hergla. *Source* Photographed by M. Hellal, July 2008

(Fig. 2) and the ancient vestiges, which give certain originality to this agglomeration, can make it a rich tourist spot. However, the mobilization of its assets for sustainable tourism development is not without constraints.

2.2 Sensitivity of the Coastal Environment and the Archaeological Sites

The village of Hergla is distinguished by its natural landscape marked by olive groves surrounding it and by its virgin beach, protected by the Madfoun Forest.[1] However, "at present, the coastline of the Madfoun Forest is showing signs of deterioration that are sometimes significant, particularly in its southern margin…The speed of erosion accelerated catastrophically especially after the establishment of the Hergla Fishing Harbor in 1982" [3].

Concerning the southern coast of the city, it is at present undergoing great degradation; the development of the fishing harbor is responsible for its degradation [4]. Thus, the cliff to the south of the harbor which is in permanent recession presents a great danger not only for the road but also for the constructions and the summer visitors who bathe at its foot. Furthermore, the retreats of the coast threaten the historical sites which constitute riches to exploit in the development of the tourism sector. To the north of the fishing harbor, "at the southern exit of the urban agglomeration of Hergla, the sea is attacking the ruins of the old city Hora Caelia. Further south, the front of the cliffs truncates, at different points, other ruins as well as the remains of ancient lime kilns [5]".

[1] It was planted in the 1960s. It presents itself as a green strip parallel to the beach. The purpose of afforestation of the site was to fix the coastal dune. It is made up mainly of acacias, eucalyptus, pines and shelters a varied fauna composed of wild boars, hares, red partridges, and hoops among others.

Due to the urban extension and the natural sense of Hergla's immediate environment caused by the lack of protection and maintenance, many archaeological sites are in danger. In a national inventory carried out in 1987 covering Hergla region, there were 41 sites which "contain ruins, many of which have disappeared today, and there are only a few walls, cisterns, columns, carved stones, and pieces of pottery left ..." (Annabi, in Fehri [6]). Mendres, which contains large carved stones and pottery pieces as well as a Roman road engulfed by the sediments of Sebkha Halk el Menjel, is among the sites in danger. In fact, the only site that is currently the object of excavations is the one located to the south of the city. There are several vestiges: cereal reservoirs covered by colored mosaics and a residential area consisting of several houses and basins covered by mosaics as well.

The intervention on the Roman quarter is not comprehensive, and it concerns only one part due to the lack of funding. According to the person in charge of the site, the intervention in the site aims only for maintenance. Indeed, maintenance is costly (weeding, restoration) seeing that the ruins previously cleared up are now in degradation. This person explains the complexity of the excavation sites and the importance of the site: "The proximity of the sea makes the task complicated especially with the harmful effects of salt on the texture of the materials. Our main target is to keep the authenticity of the site without bringing massive interventions. In reality, the Roman house which has already been cleared up and which dates back to the third century AD has specific components: mosaics, thermal baths, private port. It can be a tourist attraction without the excavation being finished. Anyway, we are not obliged to show everything to tourists.".

Finally, while natural and patrimonial heritages represent assets, their fragility is a constraint in terms of sustainable tourism development. This type of tourism must develop in interaction with pre-existing economic activities, including agriculture.

2.3 *The Evolution of the Socio-economic System in Hergla and Its Implications on the Urban Environment*

The traditional local economy of Hergla is diversified and consists of woven press baskets' crafts[2] based on alfa and pottery as well as fishing and olive growing. For a long time, the population originally has been living from the product of the home-made woven press baskets for use in the oil mills, which constitutes a certain economic and social balance by the redistribution between the populations of the induced wealth of the olive trees.

However, for at least two decades, the traditional activities of Hergla seem to go through a serious crisis. In fact, agricultural activity continues to decline because of urban development on arable land. So, traditional alfa weaving is going bankrupt due mainly to competition from nylon baskets in the 1990s and mechanical oil presses imported from Italy and Turkey during the 2000s. Moreover, the supply of alfa from

[2] An interview realised with the site manager in June 2010.

the regions of the interior has become random, insofar as these steppe territories are influenced by the advance of the desert. Also, dry land crop degradation is caused by the fragmentation of the property and the continued salinization of soils.

Hergla remained, until then, away from the Tunisian tourist area. Several causes are at the root of this exclusion; the conservatism of the Herglians who have been fiercely opposing this type of imported activity [7], the situation of Hergla on a secondary road … Previously, there was only one access to Hergla via The National Road No. 1. In fact, between Hergla and Sousse there was only one track that crossed Chott Meriem, passing through the Husseinite Bridge, which was transformed into a road after the floods of 1969. This realization enabled a relative opening-up of Hergla. The construction of the El Kantaoui Marina in 1979 and the opening of several hotels in this new resorts attracted many young people from the village to work there.

The gradual involvement of the inhabitants of Hergla in the tourist activity has contributed to the improvement of the standard of living through a fairly regular additional income from tourism. Simultaneously, the residential sector is beginning to develop in Hergla. The authenticity of the village and the quality of its beach attract mostly Sahelians to build second homes. Thus, this residential activity is combined with another type of leisure: cafés, restaurants … Moreover, there are many people who went to Hergla for a visit and who decided later to rent or to build a house. That is why at the southern entrance of the village we notice an urban boom.

The new urban network that extends southwest does not respect the local architectural style. Faced with the realization of the imported-style villas, which do not go hand in hand with its heritage and its identity bases that make its reputation, the village is gradually losing its architectural trait. In reality, the municipality is the first actor responsible for the urban management and conservation of the urban heritage of the agglomeration. Therefore, it assumes responsibility for the degradation that reaches the Medina-style tissue of the agglomeration. However, to build in the old core of Hergla, one must have the permission of the National Institute of Heritage.

Concerning the degradation at the level of the old core of the agglomeration of Hergla, the architect town planner of the municipality of Hergla does not hide that "there are several difficulties in managing the ancient urban network of the town of Hergla. Indeed, where some citizens improvise a withdrawal to benefit from a better aeration of their habitats, yet the alleyways make the originality of the medina".[3] He also highlights the difficulty of enforcing the urban planning by the law. In fact, "in Hergla, as everywhere in Tunisia, obtaining the authorization to build is not systematically a regulatory constraint. So, even after the obtaining of the authorization of the plan, we often manage, during construction, to circumvent the law. The application of urban planning regulations is particularly difficult for the well-to-do people who always try to import a foreign architectural style. However, Hergla has its specificities and colors: white for the walls, blue for the windows, doors and all the external accessories of the houses".[4]

[3] Interview conducted in June 2018.

[4] Idem.

On the ground, the urban change, which takes place through the facades of the houses, becomes an object of distinction and pride. The multiplication of styles of construction in inadequacy with the old urban network is an indication of a "crisis of identity", since the agglomeration is increasingly invested by the "foreigners". However, the authenticity of the city makes its tourist attractiveness.

Hergla suffered a change of the local economic and social system. The old community system, which is based on agriculture and handicrafts, is beginning to give way to a new system based on services (residences, leisure activity areas) that is carried by "foreign" actors. Thus, in the future, the development of tourist activity could participate in the evolution of this system.

3 The Attempts and Constraints to Realize a New Integrated Tourist Resorts in Hergla in Relation to the Sensitivity of the Site

3.1 The Stakes of the Hergla Integrated Resorts

In 1979, the first integrated resort in Tunisia was inaugurated in El Kantaoui (13,600 beds). It was intended to be an innovation in tourism development. It is organized around a marina and offers multiple choices of accommodation (hotel, residences …) [8]. The success of this project and the growing demand for residential and hotel lots prompted the Tunisian authorities to invest more under the same concept and to start a similar resort project at Hergla. This tourist project would be a few kilometers from the new Enfidha International Airport, already inaugurated in October 2009, with a capacity of 5 million passengers per year.

Officially, at the end of 1990, 13 new integrated tourism resorts were planned in Tunisia, 4 of which were planned in the Gulf of Hammamet; Hergla, Salloum I, Salloum II and Yasmine-Hammamet (already realized). The SDAT (Master Plan for the Development of the Tunisian Territory) of 1997, which confirms these achievements, projects that each of these resorts will have a golf course and a marina to culminate in a national chain of these facilities which would allow the development of recreational and golf products.

For the development of the resort between the town of Hergla and the Medfoun Forest, the Hergla Tourist Promotion Company was created. This company named after the decree of December 31, 2004: Hergla Study and Development Company (SEDH). It was created by the North of Sousse Study and Development Company (SEDSN), which has already promoted El Kantaoui, the municipality of Hergla and some Tunisian banks.[5]

---

[5] The initial social capital is 250,000 which immediately soared up to 500,000 for the financing of the studies. However, during the 1990s, the project was delayed in favor of the South-Hammamet tourist area.

The Hergla site offers, in addition to the seaside product (sand, sun, sea), cultural and landscape potentials that allow the diversification of the product and therefore the introduction of a new qualitative dimension for Tunisian tourism. In this space, there are two sensitive environmental components that are included in the development project of the tourist resorts, namely the 100-ha El Medfoun Forest along the beach and the 1300-ha Assa Djriba Salt Lake. However, the sensitive nature of the project's reception area gives a particular dimension to this new integrated resort.

The trend of sustainable tourism has contributed to the renewed reading of the "integrated resort" concept, which now places greater emphasis on the environmental dimension. Today, an integrated resort is one that integrates effectively with its urban, economic, social and natural environments. To this end, the project must be integrated with the rural landscape, which would be decor, and the agricultural activity.

The Tunisian state retains a lesson from the experience of Yasmine-Hammamet. The project of this resort, decided by a decree in 1989, can illustrate how leaving the private sector in tourist development without effective state control and without coordination between the concerned actors can lead to harmful consequences for the natural environment and the sector in general. Indeed, in this project, the state has withdrawn itself from the realization of the most expensive equipment, the marina by choosing to entrust it to an Italian promoter, Marinveste, who will benefit from the gains of such an infrastructure. This investor who designed the port and its marina, and for reasons that can be estimated mainly financially (maximum profitability on investment), increased residential and port capacity by up to 720 rings, enlarging the project by illegal encroachment on the maritime public domain.

The subsequent withdrawal of Marinevest from the completion of the project is replaced by the creation of the new SEAMHS (The South of Hammamet Marina Study and Development Company) for the realization of the project according to the initial design and the results of inherited studies without ever updating them. As a result, this orientation has obviously had immediate effects on the environment: the natural movement of algae and sand from the north to the south is prevented by the dikes to the extent that we notice a very unpleasant accumulation of these algae on the beaches of the tourist resort double damaged by sand erosion from the south coast beyond the marina. In fact, the problem of the sea functionality and environment, from which Yasmine-Hammamet suffers, is due to the disengagement of the state and the non-involvement of the territorial actors in the project. Local actors who drew attention to these problems under integration of the latter project with the natural and urban environment of the tourism project were excluded from the planning of the resort. As a result, the resort did not meet sustainable tourism standards. However, these shortcomings added insult to injury to the already existing ones related to urban choices in the resort, which ultimately resulted in a saturated tourist area during the summer season and abandoned by tourists during the rest of the year [9].

The experience of Yasmine-Hammamet provides an eloquent example of the importance of the natural environment in the success of tourism investments. Such a statement prompts tourism managers to give importance to the natural environment in the future for Hergla resort; especially since it is located in a sensitive area which has given rise to two Master Plans.

3.2 The Salloum-Hergla Sustainable Development Master Plan and the Bouficha Development Master Plan, Enfidha and Hergla

This prospective document, carried out in 1997, is decided after the programming of 4 integrated resorts in the Gulf of Hammamet; Yasmine-Hammamet, Salloum I, Salloum II, and Hergla. It assist in the adoption of management choices that will allow a coherent and well-thought territory development while favoring the aspect of environmental sensitivity as well as potential for agricultural preservation or for heritage (archaeological, urban interest …).

This SDA encourages the preservation of the major components of this sensitive system and its specificity. Indeed, "the presence of the many salt lakes "sebkhas" identified on the site and the Madfoun Forest (…) play an important role in the phenomenon of "virgin space", which characterizes this zone (in relation to the Tunisian coast)" [10, p. 31]. It also calls for the preservation of the heritage qualities and its valorization: "if the ancient remains are hardly visible, they are nonetheless very important or even unique" (idem). It emphasizes the importance of the urban specificity of the old Hergla. For this "it is advisable to have a program of safeguard and rehabilitation" (idem). Moreover, the concern for sustainability "is, therefore, the source of delimitation of the perimeter of safeguarding and prohibition of ecological sites" (idem). The SDA proposes conciliation, however difficult, between the delimitation and the safeguard of the zones of interest on the one hand and its integration with the tourist project on the other hand.

In 2006, the state updated the SDA of the Salloum-Hergla sensitive area by the SDA Bouficha, Enfidha and Hergla to look for a framework of coherence of the major projects programmed in the Gulf of Hammamet, especiallay the deep water port, the logistical zone in Enfidha, and Enfidha Airport, which are close to the future tourism project of Hergla. The SDA is justified by the socio-economic upheavals. That these projects will undoubtedly entail, the multiplication of water and other public services needs, the welcome and the housing of the employees who are of such a magnitude that, they require a global thinking on the future of this region and its management principles [11].

Moreover, Hergla is of particular importance because it is the oldest human implantation in the study area, which requires special attention to integrate it with the other programmed projects, especially that it is "The object of more or less happy transformations" [11]. The SDA encourages taking into consideration the archaeological wealth, paying particular attention to excavation and preservation, especially as these sites are situated on the southern coast, which is threatened by coastal erosion attacks.

The SDA points out "a significant degree of vulnerability that characterizes the village of Hergla as far as ChottMeriem" [11]. This finding suggests that it is "of great importance in so far as it implies certain care in the planning of coastal developments" [11]. However, he sees that "the realization of any development, even" light, "on the Medfoun Forest would prove very quickly fatal for the balance of the latter. The

SDA warns that" this risk is all the more serious when it comes to the possibility of digging the salt lake. Indeed, the artificial lake that would be developed will increase the salinity of the soils and thus jeopardize the plant species of the forest. Moreover, the opening of the salt lake on the sea will disrupt the sedimentary equilibrium and subsequently the erosion of the littoral [11].

Finally, the SDA puts a condition for the proper implementation of its proposed measures "without risk for sustainable development unless very strong coordination between all the technical partners is established in an official way" [11].

3.3 The Planning and Governance of the Hergla Integrated Resort Project

At the time of its decision by decree in 1989, the project of the tourist resort integrated into Hergla was marked by certain gigantism; It had 30,000 beds in its initial version and was to cover 1200 ha. However, the SDA of the Salloum-Hergla sensitive area imposes a reduction and the accommodation capacity is reduced to around 13,000 beds, on the basis of an average density of 80 beds per ha. Cette orientation explique le choix de développer un concept de "haut standing" et de "jardin" de station.[6] The program of the project aims to complement the regional offer to attenuate its strong seasonality and the creation of a new image by new tourist segments by "opposing the concept of Hammamet South (high-density areas exclusively reserved for seaside bathers) and preferring more integrated and more personalized areas" (Idem).

After its incorporation, the Hergla Study and Development Company carried out a Detailed Planning Plan (PAD) associated with a Feasibility Study and an Impact Study.

Through the PAD, a project has been programmed with a total area of 1200 ha, of which 800 ha are devoted to green spaces. It provides a resort with hotels, a residential area, entertainment facilities, a forest massif, a golf course … which are organized around a pole (marina, island, esplanade) that will be arranged at the Aassa Djriba Salt Lake. Through this program, we sought a certain complementarities between the programmed animation equipment and integration of the project with its natural and urban environment. The golf course, which is programmed to be located north of the resort, has a direct view over the lake and the sea. At the southern end of the development plan is a marina of 350 knots and around the marina which presents itself as the culmination of the main boulevard of the resort. All the more, "the architectural expression of the marina should be inspired by the regional architectural vocabulary" [12, p. 4]. Thus, to confirm the lake-like character, an island is located in the middle of the lake composed of a reconstituted ancient village which would delimit the marina and ensure a pedestrian promenade towards the forest.

The project impact study accompanying this TBP did not hide the fact that "the impact on the environment is not negligible since it upsets the whole of an ecosystem

[6] Municipality of Hergla and DIRASSET (Design Office), p. 5, 2008.

(…). Other effects are reported, such as the possible damage of the forest massif.” [13, p. 54]. Faced with these negative impacts, the study is limited to a few compensatory measures and the recommendation of more specific studies. It seems that certain choices of development are imposed by the state; economic concerns and the need to create jobs are strongly present in the project’s program. In such a way that the impact assessment, in view of the negative consequences of the project, considers that “these effects may seem insignificant in comparison with the formidable economic and social development that the project will bring about” (Idem).

By the time the SEDH seeks to increase its capital and purchase the rest of the lands belonging to the private sector with the help of AFT (Tourism Estate Agency) in order to realize the tourist resort, the Emirati group Emaar[7] enters online. This foreign promoter proposes to invest in the site with a cash of about 2.54 million dinars. The realization of this new resort of Hergla illustrates a great disengagement of the state since the site would be given in this case in a concession to the Emirati promoter to realize and manage the project.

In the model presented by the Emirati promoter, the project is based, in broad outlines, on the PAD ordered by the SEDH; But it is more focused on real estate, on the aquatic development and on a maximum rationalization of space. Emaar plans to build a marina and around it a “Marina of Ksours”, an 18-hole golf course, a residential complex of 1300 villas and 3350 apartments, a thalassotherapy resort, restaurants, and only 3 hotel units with a capacity of 1200 rooms. This program, which focuses on real estate to ensure maximum profitability, also aims to build for the first time in Tunisia hotels of categories of more than five stars (luxury tourism); the objective is to attract a spendthrift clientele.

This grandiose project was canceled, without the media in Tunisia giving the slightest explanation. Following the Tunisian revolution (2010–2011) and the flight of President Ben Ali to Saudi Arabia, the media have emerged from their slumber. The electronic newspaper (Invest in Tunisia) explained that the wife and brother-in-law of the deposed president, BelhassenTrabelsi, were behind this cancellation. In fact, the Emirati investor was not able to start his project without giving an important commission to the Trabelsi family. At the end of the Ben Ali era, BelhassenTrabelsi was in the habit of acting as an intermediary between the businessmen who wanted to invest in Tunisia and the former president who, in principle, used to give the permissions [14].

Finally, following the cancellation of the Emaar Project, what are the prospects for tourism development in Hergla?

[7] Emaar promoter Emaar, one of the world’s largest real estate groups. Emaar’s capital of 40 billion dollars is held at 33% by the governorate of Dubai, 34% by the group’s foundations and the remainder by shareholders on the stock exchange. Among the group achievements is the highest tower of the world “Burj of Dubai” which is 750 m high.

4 Prospects for Sustainable Tourism Development in Relation to Urban, Agriculture and Environment in Hergla

4.1 Prospects for Tourism Development in Relation to Agriculture

Even if the project of its integrated tourist resort is suspended, for several years, the tourist activity continues to develop in Hergla. Under the initiative of several actors, a "soft tourism", is developing in parallel with residential and leisure activities, in the agglomeration of Hergla. Since the 1980s, the Hergla site has been twinned with El Kantaoui restorts for tourist excursions organized by a few travel agencies; The first is an example of a typical Sahel village and the second is a reconstruction of Tunisian urban and architectural heritage. Thus, the tourist guides like the Routard incite to visit this traditional village. The authenticity of the village, which has encountered difficulties in growing because of its geographical location as an almost island and the conservatism of the Herglians by their attachment to the traditional activities, make its tourist attraction. Even if there are no hotel units in Hergla, there are 3 guest houses, which make advertisements for their rooms on electronic reservation sites. In reality, these units do not have the status of "Guest Houses", which corresponds to the specifications of the Ministry of Tourism of Tunisia, but rather they sell "guest rooms". According to our interviews carried out on August 3, 2017, within the municipality of Hergla, the owners of these units pay only the ordinary taxes on housing. Thus, two new promoters have already filed their applications with the Ministry of Tourism and the municipality of Hergla to open two real guest houses (cf. Fig. 3). Here, it is not a capital investment, as it was already in the Tunisian tourist areas, but rather a development of new tourism, which is an alternative to mass tourism. These are small accommodation units, which fit into the urban and community network of Hergla. Thus, these small constructions fit perfectly with the rural landscape of Hergla. In terms of integration with agricultural activities, these guest houses are distributed local dishes based on olive oil and other local products.

It is a new type of tourism that is beginning to enter the village without disrupting the foundations of the local economy. This type of "sustainable tourism" attempts to integrate into existing economic activities, such as handicrafts. The weaving of the alfa turns towards other designs especially touristic: the manufacture of decorative objects and souvenirs for tourists, umbrellas for restaurants and hotels, etc. Little by little, the village converts itself into touristic and para-touristic services. Today, there are more than 12 cafés, some of which are touristic, 9 restaurant-cafés and several handicraft shops based on alfa and local pottery. The residential sector is developing in parallel with activities related to tourism and leisure.

Since the 1980s, the well-off Sahelians, mostly immigrants in Europe, have begun to build luxurious villas in Hergla. In the Sahelian perception, Hergla is the "Sidi Bou Said of the Sahel" and has the most virgin beach in the region, since it is in

Fig. 3 Guest house in Hergla "Dar Khadija". *Source* https://www.tripadvisor.fr/, November 2020

front of the Medfoun Forest. The number of second homes is constantly evolving. Private and public real estate investment is booming to meet the demand of summer visitors. At the southern entrance of the city, there is a forest of buildings, which is, unfortunately, in disharmony with the old urban network of the city. This is the AFH district (Housing & Estate Agency), which is a public operator that provides land to individuals or developers for constructing buildings.

The AFH "Corniche-Hergla" Housing Project, covering an area of 40.6 ha, devotes 11.25 ha for the individual houses, 4.19 ha for the collective houses 4.8 ha for the equipment, 5.23 for the equipment and the housing and 15.1 ha for the roads (see Fig. 4).

The new districts of AFH and Ennouzha, which lie south of Hergla, are home to foreign residents. The latter are either tenants on a weekly basis or purchasers of the housing. According to our interviews with the commune of Hergla, alone in the

Fig. 4 The AFH "Corniche-Hergla" Housing Project. *Source* Photographed by M. Hellal, November 2019

district of the AFH, there are 54 foreign owners, of several nationalities, but mainly European. These residents occupy their residences mainly during the summer season.

In recent years, we have been noticing the purchase of traditional houses in the ancient network of the city by foreigners, especially French and Italians. In search of expatriation, these foreigners buy these authentic "dars". Thus, they do restoration work, and they adapt them to their needs in comfort. According to the financial service of the municipality of Hergla, there are 12 dars bought by foreigners, who reside in them for a good time of the year.

The development of the residential sector, in parallel with the leisure and tourism activities, participated in the urban development of the city. This urbanization is at the expense of agricultural land. This complicates the urban management of the city, as long as the rural landscape and the natural environment are Hergla's assets.

4.2 Urban and Environmental Management of Hergla in Relation to the Development of Tourist and Residential Activities

Hergla was erected as a commune on 26 April 1966. Hergla's first development plan was approved by Decree No. 41/67 of 10 January 1976. It was first revised and validated by Decree No. 1042/90 of 8 January 1990 and a second one approved by a decree of the Governorate of Sousse on 14 February 2003. However, the last revision of the PAU, covering an area of 337 ha decided by the decree of 9 November 2007, is not still completed. It is since the aediles awaiting the approval, especially that the city needs a document of urbanism in accordance with the regulation. The revision of the urban development plan aims at the best management of the growth of the city and the harmonization of operational planning projects (of housing and tourism) with the rest of the urban network that already has its architectural and urban specificities.

The blocking of this last revision of the PAU is explained by the competition between the private actors and the institutional actors on agricultural and lagoon fields. The former try to convert them into constructible land and the latter wish to keep them. These lands, which are naturally non-constructible, now constitute a territory invaded by the anarchic habitat. This sprawl is mainly done at the expense of productive agricultural land and the DPH (Public Hydraulic Domain) which together constitute the landscape that specifies Hergla. To this end, we note that the economic and urban dynamics have been faster than the reflex of the public authorities who have to anticipate this evolution.

The Urban Planning and Development Department of the municipality of Hergla manages the urban area through a provisional plan which does not have any legal basis. For example, the para-hotel project "Hergla Diamant" (see Fig. 5) was authorized in an area planned for this type of construction in the new PAU, but which is not yet approved. Thus, there are different zones, which have a relation to the tourist activity: a hotel accommodation area and a tourist entertainment area. The blocking

Fig. 5 The para-hotel project "Hergla Diamant." *Source* M. Hellal, November 2019

of the PAU and the lack of coordination between the municipality of Hergla and the Tourist Estate Agency (AFT), which is directly responsible for the land and the allocation of lots to tourist promoters, didn't allow a planned tourism development.

The idea to build a tourist resort resurfaces, once again, by the creation of a PIF (Perimeter of Land Intervention) decided by decree No. 2765 dated 31/12/2014. This new perimeter of an integrated tourist resort project covers part of the coastal El Madfoun Forest, the Assa Jribas Salt Lake and private land, and the procedures for their expropriation have already started. But here we wonder what the relationship between this tourism project and the principles of sustainable development is, especially if we know that the Madfoun Forest becomes an essential component of a coastal environment that is already sensitive. We recall here that the fishing port, realized in the 1980s, is responsible for the dilemma of coastal erosion in Hergla.

The problem of coastal erosion is not specific to Hergla, but it affects the northern coast of Sousse Region,[8] which is important for its tourist development. This is why the APAL (Coastal Protection and Development Agency) has embarked on a major project for the development and protection of the shoreline coastal erosion at the cost of 137 million TD. The project is funded by the state budget at 25% and by a 75% of a German grant. It is organized into 3 lots: Hergla, Chott Meriem, and Hammam Sousse. As for the Hergla lot, which is of most interest to us, the construction site, is set up between 2014 and 2017 and has several interventions:

- reassignment of the existing riding area at El Madfoun over 430 m by the addition of a carapace;
- construction of a riding area of 485 m at the foot of the El Montazah Cliff;

[8] The Sousse Region extends beyond the perimeter of the commune of Sousse to integrate the neighboring agglomerations: Hammam-Sousse, Chott Mariem, Akouda, etc.

- set up green hedges on 400 in the Madfoun Forest for the rehabilitation of the dunes.
- construction of a submerged breakwater of 150 m, in front of the Madfoun Beach.

However, we wonder whether these coastal protection operations, which are carried out independently from the tourist resort project, are sufficient to ensure sustainable development?

In any case, the Hergla "destination" is forced to adapt to the new trends of the tourism product in the world, which gives an important role to the natural environment. In fact, tourism is experiencing a new sustainable revolution [15], after those linked to the industrial revolution and the evolution of transport. Every day it gains new adepts (tourists) all over the world and new practices. This means that tourist spots evolve naturally in time and in relation to new trends. Thus, the new tourism spaces are constrained to be adapted to these changes in demand and the new requirements in terms of sustainable tourism.

5 Discussion

Hergla, which is of rural origin, has natural and heritage assets that distinguish it in the Tunisian Sahel. Over time, a socio-economic system has been established in relation to the natural resources of this territory. Here, there is the opportunity to develop tourism spaces in different ways in response to new international requirements in terms of sustainable tourism.

Although these achievements provide the opportunity to develop sustainable tourism, here, in the face of the state's disengagement, role-players, especially private ones, are turning to lucrative projects. Nevertheless, the strategic studies are against this type of planned tourism, which is at the expense of an already fragile natural environment. Although the integrated resort project has not yet been implemented, the current tourist development seems different from the mass tourism so far developed in Tunisia. It is being implemented in parallel with the development of the residential and leisure sectors under the initiative of private actors, without any control or urban planning from the public authorities. Thus, the integrated resort project, which is back, would be carried out on a new perimeter at the expense of the forest and a fragile coastal environment and farmland.

Here, we confirm the hypothesis that the state-led governance to regularize the stakes of players with divergent interests in this project would be contrary to the principles of sustainable development.

6 Conclusions

The Hergla site, through its landscape and local cultural, is quite potential for tourism development. Faced with the disengagement of the State and the weakness of local actors, the rise of private actors can lead to tourism development not in compliance with the principles of sustainable development. This observation inspires us to work more on governance and new perspectives of tourism planning, in Tunisia in general.

7 Recommendations

The management of a sustainable tourism offer in a fragile and rural territory such as that of Hergla cannot be done without good coordination between actors concerned. For that very reason we are proposing the achievement of the regional pattern of development of tourism integrating of all these actors.

References

1. Hall C-M (2019) The 2030 agenda and the managerial ecology of sustainable tourism. J Sustain Tourism 27:1044–1060
2. Despois J (1955) La Tunisie Orientale Sahel et Basse Steppe. PUF-Paris VI
3. Bada D (2007) Apport de l'analyse multi-dates à l'étude du fonctionnement, évolution et interactions des milieux géomorphologiques de la côte Nord du golfe de Hammamet - Master thesis in Geography. University of Manouba, Tunisia
4. Bourgou M (2005) Les plages : impacte des aménagements touristiques et portuaires sur leur évolution récente, exemple tunisien. CERES, Tunis.
5. Oueslati A (2005) Littoral et aménagement en Tunisie. Faculté des sciences humaines et sociales
6. Fehri A (2002) Aperçu de l'histoire de Hergla. Dar El Maaref-Sousse
7. Toumi A (1995) Mutation socio-économique dans les petites localités de Sahel: le cas de Hergla. La revue tunisienne de la géographie. 1995. N°. 14, Tunis, pp 247–258
8. Miossec J-M (1996) Le tourisme en Tunisie, un pays en développement dans l'espace touristique international. Doctoral thesis. Université François-Rabelais. UFR de droit, sciences économiques et sociales
9. Hellal M. (2015) La station touristique intégrée de Yasmine-Hammamet (Tunisie) ; enjeux, stratégies et système d'action, 1989–2009. Doctoral thesis. Université d'Angers-France
10. MEAT (Ministère de l'Environnement et de l'Aménagement de Territoire) (1997) Schéma Directeur d'Aménagement de Zone Sensible de Salloum-Hergla
11. MEHAT (Ministère de l'Equipement, de l'Habitat et de l'Aménagement de Territoire) (2007) Schéma Directeur d'Aménagement de la zone de Bouficha-Enfidha-Hergla. Phase I. Rapport définitif, Tunis
12. SEDH (Société d'Études et de Développement Hergla) (2004) Aménagement d'une station touristique intégrée à Hergla. Plan d'aménagement de détail. Rapport de deuxième phase, Tunis
13. SEDH (Société d'Études et de Développement Hergla) (2004) Aménagement d'une station touristique intégrée à Hergla, Etude d'Impact sur l'Environnement, Tunis

14. Chennoufi A (2012) Leila Ben Ali et Belhassen Trabelsi derrière l'annulation du Méga projet Al Qoussour de Hergla. www.investir-en-tunisie.net
15. Équipe MIT (2011) Tourismes 3: La révolution durable, Belin, Paris

Climate Factors Affecting Sustainable Human and Tourism Comfort in Aswan Governorate

Islam M. Jaber, El-Sayed E. Omran, and Wedad A. A. Omar

Abstract Human-biometeorology methods can be used to assess human atmospheric effects. The most significant factor for tourism among the efficient human-biometeorological complexes is the thermal effective complex. The study of the climatic characteristics aims to identify the different elements of the climate (e.g., temperature, relative humidity, and evaporation) and the contribution of various climate elements in the study of the natural factors affecting tourism based on the monthly and annual climatic conditions (Aswan). This chapter tried to address these problems in part by examining in detail how climate variables in Aswan Governorate affect sustainable tourism convenience. Several key findings emerged from an evaluation of the current studies. First outcome linked to the annual heat balance rate in the governorate of Aswan, indicating a particular heat loss trend in all governorate centers of Aswan reaching a moderate level (− 255.2 kcal/h) in the governorate of Aswan. Summer records the lowest annual rate of heat loss at 166.3 kcal/h during the year at seasonal level, while winter records the largest heat dissipation (− 255.2 kcal/h) in Aswan throughout the year. Second finding is the application of Adolf equation to the study region, showing elevated sweating rates in Aswan during the summer day at 761 g per hour, which implies fatigue. Although it is less efficient in humans than in the moist region, the elevated sweat rates during the summer. At 187 g/h, the quantity of sweat in the winter, which implies not feeling bad, makes Aswan an attractive area for winter tourism and recreation. The third finding is the application of Tom's human feeling equation, which shows that the research region

I. M. Jaber
GIS Specialist and MA Student, Department of Geography, South Valley University, Qena Branch, Qena 83523, Egypt
e-mail: islammohamed1606@gmail.com

E.-S. E. Omran (✉)
Department of Soil and Water, Faculty of Agriculture, Suez Canal University, Ismailia 41522, Egypt

Institute of African Research and Studies and Nile Basin Countries. Aswan University, Aswan, Egypt

W. A. A. Omar
Department of Geography, Faculty of Arts, South Valley University, Qena, Egypt

M. Abu-hashim et al. (eds.), *Agro-Environmental Sustainability in MENA Regions*, Springer Water, https://doi.org/10.1007/978-3-030-78574-1_17

falls within the individual's comfort zone where the coefficient is less than 21. Finally, the findings can help generate a better understanding of how weather events and heat regimes impact the behavior of tourism by integrating outcomes from this research and initiating future studies focusing on such problems.

Keywords Aswan · Climate elements · Tourism · Thermal comfort · Bioclimatology · Sustainable development

1 Introduction

Tourism has become one of the world's most notable industries and is the main driver for many countries' economies around the world. Development of tourism should be sustainable, but the issue of how this can be achieved remains the subject of discussion. Sustainable tourism is the application of sustainable development ideas to the tourism sector, that is, tourism that meets the needs of the present generation without compromising the ability of future generations to meet their own needs [1]. MENA region lacks sustainable tourism despite the huge natural and aesthetically pleasing tourist attractions scattered across the region [2]. The MENA region nations should discover measuring tools that could demonstrate their well-being (tourism) comparative benefits. This is what we are striving to do in the situation of MENA countries where we will attempt to define and determine the most appropriate aspects through sustainable development to construct an indicator that would influence growth. Tourism and the environment are very closely related as tourism depends on the quality of the natural environment. Environmental quality is the main attraction for tourists. Most tourists today are vulnerable to polluted conditions or environmental degradation. Tourism has constantly altered as a living system. This shift has influenced tourism, and it has resulted in quite varied methods. Some of these new approaches are social responsibility, health, sustainability of resources, tourism alternatives, new touristic attractions, etc.

This research related to Human Bioclimatology, which deals with the study and analysis of the impact of climate on human comfort. The most important of which is to determine the most appropriate ranges in terms of climate elements for human body. Especially with regard to the elements of temperature and relative humidity, which are playing a key role in determining the requirements of human satisfaction. This branch is one of the most recent bioclimatology, which is the most recent and most important of Applied Climatology. These applied branches of climate science emphasize two important principles: First is the principle of Dualism, which dominates most modern geographic studies, and second is the application side. The best benefit of any science is the service of humanity in general and the community in particular. Climate is the most important natural geographical factor affecting human life in general and particular their health and comfort. Its impact is not limited to one person but to all human beings without exception. Although each element of climate has direct and indirect effects on human comfort either inside building or outside, however, the temperature of air and relative humidity deserve a special pause. They are the most effective climatic elements in influencing human comfort.

The tourism industry, which is mainly an outdoor economy, can be particularly vulnerable to climate and weather effects [3]. Weather and climate variables can affect seasonal tourist arrivals, affecting financial activity's sustainability [4]. Thermal comfort describes the human satisfaction perception of the thermal environment. It relates to a number of circumstances that are comfortable for the majority of individuals. There are many areas and places with multiple tourist attractions; but they lack good climate, which reduces their importance [5]. The elements of the climate are important in determining the season and duration of the season. Boniface and Cooper [6] emphasize the importance of the natural climate despite climate can be controlled through cooling and air conditioning systems. Climate and its various components influences, directly on the human body and its social and psychological life. Because the place chosen by the tourist must have adequate climatic and environmental conditions for his movements and needs, which is different from the place where he lives, therefore, the elements of climate are one of the most important natural geographical components on which tourism is based.

Although the ideal climate for tourist recreation does not exist anywhere in the world, the atmosphere of the Aswan region in the winter is marked by moderation, making it acceptable to tourists. Most of the previous studies were about the physiological climate in Egypt or the relation between climate and tourist movement in Egypt as a whole (country level), and not in a small city because accurate data are difficult to find at large scale. This study has a key focus on assessing climate factors affecting sustainable human and tourism comfort at a large scale. So the objective of this chapter is to address some of the climate elements in the Aswan Province, and its suitability for human comfort, and the reflection on tourism activity. This study seeks to achieve the following specific objectives:

1. Analyze the temperature and relative humidity to understand the levels of comfort of the human being.
2. Linking climate elements with human comfort using biometrics and climatic standards.
3. Identify ways of measuring the thermal comfort of a human being and determine its best.
4. Determining the levels of human thermal comfort in Aswan during the months and seasons of the year.
5. Identify the best areas for human and Tourism comfort in the study area.

This chapter holds much promise and potential for environmental sustainability in MENA Regions. This chapter addresses the question of how science and technology can be mobilized to make that promise come true.

2 Climate, Thermal Comfort and Tourism

Climate change is a significant and most often overlooked parameter in the weather-to-mortality relationship. The rise in urban population and climate change have a dramatic impact on human well-being and health, changes in the frequency and intensity of weather occurrences, increases in surface temperatures and decreases in drinking water available [7].

The tourism industry, which is primarily an outdoor economy, is especially susceptible to the impacts of climate and weather [3]. Variables in weather and climate can influence tourist arrivals in season, affecting the sustainability of economic activity [4]. However, it is notable that during the summer months, extreme climatic circumstances pose an important danger to public health. The human body has no selective sensors for the perception of individual climatic parameters, but can only detect (by thermoreceptors) and provide a thermoregulatory reaction to the temperature (and any modifications) of the skin and blood flow that passes through the hypothalamus. However, these temperatures are affected by the integrated impact of all climatic parameters that influence each other in some sort of interrelation. The thermal complex involves meteorological variables such as air temperature, air humidity, and wind velocity, as well as short and long wave radiation, which affects people in indoor and outdoor environments thermo-physiologically.

Thermal comfort describes the human satisfaction perception of the thermal environment. It relates to a number of circumstances that are comfortable for the majority of individuals. In relation to wind circumstances, the skill in human comfort examines additional meteorological variables such as solar radiation, temperature, and moisture in the atmosphere. Thus, thermal comfort is achieved. Pleasant wind and thus elevated wind comfort is only one factor that contributes to an outdoor feeling of well-being. Air humidity and temperature also play a significant role in human comfort.

Improved knowledge of human responses to different environmental circumstances can assist the tourism industry to optimize both short-term tourist demand and long-term climate change risk-adaptation planning by linking to a better comprehension of the interface between tourism and weather. Industries examine these interactions and propose steps to enhance thermal comfort, and thus, residents and passers-by are well-being. Overheating of urban areas in the coming years will be a significant subject in urban planning due to climate change. So, it is possible to apply human-biometeorology techniques to obtain responses about the relationships between tourism and the atmosphere.

3 Materials and Methods

The data used in the performed analysis interpreting the climate components in tourism and recreation concern mean daily values of air temperature, relative humidity, wind speed, and precipitation for the year 2018. The climate factors are based on the General Meteorological Authority during the year 2018.

3.1 Study Area

The Geographical Location: The province of Aswan is located in the southern part of the Arab Republic of Egypt, bordered from the north by governorate of Luxor, from the south by the Republic of Sudan and from the east by the Red Sea Governorate, and from the west by the New Valley Governorate (Fig. 1). The province of Aswan occupies an estimated area of 60,923.2 km^2. The Astronomical Location: The province of Aswan lies between the latitudes of 25° 30′ 22 N and the longitudes of 23° 30′ 31 E.

Many visitors visit Aswan every year, not only for the ancient Egyptian landmarks like Philae and Abu Simbel, but also for the beauty of their landscape and cataracts. Aswan's cataracts are a wonderful place for birds migrating along the Nile and may stop for food and rest at these places. Aswan also has many traditions, the most famous of which are the Nubians. In terms of food, transport, agriculture and also in terms of religious connection, they depended on the Nile. Agriculture in Aswan is defined by a unique personality to the specific circumstances of natural variables reflecting their impacts on the largesse determination of the kinds of a crop.

Words cannot be defined about the beauty of nature and the climate in Aswan. Maybe this is the secret of this quiet city's ability to challenge other winter resorts in different areas of the globe, including endowed with its dry climate, adding to the tourism effects of medical tourism, particularly for patients with kidney and respiratory system, rheumatism, rheumatoid arthritis. Besides the intensity of ultraviolet radiation reflected throughout the year from the mountains surrounding the city, and the Nile Page, especially on the Elephantine and Isis islands.

Aswan provides a pleasant sunny weather and dry climate for most visitors. The city's climate is also known to have excellent relaxing and rejuvenating properties. May and September are the best times to visit Aswan, summers are scorching, and winter temperatures were known to achieve 27 during the day, with cold evenings. The town stays dry throughout the year and gets an average rainfall of less than 1 mm. The summers are extremely hot, however, with average temperatures around 40 °C.

3.2 Research Approaches

A scientific study follows a research curriculum that outlines it. So this study based on the geographical aspects of the physiological climate according to the following research approaches:

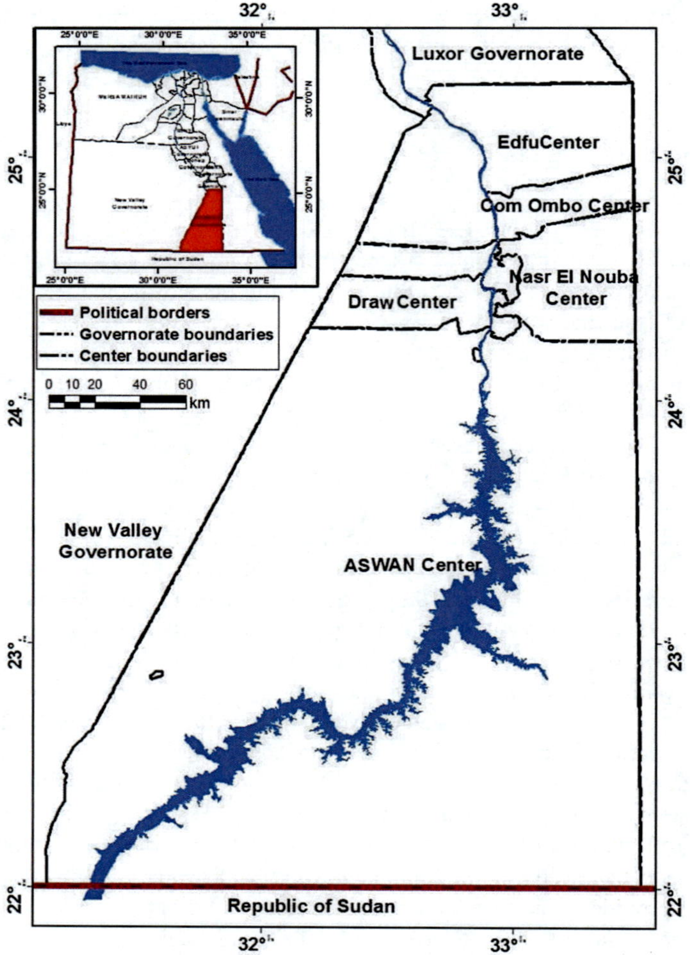

Fig. 1 The geographical and the astronomical location of the study area

1. Regional approach: Regional approach is one of the most important methods of research in geography. On the basis of which the climatic characteristics of temperature and relative humidity and their effects on human comfort were addressed within a regional framework, which includes all the lands of Aswan governorate.
2. Original approach: The geographic factors affecting air temperature and humidity in Egypt, together with the study of the elements of air temperature and relative humidity, were studied in an indigenous climate study.

3. Objective approach: The effects of heat and relative humidity on human heat in Aswan, both inside and outside buildings, will be examined objectively on the basis of physiological climatology.
4. History approach: History approach used in the current study in tracking climatic statistics over a certain period, in addition to tracking the historical (temporal) development of temperature and humidity elements and their effects on human comfort in Egypt during the study period.

The previous approaches outline the study, where the methods and means of achieving and implementing these guidelines are based on different methods, namely:

1. Descriptive method: The study of traditional geography philosophy is based on distribution, connection, and analysis between different geographical phenomena. The description method is used in descriptive interpretation of climate information and facts such as describing climatic conditions and their relations in different regions of the country, which they also have effects on thermal balance and comfort.
2. Statistical method: This method is to schedule and compile climate data to extract climatological averages and rates in preparation for the process of representing such data. Then analyze them using various statistical parameters such as the simple and multiple regression analysis equations.
3. Cartographic method: It is the translation of tables, climatic, statistical and other data into computer-based forms of biological, graphics and maps, through several cartographic programs to achieve the highest possible accuracy of maps and forms.
4. Quantitative method: One of the most important methods of analyzing climate data is a digital quantitative analysis and interpretation based on a set of physiological climatic equations and arguments for establishing the correlation between temperature and relative humidity on the one hand and human convenience on the other. The most important of which are:

   - Adulf's equations of thermal balance day and night.
   - Adolf's equation for sweat.
   - Tom's "Thom" for thermal comfort (discomfort index).

3.3 Methodology

The study of the current climate characteristics of Aswan aims at addressing the most important climatic elements, which effects directly on the human body and his social and psychological life. These elements are represented Heat, Wind, Relative humidity, Rain, and evaporation. To focus, this study includes Aswan governorate in terms of studying the temperature of air and relative humidity, and in terms of studying their effect on human comfort.

3.3.1 Effect of Temperature and Relative Humidity on Human

On the one hand, the temperature is the main climate component on which all other climatic elements depend on tourist attractions. It affects the human in terms of psychological and physical comfort. The temperature range suitable for human comfort and activity ranges between 18 and 25 °C. If the temperature is less than 18 °C or greater than 25 °C, it is a disability for human comfort [8]. On the other hand, the relative humidity is one of the most important climatic factors affecting the movement of tourism. Its height leads to an increase in the feeling of fatigue. Its lowness leads to cold feeling, which hinders the exercise of tourist activities. Humidity ranging from 40 to 60% is the most suitable for human comfort provided that it does not exceed 30%, even if the weather is cold [9]. Air is considered dry if it decreases relative humidity below 50%, while the air is considered to be the average humidity if the relative humidity is between 50 and 70% [10]. The air is high humidity if the ratio is more than 70%.

A sense of comfort is one of the many elements that attract people to recreation areas. There are many places where there is no comfort, as in the areas of ice skating, hot beaches, etc. In this regard, many quantitative studies have emerged to determine the relationship between climate and its impact on humans, in terms of comfort or distress; to determine the optimal picture of the best place for human comfort.

Relative humidity and temperature are used as indicators to measure human comfort. As relative humidity decreases and temperature rises in these areas, there is an opportunity to feel comfortable and to practice tourist activities. The temperature is 29%, and the humidity is 50% [6]. Because the study area is in the hot region, moisture is not a guide to the link between climate and human comfort. Therefore, relative heat and humidity are an appropriate guide to illustrate this relationship.

Accordingly, some quantitative methods were used to study the effect of temperature and relative humidity and its suitability for human comfort.

Discomfort Index of Tom. Tom used the Discomfort index to determine human comfort by knowing the comfort and discomfort of the population in an area, given the temperature of the air and relative humidity in that area, and calculating the presumption of comfort with Eq. 1.

$$\mathbf{THI} = \mathrm{T} + (1 - 0.1\ \mathrm{H}) + (\mathrm{T} - 14.5) \tag{1}$$

where T is the temperature, °C; H = relative humidity, %.

Tom then concluded that when the value of this index is 21, it is not annoying or comfortable. It is uncomfortable when it increases its value as being between 21 and 24. Some people feel uncomfortable when its value is 25 (50% of people feel uncomfortable). When they reach 27, most people feel uncomfortable, and if they reach 29, some people are exposed and disabled in some circles. Some laboratories stop working, especially in the United States and Europe.

In light of the outcome of the Eq. 1, Tom has classified thermal comfort as follows (Table 1).

Table 1 Levels of climatic comfort in the light of Tom's hypothesis

The output of the coefficient	How comfortable climate is
Less than 10	Exhaustion due to coldness (discomfort)
10–15	Average fatigue tends to cold (discomfort)
15–18	Relative comfort tends to cold
18–21	Comfort
21–24	Relative comfort tends to heat
24–27	Average fatigue tends to heat (discomfort)
27–29	Exhaustion due to heat (discomfort)
More than 29	Hot stress

Adolph's Equation for Sweating Rate. Other studies, such as the Adolph's equation, have been formulated. Adolph's study of the body's rate of secretion of sweat, which is used as a measure of human discomfort or discomfort from weather conditions (Eqs. 2 and 3). One of the most important criteria used to calculate how comfortable or cramped a human body is water equilibrium and especially sweat. Adolf has determined the rates of sweat in the desert conditions of a normal human being (g/h) as follows:

$$\textbf{Sweating rate}, \textbf{Day}(\mathrm{g/h}) = 720 + 41(\mathrm{T} - 33) \quad \text{For a person walking under the sun.} \tag{2}$$

$$\textbf{Sweating rate}, \textbf{Night}(\mathrm{g/h}) = 400 + 39(\mathrm{T} - 33) \quad \text{For a person walking at night.} \tag{3}$$

where: T the temperature of the air.

3.3.2 Heat Equilibrium of Human Body in Aswan Governorate

Many laboratory experiments have been conducted to estimate the thermal equilibrium values of the human body in different situations. In 2007, Adolf identified approximate primary values for human body gain and heat loss, especially in dry and dry areas. He concluded by formulating two Eqs. (4 and 5), one for thermal balance by day and the other at night, as follows:

$$\textbf{Thermal balance by day} = \mathbf{100 + 22(T - 33)} \tag{4}$$

$$\textbf{Thermal balance by night} = \textbf{20} + \textbf{18(T} - \textbf{33)} \tag{5}$$

3.3.3 Thermal Comfort and Physiological Equilibrium of Human Body in Aswan in Light of Temperature and Humidity Elements

The physiological balance of the body is the main and most important condition for thermal satisfaction of the human being, a process that has two dimensions: A thermal balance and a water balance, which, if achieved, lead to a person's feeling of rest.

Human comfort has become a new term—which includes many synonyms and various aspects of this comfort, the most important of which are: Thermal comfort (physiology), physical comfort, and sound comfort, vibrations and shocks, visual comfort, sensory comfort, patient comfort, psychological comfort… to other types of comfort.

Human comfort depends not only on heat but also on other factors such as air movement and humidity. Whereas relative humidity is 40%, the temperature of 24 °C may be suitable for the human body while with a temperature of 24 °C and relative humidity of 80%, man becomes less comfortable. Also, if air moves quickly, the object loses its heat, and the human person feels cold. We must not forget that warm air if its moisture is very low, is not suitable for humans as severe drought harms the skin and causes it to crack it as well as dry the nose and throat and increases human susceptibility to cold drops. However, the most suitable humidity is between 40 and 60%. The human body can be compared to the wet dust in terms of its effect on the temperature of the atmosphere, and it has been found that the human person begins to feel uncomfortable if the wet thermometer reaches the temperature of the 90 °C [11].

In this study, methods of determining human comfort will be applied, thermal comfort rates and comfort curves, using temperature and humidity coefficient for a moment. Its choice is because it is one of the most widespread ways to determine human comfort in the world. It also applies to the climatic conditions of Egypt, in addition to giving it accurate figures for the rates of rest translated into categories to divide the study area into regions for thermal comfort, in light of Tom's setting of his treatment.

4 Results and Discussion

4.1 *The Physiological Equilibrium of the Human Body in Aswan in Light of the Elements of Temperature and Humidity*

4.1.1 Heat Equilibrium of Human Body in Aswan Governorate

Table 2 and Figs. 2 and 3 summarize Adolf's equation results for thermal equilibrium, day.

The annual average for Adolf's day-to-day equations in Aswan refers to the relative moderation and thermal equilibrium as the average country level, with the annual average of Aswan governorate for thermal equilibrium 132.8 kilocalories/hour. The study of the results of the Adolf equation for thermal equilibrium in the day shows that only one region of the annual average for thermal equilibrium is as follows:

It is located in all governorate centers, whose stations achieved positive values kilocalories/hour, as the Draw Center records the highest temperature-gain equations of 188.66 kcal/h. The lowest heat gain values are 112.76 kgf/h in the center of Edfu. At the season level, the summer season has the highest rates of heat gain during the year, with a gain rate at Aswan governorate of 278.8 kcal/h. Winter marks the highest year-round heat dissipation rate of – 99.98 kcal/h.

Table 3 and Figs. 4 and 5 summarized Adolf's the results of the equation for Thermal Equilibrium at Night. During the night, no gain or thermal balance values are recorded during any of the seasons (Table 3). The annual rate of heat balance in Aswan governorate indicates a general trend of heat loss in all Aswan governorate centers reaching a moderate level in Aswan governorate – 255.2 kcal/h. There is a range of heat dissipation values to – 237.58 kcal/h at the center of Abu Simbel. However, at the Edfu, the value of heat dissipation is – 273.76 kcal/h. At the season level, summer records the lowest quarterly rates of heat loss during the year at 166.3 kcal/h, while

Table 2 Evaluate the day heat balance according to the Adolf equation

Station	Annual rate	Winter season	Spring season	Summer season	Autumn season
Aswan	133.44	– 72.26	150.6	295.8	159.4
Edfu	112.76	– 80.4	125.74	265.66	139.6
Kom Ombo	118.26	– 76	132.34	271.6	144.66
Nasr El-Nouba	128.16	– 106.14	181.4	289.2	147.74
Abu Simbel	115.62	– 79.74	141.14	261.26	140.04
Draw	188.66	– 185.34	201.2	289.2	163.14
Average of the governorate	132.8	– 99.98	155.4	278.8	149.1

Climate factors are based on the General Meteorological Authority during the year 2018

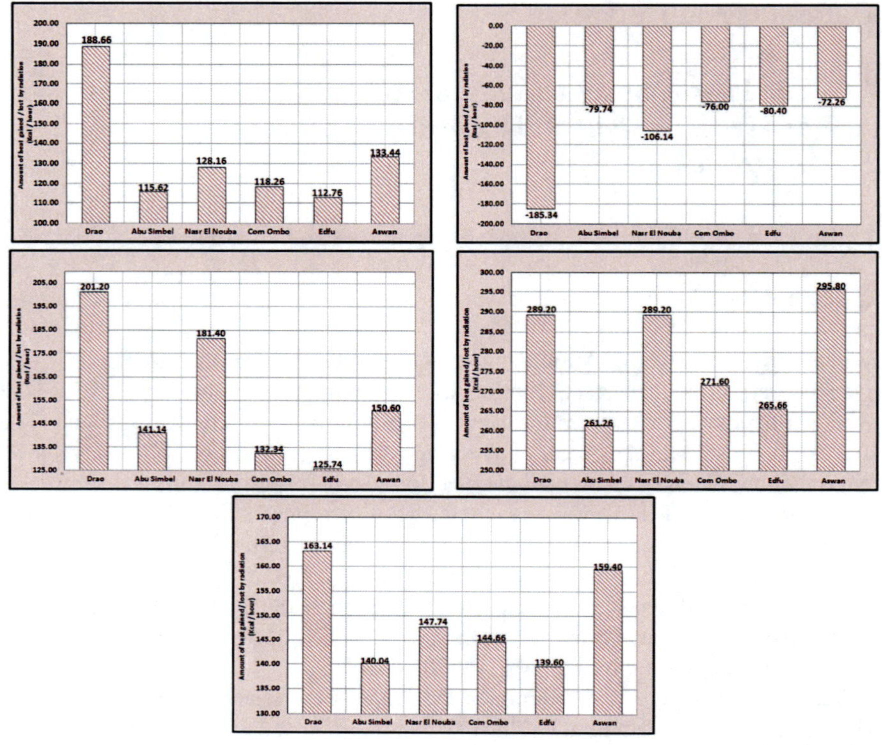

Fig. 2 Annual rate, winter rate, spring rate, summer rate, and autumn rate of the heat balance of the human body in Aswan day in the light of Adolf equation according to Table 2

winter records the highest heat dissipation equations in Aswan throughout the year − 255.2 kcal/h.

4.1.2 Water Equilibrium of the Human Body in Aswan Governorate

Applying the two previous Adolf equations to the annual and quarterly air temperature rates in Aswan, the following shows:

A. **Rate of the sweat of a person walking under the sun in Aswan**
Annual rates of sweat (Table 4) for people walking under the sun in Aswan indicate that there are rates of sweating at all stations as the annual rate of sweating reaches 368.5 g/h as the average county level. The Draw Center recorded the highest annual rate of sweating of 470.2 g/h, while the Nasr Nubia Center recorded the annual rate of sweating 27.4 g/h. In the summer season, the highest annual rate of sweating is 594.3 g/h as the average county level, while the winter recorded the lowest annual rate of sweat 82.2 g/h (Figs. 6, 7, 8 and 9).

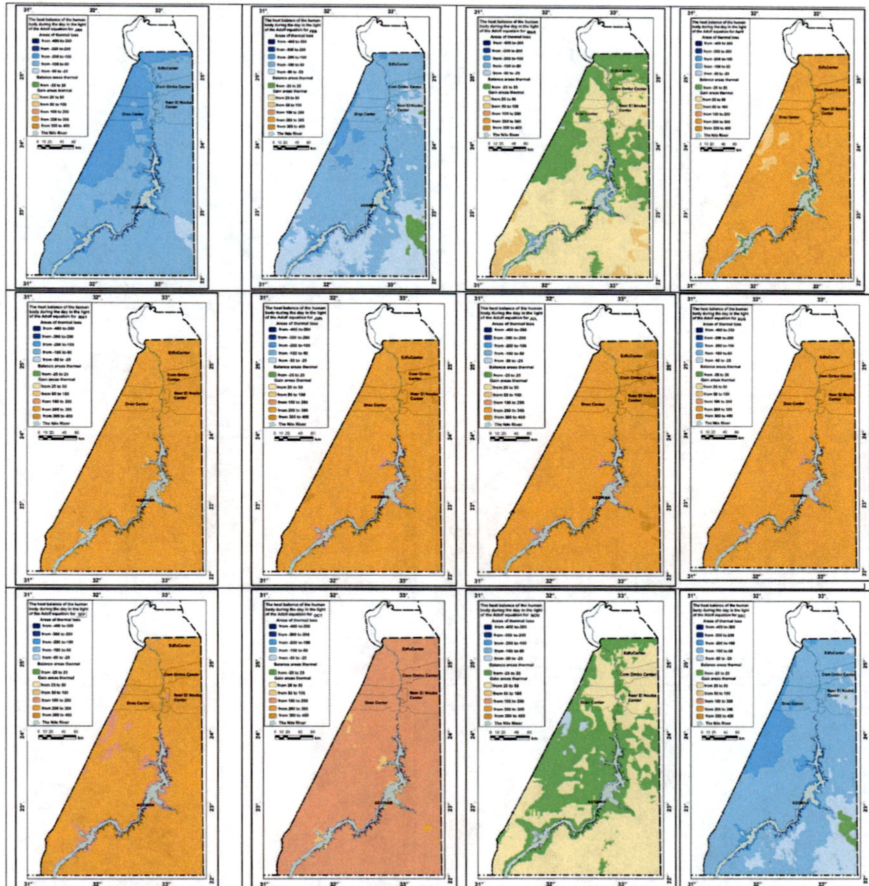

Fig. 3 The annual rate of body secretion of sweat at night, gm/h based on the Adolf equation according to Table 2

B. **Rate of sweating for a person walking at night in Aswan**

Annual rates of sweating for people walking at night in Aswan indicate that there are rates of sweating at all stations as the annual rate of sweating reaches 65.6 g/h as the average county level. The Draw Center has recorded the highest rates of sweating (162.43 g/h), while the Nasr El-Nouba Center recorded the annual sweating rates – 258.78 g/h. In the summer season, the highest annual rate of sweating is 280.4 g/h as the average county level, while the winter recorded the lowest annual rate of sweating – 206.7 g/h.

By applying these equations to the study area, Table 4 was prepared to show the relationship between the average and the maximum temperature and the rate of the body's secretion of sweat.

Table 3 Evaluate the heat balance at night according to the Adolf equation

Station	Annual rate	Winter season	Spring season	Summer season	Autumn season
Aswan	− 258.46	− 121	61.4	180.2	68.6
Edfu	− 273.76	− 127.6	41	155.6	52.4
Kom Ombo	− 265.48	− 124	46.4	160.4	56.6
Nasr El-Nouba	− 247.48	− 148.6	86.6	174.8	59
Abu Simbel	− 237.58	− 127	53.6	152	52.76
Draw	− 248.74	− 213.4	102.8	174.8	71.6
Average of the governorate	− 255.2	− 143.6	65.3	166.3	60.2

Climate factors are based on the General Meteorological Authority during the year 2018

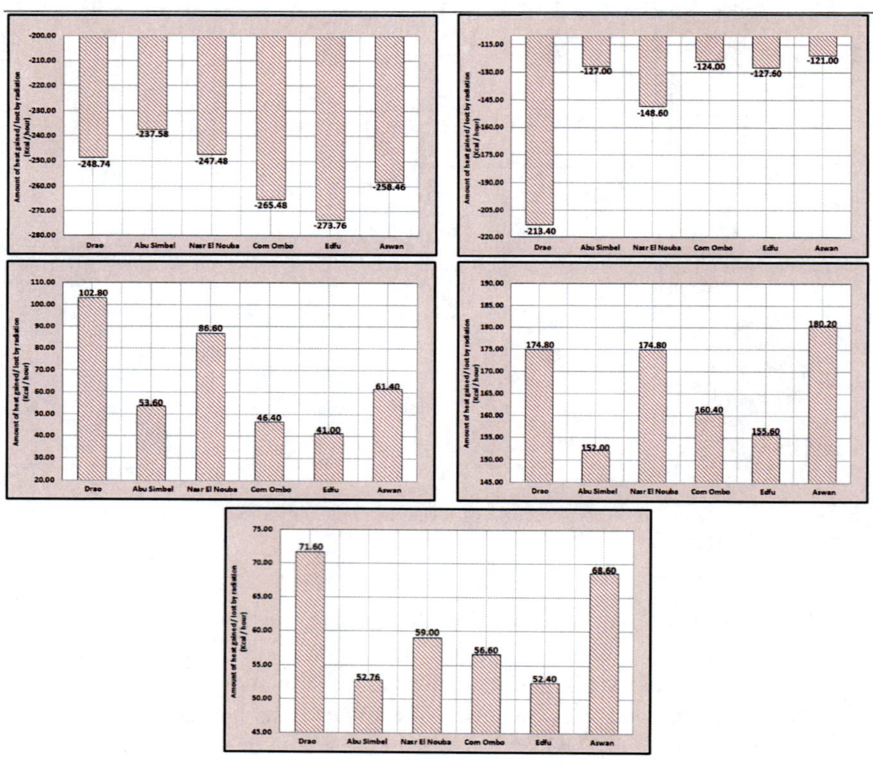

Fig. 4 Annual rate, winter rate, spring rate, summer rate, and autumn rate of the heat balance of the human body in Aswan day in the light of Adolf equation according to Table 3

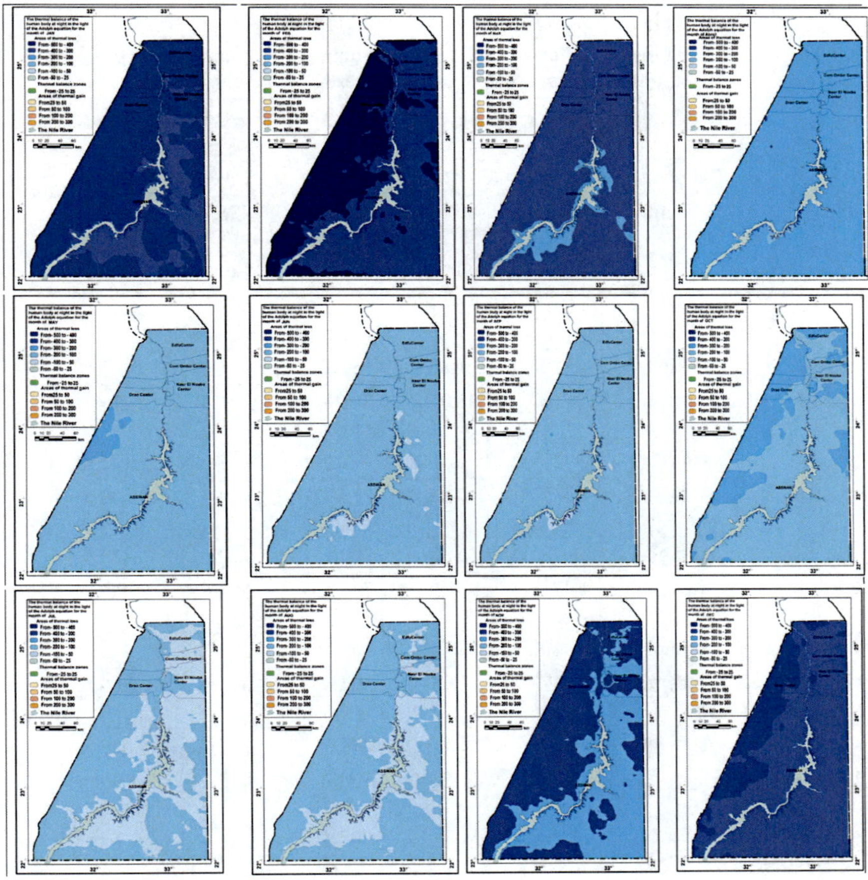

Fig. 5 Annual rates of perspiration for a person walking under the sun in the light of the Adolf equation for sweating

It is noted that high sweating rates in the summer day, at 761 g/h in Aswan, which means fatigue, but the body resists this increase by adjusting the weather conditions, using the means of cooling and air conditioning. Although the high rates of sweat during the summer, it is less effective on humans than in the wet region. Dry air increases evaporation from the human body, smoothing the temperature of the body, unlike air in the wet region. The amount of sweat in the winter at 187 g/h, which means not feeling bad, makes Aswan an area of attraction for winter tourism and recreation. This indicates the existence of a positive relationship between them. This method is defective because it is limited to the element of heat, and the rest of the climate is neglected.

Table 4 Results of applying the Adolf equation for sweating day and night

Station	Sweating rate (g/h)	Winter season	Spring season	Summer season	Autumn season	Annual rate
Aswan	Under the sun	105	480.8	761	520.5	466.83
	At night	– 185	172.5	439	210.2	159.2
Edfu	Under the sun	46.2	404.3	677.63	456.2	396.1
	At night	– 240.9	99.7	359.7	149.1	91.9
Kom Ombo	Under the sun	62.6	422.1	669.78	396.1	410.45
	At night	– 225.3	116.6	352.23	91.9	105.5
Nasr El-Nouba	Under the sun	5.2	85.9	20.267	– 1.6	27.4
	At night	– 279.9	– 203.2	– 265.6	– 286.4	– 258.78
Abu Simbel	Under the sun	87.2	461.7	710.43	499.9	439.8
	At night	– 201.9	154.3	390.9	190.7	133.5
Draw	Under the sun	187	551.9	726.83	506.8	470.2
	At night	– 107	240.1	406.5	197.2	162.43
Average of the governorate	Under the sun	82.2	401.1	594.3	396.3	368.5
	At night	– 206.7	96.7	280.4	92.1	65.6

Climate factors are based on the General Meteorological Authority during the year 2018

4.2 *The Thermal Comfort of a Human Body in Aswan in Light of the Elements of Temperature and Humidity*

Table 5 and Fig. 10 show the results of the application of Tom's equation to Aswan. By applying Tom's equation to Aswan, it became clear that the annual rate of climate comfort levels in light of Tom's presumption is 32.8 (hot stress), and indicates that the levels of comfort in Aswan are limited to two levels of eight set by Tom. They range from relative comfort that tends to cool in some areas to heat stress in others.

A. The relative comfort region that tends to cool (15–18). The province is located in one station and represents the central Nasr El-Nouba, where 16.1 is recorded.
B. Hot exhaustion region (more than 29). This region represents the rest of Aswan governorate's stations, namely the Draw Center, Aswan Center, the Abu Simbel Center, the Kom Ombo Center, and the Edfu Center at rest rates of 37.8, 37.6, 36.3, 34.9, and 34.1, respectively.

Table 5 shows the state of human feeling comfortable in Aswan according to the coefficient of Tom. By applying the "Tom" scale to the study area, it is clear that the

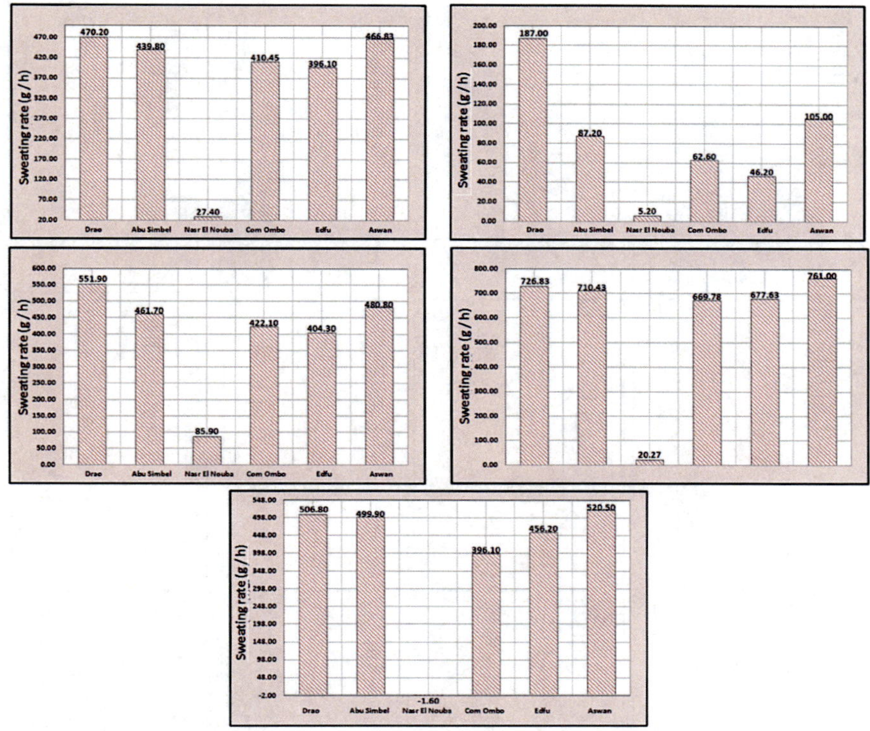

Fig. 6 Annual rate, winter rate, spring rate, summer rate, and autumn rate of perspiration for a person walking under the sun in the light of the Adolf equation for sweating in Aswan day in the light of Adolf equation according to Table 4

study area falls within the comfort zone of individuals, where the coefficient is less than 21. The results of Tom cannot be considered realistic and correct, as the study of man is confined to his home while ignoring other factors of the climate.

5 Conclusions and Future Direction

To better comprehend how tourism populations can become more resilient to climate change, society must first know better how tourists today are affected by weather and climate variables. This study attempted to tackle these issues in part by examining in detail how climate factors affecting sustainable tourism comfort in Aswan Governorate. Several main results arose on the basis of an assessment of the present research.

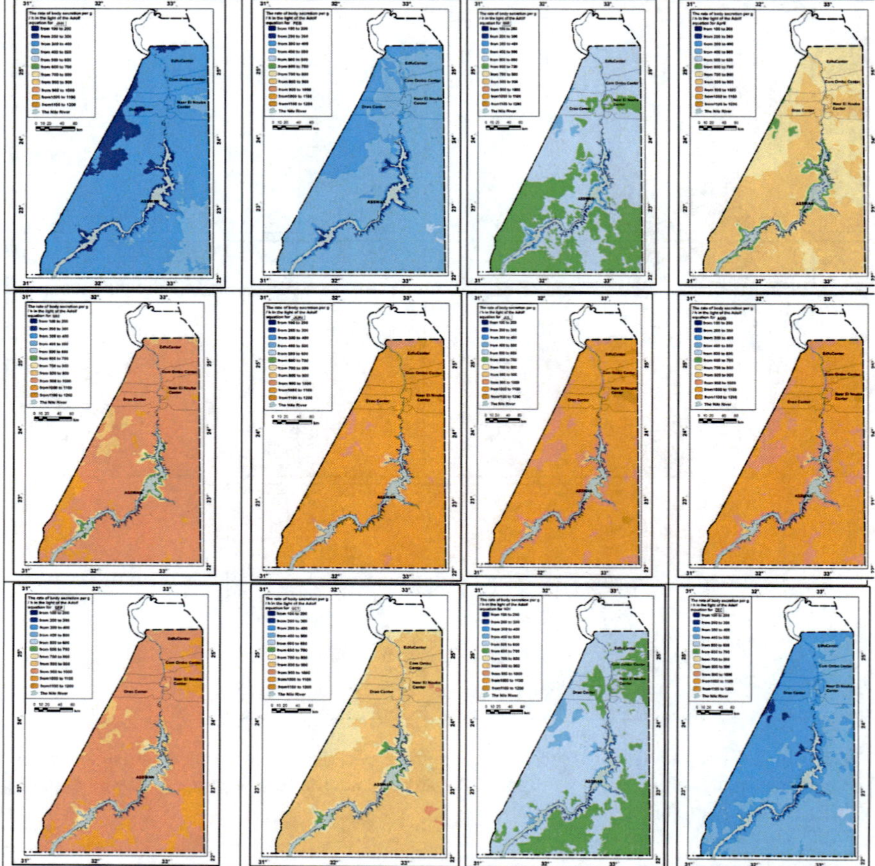

Fig. 7 Annual rates of perspiration for a person walking in the night in the light of the Adolf equation for sweating

1. The issue of the effect of heat and humidity on human well-being in Aswan Governorate in general, as reflected in this study, is still considered new to the Egyptian Geographic Library.
2. The importance of the topic of human well-being in Aswan governorate, where there are climatic differences effects on human well-being.
3. The development of meteorology in Aswan governorate and the availability of modern climate data for different climate elements throughout the year, thus enabling the use and adaptation of such data to serve geographical research.

The results can assist create better knowledge of how weather events and heat regimes affect tourism conduct by incorporating results from this study and initiating future studies with a focus on such issues. In the short term, such results can assist to better educate people and tourists about locations of participation. In a longer time

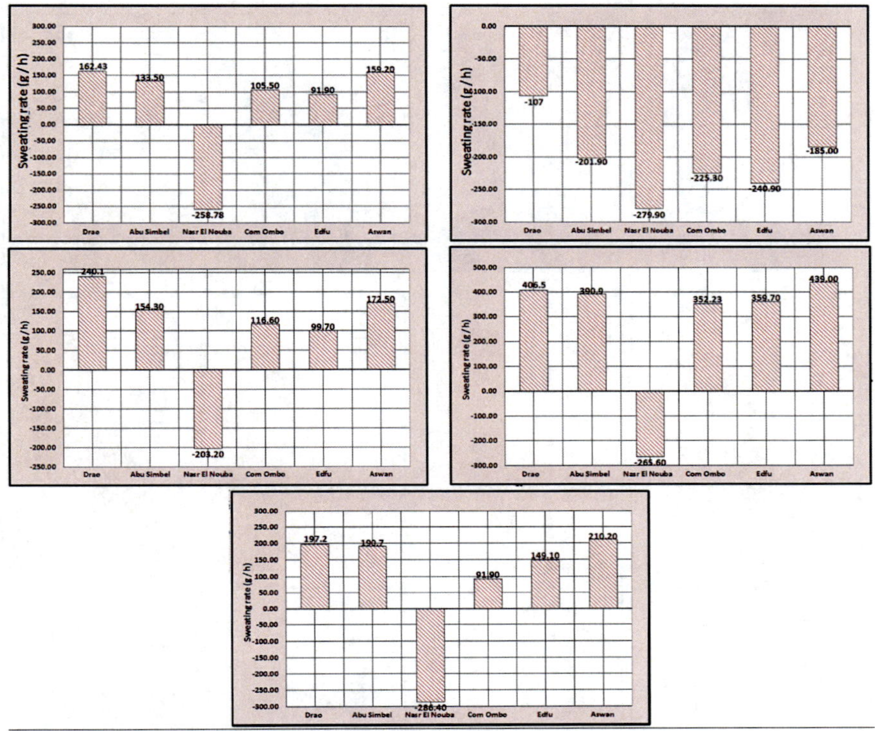

Fig. 8 Annual rate, winter rate, spring rate, summer rate, and autumn rate of perspiration for a person walking in the night in the light of the Adolf equation for sweating in Aswan night according to Table 4

frame, a better knowledge of the reaction of tourism to outdoor heat circumstances and variations in reactions across different places can serve as a basis for a more precise future modeling of attendance-response. Such a framework can only assist enhance our assessment of the prospective effect of wider climate change on tourism and contribute to a more strict and resilient planning of long-term tourism. Ultimately, this can enhance profits.

Mass tourism is a thing of the past for the tourist countries that wish to develop their tourism in a well-considered way. The 2030 Agenda for Sustainable Development, along with the 17 Sustainable Development Goals (SDGs), calls on all stakeholders to act in collaborative partnerships. Among the SDGs, tourism is featured as a target in Goals 8, Goal 12, and 14. For this reason, tourist countries, including Egypt, need to develop a new model of tourism that should also be based on so-called alternative tourism. Tourism has been turning more and more towards new concepts. We are no longer talking about seaside tourism alone but also about other concepts. Alternative tourism is a generic concept encompassing various forms of tourism, such as cultural tourism, Ecotourism, agro-tourism, rural tourism, agritourism, and green tourism.

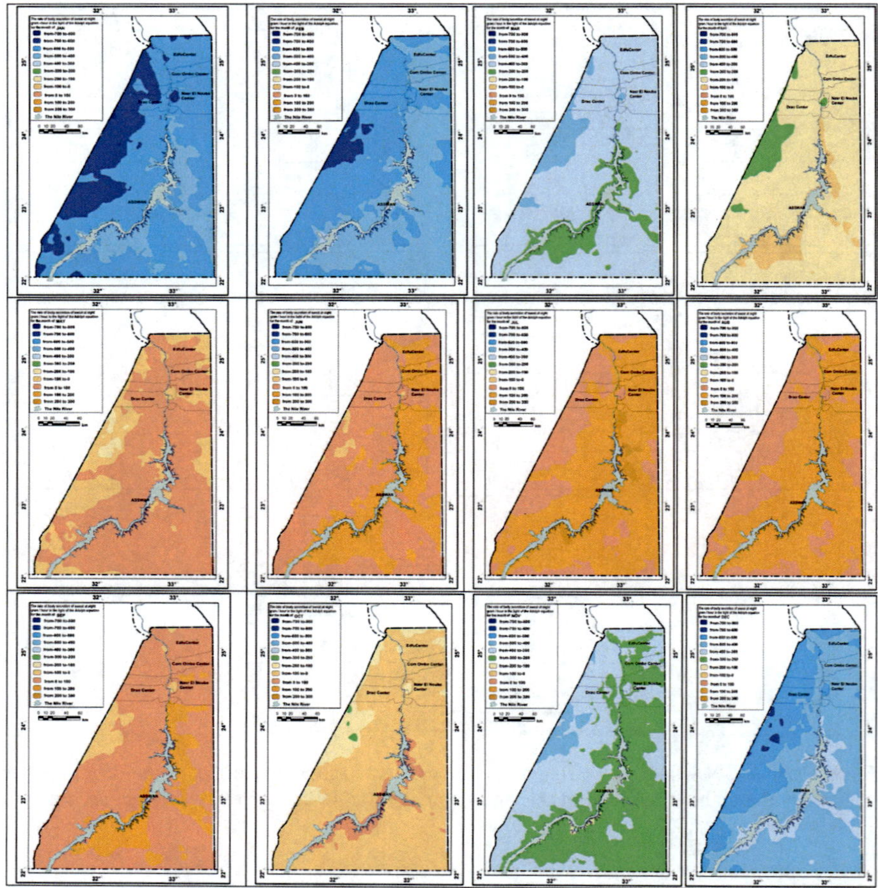

Fig. 9 Annual rate of heat balance of the human body in Aswan day in the night of Adolf equation

1. Identify the climatic characteristics of the air temperature and relative humidity elements in Aswan in great detail and depth because the relative humidity element did not receive a detailed independent study.
2. The need to select and apply some of the global climate equations, models, and curves to Aswan governorate that have not been applied before in the physiological climate studies on Aswan to modeling and classification.
3. Identify the boundaries of the physiological balance and comfort of the human body in Aswan throughout the year.

Table 5 Results of applying Tom's equation on Aswan

Station	DEC	JAN	FEB	Win	MAR	APR	MAY	Spr	JUN	JUL	AUG	Sum	SEP	OCT	NOV	Aut	Annual rate
Aswan	37.6	16.22	20.71	18.72	28.89	39.82	48.55	39.1	51.7	53.4	52.9	52.6	48.9	41.3	30.7	39.9	19.23
Edfu	16.24	13.31	17.73	15.76	25.96	35.59	44.61	35.4	48.3	48.8	48.6	48.6	44.7	38.8	27.8	36.7	34.1
Kom Ombo	16.96	14.4	18.55	16.6	27.16	36.66	45.02	36.3	49.1	48.9	49.5	48.2	45.1	39.5	28.7	33.9	34.9
Nasr El-Nouba	10.64	14.1	16.53	13.7	20.36	20.39	18.81	19.8	18.1	15.64	15.8	16.5	16.7	15.2	12.4	14.4	16.1
Abu Simbel	18.19	15.24	20.21	17.9	28.81	38.77	46.76	38.1	50.2	49.91	50.1	50.1	472	41.5	29.3	39	36.3
Draw	33.36	15.2	19.75	22.7	28.96	38.46	46.82	42.6	51.1	51.1	50.9	51	46.9	41.5	30.5	39.3	37.8
Average of the Governorate	14.7	33.9	18.9	17.6	26.7	34.9	41.8	35.2	44.7	44.6	44.6	44.5	41.6	36.3	26.6	19.1	32.8

Climate factors are based on the General Meteorological Authority during the year 2018

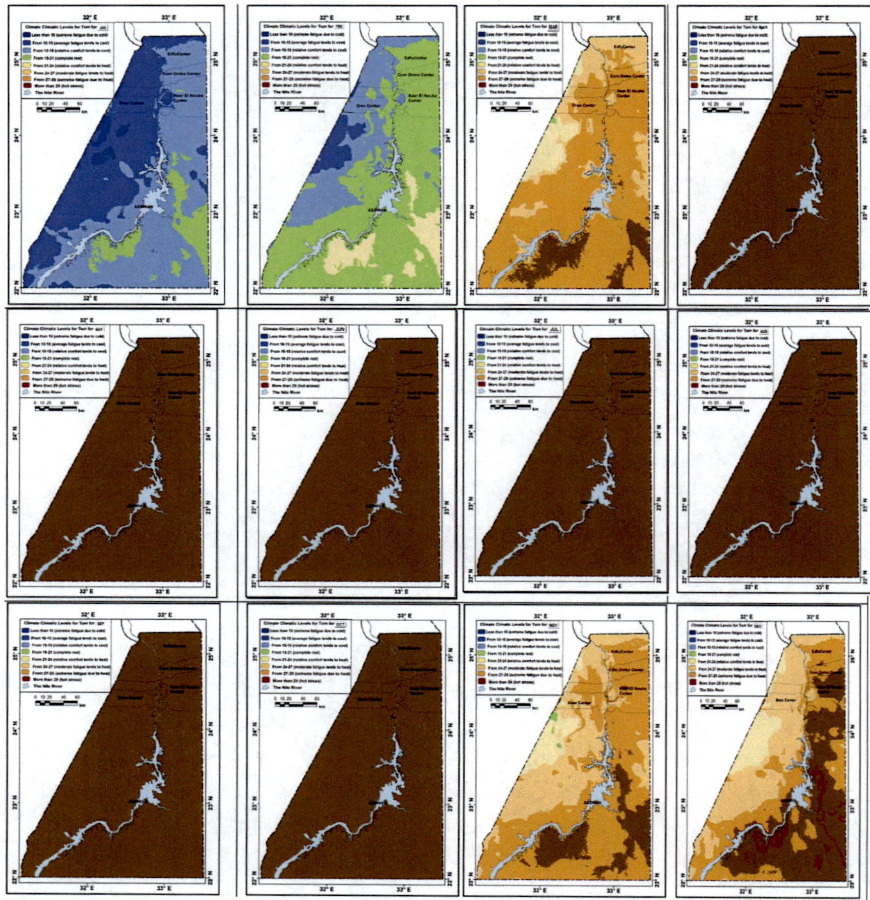

Fig. 10 Annual rate of climatic comfort with Tom's hypothesis

References

1. Weaver D (2006) Sustainable tourism: theory and practice. Butterworth-Heinemann
2. Muhanna E (2006) Sustainable tourism development and environmental management for developing countries. Probl Perspect Manage 4:14–30
3. Perkins D (2016) Using synoptic weather types to predict visitor attendance at Atlanta and Indianapolis zoological parks. Int J Biometeorol
4. Shakeela A, Becken S (2015) Understanding tourism leaders' perceptions of risks from climate change: an assessment of policy-making processes in the Maldives using the social amplification of risk framework (SARF). J Sustain Tour 23:65–84
5. Abdul Hakim MS, Al-Dib HA (2012) Geography of tourism. The Egyptian Anglo Library, Cairo, p 348
6. Boniface BG, Cooper CP (2009) Worldwide destination case book: the geography of travel and tourism, 2nd edn. Butterworth-Heinemann, New York

7. Costello A, Abbas M, Allen A (2009) Managing the health effects of climate change. Lancet and University College London Institute for Global Health Commission. Lancet 373:1693–1733
8. Huebner GM, Hamilton I, Chalabi Z, Shipworth D, Oreszczyn T (2018) Comparison of indoor temperatures of homes with recommended temperatures and effects of disability and age: an observational, cross-sectional study. BMJ Open 8:e021085
9. Ewelina D, Bożena B, Robert S (2018) Analysis of temperature, air humidity and wind conditions for the needs of outdoor thermal comfort. E3S Web Conf 44
10. Wolkoff P (2018) Indoor air humidity, air quality, and health—an overview. Int J Hyg Environ Health 221:376–390
11. Schulte P, Brenda J, Jon W, Nina T, Aitor C, Jung-Hyun K, Kristin M (2013) Criteria for a Recommended Standard: Occupational Exposure to Heat and Hot Environments. Department of Health and Human Services, Centers for Disease Control and Prevention, National Institute for Occupational Safety and Health

Agri-Environmental Policies: Comparison and Critical Evaluation Between EU and Egyptian Structure

Elisa Mutz, Benjamin O. Emmanuel, Fadi Abdelradi, and Johannes Sauer

Abstract This chapter outlines selected agri-environmental policies in both Egypt and the European Union (EU). It focuses on legislation introduced since the year 2000 on water and agricultural conservation, specifically ecosystem, and compares the findings to the United Nations (UN) 2030 Agenda for Sustainable Development guidelines. The 2030 Agenda serves as a framework for both Egypt and the EU and as a standard for international comparison. The United Nations framework, Egyptian as well as EU legislation, respectively, have been analysed with the Qualitative Data Analysis Software ATLAS.ti. Results show a graphical representation of our analysis including overlaps and gaps between the United Nations guidelines and Egyptian as well as EU legislation. Egypt and the EU have introduced legislation that covers certain aspects of the UN guidelines, however, neither of them in the whole scope. Discussing the results with respect to the conducted literature review points to possible weaknesses of both the Egyptian and EU legislation since the year 2000.

Keywords Agri-environmental policies · Qualitative analysis · Egypt · Water resources · Agriculture

E. Mutz
Agricultural and Horticultural Science, Technical University of Munich, Munich, Germany
e-mail: elisa.mutz@tum.de

B. O. Emmanuel · J. Sauer
Production and Resource Economics, Technical University of Munich, Munich, Germany
e-mail: emmanuel.benjamin@tum.de

J. Sauer
e-mail: jo.sauer@tum.de

F. Abdelradi (✉)
Department of Agricultural Economics, Faculty of Agriculture, Cairo University, 3 Gamma St., Giza, Egypt
e-mail: fadi.abdelradi@agr.cu.edu.eg

M. Abu-hashim et al. (eds.), *Agro-Environmental Sustainability in MENA Regions*, Springer Water, https://doi.org/10.1007/978-3-030-78574-1_18

1 Introduction

The world today is facing an uncertain future. How do we deal with climate change and the destruction of the environment, growing population and migration? People and governments often do not seem to know how to adequately tackle some of these challenges, not even talking about all of them. The term "sustainability" is omnipresent, especially when it comes to environmental issues, but how can it be applied to governance? There is indeed need for action and therefore guidance might play a central role. "The world's best plan to build a better world for people and our planet by 2030", is how the United Nations (UN) advertise their 2030 Agenda for Sustainable Development [1]. It has been adopted by all UN member states in 2015. The Sustainable Development Goals (SDGs) included in this agenda are a call for action by all countries—"to promote prosperity while protecting the environment". They consist of 17 goals and 169 targets and the progress towards their achievement is being reviewed annually [1]. The SDGs build upon the 2015 Agenda and the Millennium Development Goals (MDGs) and now try to combine the economic, social and environmental aspects of sustainable development [2].

Everyone and every country contribute to today´s challenges and the achievement of the 2030 Agenda. For example, "the way we produce and consume food is causing major environmental and human health issues. Transforming the agri-food system has become a priority for achieving the Sustainable Development Goals. Food must be produced, distributed, consumed and disposed of in ways that optimize resource use, minimize greenhouse gas emissions, avoid chemicals that harm ecosystems or human health, and stop further loss of biodiversity", states the United Nations Environment Programme (United Nations Environment Programme, 2019). Agricultural transformation is believed to be vital for achieving a sustainable future and being able to contribute to the achievement of several SDGs [3]. Agriculture uses a lot of lands—37% of the total land area are being used for agriculture worldwide [4] 70% of all freshwater withdrawals go to agriculture for irrigation [3] and many more resources are involved. Thus, it is certainly true, that agriculture can cause damages and induce negative impacts. On the other hand, however, agriculture produces what we all need for living: food. It employed 27% of the world´s population in 2016 [4] and is thus the main source of income for a lot of people worldwide.

In the European Union (EU), 39% of the total land area was being used for agriculture in 2016, yet only 4.2% of the population worked in the agricultural sector at the same time (eurostat. 2018). In 2017, agriculture contributed to just 1.7% of the gross domestic product (GDP).

In Egypt however, agriculture is the largest employer providing income-generating activities to more than 30% of the workforce [5]. The agricultural sector plays a central role in the Egyptian economy as it accounts for 14.5% of the GDP [5]. At the same time, the estimated land under agricultural use consists of only about 2.5% of the country´s total area (24,960 km^2 of 1 million km^2).

For the provision of food and employment of people, agriculture relies on water, functioning ecosystems, resources and land. Therefore, the protection of ecosystems and water is vital for agricultural production. Particularly since the Millennium Ecosystem Assessment (MEA) in 2005, initiated by the UN, acknowledged that the earth´s ecosystems are degrading, mostly since the middle of the twentieth century [6]. This happened mainly due to growing demands for food, timber, freshwater or fuel [6] In the year 2000, the MDGs involved environmental sustainability (MDG 7) in their agenda: reversing the loss of environmental resources, reducing biodiversity loss, expanding access to safe drinking water and sanitation—and the world agreed on contributing to that [7].

Today, however, the SDGs—building upon the MDGs—are the relevant global goals and framework we want to focus on. This study will first evaluate what the UN claim in their 2030 Agenda, concerning ecosystems and water. Then we shall investigate whether Egypt and the EU have introduced this agenda in their legislation since the year 2000, esp. with regards to ecosystems and water. The aim is to find an answer to the following questions: was there new legislation introduced for water and ecosystem conservation since the launch of the MDGs and now the SDGs in both Egypt and the EU? What aspects of the UN 2030 Agenda are already enacted in national polices?

For the ecosystem part, the focus is on several agri-environmental laws to see, if and in what way Egypt and the EU policies involve agriculture in the protection of ecosystems since the year 2000. For the water section, the focus lies on the overall water legislation since 2000 to find out how water is protected, not only with regards to agriculture but from all sources. For this purpose we first give an overview of the UN 2015 Agenda, 2030 Agenda, the MDGs and SDGs as well as their approach to global governance. Finally, we critically analyze Egyptian and EU legislation regarding the subject matter. We then outline current developments concerning ecosystems and water worldwide, with a special focus on the current state and problems in Egypt and in the EU.

The UN framework, Egyptian and EU laws have been analyzed with the Qualitative Data Analysis Software ATLAS. The analysis is detailed in the subsequent sections of this study which include review literature as well as results. The evaluation and discussion build the final part of the study.

2 Literature Review

In the following, the United Nations framework will be described, followed by an overview of water and ecosystem governance in Egypt and the EU. Thereafter, water and ecosystem developments worldwide, in Egypt and in the EU will be outlined.

2.1 United Nations Framework

The first part of the study deals with the United Nations approach to global goal setting—both the 2015 and the 2030 Agenda—as well as their approach to water and ecosystem development.

2.1.1 Millennium Development Goals - 2015 Agenda

In September 2000, the United Nations Millennium Declaration was adopted in New York Eight time-bound targets—with a deadline of 2015—were deduced, to be known as the Millennium Development Goals: Eradicating extreme poverty and hunger (Goal 1), achieving universal primary education (Goal 2), promoting gender equality and empower women (Goal 3), reducing child mortality (Goal 4), improving maternal health (Goal 5), combating HIV/AIDS, malaria and other diseases (Goal 6), ensuring environmental sustainability (Goal 7) and developing a global partnership for development (Goal 8) For assessing progress, relevant indicators were selected and regular report on the progress, based on data on the selected indicators, was given by the Secretary-General [8].

In 2015, the final MDGs report was published, giving an overview of the achievements and addressing the still existing gaps between targets and reality [2]. Former UN Secretary-General Ban Ki-moon resumes: "Experiences and evidence from the efforts to achieve the MDGs demonstrate that we know what to do. We need to tackle root causes and do more to integrate the economic, social and environmental dimensions of sustainable development. The emerging post-2015 development agenda, including the set of Sustainable Development Goals, strives to reflect these lessons".

Since the first, second, and third United Nations development decades (the 1960s, 1970s, and 1980s) various development goals have been set, but there was still an innovation to the MDGs: They were adopted by an unprecedented assembly of the world's heads of state and they put human development at the center of the global development agenda Also, the MDGs defined concrete goals that could be monitored and were therefore accountable and measurable In addition, development outcomes as well as inputs from rich countries were addressed [9].

While the UN clearly see the connection between MDGs and progress [2] and "it is widely believed that the MDGs mobilized action" (Fukuda-Parr S. From the Millennium Development Goals to the Sustainable Development Goals: shifts in purpose, concept, and politics of global goal setting for development. Gender & Development (2016). Some researchers argue: even though since launching the MDGs there has evidently been fast progress in the improvement of the quality of life in developing countries, it can´t be clearly stated in how much the MDGs contributed to this trend [10]. In terms of their extent, the MDGs are being criticized because they did not address the root causes of poverty [11]. In addition their targets were judged to be under-ambitious or irrelevant to current challenges, leaving out the most important issues of today: e.g. inequality, climate change and migration [11].

2.1.2 Sustainable Development Goals—2030 Agenda

15 years after launching the MDGs, the UN member states adopted the 2030 Agenda for Sustainable Development in September 2015, the core part of it are the 17 Sustainable Development Goals (SDGs)—time bound like the MDGs with a deadline of 2030 [1]. At the United Nations Conference on Sustainable Development (Rio + 20) in Rio de Janeiro, Brazil, in June 2012, UN member states decided to initiate the development of a set of SDGs to build upon the MDGs and to establish the UN High-level Political Forum on Sustainable Development [1]. While the MDGs were mainly elaborated within the UN Secretariat itself, about 70 governments and representatives of civil society were involved in the development of the SDGs [12]. After a negotiation process, the 2030 Agenda for Sustainable Development was adopted in September 2015 [1]. The annual High-level Political Forum on Sustainable Development is now the central UN platform for the follow-up and review of the SDGs [1].

In the publication Transforming our World: The 2030 Agenda for Sustainable Development the UN defines sustainable development as having three dimensions: economic growth, social inclusion and environmental protection. The 17 SDGs are displayed, each of them including several targets, leading to 169 targets as a whole: No poverty (Goal 1), zero hunger (Goal 2), good health and well-being (Goal 3), quality education (Goal 4), gender equality (Goal 5), clean water and sanitation (Goal 6), affordable and clean energy (Goal 7), decent work and economic growth (Goal 8), industry, innovation and infrastructure (Goal 9), reduced inequalities (Goal 10), sustainable cities and communities (Goal 11), responsible consumption and production (Goal 12), climate action (Goal 13), life below water (Goal 14), life on land (Goal 15), peace, justice and strong institutions (Goal 16), partnerships for the goals (Goal 17) [13]. The global indicator framework for the SDGs is defined in the Resolution adopted by the General Assembly in July 2017 [14].

In contrast to the MDGs and their particular relevance for developing countries, the SDGs and their targets apply to every country (Fukuda-Parr S. From the Millennium Development Goals to the Sustainable Development Goals: shifts in purpose, concept, and politics of global goal setting for development.). The MDGs focused on poverty alleviation through meeting basic needs whilst the 2030 Agenda is broader in scope, including more goals and targets, and incorporates the economic, social and environmental side [11].

2.1.3 Sustainable Development Goals—The Way of Implementation

According to [12] the SDGs can be regarded as a new way of global governance that bases on non-legally binding goals. Governments do not have to follow any legal obligation to transfer the SDGs into national law, as opposed to e.g. legally binding treaties. Governance through goals rests upon fragile institutional agreements at the intergovernmental level and also permits much leeway to countries in terms of interpretation and implementation of the goals [12]. Fukuda-Parr [11] argues that there is the possibility of neglecting some of the most transformative goals and targets through selectivity and simplification of the goals and through the national adaptation process. This could also be induced by the fact that the SDGs contain 169 targets, not all of them being measurable but instead complex concepts [11]. However, [12] emphasize the potential of the SDGs and the absence of alternatives that respond to today's challenges.

2.1.4 Approaches to Water and Ecosystems

According to the World Water Development Report 2015, "water flows through the three pillars of sustainable development—economic, social and environmental. Water resources, and the essential services they provide, are among the keys to achieving poverty reduction, inclusive growth, public health, food security, lives of dignity for all and long-lasting harmony with earth's essential ecosystems" [15]. The UN identifies water as being at the core for achieving sustainable development as it is essential for the function of ecosystems and also ensures economic development [15].

The UN has already addressed water issues in the 2015 Agenda with goal 7 (ensuring environmental sustainability): "Halve, by 2015, the proportion of the population without sustainable access to safe drinking water and basic sanitation" (target 7.C) (United Nations, 2015) The UN reports that the target has been met concerning drinking water but missed when it comes to sanitation [2]. In contrast however, [16] perceives that the drinking water target has never been met and the success has been overestimated given that the indicator framework was not well-chosen and didn't consider e.g. water quality or affordability [16]. In the World Water Development Report 2015, the UN also reflects the fact that no specific targets for sustainable water resources management, water quality, wastewater treatment or for maintaining ecosystems were included in the MDGs [15].

In March 2019, the UN General Assembly declared 2021–2030 to be the UN Decade on Ecosystem Restauration—calling for global action to reverse the degradation of ecosystems, necessary for achieving the SDGs (United Nations Environment Programme. New UN Decade on Ecosystem Restoration offers unparalleled opportunity for job creation, food security and addressing climate change. 2019). Throughout the MDGs, the UN addressed the reduction of biodiversity loss (target 7.B: "Reduce biodiversity loss, achieving, by 2010, a significant reduction in the rate of loss") [7], but ecosystems themselves were not mentioned in the Millennium Declaration [17].

To build up upon the MDGs, in the 2030 Agenda, the UN emphasizes the importance of environmental protection and continue, or rather strengthen, their focus on water and ecosystems. Goal 6 (clean water and sanitation—ensure availability and sustainable management of water and sanitation for all) is entirely devoted to water. It aims at achieving access to drinking water (6.1) and sanitation (6.2), improving water quality and recycling (6.3), increasing water-use efficiency, addressing water scarcity (6.4), implementing integrated water resources management (6.5) and protecting water-related ecosystems (6.6). Also, the UN urges that the release of all wastes and chemicals to water should be reduced (12.4). Goal 14 (life below water—conserve and sustainably use the oceans, seas and marine resources for sustainable development) aims at preventing and reducing marine pollution (14.1) and sustainably managing and protecting marine and coastal ecosystems (14.2). Goal 15 (life on land—protect, restore and promote sustainable use of terrestrial ecosystems, sustainably manage forests, combat desertification, and halt and reverse land degradation and halt biodiversity loss) contains the target of the conservation and sustainable use of terrestrial and inland freshwater ecosystems (15.1). In addition, the SDGs aim on the sustainable management of forests (15.2), the achievement of a land degradation-neutral world (15.3), the conservation of mountain ecosystems (15.4) and the protection of natural habitats and biodiversity (15.5). Also, the UN addresses the necessity of the utilization of genetic resources (15.6), the conservation of protected species (15.7), the introduction of invasive species (15.8) and integrating ecosystem and biodiversity values into development and strategies (15.9) [13].

In Transforming Food and Agriculture to Achieve the SDGs, the Food and Agriculture Organization of the UN (FAO) describes key policies and practices for engaging agriculture in achieving the SDGs. They suggest conserving genetic resources for agriculture to preserve biodiversity, applying integrated pest management and setting payments for providing ecosystem services. In addition, water and soil should be managed in a sustainable way, fertilization should be judicious [3].

Integrated pest management (IPM) is being strongly promoted by the FAO and is their preferred approach to crop protection as well as pesticide risk reduction in the face of sustainable agricultural intensification [5]. Typically, the IPM approach only uses pesticides when prevention measures and non-chemical methods did not succeed and the use of pesticides is economically justified. Thus, pesticides are used in the minimum possible amount and with least side effects on the environment, humans and non-target organisms [5].

Payments for ecosystem services (PES) aim atprotecting and conserving ecosystem services [18]. Farmers can voluntarily decide to go with production practices that positively affect ecosystem services and receive payments for that, for example from the government [18]. The FAO itself identifies some difficulties of their implementation, for example valuing ecosystem services in a monetary way or identifying the actions that are providing ecosystem services as they are very complex systems [18]. In literature, PES programs have been criticized because "capitalist markets cannot be the answer to their ecological contradictions" [19] and have been judged as "commodity fetishism" [20]. While others emphasize also their potential, especially if payments are not the only approach but one amongst others and critically thought of before implementation [21]. For low-income landholders in developing countries, ecosystem service markets and systems were found to present many potential opportuniti Klicken oder tippen Sie hier, um Text einzugeben. es and to contribute to poverty alleviation at the global level [21].

Governance of Water and Ecosystems

The subsequent section describes Egyptian and EU legislation concerning water and ecosystems.

2.1.5 Legislation in Egypt

The Ministry of Environment, together with the Egyptian Environmental Affairs Agency (EEAA), are the institutions responsible for governing the environment in Egypt [22]. They are responsible for protecting the environment, defining policies and preparing laws [22].

The main institutional body for governing water resources in Egypt is the Ministry of Water Resources and Irrigation (MWRI) [23]. It has the legal responsibility of protecting water resources and ensuring water quality Water Resources and Agriculture in Egypt. Also, other ministries are connected to the Egyptian water management, especially the Ministry of Agriculture and Land Reclamation (MALR) given that agriculture is the number one consumer of water [23].

In 1983, the Nature Protection Law 102/1983 was implemented (Arab Republic of Egypt. Nature Protection law, 2019). Law No. 4 was implemented in 1994 and amended in 2009, being Egypt's overall law for the protection of the environment (Arab Republic of Egypt. Law 4 for the Protection of the Environment Amended by Law 9/2009.) In 2017, Decree No. 974 of 2017 was introduced, concerning the registration, handling and use of pesticides (Arab Republic of Egypt. Decree No. 974/2017. 2017). The Decrees No. 225/2004 and No. 1423/1990 are related to fertilizer application [24].

To regulate irrigation, water distribution, establishment and maintenance, and the groundwater process in the Nile region, Law No. 12/1984 was established in 1984 [23]. To further protect the river Nile, its canals and water passages, law No. 48/1982 was introduced addressing water pollution Decree No. 92/2013 is an amendment to law No. 48/1982 and concerns the issuance of permits for disposing of liquid wastes in the water passages of the Nile as long as certain requirements are being met [23]. Decree No. 99/2000 and Decree No. 603/2002 concern the reuse of treated wastewater in agriculture [25].

Tayie and Negm [23] argue that the MWRI does not "display major concern" regarding the quality of water even though this is one of its core responsibilities. Also, they state that not enough attempts have been made when it comes to water efficiency and productivity [23]. However, MWRI started to pay great attention to these issues.

2.1.6 Legislation in the EU

In the EU, three main institutions are involved in legislation: the European Parliament, the Council of the European Union and the European Commission [26]. While regulations are binding legal acts and must be applied entirely among the member countries, directives set out goals that must be achieved across the EU [26]. Therefore, the member countries have to set up their legislation in order to meet the requirements [26]. In the EU, several laws for the protection of the environment and water exist. A selection is presented below:

For the protection of the environment and ecosystems, the Council Directive 92/43/EEC on the conservation of natural habitats and of wild fauna and flora [27] and the Directive 2009/147/EC on the conservation of wild birds set up standards for establishing conservation and protection areas in Europe, the so-called Natura 2000 network [28]. At this point, however, the focus lies more on agri-environmental laws.

The Common Agricultural Policy (CAP) of the EU consists of financial support for farmers [29]. Entailed in the so-called "second pillar", the Regulation (EU) No 1305/2013 on support for rural development includes agri-environment schemes (AES), which are payments for voluntary agri-environment commitments [29]. These were first introduced in the 1980s and are a compulsory part for member countries since 1992 [29]. They exist in addition to the so-called "first pillar" with its cross-compliance requirements, being the basic part of environmental requirements that farmers must meet in order to get CAP funding [29]. Batáry et al. [30] argue that AES must be easy to implement, possible on a large scale, and attractive to farmers. They also judge AES to be expensive and need to be carefully designed and targeted to make a difference [30].

For pesticides, a legislative package was introduced in 2009, consisting of Directive 2009/128/EC establishing a framework for Community Action to achieve the sustainable use of pesticides, Regulation 2009/1107/EC concerning the placing of plant protection products on the market, Directive 2009/127/EC with regard to machinery for pesticide application and Regulation 2009/1185/EC concerning statistics on pesticides [31]. The legislation concerning the sustainable use of pesticides is regarded as significantly driving IPM in the EU. European legislation "calls for the rapid and widespread mainstreaming of IPM" [31].

EU water legislation is rather comprehensive. Directive 2000/60/EC establishing a framework for Community Action was established in 2000, including the protection of water bodies, their related ecosystems and sustainable use of water [32]. This Directive is also being called Water Framework Directive (WFD) and is the most substantial part of the EU water legislation [33].

In 1991, Directive 91/271/EC was introduced, concerning urban wastewater treatment, to protect the environment from wastewater, originating also from industries [34]. In the same year, Council Directive 91/676/EEC was established, concerning the protection of waters against pollution caused by nitrates from agricultural sources [35]. In 1998, Council Directive 98/83/EC set standards for drinking water [36]. Directive 2006/7/EC concerning the management of bathing water quality strengthens the rules of the WFD and supplements it on water protection and management [37]. Directive 2006/118/EC on the protection of groundwater against pollution and deterioration is related to the WFD and sets out measures for protecting the groundwater from pollution [38]. Directive 2005/35/EC addresses marine pollution caused by ships [39]. In May 2018, the European Commission proposed a regulation on minimum requirements for water reuse that has not entered into force yet [40].

Researchers have long perceived the WFD's approach to be innovative and a role model for other water policies worldwide [41]. As the Directive's goals were not met by its deadline in 2015 and the improvement of water status is being slower than intended, stronger policy integration across different sectors is needed and implementation weaknesses need to be eliminated [41]. Voulvoulis et al. [42] also argue that current implementation efforts need to be reviewed or revised for reaching the potential of the WFD among the EU members. They designate the WFD as the most ambitious piece of EU environment legislation so far [42].

2.2 Water and Ecosystem Developments

This part of the study deals with water and ecosystems and their services in relation to agriculture—worldwide, in the EU and Egypt. Also, it describes current developments and threats.

2.2.1 Water and Ecosystems Around the Globe

Unsustainable development pathways and governance affect water quality and availability worldwide. Demands for freshwater are growing—the global water demand is going to increase by 55% until 2050—as projected by the UN, influenced by population growth, urbanization, industrialization and increases in production and consumption [15]. At the same time, these processes also lead to the pollution of waters and the reduction of its accessibility. Increasing demand would likely induce competition and conflicts. Already, negative impacts on the most vulnerable people and ecosystems can be identified [15].

Climate change has large impacts on several aspects of the hydrological cycle—leading to the increase of extremes—and therefore makes the management and use of water resources more difficult [43].

Whilst groundwater plays an important role in water supply, an estimated 20% of the world's groundwater bodies is being over-exploited contributing to land subsidence and saltwater intrusion [15]. Especially around mega-cities and intensely used agricultural areas, groundwater levels are decreasing while providing 43% of all water used for irrigation in agriculture [15]. Not only growing water demand but also water pollution affects water availability.

How our water resources are being managed also has an impact on the world´s ecosystems. Polluted or contaminated waters can degrade connected ecosystems while residues of nutrients from fertilization lead to the eutrophication of water-related ecosystem [15]. These ecosystems provide essential services and contribute to water purification or flood control [15]. Ecosystems and biodiversity also provide services, that are essential for agriculture e.g. pollination, healthy soils and nutrient cycling, biological control of pests or diseases, provisioning as well as regulating services –food, bioenergy, and carbon storage [44].

Processes within agricultural systems, however, can affect ecosystem services and lead to disservices like pesticide poisoning, water pollution and loss of biodiversity [44]. A positive or negative influences on ecosystem services depends on how agroecosystems are being managed [44] For example, ecosystem services like pest control provided by natural enemies have widely been replaced by pesticides which has degenerated the capacity of agro ecosystems to provide pest control [6]. In some systems however, applying IPM enhances pest control provided by natural enemies [6].

Worldwide, fertilizer consumption is still growing since 2000 and projected to rise further [45]. However, nitrogen use efficiency for cereal production is only about 33 percent while the remaining percentage is leached or emitted in the form of greenhouse gases [3] and potentially degrade the ecosystem.

The Millennium Ecosystems Assessment, initiated by the UN in 2005, found that more than 60% of ecosystem services under consideration are being degraded or transformed worldwide. The major changes occurred since the middle of the twentieth century to meet the growing demand for food, freshwater, timber, fibre and fuel [6]. Such degradation and losses are mostly irreversible [6].

In the face of such trend, by the year 2050, agricultural food production will still need to increase by about 60% globally and by ca. 100% in developing countries [15]. Whereas, water and intact ecosystems are essential for this development.

2.2.2 Water and Ecosystems in Egypt

Egypt is located in a semi-arid climate zone characterized by hot dry summers and mild winters with very little rainfall [46]. Annual rainfall ranges between a maximum of about 200 mm along the northern coast to a minimum of nearly zero in the south [5]. However, the average rainfall per year is only about 10 mm [47].The majority of the country is desert land whileSinai obtains more rainfall than the other desert areas leading to numerous wells and oases in this region [5]. The Nile Valley and Delta account for about 4 percent of the total country area [5]. According to [47], "nearly 95% of the freshwater requirements in Egypt are covered by the River Nile. Moreover, any development (i.e., agriculture, industrial, cultural) mainly depends on the availability of Nile water". As a result, about 95% of Egypt´s population lives in the Nile Valley and Delta [5]. The population is estimated to reach about 100 million by 2025 and 140 million by 2050 [47]. In the near future, Egypt's per capita share of water is estimated to be reduced by half even without considering the impact of climate change [46].

In Egypt, the main water consumer is the agricultural sector, accounting for the use of about 85% of the available freshwater and returns about 25% to the system via drainage and deep percolation [47]. Drainage water often does not fulfill the required quality and contains dissolved salts and residues of pesticides and fertilizer.

This, together with industrial and domestic wastewater, contributesto the growing pollution of the Nile [48].

In case of water resources, Egypt has a fixed amount of water supply of ca. 57 Billion Cubic Meters (BCM) per year, of which 97% comes from the River Nile and the rest from precipitation and underground water [49] On the demand side, it is estimated that 72 BCM per year is required with 80% going for agriculture [49]. Excess demand is accommodated by tapping groundwater (6 BCM) and non-renewable aquifers (1 BCM), plus reuse of agricultural drainage water (about 7.5 BCM) and treated municipal and industrial wastewater (variously estimated at 0.7–3 BCM) [49]

There is already a gap between Egypt´s freshwater resources and its demand [47]. This gap is likely to grow as Egypt´s water requirements increase not only due to population growth and higher living standards, but also due to agricultural expansions [47]. Furthermore, there are uncertainties related to the main water resource, Nile River (see—The Nile Water Agreement with Sudan provided by the Aswan High Dam), due to shifts in the regional politics [47]. In the unstable political period from 2011 to 2014, Egyptian negotiation of its water rights with Sudan and Ethiopia in the face of the Ethiopian Renaissance Dam was not successful [23].

Also, climate change makes the management and use of water resources more difficult [43] The rise of temperature associated with climate change will likely lead to increasing evaporative losses from the Nile River [46]. In addition to that, higher temperatures also increase agricultural water demands [46]. Thus, groundwater exploitation is undoubtedly on a rise [47]. In numerous parts of the country extensive water extraction does not pay enough regards to the recharging capacities of aquifers [47].

As conventional water resources are limited, usage of unconventional water resources, like treated wastewater and drainage water or desalinated water is an important approach to increase available water resources [50].). In achieving this goal, the infrastructure for wastewater treatment needs to be expanded according to [51]. Currently, untreated or partially treated wastewater is used for agricultural irrigation at some places with possible negative impacts on the environment and human health [51].

As earlier mentioned, the arable land accounts for a small percentage of the whole country´s area in Egypt. The ecosystem of the Nile Delta´s coastal zone is an important aspect of agricultural production and is affected by human interventions and land-use changes [52]. These developments cause seawater intrusion, erosion as well as possible disappearance of natural ecosystems in that area [52]. Major land-use changes in the western Nile Delta were due to transforming barren land into agricultural land resulting in land degradation [53]. Whereas, climate change will also have an impact on the Nile Delta. Sea level rise will affect water quality and lead to groundwater salination leading to a projected loss of arable land and damaging of natural habitats [54].

2.2.3 Water and Ecosystems in the EU

While Egypt consists of mostly deserts, the EU's 28 (United Kingdom included) member countries are partly located in the moist mid-latitudes and partly in the most subtropical mid-latitudes, i.e. the Mediterranean area (Schultz and Ökozonen, 2008). Throughout these climate zones have at least 6 humid months per [55]. In Germany, the average annual precipitation is 700 mm, in Norway 1400 mm, in Portugal 850 mm and Greece 650 mm [5] Cyprus, Bulgaria and Malta are countries confronted with water stress in Europe [56].

Projected rising temperatures and changes in precipitation patterns [57] can be a challenge for agriculture also in Europe. The above-average hot and dry summer of 2018 has already led to poor yields across the EU. In Germany, cereal yields were about 20% lower than previous three year´s average [58]. While in Germany and Norway, the main water consumer is industry, in Portugal and Greece, most water is being used for agricultural purposes [5].

According to the European Commission, 60% of all surface water bodies did not achieve good ecological status as defined in EU´s water legislation [40]. While 38% of all surface water bodies and 35% of the area of groundwater bodies have been found to be affected by diffuse source pollution i.e. pollution resulting from many sources [59]. This pollution is as a result of agricultural emissions of nutrients—nitrogen and phosphorus and chemicals—pesticides [59]. Thus, the pollution of ground water can be linked to agriculture and drainage of peat lands especially in northern Europe [33].

There are also declining groundwater levels in southern Europe (Spain and Greece) due to extraction for irrigation [33]. Water withdrawals for public, agricultural and industrial supply are the main reasons for failing to achieve the required quantitative status [59]. Overall, 89% of the area of groundwater bodies achieve EU requirements [59].

Presently, treated urban wastewater that is being reused is less than 0.5% of the amount of annual freshwater withdrawals across the EU [40]. While Cyprus reuses more than 90% of its wastewater and Malta 60%, Greece, Italy and Spain reuse between 5 and 12% of their wastewater [40]. The UN state that the significant potential for wastewater reuse in EU is inter alia restricted by missing standards and safety concerns [15]. The potential for water reuse in Europe, however, is very high and a significant number of water reuse projects are in an advanced planning phase [60].

In the EU, more than half of the landscape is under agricultural use, thus, ecosystems in Europe depend on agricultural management but also show on going declines [30]. According to Kløve et al. [33] a major threat to ecosystem in the EU is the use of fertilizer and pesticides in agriculture. This pollutes the groundwater which essentially contributes to various terrestrial and aquatic ecosystems [33]. Furthermore, changes in land-use practices and water withdrawals are endangering the ecosystem [33].

3 Material and Methods

In this section of the study literature as well as law documents were analyzed. In addition, we describe how the software ATLAS was used for the elaboration and depiction of own results.

3.1 Literature

The literature search for this study was conducted with google scholar and Scopus. UN documents have been derived from the UN website or related websites, e.g. FAO. The 2030 Agenda for Sustainable Development is the overall framework and serves as a blueprint for both the Egypt and EU in the fields of water and ecosystems. Concerning ecosystems and ecosystem services, the publication Transforming Food and Agriculture to achieve the SDGs by the FAO, was reviewed inter alia, how to involve agriculture in the protection of ecosystems according to the 2030 Agenda. Soil and genetic resources as well as targets 6.1 and 6.2 concerning drinking water and sanitation were not included in the research as they are outside the scope of this study.

The selection of relevant Egyptian and EU laws was contained by the criterion that they must not have been implemented before the year 2000. The aim was to find legal acts, i.e. laws or decrees, directives and regulations or their amendments, implemented to tackle issues related to water and ecosystems at the turn of the millennium. For water, we took into account laws that do not necessarily address water use in agriculture alone but other sectors. For ecosystems, however, only legal acts which are directly linked to agriculture have been selected.

The EU laws were received from the EUR-Lex website (https://eur-lex.europa.eu/homepage.html). The webpage contains all EU treaties, legal acts, international agreements and other related documents. These documents are available in English, as well as in all member state´s languages.

It was more difficult to find and retrieve the relevant Egyptian laws on-line. The Ministry of Environment's website (http://www.eeaa.gov.eg/en-us/home.aspx) was accessible and available in English. The Ministry of Agriculture and Land Reclamation's website (http://www.agr-egypt.gov.eg) could not be accessed. The Ministry of Water Resources and Irrigation´s website (https://www.mwri.gov.eg/en/index.php/ministry-8#) could be accessed in English but did not respond to inquiries. Thus, for Egyptian legislation, it was necessary to use summaries of legislation.

3.2 ATLAS.ti—Coding Strategy

ATLAS is a Qualitative Data Analysis-Software and a tool for detecting and summarizing relevant information in vast data sets. The software can be downloaded online on the ATLAS website (atlasti.com) of which version ATLAS s7.5.7 was used for the study. With ATLAS it was possible to create a suitable basis for comparing the UN framework and/or guideline with Egyptian and EU law. The software also allows a graphical presentation of the final analysis.

In this section, the process of working on this chapter with ATLAS is being described. To begin working with ATLAS, it is necessary to import documents to the so-called Hermeneutic Unit (HU). The HU is the whole research plat form and consists of imported documents, selected quotations and assigned codes. The imported documents are the Primary Documents (PD). These documents can be organized in the Primary Doc Manager i.e. renamed or re-grouped. For this study, the documents have been imported to ATLAS and grouped to the three PD Families "United Nations framework", "EU laws" and "Egyptian laws".

Relevant sections of the documents concerning water and ecosystems were highlighted and quoted. These quotations were then organized in the Quotation Manager. To every quotation, one or more codes have been assigned. The codes capture the idea of the quotation. As coding is a form of qualitative and therefore subjective analysis, a coding strategy was developed in the coding process. This coding strategy and process is described in Table 1 or transparency and comprehension. The codes are then organized in the Code Manager after which the analysis of the documents was

Table 1 Coding Strategy for the UN framework and/or guideline for water and ecosystems

Code	Assigned to
For water	
Improving/maintaining water quality	Quotations about minimizing or preventing water pollution and contamination, maintaining or improving water quality
Efficient/sustainable water use/abstraction	Quotations about ensuring water productivity and water-efficient and/or sustainable use, about sustainable withdrawals and abstraction
Water recycling and reuse	Quotations about ensuring reuse and recycling of water through e.g. wastewater treatment
For ecosystems	
Payments for agri-environment commitments	Quotations about realising payments for agri-environment commitments by farmers
Promoting integrated pest management	Quotations about promoting integrated pest management
Adjusting fertilization	Quotations about judicious fertilizer use and adjusting fertilization measures

done. The results of the analysis are represented in graphic form using the Network View Manager of ATLAS.

4 Results

The results gained from the analysis in ATLAS show the focus of the UN concerning water usage as well as the UN approach for engaging agriculture in the protection of ecosystems. They also display, the extent Egypt and the EU have implemented into legislation and/or laws the UN framework and/or guidelinesince the year 2000, revealing progress and possible gaps. The following part outlines the codes that have been assigned and consists of figures that display the results.

4.1 UN Water Approaches—Egyptian and EU Law Implementation

Coding the 2030 Agenda, in particular the SDGs, revealed the following UN focus on water and led to three major codes:

- Efficient/sustainable water use/abstraction
- Improving/maintaining water quality
- Water recycling and reuse

For Egypt, the code improving/maintaining water quality was assigned to the amendment of the Environment Law in 2009 and Decree No. 92/2013, an amendment to law No. 48/1982. Although the Environment Law with regards to water only deals with protecting marine waters. For Decree No. 99/2000 and Decree No. 603/2002, from 2000 and 2002 respectively, the code water recycling and reuse were assigned. The code efficient/sustainable water use/abstraction was not assigned. Figure 1 shows the UN water approaches in the 2030 Agenda and their implementation in Egypt since the year 2000.

For the EU, the Water Framework Directive, established in 2000, the codes efficient/sustainable water use/abstraction and improving/maintaining water quality were assigned. The code of water recycling and reuse could not be assigned.

Figure 2 shows the UN approaches to water and their implementation to EU legislation since 2000.

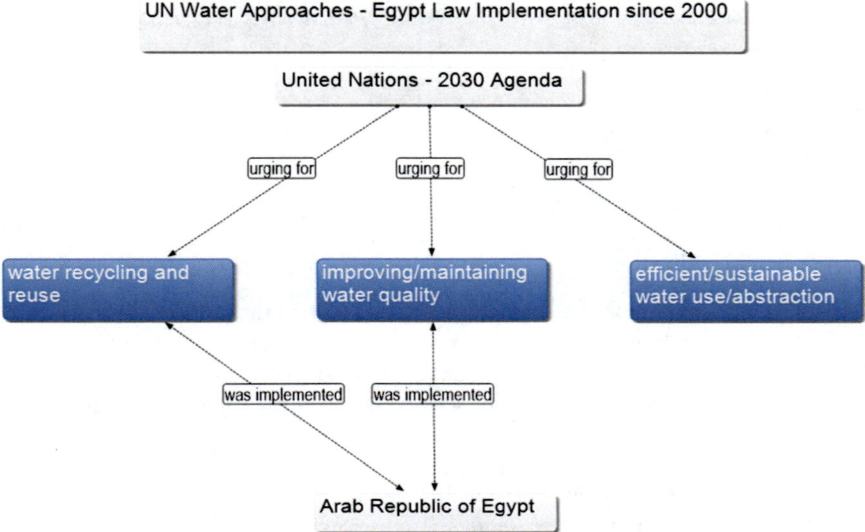

Fig. 1 UN-water approaches—Egypt law implementation (own figure)

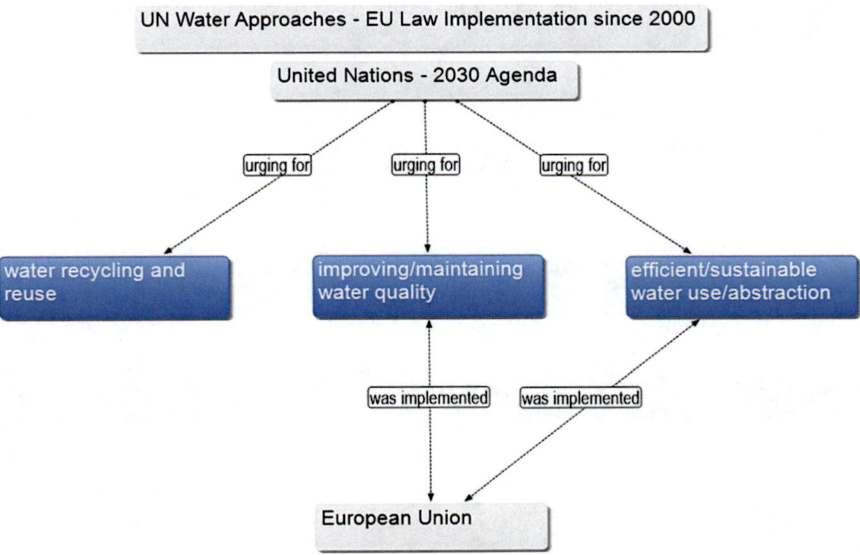

Fig. 2 UN water approaches—EU law implementation (own figure)

4.2 UN Ecosystem Approaches—Egyptian and EU Law Implementation

Coding the FAO document Transforming Food and Agriculture to achieve the SDGs led to three major codes:

- Promoting integrated pest management (IPM)
- Payments for agri-environment commitments
- Adjusting fertilization.

To the Egyptian legislation since the year 2000, two of these codes were assigned.

The code promoting integrated pest management was assigned to the Decree No. 974 of 2017. The code adjusting fertilization was assigned to Decree 225/2004.

Figure 3 shows the UN ecosystem approaches as well as the implementation in Egypt

For the EU, the code promoting integrated pest management could be assigned to the Directive 2009/128/EC establishing a framework for Community action to achieve the sustainable use of pesticides. To the Regulation (EU) No 1305/2013 on support for rural development the code payments for agri-environment commitments was assigned. The code adjusting fertilization was not assigned. Figure 4 displays the UN ecosystem approaches according to the 2030 Agenda and implementation in the EU.

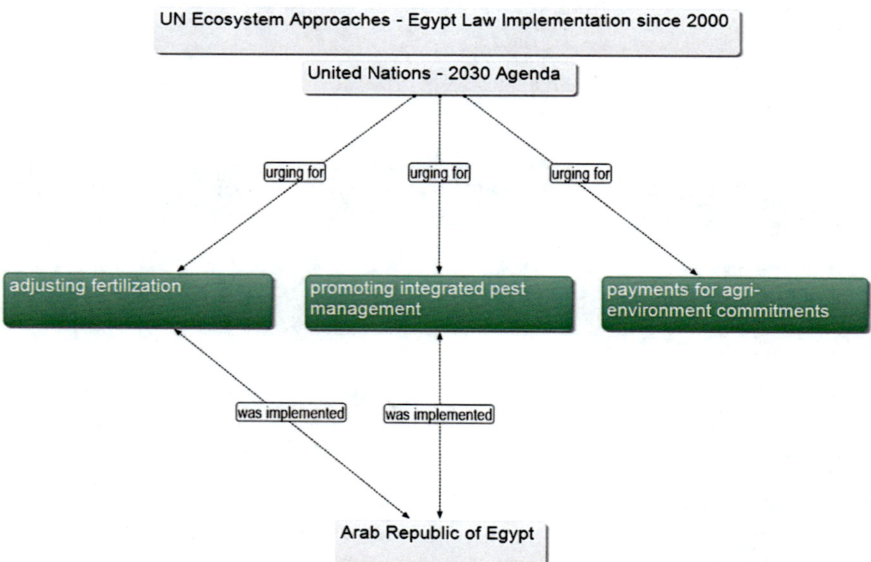

Fig. 3 UN ecosystem approaches—Egypt law implementation (own figure)

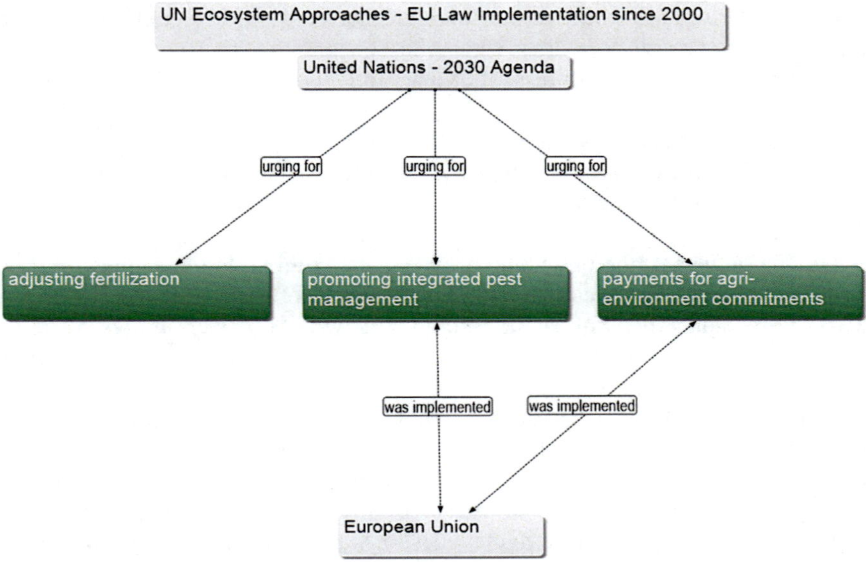

Fig. 4 UN ecosystem approaches—EU law implementation (own figure)

5 Discussion

In the section below, the own results will be evaluated with respect to the existing literature review.

5.1 *UN Water Approaches—Egyptian and EU Law Implementation*

Results show that the UN focus on water is based on the maintenance or improvement of water quality, sustainable and efficient water abstraction and use, as well as water reuse and recycling. The UN therefore took an important step ahead from the MDGs, where drinking water and sanitation were the only topics addressed. Thus, along with the SDGs, the UN address water in a broader scope which is indeed necessary regarding the importance of water and the current development. As the population is growing, and accordingly the demand for freshwater [15]—not to forget, the impact of climate change [43–46, 48–57]. It is of great importance that water is being protected from overexploitation and pollution and that safe reuse and recycling are being promoted.

According to own results, Egypt and the EU both implemented legislation that entails parts of the UN approaches since 2000.

It can be derived from the reviewed literature that there is an essential need in Egypt to protect existing water resources which must encompass efficient use and sustainable abstraction measures. In respect of that, no legislation has been found to have been introduced since the year 2000. Accordingly, [23] perceive that water efficiency is not being sufficiently considered in Egypt. Regarding the gap between Egypt's freshwater resources and demand which is likely to grow further [47] adequate legislation ensuring water efficiency and sustainable withdrawals can be perceived as of major relevance.

The importance of the protection of Nile water from pollution can possibly not be overestimated. The amendment of law No. 48/1982 by Decree No. 92/2013 in 2013 and the amendment of The Environment Law in 2009 were found to address water quality. However, [23] describe the Ministry of Irrigation as not adequately contributing to ensuring water quality.

Using unconventional water sources like treated wastewater, drainage water and desalinated water is considered to be an important approach for Egypt [50]. Although in 2000 and 2002, as own results show, new legislation concerning water reuse has been introduced, research voices say that capacities for wastewater treatment need to be increased [51]. Present on-farm practices for water reuse (e.g. using non or partly treated wastewater for irrigation) [51] lead to the conclusion that either legislation itself or the implementation is too lax when it comes to water reuse.

In the EU, results show that the WFD is regulating sustainable water abstraction measures. There are declining groundwater levels in southern Europe [33] but, overall, most of the groundwater bodies are in good quantitative status as defined by the WFD itself [59]. With possible precipitation changes in the future (IPCC, 2014) it is important, that sustainable abstraction is further being warranted across the EU.

The WFD also includes maintaining or improving water quality and has been referred to as the most ambitious piece of European environmental legislation so far [42]. It surely must be acknowledged that the EU adopted it already 15 years before the SDGs and their call for ensuring water quality were developed. However, water quality has not fulfilled the standards set in the WFD for 60% of surface water bodies by the deadline of 2015 [59]. Agricultural pollution affects more than one-third of surface and groundwater bodies [59]. Researchers suggest stronger policy integration across sectors [41] and reviewing and changing the way of WFD implementation [42].

The amount of reused wastewater across the EU can almost be neglected in comparison to freshwater withdrawals [40], although the potential is being seen as very high [60]. At present, there is no EU-wide legislation for water reuse, but a proposal for a regulation that sets minimum requirements for water reuse has been made public on May 2018 [40].

Egypt´s per capita share of water is strongly decreasing [46]. This implies that deliberate water legislation should be a priority for Egypt. However, Egypt has not implemented legislation covering all the UN approaches as it has not introduced new legislation concerning sustainable water withdrawal and use. Also, not much can be drawn on the quality of the existing water-related laws or extent of their amendments. Water wise, the EU is not facing as many constraints as Egypt and has, with the WFD, an innovative piece of water legislation that, however, needs to be properly implemented to unfold its potential. Water reuse is important to be promoted and represents a current gap in view of the UN approaches.

5.2 UN Ecosystem Approaches – Egyptian and EU Law Implementation

Results show that to engage agriculture in the protection of ecosystems, the UN suggests promoting IPM, setting payments for agri-environment commitments and adjusted fertilization. Along with the MDGs, the UN did not address environmental protection to the same extent as in the 2030 Agenda where several goals and targets address ecosystems in different ways. The 2030 Agenda emphasizes the importance of ecosystem conservation [13] while in the United Nations Millennium Declaration the term ecosystem was not yet mentioned throughout the whole document [17]. Likewise, the fact that the UN introduced the UN Decade on Ecosystem Restoration from 2021 - 2030 shows the new-found courtesy. Given that ecosystem services are so important yet degrading makes it indeed necessary to introduce measures to protect them. The UN also recognizes the role agriculture is playing in both ecosystem degradation but also in ecosystem conservation [3], as agricultural practices can affect ecosystem services [44]

The own results show that both Egypt and the EU have introduced legislation that partly covers the UN framework.

Egypt has introduced legislation that entails the promotion of IPM since the year 2000. As the Ministerial Decree No. 974 of 2017 is rather new, future will have to show if its establishment has positive impacts on e.g. the amount of pesticide residues that can be found in Egyptian waters today.

According to Decree 225/2004 fertilization is being restricted with fixed shares of fertilizer going to the farmers. However, residues of fertilizer in drainage waters contribute to the pollution of the Nile [48] and thus also to the degradation of ecosystems.

Egypt was not found to have introduced PES to their legislation. It has however been proposed by [61] also see [62], as a way of reducing the inequality in the distribution of cost and benefit among farmers resulting to a win–win situation in Egypt.

The EU promotes IPM in the Directive 2009/128/EC, implemented in 2009. The European pesticide legislation is judged to be a significant boost for IPM. In addition, pesticides availability is also shrinking due to stricter regulations [31]. As pesticides residues can be found in EU waters [59], pesticide legislation is important to be carefully implemented in the member countries.

Payments for agri-environment commitments have been found to be covered along with the so-called "second pillar" in the EU, with Regulation (EU) No 1305/2013 on support for rural development which is an amendment to earlier legislation. AES has a long history in EU and it has been stated that they need to be carefully designed and targeted to make a difference [30].

Even though results show, that the EU has not introduced legislation concerning fertilization practices, the WFD sets maximum levels of nutrients that are allowed be found in waters. So, member countries must introduce their laws to meet these requirements. Therefore, it can be concluded that the EU legislation is in that way partly contributing to adjusted fertilization practices. Aas stated above, it was found that the EU member countries do not meet these water quality requirements throughout [59] which contributes to the degradation of ecosystems [33].

Literature shows that in Egypt, the Nile Valley and Delta represent a significant share of the limited amount of fertile land. Parts of this land area are degrading and agricultural production is contributing to that [52]. Also, the ecosystems in Europe show declines [30]. According to [30] they depend on agricultural management as the area under agricultural use is considerably large in the EU. Their long history of payments for agri-environment commitments shows that the EU member countries have been sensitized for engaging agriculture in ecosystem conservation for quite some time. In view of the 2030 Agenda and current developments, both Egypt and the EU have their gaps to close either in legislation setting or implementation of already existing agri-environmental laws.

6 Conclusion

This study reviews the 2030 Agenda as a blueprint for water and ecosystem services in the context of agriculture. Since it has been agreed on and adopted by the UN member states. Despite the possible weaknesses of the 2030 Agenda, the potential of a common plan and framework in the world today should not be neglected. For both Egypt and the EU, implementation of this international consensus could lead to improved water handling and agricultural practices. Reviewing already existing legislation or ensuring its implementation can also be regarded as an important approach.

Our results show that the 2030 Agenda with its ambitious targets has been adopted and to some extent already passed into law by the EU member states and Egypt. However, these laws and decrees have to be adequately enforced for some of the targets to be achieved by the EU member states and Egypt. Results also show that there are still gaps to close, especially in areas where no new legislation has been introduced in the last two decades.

Today, the whole world is in the same boat when it comes to the sustainable development pathway—with or without the 2030 Agenda. To put it simple: The UN member states committed to the SDGs and to providing voluntary reports about their progress. Thus, a development towards the achievement of the 2030 Agenda's targets should only be a question of time—and time is of the essence.

7 Recommendation

Based on the results there are a number of things that policy makers can do to improve the current available water resources. First, finding a solution to change the main irrigation system in Egypt from flood system to a more efficient system that takes into consideration the salinity problem of soils. Second, finding unconventional sources of water such as using treated wastewater for agricultural purposes since agriculture sector takes 85% of the freshwater. Fourth, develop new legislations that monitors and protects current water resources in Egypt. Fifth, minimize the evaporation from Lake Nasser which is the world largest artificial lake that exceeded the earlier estimated amounts. Finally, developing more water efficient cropping patterns. Future research will not only focus legislative and policy frameworks, which is a component of the support ecosystem but as well will investigate the water user across specific agricultural value chains and develop more water efficient crop mixes.

References

1. United Nations (2019) Sustainable development goals. https://sustainabledevelopment.un.org/. Accessed 31 May 2019
2. United Nations (2015) The millennium development goals report. New York, United Nations
3. FAO (2018) Transforming food and agriculture to achieve the SDGs: 20 interconnected actions to guide decision-makers. Rom
4. FAO. FAOSTAT (2019). http://www.fao.org/faostat/en/#country. Accessed 5 Jul 2019
5. FAO (2019) Integrated pest management. http://www.fao.org/agriculture/crops/thematic-sitemap/theme/pests/ipm/en/. Accessed 5 Aug 2019
6. Millennium Ecosystem Assessment (2005) Ecosystems and human well-being: synthesis; a report of the millennium ecosystem assessment. Island Press, Washington, DC
7. United Nations (2015) United Nations millennium development . https://www.un.org/millenniumgoals/. Accessed 4 Jul 2019
8. United Nations Statistics Division (2008) Millennium development goals indicators—the official united nations site for the MDG Indicators. http://mdgs.un.org/unsd/mdg/Host.aspx?Content=Indicators/About.htm. Accessed 29 May 2019
9. Millennium F-P, Goals D (2004) Why they matter. Glob Gov 10:395–402
10. Kenny C, Sumner A (2011) More money or more development: what have the MDGs achieved? SSRN J
11. Fukuda-Parr S (2016) From the millennium development goals to the sustainable development goals: shifts in purpose, concept, and politics of global goal setting for development. Gend Dev 24:43–52
12. Biermann F, Kanie N, Kim RE (2017) Global governance by goal-setting: the novel approach of the UN sustainable development goals. Curr Opin Environ Sustain 26–27:26–31
13. United Nations (2015) Transforming our world: the 2030 agenda for sustainable development. New York
14. United Nations (2017) Resolution adopted by the general assembly on 6 July 2017: work of the statistical commission pertaining to the 2030 agenda for sustainable development
15. UNESCO (2015) Water for a sustainable world. UNESCO, Paris
16. Martínez-Santos P (2017) Does 91% of the world's population really have "sustainable access to safe drinking water"? Int J Water Resour Dev 33:514–533

17. United Nations (2000) Resolution adopted by the general assembly: united nations millennium declaration
18. FAO (2007) Paying farmers for environmental services. Rome
19. Büscher B (2012) Payments for ecosystem services as neoliberal conservation: (reinterpreting) evidence from the Maloti-Drakensberg, South Africa. Conserv Soc 10:29
20. Kosoy N, Corbera E (2010) Payments for ecosystem services as commodity fetishism. Ecol Econ 69:1228–1236
21. Benjamin EO (2013) Adverse selections and microfinance in rural Africa: signalling through environmental services. Enterprise Dev Microfin 24:28–39
22. Arab Republic of Egypt (2019) Ministry of environment website. http://www.eeaa.gov.eg/en-us/home.aspx. Accessed 6 Jul 2019
23. Tayie MS, Negm A (2019) Administrative context and the legal framework governing water resources and agriculture in Egypt. In: Negm AM (ed) Conventional water resources and agriculture in Egypt. Springer International Publishing, Cham, pp 101–124
24. MALR (2019). Central administration of agriculture economics: ministry of agriculture and land reclamation
25. AbuZeid K, Elrawady M (2014) 2030 strategic vision for treated wastewater reuse in Egypt", Water Resour Manage Prog
26. European Union (2019) European Union official website. https://europa.eu/european-union/index_en. Accessed 4 Jul 2019
27. European Commission (1992) Council Directive 92/43/EEC. https://eur-lex.europa.eu/legal-content/EN/TXT/?uri=CELEX%3A31992L0043. Accessed 12 Aug 2019
28. European Commission (2009) Directive 2009/147/EC. https://eur-lex.europa.eu/legal-content/EN/TXT/?uri=CELEX%3A32009L0147. Accessed 12 Aug 2019
29. de Krom MPMM (2017) Farmer participation in agri-environmental schemes: regionalisation and the role of bridging social capital. Land Use Policy 60:352–361
30. Batáry P, Dicks LV, Kleijn D, Sutherland WJ (2015) The role of agri-environment schemes in conservation and environmental management. Conserv Biol J Soc Conserv Biol 29:1006–1016
31. Barzman MS, Bertschinger L, Dachbrodt-Saaydeh S, Graf B, Jensen JE, Joergensen LN, et al (2014) Integrated pest management policy, research and implementation: European initiatives. In: Peshin D (Hg.) 2014—experiences with implementation, global overview, pp 415–428
32. European Commission (2000) Directive 2000/60/EC. https://eur-lex.europa.eu/legal-content/EN/TXT/?uri=CELEX:32000L0060. Accessed 12 Aug 2019
33. Kløve B, Allan A, Bertrand G, Druzynska E, Ertürk A, Goldscheider N et al (2011) Groundwater dependent ecosystems. Part II. Ecosystem services and management in Europe under risk of climate change and land use intensification. Environ Sci Policy 4:782–793
34. European Commission (1991a) Directive 91/271/EC. https://eur-lex.europa.eu/legal-content/EN/TXT/?uri=CELEX:31991L0271. Accessed 12 Aug 2019
35. European Commission (1991b). Directive 91/676/EEC. https://eur-lex.europa.eu/legal-content/en/ALL/?uri=CELEX:31991L0676. Accessed 12 Aug 2019
36. European Commission (1998) Council Directive 98/83/EC. https://eur-lex.europa.eu/legal-content/EN/TXT/?uri=CELEX%3A31998L0083. Accessed 12 Aug 2019
37. European Commission (2006a) Directive 2006/7/EC. https://eur-lex.europa.eu/legal-content/EN/TXT/?uri=CELEX:32006L0007. Accessed 26 Aug 2019
38. European Commission (2006b). Directive 2006/118/EC. https://eur-lex.europa.eu/legal-content/EN/TXT/?uri=CELEX:32006L0118. Accessed 26 Aug 2019
39. European Commission (2005) 2005/35/EC. https://eur-lex.europa.eu/legal-content/EN/TXT/?uri=CELEX%3A32005L0035. Accessed 26 Aug 2019
40. European Commission (2018) Environment. http://ec.europa.eu/environment/water/reuse.htm. Accessed 17 Jul 2019
41. Carvalho L, Mackay EB, Cardoso AC, Baattrup-Pedersen A, Birk S, Blackstock KL et al (2019) Protecting and restoring Europe's waters: an analysis of the future development needs of the water framework directive. Sci Total Environ 658:1228–1238

42. Voulvoulis N, Arpon KD, Giakoumis T (2017) The EU water framework directive: from great expectations to problems with implementation. Sci Total Environ 575:358–366
43. Trenberth KE (2011) Changes in precipitation with climate change. Clim Res 47:123–138
44. Power AG (2010) Ecosystem services and agriculture: tradeoffs and synergies. Philos Trans R Soc Lond B Biol Sci 365:2959–2971
45. FAO (2017) World fertilizer trends and outlook to 2020: summary report
46. Khedr M (2019) Challenges and Issues in water, climate change, and food security in Egypt. In: Negm AM (ed) Conventional water resources and agriculture in Egypt. Springer International Publishing, Cham, pp 229–243
47. Meguid MA (2019) Key features of the egypt's water and agricultural resources. In: Negm AM (ed) Conventional water resources and agriculture in Egypt. Springer International Publishing, Cham, pp 39–99
48. Shamrukh M, Abdel-Wahab A (2011) Water pollution and riverbank filtration for water supply along River Nile, Egypt. In: Shamrukh M (ed) Riverbank filtration for water security in desert countries. Springer Science+Business Media B.V, Dordrecht, pp 5–28
49. Misheloff R (2010) Integrated water resources management II, feasibility of wastewater reuse: report no. 14
50. Negm AM, Omran E-SE, Awaad H, Abu-hashim M (2019) Update, conclusions, and recommendations for sustainability of the agricultural environment in Egypt: the soil–water–food nexus. In: Negm AM, Abu-hashim M (eds.) Sustainability of agricultural environment in Egypt: part I. Springer International Publishing, Cham, pp 351–369
51. Elbana TA, Bakr N, Elbana M (2019) Reuse of treated wastewater in egypt: challenges and opportunities. In: Negm AM, Abdelrahman MAM (eds) Unconventional water resources and agriculture in Egypt. Springer, Cham, pp 429–453
52. Ali EM, El-Magd IA (2016) Impact of human interventions and coastal processes along the Nile Delta coast, Egypt during the past twenty-five years. Egypt J Aquatic Res 42:1–10
53. Abd El-Kawy OR, Rød JK, Ismail HA, Suliman AS (2011) Land use and land cover change detection in the western Nile delta of Egypt using remote sensing data. Appl Geogr 31:483–494
54. El-Nahry AH, Doluschitz R (2010) Climate change and its impacts on the coastal zone of the Nile Delta, Egypt. Environ Earth Sci 59:1497–1506
55. Schultz J (2008) Die Ökozonen der Erde, 4th edn. Ulmer, Stuttgart
56. Sofroniou A, Bishop S (2014) Water scarcity in Cyprus: a review and call for integrated policy. Water 6:2898–2928
57. IPCC (ed.) (2014) Climate change 2014: synthesis report. Intergovernmental Panel on Climate Change, Geneva, Switzerland
58. BMEL (2019) Bundesministerium für Ernährung und Landwirtschaft. https://www.bmel.de/DE/Landwirtschaft/Pflanzenbau/Ackerbau/_Texte/Ernte2018.html;nn=374460. Accessed 17 Jul 2019
59. European Environment Agency (2018) European waters: assessment of status and pressures 2018. Luxembourg
60. Angelakis AN, Gikas P (2014) Water reuse: overview of current practices and trends in the world with emphasis on EU states 2014:67–78
61. Slootweg R, Rajvanshi A, Mathur VB, Kolhoff A (2009) Biodiversity in environmental assessment. Cambridge University Press, Cambridge
62. Benjamin EO, Ola O, Buchenrieder G (2018) Does an agroforestry scheme with payment for ecosystem services (PES) economically empower women in sub-Saharan Africa? Ecosyst Serv 31:1–11

Conclusions

Update, Conclusions, and Recommendations for "Agro-Environmental Sustainability in MENA Regions"

Mohamed Abu-hashim, El-Sayed E. Omran, Faiza Khebour Allouche, and Abdelazim Negm

Abstract The present situation in the Middle East and North Africa (MENA) regions is framed by triple environmental sustainability-related challenges. Natural resources are at the heart of sustainable development in MENA regions and are critical to socio-economic growth. This chapter captures the agro-environmental sustainability in MENA Regions (in terms of findings and suggestions) and provides ideas extracted from the volume cases. Besides, some (update) findings from a few recently published research works related to the agro-environmental sustainability covered themes. This chapter provides the present challenges faced by the agro-environmental sustainability in MENA Regions with a set of recommendations to safeguard the resources. Environmental sustainability is concerned with the possibility of protecting and maintaining environmental resources for future generations. Developing countries are looking for growth while developed nations are looking for instruments for post-growth and intellectual development to think about sustainable development (economically effective, socially equitable and environmentally sustainable). For any economy that needs progress and development, growth has always been a significant goal. It is based mainly on the growth of factors of production owing to the enhanced use of available resources. However, the MENA region faces a triple challenge: Emerging climate patterns herald a future where water

M. Abu-hashim
Department of Soil and Water Science, Faculty of Agriculture, Zagazig University, Zagazig, Egypt
e-mail: mabuhashim@zu.edu.eg

E.-S. E. Omran
Department of Soil and Water, Faculty of Agriculture, Suez Canal University, Ismailia 41522, Egypt

Institute of African Research and Studies and Nile Basin Countries. Aswan University, Aswan, Egypt

F. K. Allouche
Department of Horticultural Sciences and Landscape, High Institute of Agronomic Science-Chott Meriem, University of Sousse, ISA CM, BP 47, 4070 Sousse, Tunisia

A. Negm (✉)
Department of Water and Water Structures Engineering, Faculty of Engineering, Zagazig University, Zagazig 44519, Egypt
e-mail: amnegm@zu.edu.eg

M. Abu-hashim et al. (eds.), *Agro-Environmental Sustainability in MENA Regions*, Springer Water, https://doi.org/10.1007/978-3-030-78574-1_19

resources lie below sustainable levels; rapid population growth is threatening to imperil food security; and over-reliance on oil is curtailing governments' ability to act. Nevertheless, water crisis is more acute than in the MENA region. In addition to food security, water scarcity is the biggest threat to human and environmental security in the region. Droughts, soil salinity and pollution, land subsidence and rural exodus have been triggered by the lack and inefficient use of resources. It has also helped trigger conflicts. Agriculture and food production is a big culprit behind the water shortage in the MENA region. The situation needs new thoughts on sustainability, successful technology deployment and radical agricultural transformation. This book holds much promise and potential for agro-environmental sustainability in MENA Regions. This book addresses the question of how science and technology can be mobilized to make that promise come true.

Keywords Water resources · Agriculture · Assessment · Sustainability · Groundwater · Deserts · Environment · Egypt · MENA regions · Food security

1 Introduction

Environmental sustainability is concerned with the possibility of protecting and maintaining environmental resources for future generations. Developing countries are looking for growth while developed nations are looking for instruments for post-growth and intellectual development to think about sustainable development (economically effective, socially equitable and environmentally sustainable). For any economy that needs progress and development, growth has always been a significant goal. It is based mainly on the growth of factors of production owing to the enhanced use of available resources. The MENA region, however, faces a triple challenge: Emerging climate patterns herald a future where water resources lie below sustainable levels; rapid population growth is threatening to imperil food security; and over-reliance on oil is curtailing governments' ability to act. Nowhere else is the water crisis more acute than in the MENA region. In addition to food security, water scarcity is the biggest threat to human and environmental security in the region [1, 2]. Droughts, soil salinity and pollution, land subsidence and rural exodus have been triggered by the lack and inefficient use of resources. It has also helped trigger conflicts. A major culprit behind water scarcity in the MENA region is agriculture and food production. The situation needs new thinking on sustainability, efficient implementation of technology and radical transformation of agriculture.

This book holds much promise and potential for agro-environmental sustainability in MENA Regions. This book addresses the question of how science and technology can be mobilized to make that promise come true. Therefore, the intention of the book is to improve and address the following main theme.

- Climate change and Water Management Practices.
- Diagnosis and New Farming Technologies.
- Practices for Sustainable Plant and Soil Production.

- Industrial, Landscape, Touristic and political Approaches for Agro-environment sustainability.

The next section presents a brief of the important findings of some of the recent (updated) published studies on the agro-environmental sustainability in MENA Regions, then the main conclusions of the book chapters in addition to the main recommendations for researchers and decision makers. The update, conclusions, and recommendations presented in this chapter come from the data presented in this book.

2 Update

The following are the major update for the book project based on the main book theme:

2.1 Climate Change and Water Management Practices

Four chapters are identified in the book related to water management practices. The first study considers climate change impacts on water balance in Egypt and opportunities for adaptations. The MENA is known as one of the world's most water-scarce regions. The availability of freshwater declined from 990 m^3 in 2005–800 m^3 in 2015 [3] and is expected to reach 600 m^3 per capita by 2050 well below the indix of poverty (1000 m^3/capita) [4]. The region's water shortage will be compounded by the dramatically growing population, as it is expected to double to almost 450 million by 2050 [5]. According to the World Bank [6], most of MENA's agricultural areas rely primarily on rain-fed and 60% of the population lives in surface water-scarce areas, which means the country is highly vulnerable to changes in temperature and precipitation. Therefore, recognizing climate change threats is key to formulating policies and rectifying approaches to reduce the risks. With its limited water supplies and rapidly increasing population, Egypt is facing major challenges. Per capita water consumption has declined from 2200 m^3/capita/year in the 1960s to 570 m^3/capita/year in 2018, and is estimated to be only 324 m^3/capita/year in 2050. Across the world, climate change is projected to trigger much greater concerns, with significant environmental, social, and economic implications.

The second study gives an overview of the decentralized wastewater treatment by using biofilm technologies in the MENA region. Decentralized wastewater treatment used Biofilm technologies. Biofilm processes require less space as they consistently keep the active sludge concentration in the biological reactor at a high level. This is especially of great importance for applications where the type of waste implies the risk of high sludge volume indices. The increase of the demands on efficiency and cost of wastewater treatment leads to new interest in biofilm technology, which is lead to an important positive impact that can affect the treatment cost, operation, and treatment efficiency in several countries in the region. Also, the effect of climate change on water resources in the MENA region is expected to be significant as a result of a decrease in precipitation and water technologies changes in its spatial and temporal distribution. Furthermore, to demonstrate the application of several biofilm technologies with the high-efficiency treatment of municipal wastewater in a small and large scale with respect to the application of low-cost wastewater treatment and also to reuse the treated wastewater for irrigation purpose as successful case studies from and out of the MENA region.

The third study concern the estimation of the olive orchards' water requirements using climatic and physiological methods: Case study Tunisian Semi-arid. Irrigation water management should be done with methods to save water and maximize productivity. Water productivity increases with deficit irrigation. This technique aims to save water and control vegetative growth in orchards without any adverse effects on production. During the first years after planting, irrigation should compensate for the water requirements needed for tree growth to establish as soon as possible [7]. Subsequently, a deficit irrigation strategy is the best option for most olive orchards [7–9]. Whenever there is a shortage of water for irrigation, supplementary or supplementary irrigation is sufficient to significantly increase crop performance.

2.2 Diagnosis and New Farming Technologies

Five approaches are used to delineate and discussed the new farming technologies. The first approach is the agrarian system diagnosis in Kerkennah Archipelago, Tunisia. The application of a detailed diagnosis on the agrarian systems in MENA region, more precisely in a Tunisian archipelago through a social survey allowed us to describe the characteristics of cropping systems in general and detailed farming systems. The interaction between different agricultural and social indicators helped to identify gaps in the evolution of such system. Tunisia needs to develop the agricultural practices of its producers. The purpose of diagnostic analysis of the agrarian system in a small region is to identify the factors that influence the evolution of agricultural production systems. "Agrarian system is the theoretical expression of a historically constituted and geographically located type of agriculture, consisting of a characteristic cultivated ecosystem and a defined productive social system. This allows sustainable exploitation of the corresponding cultivated ecosystem".

The second approach is precision farming technologies to increase soil and crop productivity. In precision farming (PF) or Site-specific land management (SSLM), the farm field is classified into "site-specific management zones" depending on soil pH, yield rates, pest invasion, and other factors that affect soil and crop production. Management decisions become in need for each zone and SSLM tools, for example, remote sensing, GPS, and GIS are used to observe zone variable-rate inputs. This method differs from traditional farming due to traditional farming used a "whole field" approach where the field is considered as a homogeneous area, but SSLM method classifies the field into zones based on the field variability [10, 11]. The application of new advanced technology in order to increase agricultural productivity becomes an urgent solution to meet the food demand of the growing population. Remote sensing technique is an important factor in precision farming [12]. Different components of technologies used in precision farming include remote sensings such as satellites, aerial photography, UAVs, GIS, GPS, the variable rate application, and geostatistics. UAVs have been normally used for image surveys such as mapping weed in coffee farms, finding the anomaly in the fertilization delivery system and determining maturity analyses [13].

The third potential practice is implementing an environmental information system in data-scarce countries: issues and guidelines. Integrating the environmental dimension into socio-economic development strategies has become a requirement for sustainable development. To reach this goal, it is necessary to monitor the state of the environment in order to identify, manage and supervise the different environmental issues and to determine the appropriate responses for them. This objective cannot be achieved without organizing environmental data in an information system that can be used and updated by all the actors and partners concerned by the environmental questions. The aim of implementing an Environmental Information System (EIS) is to develop an integrated framework for the storage, production, management and exchange of environmental information in a decision-making perspective. However, the availability and scarcity of data remains a major obstacle for realizing such systems, especially in developing countries. Society responds then by trying to provide solutions to these changes by adopting environmental, economic and social policies (Responses) [14]. To overcome the problems of access and availability of data, other alternatives are to be explored. Earth observation data captured from satellites is one of the promising options in this regard. Indeed, with the free and open access to satellite data of many types and resolutions, these databases become a valuable source that countries suffering from data availability problems can exploit and glamorize [15]. Big data techniques present three significant advantages for the implementation of information systems. Firstly, they provide solutions for the storage of mass data that evolve over time, such as those concerning the environment. They provide algorithms and methods for processing and analyzing extensive data coming from various sources. Finally, they provide tools and interfaces to display information in many ways that make it easy to understand and interpret the phenomena being studied [16].

The fourth approach is related to the green spaces for residential projects as a commitment to environmental concerns and a sustainable development initiative: Design of a periurban park in Casablanca, Morocco. Morocco has embarked on a long process of modernization of the territorial administrative organization and its adaptation to the current political, economic and social context while gradually introducing a decentralization that will allow a rapprochement between the administration and the local authorities. The need for green space is no longer to be demonstrated in urban areas; It is not only a question of environmental and aesthetic stakes but also the place of founding experiences and social openness. The presence of this vital space is essential for the well-being of human populations because its absence presents itself as a phenomenon to be analyzed. Casablanca is a city that is much better known for its buildings than for its green spaces. Places of relaxation in the White City are numerous, but they are poorly maintained. This will make Morocco one of the most emerging countries in the MENA region. The processes of territorial recomposition in Morocco are characterized by a continuous dynamic, reflecting the major changes that affected the different sectors of the country during the twentieth century.

The fifth approach is related to the environment and sustainable development in front of the artificialisation of the coastlines: Coasts of Tunisia, Morocco and Algeria. The traditional economy of the Sahel region was mainly based on agriculture and, in particular, olive growing, livestock farming and vegetable crops. This was also the case for all of Tunisia, which had just emerged from decades of colonialism. Thus, at that time, the coastal landscapes of the Sahelian region, in general, and the Sousse region, in particular, were formed, essentially, as the following figures show, by a succession of traditional vegetable crops plots (called chatts) and olive groves. The latter were particular and occupied most of the fertile coastal lands of the region [17]. The impact of coastal artificialisation, notably on the environment and sustainable development, is found not only in the Sahelian region of Tunisia but also in several countries of the Mediterranean basin, such as: Morocco, where tourist and resort urbanization has taken place in the form of a ribbon along the beaches and on former coastal agricultural land, and Algeria, where diffuse urbanization generated in particular by the most widespread form of dwelling which is the form of individual dwelling, has invaded coastlines such as those of the city of Oran. This development of coastal tourism infrastructure has therefore not been made for the sake of sustainability, given that we have noticed that, in recent decades, tourism has been turning more and more towards new concepts. We are no longer talking about seaside tourism alone (this concept is the main reason for the coastal polarization of tourist infrastructures) but also about other concepts among them cultural tourism, green tourism and agritourism. Tunisia has important potentials that could be essential in promoting these new concepts. Among these potentials: coastal agriculture. Indeed, the Sahel region is characterized not only by its tourism but also by its agriculture located on the seafront and which is entirely productive. This agriculture could play, in addition to its primary function of food production, other functions that can be assets for the promotion of tourism respectful of the environment and perfectly in agreement with sustainable development of the region. It could also be one of the links in the chain that would lead to environmental protection and sustainable development of this region.

2.3 Practices for Sustainable Plant and Soil Production

Six potential practices for Sustainable Soil Production are updated. The first potential practice is sustainable agriculture in some Arab Maghreb countries (Morocco, Algeria, and Tunisia). Agriculture has changed fundamentally and over time has become the main economic sector responsible for multiple environmental impacts. The environmental impact of a pesticide depends on the degree of exposure (resulting from its dispersion and concentration in the environment) and its toxicological characteristics. Intensive agriculture exposed Arab Maghreb countries (Marocco, Algeria, Tunisia) to chemical pollution in irrigated areas: contamination and degradation of water ground, watercourses and soil, mainly due to the use of mineral fertilizers and crop protection products [18]. As many countries in this planet are exposed to climate change, due to the bad management of this fragile sector, the reduction of yields in Arab Maghreb countries depends on climate changes [18, 19]. Without forgetting the lower levels of the agriculture users, they had the hard effects on agrosystems then on the environment [19].

The second approach is the possibilities of mineral fertilizer substitution via bio and organic fertilizers for decreasing environmental pollution and improving of sesame (*Sesamumindicum* L.) vegetative growth. Nitrogen is an essential element for other vital processes like carbohydrates, proteins, metabolism, and (reinforcing both stages of cell division and cell enlargement). Potassium is playing a vital role in stimulating the enzymes and fight biotic and abiotic stresses such as cold, diseases, waterlogging, other opposite conditions [20], pests, drought, salinity. Mineral fertilizers are expensive, not easily available, and increasing its doses about plant needs lead to environmental pollution. Possibilities of partial or complete substitution of mineral fertilizer via bio and organic fertilizers were assessed for improving the characteristics of sesame vegetative growth and contributing to decreasing environmental pollution. The sesame plant is an important oilseed crop, which has medical and pharmaceutical uses. Its seed oil contains some of vitamins and minerals such as copper, calcium, high content of phosphorus, iron, magnesium, manganese, zinc, and vitamin B1.

The third potential practice gives an overview of the animal and Rangeland resources in Shalatin—AbouRamad—HalaibsTriangle Region (SAHTR), Red Sea governorate, Egypt. Shalatin—AbouRamad—HalaibTriangle Region (SAHTR) has vital and strategic importance to Egypt. Due to harsh climatic conditions, low and erratic rainfall precipitation and lack of knowledge, there is no any traditional agriculture activities [21]. The major income and food sources have been traditionally through nomadic animal husbandry and trading. A part from livestock, other traditional sources of income is charcoal, wild honey, medicinal herbs, and seed collection. A combination of severing long–lasting drought and loss of access to traditional grazing areas has dramatically undetermined the continued viability of their nomadic pastoralist way of life. Most of them loose their animals and become extremely poor and food insecure, with high levels of malnutrition amongst women and children. The chapter focuses animal and natural range resources in Shalatin—Abou Ramad—Halaib Triangle Region (SAHTR) to identify the potentialities of animal and rangeland resources; main constraints and problems that would help in planning specific strategies for developing animal production in the region and enhancing local Bedouin's welfare.

2.4 Industrial, Landscape, Touristic and Political Approaches for Agro-Environment Sustainability

Six sustainable industry approaches are identified. The first approach is the Chemical Industry Vis-A-Vis Sustainability In MENA Regions: The Case Of Tunisia. While the second approach is the sustainable mining site remediation under (Semi) Arid Climates in the Middle East and in Northern Africa. The Djebel Ressas Mine in Tunisia as an example of the Orphaned Mines Issue. Orphaned mines after coal, metals, oil shale, and other extracting, are a universal problem all over the world, in different cultures [22]. As primary effects, if the underground or open cast mining is stopped, abandoned mines can cause unstable underground structures or a heightened exposure of minerals releasing chemicals, for example. As secondary effects, deforestation, erosion and groundwater contamination from isolating metals can follow, among others. As a tertiary effect, areas near open-cast mines can become inhabitable. For example, as mining often affects the groundwater levels via pumping, the stopping of the mining activities does not only restore the former groundwater levels. It may also create an artificial hydrogeological situation where the underground aquifers are even destroyed. Even if the site remediation addresses the groundwater levels' restoration, the water quality could be deteriorated by heavy metals and other pollutants such as radioactive substances. So for agriculture and human habitats, the availability of fresh, unpolluted water is a key issue. The semi-arid climate limits the water availability in the MENA area, so water issues will be among the top priority future issues. For the third approach is the steppe ecological systems dynamics: between conservation of diversity and adaptation to agropastoral production systems.

The second approach is the agricultural landscape of Sahel Bizertin: A Heritage in Peril. The knowledge of landscapes as a value has become very distinctive and topical in the last decades. The landscape with harmonious and consistent relations between human activity and preserved nature is an essential competitive advantage. A recognizable outstanding landscape presents above all the identity of a certain area at different levels. The treatment of the landscape also depends on the attitude of society towards it, thus influencing the state and image of landscapes and the quality of our life. The Sahel of Bizerte is rich in heritage and landscape identity reflecting the evolution of the human occupation of the territory. The heritage value of the landscape is mainly due to the time.

The third approach is the urban tree: a key element for the sustainable development of Tunisian cities. A number of programs and projects aiming to improve the quality of life in urban and rural areas, have been developed by the Tunisian Ministry of the Environment and implemented. We can cite the program of the promotion of urban aesthetics (green spaces, urban parks, environmental boulevards, boulevards of the earth, strategic routes and entrances of cities …), besides the program of assistance to the preparation and implementation of Agenda 21 and urban development strategies. Green spaces are an essential element for the aesthetics, setting and quality of life of a city. They help to ventilate the cities and must be considered as the lungs of the city. They are places of relaxation, walk, rest, games for the youngest, sports fields for all ages. It was necessary to return to the role of the tree in the city, a role that merges with the three pillars of sustainable development because of its benefits for the territory, for man and society. Then, it was about to see the short history of the introduction of the tree in urban areas in Tunisia and especially in the city of Tunis. The tree that was having a role only food now is having an ecological ornamental role and even symbolic.

The fourth approach is related to the question: What prospects for sustainable tourism development in Hergla (Tunisia)? Hergla is an old farming village located on the eastern coast. It is located to the north of the integrated El Kantaoui Tourist Resorts, in the governorate of Sousse. Faced with the natural and cultural specificities of the site, the authorities are compelled to adjust tourism spaces otherwise in response to new international standards in terms of sustainable tourism. Tourism is affected by a tight transformation aimed at a development that is inevitably more sustainable [23]. To achieve the principles of sustainable development, the Charter initiates the creation of new partnership relationships between the main actors involved in this activity, in particular, the territorial actors, to forge the hope of developing more responsible tourism in regard with our common heritage. Moreover, tourism development is part of a global land management, which integrates local actors. Indeed, "the concept of sustainable land management (SLM) offers solutions that go beyond technologic recommendations by including aspects of social participation and policy dialogue".

The fifth approach is the climate factors affecting sustainable human and tourism comfort in Aswan Governorate. Tourism has become one of world's most notable industries and is a main driver for many countries' economies around the world. The development of tourism should be sustainable, but the issue of how this can be achieved remains the subject of discussion. Sustainable tourism is the application of sustainable development ideas to the tourism sector, that is, tourism that meets the needs of the present generation without compromising the ability of future generations to meet their own needs [24]. The tourism industry, which is mainly an outdoor economy, can be particularly vulnerable to climate and weather effects [25]. Weather and climate variables can affect seasonal tourist arrivals, affecting financial activity's sustainability [26]. Climate and its various components influence, directly on the human body and its social and psychological life. Because the place chosen by the tourist must have adequate climatic and environmental conditions for his movements and needs, which is different from the place where he lives, therefore, the elements of climate are one of the most important natural geographical components on which tourism is based.

The sixth approach is the Agri-environmental policies: Comparison and critical evaluation between EU and Egyptian structure. Everyone and every country contribute to today´s challenges and the achievement of the 2030 Agenda. For example, "the way we produce and consume food is causing major environmental and human health issues. Transforming the agri-food system has become a priority for achieving Sustainable Development Goals. Food must be produced, distributed, consumed and disposed of in ways that optimize resource use, minimize greenhouse gas emissions, avoid chemicals that harm ecosystems or human health, and stop further loss of biodiversity", states the United Nations Environment Programme [27]. Agricultural transformation is believed to be vital for achieving a sustainable future and being able to contribute to the achievement of several SDGs [28]. Agriculture uses a lot of land—37% of the total land area are being used for agriculture worldwide [29]. 70% of all freshwater withdrawals go to agriculture for irrigation [28] and many more resources are involved. Thus, it is certainly true, that agriculture can cause damages and induce negative impacts. On the other hand, however, agriculture produces what we all need for living: food. It employed 27% of the world's population in 2016 [29] and is thus the main source of income for a lot of people worldwide. In the European Union (EU), 39% of the total land area were being used for agriculture in 2016, yet only 4.2% of the population worked in the agricultural sector at the same time [30]. In 2017, agriculture contributed to just 1.7% of the gross domestic product (GDP) [30]. In Egypt, however, agriculture is the largest employer providing income-generating activities to more than 30% of the workforce [31]. The agricultural sector plays a central role in the Egyptian economy as it accounts for 14.5% of the GDP [31]. At the same time, the estimated land under agricultural use consists of only about 2.5% of the country's total area (24,960 km^2 of 1 million km^2) [32].

3 Conclusions

Throughout the course of the present book project, several conclusions drawn from this book were reached by the editorial teams. In addition to methodological ideas, the chapter draws important lessons from the book cases, in specific the agro-environmental sustainability covered themes. This book provides the present challenges faced by the agro-environmental sustainability in MENA Regions. In order to improve the sustainable environment in MENA Regions, these findings are crucial. Based on the materials described in all chapters of this volume, the following findings could be indicated:

1. The expected impacts of climate change on the water resources in Egypt are related to rainfall patterns over the Nile Basin and evaporation rates over open water bodies. Therefore, some of the available climate change impact studies for the Nile River Basin and Egypt were reviewed in this article to provide helpful literature for future climate change studies in Egypt. There is a consensus among the regional climate projections performed in the literature indicating a general increase in precipitation during the main rainy season (JJAS) in the Ethiopian Plateau, where both the Blue Nile and the Atbara River have their sources (the main freshwater source in Egypt). For other seasons, no significant model agreement was found. For the Nile Equatorial Lakes region, however, most of the models indicate a significant drying during the summer and a general increase in precipitation during the winter. Furthermore, most of the studies concluded that the uncertainties in predicting the impact of climate

change on Nile flows are very high, which complicate the proposed plans of water resources in basin countries.

2. The applications of wastewater treatment by biofilm technologies were successful in improving wastewater treatment efficiency in the MENA region. In the two previous pilot projects in Jordan and in Tunis, the treated wastewater was usable for irrigation after disinfection or discharging to the appointed recipient according to the local wastewater disposal standards. The generated excess sludge in the two above pilot projects was exposed for drying in sludge drying bed for composting and sludge reduces from the wastewater treatment stages, and also the sludge as an end product. This is after ensuring that it meets the local reuse standard which can be reused safely in agriculture or for any other sludge disposal purposes.
3. Water has particular importance in the agricultural sector, particularly in arid and semi-arid regions, but water resources are increasingly scarce. Having reached a high level of mobilization of these hydraulic resources, Tunisia is now confronted with the imperative to better manage and valorize them. The study of the olive water needs allowed more efficient management of the water supplies and saving in irrigation water and this by an estimation of the needs in water by using more precise methods integrating the maximum of parameters from the Soil–Plant-Atmosphere continuum. Each of these methods is based on a set of climatic, edaphic and/or physiological parameters of the olive tree. The study of the water requirements of the olive tree in a semi-arid region has allowed us to manage water supplies more efficiently and save on irrigation water using the climatic method and/or the physiological method. The evaluation of the different irrigation doses applied according to the climatic (T2 = 60% ET0) and physiological (T1 = 100% sap flow) methods and in relation to the transpiration of the trees measured by the sap flow method, allows to note that the higher the irrigation dose is, the highest is loses by transpiration.
4. Some of our recent results in the context of the stochastic hydrological processes modeling and prediction in case of time-varying linear systems are presented. It concerns water resources in the northern part of Algeria, where discrete Kalman Filter technique has been applied to the modeling and prediction of Stream flows and rainfalls in a number of sites simultaneously (multi-site) for each of the monthly and annual scales. The developed operators have the particularity of automatic self-adapting as soon as a new measure becomes available. This is an advantage of the KF algorithm recursive character that can be used in real-time predictions. As a result, optimal stream flow and rainfall predictions are obtained considering time variations of the underlying hydrological generating processes, as well as their stochastic character. The obtained predictions can be appreciated from a temporal point of view, where observations and predictions in a single site are obtained during a period of time, but can also be the extent to any further step where observation is available. These predictions can also be appreciated from a spatial point of view, where observations and predictions in all considered sites are obtained during a period of time, but can also be the extent to any other site

where observation is available. One of the most important advantages of the developed operators is to provide the error prediction covariance matrix with certainty at each calculation step. This is of great interest because it constitutes a measure of the prediction accuracy at each step calculation. The accuracy of the predictions, consequently the suitability of the operators, has also been checked by the prediction relative error in percent whose overall average value is less than 10% which is highly acceptable. Another advantage of the developed operators is that the algorithm may be initiated with very little objective information and the prediction is obtained in the time-domain. This is of great interest because it offers a real time forecasting possibility. The developed operators are interesting because they can help policy and decision-makers in water resources in Algeria acting efficiently for a better management and sustainable development of such resources in the country.

5. There is a clear need for agricultural extension services to support the education and awareness of breeders. In the Archipelago, a dominance of ovine breeding (73%) based on an aboveground farming system is noted. The majority of the breeders use the hay, the concentrated, and the agropastoral food as basic ration completes what establishes a constraint. Furthermore, the analysis of the ovine reproduction system shows that 94% of the breeders do not practice mating against the season, which explains the lack of knowledge of the breeding conduct. Also, more than half of the sheep and goat breeders do not apply the techniques of flushing, destemming and of creep-feeding.
6. Land degradation is threatening the livelihoods of some Two billion people living in the world's drylands. Remote sensing data and Geographic information systems provide essential base information for integrated, region- or problem-specific approaches combining satellite data products with specific GIS interpretation and modeling approaches for land degradation assessment. There are many products of remote sensing data with different spatial and temporal scales. These products are very useful for deriving spectral indices that contribute in land degradation assessment. Also, the geographic information system (GIS) is a database management system that deals primarily with spatial and temporal data and gives the ability to model land degradation derivatives and processes. Some land degradation assessment frameworks were developed under the umbrella of remote sensing and GIS. Those land degradation assessment frameworks are more accurate especially in the way of determining the land degradation indicators and derivatives, and also in the monitoring of the processes of land degradation and how they are changed spatially and temporally. This chapter focused very briefly on the role of remote sensing and geographic information system in the assessment of land degradation, with some attention to the research of assessing land degradation in Egypt using remote sensing and GIS.
7. Precision farming or precision agriculture or site-specific land management is relatively a new farming technology for increase soil and crop production. However, precision farming technology includes remote sensing, GIS, DGPS,

yield map, variable rate application, soil sampling, and site-specific management zones. The main aim of precision farming include a good understanding of soil and crop variability within-field and follow up by farm management inputs depending on the main source of the variability. In the last two decades, precision farming technology has been more and more developed, including mechanization and automation. Site-specific management has been an important sector of the agricultural system of the developed countries, but the implementation of this high technique of precision farming to farm through the farmers in developing countries is still very slow and needs a lot of extension efforts and farmer motivation. Due to most farmers do not know new information technologies, e.g. remote sensing, GIS, and GPS. The government should be made to increase investments to educate farmers for the application of new methods and technologies in agriculture. The size of the agricultural land is another problem for adopted precision farming technology in some countries like Egypt. Egyptian farmers and experts should use the PF experiences that have been used by developed countries such as USA, Germany, and Japan.

8. Today's context, triply marked by economic globalization, socio-technological changes and ideological and political uncertainties, poses new challenges for developing countries. Local governance consists of a set of institutions and mechanisms. Processes that allow citizens and citizens' groups to express their interests and needs, to settle their differences and to exercise their rights and obligations at the local level. Today, we talk about the attractiveness and competitiveness of territories, territorial anchoring of companies, industrial districts, territorial projects, etc. However, the question is no longer simply to initiate local projects but to ensure those synergies are created between the actors and accelerate the local dynamics observed in territories that remain to be built. What is the real contribution of these projects to problem-solving at the local, regional and national levels? How to act effectively and accelerate the dynamics of growth and development? In other words, how to move from localized action to a territorial approach to development problems? These are the questions facing Morocco today regarding local development.
9. The coasts of the Sahelian Tunisian region have undergone spectacular transformations throughout the decades that followed the independence of the country. Indeed, from an open green space, they have become the capital of tourist and resort urbanization. These transformations have not been without consequences including social, landscape and environmental. The installation of this urbanization has led to an artificialization of the coast; a barrier seems to have been erected between the coast and the hinterland. This barrier, formed by a narrow cordon of urbanization, has relegated coastal agriculture to the hinterland and has also given rise to a form of social segregation since most of the coastal territory has been invested by the tourist sector which privatizes the beaches. All these transformations have not been without environmental consequences. The precarious balance of this portion of the territory has been weakened by the large consumption of space of tourist and resort urbanization and by the acceleration of coastal erosion, mainly due to: the location of

the urbanization in front of the sea, the destruction of the coastal dune, the degradation of spontaneous plantations by the discharge of swimming-pool water, etc. However, until the end of the 1970s, coastal and traditional agriculture occupied most of this territory and, in addition to its social and landscape functions provided an environmental balance for this fragile area. The urbanization of the coastlines and its social, landscape and environmental impacts are a phenomenon that affects the majority of countries in the Middle East and North Africa and, in particular. Morocco where coastal farmland is giving way to tourist urbanization.

10. The Arab Maghreb countries (Morocco, Algeria and Tunisia) are among the growing countries whose economy is based on natural resources, agriculture and industry. Agriculture is practiced in a traditional and modern way, with large areas, irrigated and non-irrigated, which have absorbed a lot of work to ensure the availability of food. Since the independence of these countries have changed the policies, plans and programs for the recovery of this sector, but so far have not achieved the desired objectives, which are demonstrated by the large volume of the bill of the importation of basic foodstuffs. The geographical situation of the Arab Maghreb countries in the South of the Mediterranean sea, which is characterized by a semi and arid climate, lack rainfall generally, in addition to the random and excess use of water in the agricultural system, and even in drinking make these countries exposed to water shortage crises several times in their history, and this has had a negative impact on their environment in general: on soil structure, on vegetable biodiversity, and even on climate change; and these changes have also had a negative impact on the agro-socio-economic system. The impact studies available in these countries, though rare, have been shown to be intensive agriculture using many chemicals (fertilizers and pesticides) plus the human factor (unskilled) leads to soil pollution, water and air, as well as human exposure to health problems. Sustainable agricultural production and conservation of the environment is a dream for these countries comes after the dream of food insurance for lack of qualified human potential and material possibilities.

11. It can be concluded that the mineral fertilizer was partially replaced by the treatment of 50% of mineral in interaction with 50% of compost gave the highest values of the dry weight of whole plant (g), the dry weight of shoot (g), absolute growth rate (mg day^{-1}), and unit shoot rate (mg mg^{-1} day^{-1}) in the first season. Also, the treatment of 50% of mineral in interaction with 50% of compost attained a maximum value of absolute growth rate (mg day^{-1}) and unit shoot rate (mg mg^{-1} day^{-1}) in the second season and in both seasons. Treatment of 100% of bioformulations recorded the highest value of the dry weight ratio of shoot/root in the first season and in both seasons. Treatment of 100% of bioformulations in combination with 50% of mineral achieved a maximum value of total chlorophyll content (%) in the first season and in both seasons. Treatment of 100% of bioformulations combined with 33% of compost and 33% of mineral gave a maximum value of the dry weight of root (g) in the first season and in both seasons.

12. As a result of the factors mentioned above, there was a significant effect by the combination of K-feldspar at different diameters and vermicomposting with SDB on yield, NPK-Si-uptake of wheat plants, available potassium and silicon in the soil. The application of K-feldspare > 45 μm combined with Ver in the presence of SDB gave the highest values of straw and grains yield of plants, the weight of 1000 grain, biological yield, protein content, NPK and Si-uptake, available K and Si in the soil. The available potassium and silicon were remarkably increased after 75 days at all treatments by application of various diameters of K-feldspar at different diameters and vermicomposting with SDB.
13. Shalatin—AbouRamad—Halaib Triangle Region (SAHTR) has vital and strategic importance to Egypt. It is characterized by broad biodiversity of both plant and animal resources. Livestock production on the native ranges of Shalatain—AbouRamad—Halaib Triangle Region. Red Sea Governorate is facing several major constraints mainly: drought, feed and water shortage, diseases, and environmental stresses.
14. The use of tree or shrub species that are capable of developing under conditions of extreme aridity and absolute marginality of soils represent a significant solution to the problem of desertification in the Mediterranean basin. Among these species, *Opuntiaficus-indica* is a potential candidate in rehabilitation programs besides to its food and forage properties. Our findings showed that after the establishment of *Opuntiaficus-indica* in arid and semi-arid steppes. There is an improvement in the environment, particularly in soil organic matter due to the accumulation of litter and vegetation cover. These plantations act as a refuge to plants or resource-plant for the sequestration of seeds that cannot be established elsewhere, and thus creating a favorable microhabitat. In addition, this species is mycotrophic (mycorrhizal colonization > 80%) that can contribute to the improvement of soil quality. All of these statements require certainly more elaborative and exhaustive studies to reinforce the inferences and conclusions: such as soil study at the foot of the shrub instead of interspacing or investigating soil enzymes that are considered as consistent indicators of any change in soil condition. The expansion of PPP is emerging as a promising strategy for raising resource dynamics and restoring vegetation in arid and semi-arid environments. The evaluation of the rehabilitation actions can be developed for producing aided decision tools for programs of combating desertification. Long-term rehabilitation with this allochtone species is possible because this cactus is a long-term soil stabilizer and without ecological requirements.
15. The soils of the Tadjmout perimeter belong to the arid zone. The analytical results show that these soils are at alkaline pH Cation exchange capacity (CEC) relatively average, which does not exceed 15 Cmol/kg; Low to medium and sometimes high salinity (EC); Exchangeable phosphorus: the values are between 26.62 and 34.87 ppm. This high content of exchangeable phosphorus could be due to phosphate fertilizers. Exchangeable potassium has average to low values. These soils require fertilization to improve their chemical fertility. The thematic maps allow us to manage a fertilization plan to improve the

chemical fertility of these soils, in order to increase agricultural production. It is imperative for the agricultural exploitation of arid soils in the context of sustainable development, further research based on experimentation and monitoring over a long period.

16. The main areas of rangelands are distributed over the Northwest Coast region, the Sinai Peninsula and the Halayeb-Shalateen region in the southeast corner of Egypt bordering the Red Sea. The increase of the number of palatable perennial forage species indicates that the pasture condition in the study area is generally good. Also, the vegetation potential has not been destroyed, and that there are some chances of improvement and development. The studied species can be ranked depending on its palatability and the percentage of DCP, TDN and water content as follows:*Alhagigraecorum. Asphodelusramosus,Panicum turgidum, Lyciumshawii,* new sprouts of*Pharagmitisaustralis,Gymnocarpusdecander,* fruits of *Acacia tortilis,* and new sprouts of*Juncusrigidus.* The productivity of *Alhagigraecorum,* Pharagmitis*australis,* and *Juncusrigidus* is high and these plants cover a wide area in the Egyptian Oases.These species vary in their chemical composition that may meet the requirements of the grazing animals. Most of the palatable shrubs had a low value of density due to these species subjected to heavy grazing. The grazing and burning had a negative effect on *Pharagmitisaustralis*particularly on productivity and palatability. The burning improves the palatability of *Juncusrigidus.* Most of carrying capacity of the eight range plants are reported are high.
17. Humanity is forced to adopt the concepts and requirements of sustainable development to handle the fragility of ecological equilibrium and the degradation of the environment as a result of the imperatives of socio-economic development. One of the pillars of this requirement is to continuously monitor the state of the environment through an operational information system that can assess and support the actions and policies of society towards the environment. The implementation of such systems encounters difficulties in developing countries due mainly to problems of availability and scarcity of data. This study provides an overview of the different issues concerning this topic and presents a list of suggestions to overcome these obstacles.
18. Throughout this article, we have shown that the interest of the agricultural landscapes of the Sahel of Bizerte is not limited to the fact that they offer a remarkable aesthetic beauty.We have also been able to show that the key to the preservation of an agro-biodiversity an overall meaning and present a valuable cultural heritage, but they also provide multiple sustainable solutions related to work, to food and the well-being of farmers. Although in most countries of the world, modernity has been characterized by a process of cultural homogenization and economic, in many rural areas, specific cultural groups are still associated with a geographical context and society in which there are special forms of agriculture. Dynamic conservation of these sites and their cultural identity may be the starting point for territorial development and social and cultural renewal.

19. For post-mining site remediation, the finding and application of environmental quality targets is feasible and can be modeled after projects like "Saumon 2020" or its predecessor "Saumon 2000". Redevelop the current nature reserve around the Djebel Ressas mountain top into an area in which *Macacasylvanus*is to be resettled into a habitat suitable for that species. That can be implemented if the growing cement works still allow such a measure. For Phytoextraction, Phytostabilization of the affected soil by adapted plants, Alfalfa could be used. However, as the Barbary macaque needs trees such as in "high cedar forests *(Cedrusatlantica),* cedar/holm oak *(Quercus ilex)* mixtures, pure holm oak forests and cliffs and gorges dominated by scrub vegetation.", the fauna must be adopted: The dietary habits of the ape species need to be considered: "Its diet is primarily composed of cedar *(Cedrusatlantica)* and the oak *(Quercus sp.),* which make up over 50% of its total intake. It eats fruits (33% of its intake), tree leaves (16%), and other plant parts (24%)." The local population should be fenced, in a natural way by trees and bushes, from the mining deposits' flue dusts so that the exposure to heavy metals via inhalation and ingestion is closer to international WHO levels than under the current conditions. Also, under a site remediation scheme, jobs in local projects can be created to ensure high acceptance levels among the inhabitants of Djebel Ressas village and the surrounding area.
20. In Tunisia and in the largest Tunisian cities, green spaces were born with the extension of the European city outside the medinas where this notion is totally illusory. Tunisia has also become, over the years, an attractive tourist destination, since then, the public authorities have tried to promote the creation of green spaces by multiplying the national programs of the creation of urban parks (PNPU), development of city entrances, creating boulevards of the environment in all Tunisian cities. The creation of green spaces in urbanized areas has become a recurring message in messages conveying an environmental policy, especially after adhering to the precepts of "sustainable development". Green spaces have been rationalized as urban equipment as well as cultural, sports or health facilities. They have been standardized in the form of m^2 per capita (the public authorities have set a minimum of 10 m^2 green space/inhabitant). This minimum standard was required for any new development project. Tunisian green policy should be more in the direction of a green infrastructure aimed at restoring connections between existing natural spaces and improving the ecological quality of the environment as a whole. For this, it may include several natural components (protected areas, wetlands, forests, etc.) as well as multifunctional areas in which priority is given to land uses that maintain or restore healthy and diverse ecosystems rather than other harmful activities. This could support the construction of "villedurable" while taking into account the preservation of biodiversity in current urban planning.
21. Several main results arose on the basis of an assessment of the present research. The issue of the effect of heat and humidity on human well-being in Aswan Governorate in general, as reflected in this study, is still considered new to the Egyptian Geographic Library. The importance of the topic of human well-being

in Aswan governorate, where there are climatic differences effects on human well-being. The development of meteorology in Aswan governorate and the availability of modern climate data for different climate elements throughout the year, thus enabling the use and adaptation of such data to serve geographical research.

22. This study reviews the 2030 Agenda as a blueprint for water and ecosystem services in the context of agriculture. For both Egypt and the EU, implementation of this international consensus could lead to improved water handling and agricultural practices. Reviewing already existing legislation or ensuring its implementation can also be regarded as an important approach. Our results show that the 2030 Agenda with its ambitious targets has been adopted and to some extent, already passed into law by the EU member states and Egypt. However, these laws and decrees have to be adequately enforced for some of the targets to be achieved by the EU member states and Egypt. Results also show that there are still gaps to close, especially in areas where no new legislation has been introduced in the last two decades. Today, the whole world is in the same boat when it comes to the sustainable development pathway—with or without the 2030 Agenda. To put it simple: The UN member states committed to the SDGs and to providing voluntary reports about their progress. Thus, a development towards the achievement of the 2030 Agenda's targets should only be a question of time—and time is of the essence.
23. The Hergla site, through its landscape and local cultural, is quite potential for tourism development. Faced with the disengagement of the State and the weakness of local actors, the rise of private actors can lead to tourism development not in compliance with the principles of sustainable development. This observation inspires us to work more on governance and new perspectives of tourism planning, in Tunisia in general.

4 Recommendations

The capacity to adapt to future problems is a main element of agro-environmental sustainability in MENA Regions. We contend that to accomplish this objective, agro-environmental sustainability needs integrated flexibility. The editorial teams observed certain aspects that could be explored for further enhancement throughout the course of this book project. Based on the results and conclusions of the contributors, this chapter provides a number of recommendations that provide suggestions for future scientists to exceed the limited scope of the book to the future unlimited space of the topic.

1. The following are suggestions for possible future researches related to climate change impacts on water resources in Egypt: Evaluating the updated IPCC [33] climate change scenarios to have a comprehensive assessment of climate change impacts on different hydrometeorological processes (e.g., evapotranspiration and precipitation) in Egypt, updating Regional Climate Models

capable of predicting the impact of climate change on the local (Egypt) and regional (Nile Basin) water resources and studying the basic characteristics of stream networks that contribute to flash floods by using advanced tools (e.g., GIS and remote sensing). Investigating the impact of change in both land use and climate on hydrological processes and water resources and conducting comprehensive research on the joint impact of climate change and new water projects on the water resources of the Blue Nile and assessing climate change impacts on food security, energy, and water resources.

2. Biofilm system applications normally faced a several operations problem in the region. Some recommendations have to be taken into consideration. Fats, oils, Solids and Suspended solids, and greases must be removed or minimized in the influent by very good pre-treatment efficiency to protect the biofilm and its specific surface area. This leads to an increase pf the Biofilm treatment efficiency. If there is a high concentration of BOD5 and TSS in inflow, then some pre-treatment may be required to reduce it. Researches demonstrate that biofilm technologies should be designed based on surface area loading rate (g COD/m^2d), and the Media material and type play an important role in such treatment process. For tricking filter: Pressure distribution should be required for the design option in order to strive for equal distribution of effluent at the design organic loading rate and also to keep the attached biofilm wet. Some methods of effluent quality monitoring may be necessary. An intermediate clarifier tank is useful to build between treatment stages to draw out the sludge produced and to keep the attached bacteria working in its full capacity; this will increase the treatment efficiency.
3. Deficit irrigation strategies, applying climatic or physiological methods, allow better water use efficiency for olive orchards without affecting yield or fruit and oil quality. The physiological method (T1), equivalent to 32% of the ET0, allowed us a saving in water of the order of 50% in comparison with the climatic method. So the physiological method allows for more efficient management of water resources. To apply the physiological method in different bioclimatic stages, the use of transpiration modelingin relation to climatic parameters is necessary to meet the needs of the economy and the management of water supplies.
4. Precision farming systems can be implemented in developing countries through collect information about soil and crop using new information technology. Collecting data on the spatial variation in soil and crop features requires accurate position determination in the field, using GPS. Differential action. In response to spatial variability, farming operations, such as sowing rate, fertilizer, pesticide and lime application, tillage and water use, can be varied in real-time across a field. Variation in treatment corresponds to the mapped variation in the field attributes measured using remote sensing technology, GPS and GIS. Soil and crop monitoring. Soil and crop attributes are monitored on a finite scale. When observations are targeted with GPS, and remote sensing they provide data on the spatial variability of the attributes within a

field. Values for soil and crop attributes are predicted for unsampled locations across a field using GIS. This enables a detailed representation of the spatial variability within an entire field through the creation of a smoothed map. Knowledge about the effects of field variability on crop growth—and the suitable agronomic responses—can then be combined to formulate differential treatment strategies (http://www.fao.org/3/a0869t/a0869t04.pdf).

5. We recommend decision-makers to use the agrarian diagnosis method merged to new agronomic technologies for planning a future strategic plan. [10] Has proposed the introduction of this technology to improve the quality production, with the concern to save water in irrigable areas and to cultivate varieties adapted to the weakness and the irregularity of the pluviometry in rainfed areas, using energy-saving techniques energy and inputs from oil. The recommendation can be applied to different MENA agrarian systems.
6. First, the "urban and landscape soils" will recognize the quality of "urban and landscape terroir" in the periurban areas in which, the communes, departments, regions, and the state wish to carry out voluntary, concerted and contractual action. This label would be concretized by: the signing of a charter applicable to a territory; the definition of social, urban and agricultural objectives to respect to manage the space and requalify it when it is degraded; the mobilization of cross-financing State-local authorities. Second, develop the land policies of local authorities by: Including an agricultural and landscaping component in city contracts in order to treat peri-urban space as a heritage and dynamic component of the city; Reconsidering the question of property assessments, in particular, to enable municipalities to purchase land if they so wish and facilitate the maintenance of agriculture; Third, create regional environmental public institutions: These agencies would ensure the protection of peri-urban green and agricultural areas under; Allocate them a share of the proceeds of the departmental tax of sensitive natural areas.
7. The tourist development of the Tunisian Sahel region has, of course, had major repercussions on the economy of the region, which has become one of the most important Tunisian tourist polities, but this has not been without consequences both socially and environmentally. Moreover, today, tourist demand has evolved towards other centers of interest. Similarly, new concepts applied to the tourism industry have promoted its sustainability through the development of a "Sustainable Tourism" based on the respect, preservation and development of environmental aspects economic and sociocultural countries. Tunisian, in general, and Sahelian tourism, in particular, must set out new directions for its development. It is, therefore a question of stopping the increasing densification of the tourist zones and of easing the pressure which is exerted on several sites, including, in particular, the natural and agricultural sites of the Sahelian littoral. These should be protected and introduced as an integral part of a new tourism offer, based on the multifunctionality of agriculture and in accordance with the rules of sustainable development. The Tunisian Sahel is also the site of traditional coastal agriculture. The latter could play an important role, within the framework of this new tourist offer, on the one hand: its singularity, the

quality of its products, its contribution in the fight against the erosion of the beaches, its maintenance as an open space to inside the coastal urbanization, etc. …

8. Water is a natural resource; where there is there is a life, balance and sustainability. It is important to these Maghreb countries to study how to exploit this dear gold to their life, eventually for their food security: Storage of rainwater in dams and distribute tightly in agriculture and drinking. Establish strict laws prohibiting water loss. It is best to irrigate cultures by drip system. Cultivate varieties in appropriate places according to their environmental needs (soil, water and temperature). Uses desalinate water of the sea or the ocean. It is necessary to fertilize the poor soil, and to control plant diseases with bioproducts to avoid negative impacts of chemical products. Climate change, always followed by special attention to the soil and the cultivation of plant varieties adapted to the new climate. Most important for preserving safety agriculture without impact on the environment; It is the human factor that practices agriculture; it needs to be structured to understand what it is for sustainable agriculture.As well as coordination between all actors in the agricultural sector.
9. The recent major problem facing the farmers is the high coast of chemical fertilizers. The alternative depending on expensively imported fertilizers is to exploit indigenous resources such as K-bearing minerals. Making use of such minerals is meaningful in increasing crop yield and protecting the ecological environment. The main source of K and silicon for plants growing under natural conditions comes from the weathering of K minerals and organic K-sources such as composts and plant residues. The application of K-feldspare at different diameters combined with vermicomposting at the rate of 2% (20 Mg ha^{-1}) in the presence of silicate dissolving bacteria increases yield, nutrient uptake of wheat plants and available nutrients in sandy soils. Finally, From an economic point of view, this approach of using the naturally deposited materials (K-feldspar) instead of chemical fertilizers or combination together or mixed with vermicomposting will be very beneficial for the farmers who subsidize the costs of chemical fertilizers (potassium sulphate).
10. All efforts should be directed towards setting up a sustainable national strategy to improve both rangelands and livestock production which has the following tasks: Optimum utilization of the available local feed and drinking water resources—applying proper approaches for native range management and conservation. Development of watering points combined with proper control of grazing. Utilizing the available non-conventional feed resources efficiently. Genetically constitution improvement, particularly through the selection. An effective program for the improvement of feeding, disease control, and management should precede any attempt at genetic improvement. They were providing proper assistance and advice to local Bedouins or investors to best utilize available resources, strengthening the cooperation between scientists, specialists, policymakers and Bedouins.
11. The experiences of the countries with PPP are quite developed. PPP show its importance in the valorization of degraded soils, in the use of cladodes as fodder

for animals, and in human consumption (fruits and extraction of essential oils). It would be wise to develop this culture for agricultural and rural renewal in the arid and semi-arid zones: (i) to develop the agroforestry approach and associate the cultivation of *Opuntia* spp. with the suitable crops in particular olive, pistachio and legume trees, (ii) initiate research projects on improving planting techniques to improve yields (study the effect of interline spacing, planting in pots or inline), (iii) sensitize rural populations on the different uses and ecological functions and services of this crop such soil protection, agricultural, industrial, cosmetic, culinary uses, and (iv) develop the role of rural women by integrating them into all income-generating activities.

12. The mapped soils require phospho-potassium and nitrogen fertilizer maintenance to improve their chemical fertility. Before the planting of each crop, the analysis of nitrogen and phosphorus in the soil is necessary in order to avoid pollution of groundwater.
13. Using the reported species as forage resources particularly in the dry season. Complete protection for at least five consecutive years in some sites of the communities of *A. ramosus* and *L. shawii*to increase their regeneration and productivity. It would perhaps be advisable to reseed, or established as mother plants the endangered high palatable species e.g. *L. shawii*, *Gymnocarposdecander*especially in the soil deep enough and soil moisture content is high. It would be useful to utilizating high techniques of water harvesting in the growing area of the reported species for cultivation of these species and to promote the seedling and propagation of these plants. Using the palatable halophytic species e.g. *Alhagigraecorum and Panicum turgidum* in the rehabilitation of the salt affected lands and increase their productivity. They were promoting the fire technique in the habitat of *J. rigidus* to manage these grasses for forage. Improve the acceptability of the medium and low palatable species for animal through mix these species with the other acceptable species.
14. To implement a successful and useful environmental information system, the authors recommend taking into consideration all the elements of the guideline presented in this study. They highlight the need to pay special attention to the problem of lack or scarcity of data. For that, they also recommend studying and analyzing the new opportunities offered by remote sensing techniques and big data technologies.
15. In order to increase the efficiency of GHs, other systems are integrated with the GH, such as desalination, solar power generation and heating systems. In these systems, the extrasolar radiation is used to generate electricity via PVs or to produce water via desalination systems. Also, an Aquifer Coupled Cavity Flow Heat Exchanger System (ACCFHES) is integrated with the GH for its cooling and heating.
16. It is essential today to involve the local community in local government. With the new direction taken by the Tunisian government in 2018, promoting decentralization, it is necessary to involve the population for a participatory approach. The local government to support by the participation of the population can prove the solution for the future and the safeguarding of the agricultural lands

of Ghar El Melh. A site remediation project around Djebel Ressas village should ideally satisfy all three aspects of sustainable development as defined by the Rio AGENDA 21: economic factors, social factors, and environmental factor.

17. For future vision, tourist countries, including Egypt, need to develop a new model of tourism that should also be based on so-called alternative tourism. Alternative tourism is a generic concept encompassing various forms of tourism, such as cultural tourism, Ecotourism, agro-tourism, rural tourism, agritourism, and green tourism. Identify the climatic characteristics of the air temperature and relative humidity elements in Aswan in great detail and depth because the relative humidity element did not receive a detailed independent study. The need to select and apply some of the global climate equations, models, and curves to Aswan governorate that have not been applied before in the physiological climate studies on Aswan to modeling and classification. Identify the boundaries of the physiological balance and comfort of the human body in Aswan throughout the year.
18. The management of a sustainable tourism offer in a fragile territory such as that of Hergla cannot be done without good coordination between actors concerned. For that very reason we are proposing the achievement of the regional pattern of development of tourism integrating of all these actors.

Acknowledgements Mohamed Abu-hashim and Abdelazim Negm acknowledge the support of the Science, Technology, and Innovation Authority (STIFA) of Egypt in the framework of the grant no. 30771 for the project titled "a novel standalone solar-driven agriculture greenhouse—desalination system: that grows its energy and irrigation water" via the Newton-Mosharafa funding scheme.

References

1. Omran ESE, Negm A (2018) Environmental impacts of AHD on Egypt between the last and the following 50 years. In: Negm A, Abdel-Fattah S (eds.) Grand Ethiopian Renaissance Dam versus Aswan High Dam. The Handbook of Environmental Chemistry, vol 79. Springer, Cham
2. Omran ESE, Negm A (2018) Environmental impacts of the GERD project on Egypt's Aswan High Dam Lake and mitigation and adaptation options. In: Negm A, Abdel-Fattah S (eds.) Grand Ethiopian Renaissance Dam versus Aswan High Dam. The Handbook of Environmental Chemistry, vol 79. Springer, Cham
3. AFED (2017) Arab Environment in 10 Years. Annual report of Arab forum for environment and development. In: Saab N (ed.) Technical Publications, Beirut, Lebanon
4. Abouelnaga M (2019) Why the MENA region needs to better prepare for climate change. https://www.atlanticcouncil.org/blogs/menasource/why-the-mena-region-needs-to-better-prepare-for-climate-change. Access date 7 May 2019
5. World Bank (2019) Database. https://data.worldbank.org/region/middle-east-and-north-africa?view=chart
6. World Bank (2018) Beyond scarcity: water security in the Middle East and North Africa. https://openknowledge.worldbank.org/handle/10986/2168
7. Fernández JE, Diaz-Espejo A, Romero R, Hernandez-Santana V, García JM, Padilla-Díaz CM, Cuevas MV (2018) Chapter 9—precision irrigation in olive (*Oleaeuropaea L.*) Tree

Orchards. Water scarcity and sustainable agriculture in semiarid environment. Tools, strategies, and challenges for woody crops, pp 179–217

8. Hernandez-Santana V, Fernández JE, Rodriguez-Dominguez CM, Romeroa R, Diaz-Espejo A (2016) The dynamics of radial sap flux density reflects changes in stomatal conductance in response to soil and air-water deficit. Agric Meteorol 218–219(2016):92–101
9. Padilla-Díaza CM, Rodriguez-Dominguez CM, Hernandez-Santana V, Perez-Martin A, Fernandes RDM, Montero A, García JM, Fernández JE (2018) Water status, gas exchange and crop performance in a super high density olive orchard under deficit irrigation scheduled from leaf turgor measurements. Agric Water Manage 202(2018):241–252
10. Miao Y, Mulla DJ, Pierre C, Robert PC (2018) An integrated approach to site-specific management zone delineation. Front Agr Sci Eng. https://doi.org/10.15302/J-FASE-2018230
11. Sao Y, Singh G, Jha MK (2018) Site specific nutrient management for crop yield maximization using two soil types of Bilaspur District of C.G. on grain and straw yield. J Pharmacogn Phytochem 7(1):08–10
12. Abdullahi HS, Mahieddine F, Sheriff RE (2015) Technology impact on agricultural productivity: a review of precision agriculture using unmanned aerial vehicles. In: Pillai et al (eds) Wireless and satellite systems 7th international conference, WiSATS 2015 Bradford, UK, July 6–7, 2015 Revised selected papers, pp 388–400. https://doi.org/10.1007/978-3-319-25479-1. Springer, Cham, Heidelberg, New York
13. Mogili UR, Deepak BV (2018) Review on application of drone systems in precision agriculture. Proc Comput Sci 133:502–509; International conference on robotics and smart manufacturing (RoSMa2018)
14. Neri AC, Dupin P, Sánchez LE (2016) A pressure-state-response approach to cumulative impact assessment. J Clean Prod 126:288–298. https://doi.org/10.1016/j.jclepro.2016.02.134
15. Guo H, Wang L, Liang D (2016) Big earth data from space: a new engine for earth science. Sci Bull 61:505–513. https://doi.org/10.1007/s11434-016-1041-y
16. Kamilaris A, Kartakoullis A, Prenafeta-Boldú FX (2017) A review on the practice of big data analysis in agriculture. Comput Electron Agric 143:23–37. https://doi.org/10.1016/j.compag.2017.09.037
17. Ben Attia O (2016) Interactionsbetween the evolution of Tunisian coastal agriculture and the development of urbanization and tourism. The case of peri-urban areas north-east of Sousse and southeast of Monastir
18. OTEDD (2016) Rapport national surl'état de l'environnement et du développement durable de l'année2015.P. In: Lerin F, Mezouaghi M (coord.) Perspectives des politiquesagricoles en Afrique du Nord. Paris : CIHEAM, pp 51–91 (Option s Méditerranéennes : Série B. Etudes et Recherches ; n°64)
19. Si-Tayeb H (2015) «Les transformations de l'agriculture Algériennes dans la perspective d'adhésion à l'OMC», thèse de doctorat en science agronomique, option économie rurale, université Mouloud Mammeri de TiziOuzou, p 282
20. Jat R, Naga SR, Choudhary R, Jat S, Rolaniya MK (2017) Effect of potassium and sulphur on growth, yield attributes and yield of sesame (*Sesamumindicum L.*). Chem Sci Rev Lett 6(21):184–186
21. El-Shaer HM, El-khouly AA (2016) Natural resources in saline habitats in South East of Egypt. Natural resources in saline habitats in South East of Egypt. Academy of Scientific Research and Technology, 168 p
22. García A, Álvarez E (2008) Soil remediation in mining polluted areas. ©Macla. Revista de la sociedadespañola de mineralogía 10:76–84
23. Scol J (2012) Équipe MIT: Tourismes 3 : La révolution durable. Territoire en mouvement Revue de géographie et aménagement, pp 14–15
24. Weaver D (2006) Sustainable tourism: theory and practice. Butterworth-Heinemann
25. Perkins D (2016) Usingsynopticweather types to predictvisitorattendanceat Atlanta and Indianapolis zoologicalparks. Int J Biometeorol
26. Shakeela A, Becken S (2015) Understanding tourism leaders' perceptions of risks from climate change: an assessment of policy-making processes in the Maldives using the social amplification of risk framework (SARF). J Sustain Tour 23:65–84

27. United Nations Environment Programme (2019) A new deal for nature—change the way we produce and consume food
28. FAO (2018) Transforming food and agriculture to achieve the SDGs: 20 interconnected actions to guide decision-makers. Rom
29. FAO (2019) FAOSTAT. Retrieved from http://www.fao.org/faostat/en/#country
30. Eurostat (2018) Agriculture, forestry and fisherystatistics, 2018 edn.
31. FAO (2016) AQUASTAT Website. Retrieved from http://www.fao.org/aquastat/en/
32. Meguid MA (2019) Key features of the Egypt's water and agricultural resources. In Negm AM (ed.) The handbook of environmental chemistry. Conventional water resources and agriculture in Egypt, vol 74, pp 39–99. Springer International Publishing, Cham
33. Setiawati TC, Mutmainnah L (2016) Solubilization of potassium containing mineral by microorganismsfrom sugarcane rhizosphere. Agric Agric Sci Proc 9:108–117